Mar 98

D1807314

Broadband Communications

Visit the IT & Applied Computing resource centre
www.IT-CH.com

IFIP – The International Federation for Information Processing

IFIP was founded in 1960 under the auspices of UNESCO, following the First World Computer Congress held in Paris the previous year. An umbrella organization for societies working in information processing, IFIP's aim is two-fold: to support information processing within its member countries and to encourage technology transfer to developing nations. As its mission statement clearly states,

> IFIP's mission is to be the leading, truly international, apolitical organization which encourages and assists in the development, exploitation and application of information technology for the benefit of all people.

IFIP is a non-profitmaking organization, run almost solely by 2500 volunteers. It operates through a number of technical committees, which organize events and publications. IFIP's events range from an international congress to local seminars, but the most important are:

- the IFIP World Computer Congress, held every second year;
- open conferences;
- working conferences.

The flagship event is the IFIP World Computer Congress, at which both invited and contributed papers are presented. Contributed papers are rigorously refereed and the rejection rate is high.

As with the Congress, participation in the open conferences is open to all and papers may be invited or submitted. Again, submitted papers are stringently refereed.

The working conferences are structured differently. They are usually run by a working group and attendance is small and by invitation only. Their purpose is to create an atmosphere conducive to innovation and development. Refereeing is less rigorous and papers are subjected to extensive group discussion.

Publications arising from IFIP events vary. The papers presented at the IFIP World Computer Congress and at open conferences are published as conference proceedings, while the results of the working conferences are often published as collections of selected and edited papers.

Any national society whose primary activity is in information may apply to become a full member of IFIP, although full membership is restricted to one society per country. Full members are entitled to vote at the annual General Assembly, National societies preferring a less committed involvement may apply for associate or corresponding membership. Associate members enjoy the same benefits as full members, but without voting rights. Corresponding members are not represented in IFIP bodies. Affiliated membership is open to non-national societies, and individual and honorary membership schemes are also offered.

Broadband Communications

The future of telecommunications

IFIP TC6 / WG6.2 Fourth International Conference on
Broadband Communications (BC'98)
1st-3rd April 1998, Stuttgart, Germany

Edited by

Paul J. Kühn
University of Stuttgart - IND
Stuttgart
Germany

and

Roya Ulrich
SUN Microsystems Inc.
Stuttgart
Germany

Published by Chapman & Hall on behalf of the
International Federation for Information Processing (IFIP)

CHAPMAN & HALL

London · Weinheim · New York · Tokyo · Melbourne · Madras

**Published by Chapman & Hall, an imprint of Thomson Science, 2–6 Boundary Row,
London SE1 8HN, UK**

Thomson Science, 2–6 Boundary Row, London SE1 8HN, UK
Thomson Science, 115 Fifth Avenue, New York, NY 10003, USA
Thomson Science, Suite 750, 400 Market Street, Philadelphia, PA 19106, USA
Thomson Science, Pappelallee 3, 69469 Weinheim, Germany

First edition 1998

© 1998 IFIP

Thomson Science is a division of International Thomson Publishing I(T)P˙

Printed in Great Britain by Athenæum Press Ltd, Gateshead, Tyne & Wear

ISBN 0 412 84410 9

All rights reserved. No part of this publication may be reproduced, stored in a retrieval
system or transmitted in any form or by any means, electronic, mechanical, photocopying,
recording or otherwise, without the prior permission of the publishers. Applications for
permission should be addressed to the rights manager at the London address of the publisher.
 The publisher makes no representation, express or implied, with regard to the accuracy of
the information contained in this book and cannot accept any legal responsibility or liability
for any errors or omissions that may be made.

A catalogue record for this book is available from the British Library

∞ Printed on permanent acid-free text paper, manufactured in accordance with
ANSI/NISO Z39.48-1992 (Permanence of Paper).

CONTENTS

PREFACE

New Services such as for Internet data and multimedia applications, have caused a fast growing demand for broadband communications. The fundamental technologies for the integration of these services have been developed in the last decade: optical communications, photonic switching, high speed local area networks, Asynchronous Transfer Mode (ATM), ISDN and B-ISDN, Internet packet networks and mobile communications. The development was possible through the dynamic progress in communication and computer technologies and through worldwide standardization activities within ITU-T, the ATM Forum, the IETF, IEEE, ANSI, ETSI and other bodies. These developments have been supported by research and field trial programmes. Past developments, such as about LAN, Internet or ISDN networking technologies, have shown that it needs a time span of 10 years for a new technology from its research stage to its full application. Broadband Communications is just at its onset for full deployment. It will have a dramatic effect not only on the networking situation but on the whole development of information technology throughout our social and economic life, which is expressed by the conference theme „The Future of Telecommunications".

The Broadband Communications conference series of IFIP WG 6.2 addresses the fundamental technical and theoretical problems related with these technologies. BC '98 is the fourth meeting in a series on conferences being held in Stuttgart, Germany. The previous confernces were held in Estoril, Portugal, in 1992, in Paris, France, in 1994, and in Montreal, Canada, in 1996. Key topics of BC' 98 are Access Networks, Broadband Mobile Networks, next generation Internet and related issues with the engineering, control, operation and management of such networks: traffic modeling, traffic control (connection admission, flow and overload control, routing, multicast), network planning, network management (including charging), tests, measurements and field trials. As in the previous BC conferences, the programme is a cross-section of topics from technology, applications and theoretical issues and profits quite much from the interaction between them.

Except for the four plenary sessions, the conference is subdivided in two parallel tracks: track A focuses on „Broadband technologies, architectures and network design", while track B concentrates on „Traffic control, modelling and analysis".

The organizers are thankful to all authors who have contributed by their submitted papers, to the members of the Scientific and Organization Committees and to all reviewers for their support, to the invited, tutorial and panel speakers and to the sponsoring companies. Special thanks are due to IFIP WG 6.2 for the encouragement and support of BC '98, to the Information Technology and Computer Societies of Germany (ITG, GI), to the University of Stuttgart and to the staff of the Communication Networks and Computer Engineering Institute (IND) for their support in the organization of the conference.

Paul J. Kuehn   Conference chairman   March 1998
Roya Ulrich   Conference co-chair

BROADBAND COMMUNICATIONS '98

General Chair:
P.J. Kuehn, University of Stuttgart

Co-Chair:
Roya Ulrich, SUN Microsystems

Organizing Committee

Christ, P.	Reg. Computing Center Stuttgart, Germany
Frohberg, W.	Alcatel/SEL, Germany
Hohlfeld, B.	Daimler Benz AG, Germany
Kuehn, P.J.	University of Stuttgart, Germany
Nietz, R.	Deutsche Telekom Darmstadt, Germany
Paterok, M.	IBM/European Networking Center, Germany
Rathgeb, E.	Siemens AG, Germany
Schroeder, J.	Bosch Telecom GmbH Backnang, Germany
Ulrich, R.	Sun Microsystems, Germany

Scientific Programme Committee:

Ajmone-Marsan, M.	Politecnico di Torino, Italy
Albanese, A.	Int. Computer Science Inst., U.S.A.
Blondia, Ch.	University of Antwerp, Belgium
Butscher, B.	DeTeBerkom/GMD, Germany
Casaca, A.	IST/INESC, Portugal
Casals, O.	UPC, Spain
Christ, P.	Reg. Comp. Center Stuttgart, Germany
Cuthbert, L.	QMW College London, UK
Denzel, W.	IBM Rueschlikon, Switzerland
Drobnik, O.	University of Frankfurt, Germany
Eberspaecher, J.	Techn. University of Munich, Germany
Frohberg, W.	Alcatel/SEL, Germany
Gallassi, G.	Italtel, Italy
Guerin, R.	IBM Yorktown Heights, U.S.A.
Hebuterne, G.	INT, France
Hohlfeld, B.	Daimler-Benz AG, Germany
Iversen, V. B.	Technical University of Denmark
Kawashima, K.	NTT, Japan
Killat, U.	Techn. Univ. of Hamburg-H., Germany
Koerner, U.	University of Lund, Sweden
Kofman, D.	Telecom Paris, France
Kuehn, P.J.	University of Stuttgart, Germany
Lehnert, R.	Dresden Univ. of Technology, Germany
Leopold, H.	Alcatel, Austria

Mason, L. G.	INRS, Canada
Nunes, M.S.	INESC, Portugal
Paterok, M.	IBM/Europ. Network. Center, Germany
Pettersen, H.	Telenor R&D, Norway
Rathgeb, E.	Siemens AG, Germany
Roberts, J.W.	CNET, France
Rosenberg, C.	Nortel Imperial College, UK
Schroeder, J.	Bosch Telecom GmbH, Germany
Stuettgen, H.	NEC Europe Ltd., Germany
Spaniol, 0.	RWTH Aachen, Germany
Takahashi, Y.	Nara Inst. of Science and Techn., Japan
Tohme, S.	Telecom Paris, France
Tran-Gia, P.	University of Wuerzburg, Germany
Tsang, D.H.K	HKUST, China
Ulrich, R.	SUN Microsystems, Germany
van As, H.	Vienna University of Techn., Austria
Walke, B.	RWTH Aachen, Germany
Wolisz, A.	Techn. Univ. of Berlin/GMD, Germany
Zitterbart, M.	Techn. Univ. of Braunschweig, Germany

Reviewers

Aarstad, E.
Abu-Amara, H.
Ahmad, I.
Ajmone-Marsan, M.
Albanese, A.
Aldejohann, A.
Andersen, A. T.
Andreassen, R.
Aramomi, J.
Atmaca, T.
Autenrieth, A.

Barcelo, J. M.
Baumann, M.
Bengi, K.
Bensaou, B.
Berghoff, J.
Beylot, A. L.
Bianco, A.
Blaabjerg, S.
Blondia, C.
Boel, R.
Bonnaventure, O.
Butscher, B.

Carle, G.
Casaca, A.
Casals, O.
Casetti, C.
Cerda, L.
Chan, A.
Cheung, L. C. C.
Chiasserini, C.-F.
Christ, P.
Cosmas, J. P.
Cseh, C.
Cuthbert, L.

Daduna, H.
Daniels, T.
Decreusefond, L.
Denzel, W.

Drobnik, O.
Droz, P.

Eberhardt, R.
Eberspächer, J.

Festas,
Floris, T.
Francini, A.
Franz, W.
Frings, J.
Frohberg, W.

Gagnaire, M.
Gallassi, G.
Garcia, J.
Girard, A.
Guerin, R.
Gurtler, J

Hadingham, R. G.
Halberstadt, S.
Hebuterne, G.
Hiramatsu, Y.
Hjalmtysson, G.
Hohlfeld, B.
Hoch-Hallwachs, J.

Iliadis, I
Iselt, A.
Itoh, A.
Iversen, V. B.

Jukan, A.

Kabota, M.
Karabek, R.
Kasahara, S.
Karsten, M.
Kawashima, K.
Kenji, L
Killat, U.

Körner, U
Kofman, D.
Kouichi, G.
Kuehn, P. J.

Lagrange, X.
Lehnert, R.
Leonardi, E.
Leopold, H.
Lingnau, A.
Lo Cigno, R.

Mason, L. G.
Mönch, C.
Müller, T.

Najm, E.
Neri, F.
Nunes, M. S.
Nyberg, C.

Orlamünder, H.

Paglino, R.
Panken, F.
Paterok, M.
Peeters, S.
Pettersen, H.
Pinto, P.
Pioro, M.
Pitts, J. M.
Plasser, E.
Prögler, M.

Rathgeb, E.
Ratke, B.
Reichl, P.
Rigault, C.
Ritter, M.
Roberts, J,
Rose, O.
Rosenberg, C.

Roth, R.

Sarmento, H.
Schödl, W.
Schollenberger, W.
Schormans, J. A.
Schott, W.
Schröder, J.
Schuba, M.
Schütt, T.
Sigle, R.
Smirnow, M.
Spaey, K.
Spaniol, O.
Staalhagen, L.
Stoer, M.
Sviunset, I.

Takahashi, Y.
Thome, S.
Toyoizumi, H.
Tran-Gia, P.
Tsang, D. H. K.
Tsi-Mei, K. O.
Tutschku, K.

Ulrich, R.

Van As, H.
Van Moudt, B.
Vicari, N.

Walke, B.
Wallmeier, E.
Wolf, M.
Wolisz, A.

Yaiche, H.

Zitterbart, M.

New Applications and Internet Technologies

1

New applications of broadband networks: A vision based on the state of the art in networking

B. Plattner
Computer Engineering and Networks Laboratory, ETH Zurich
ETH-Zentrum, CH-8092 Zurich
Phone +41 1 632 7000, Fax ext. 1035, plattner@tik.ee.ethz.ch

Abstract

In this paper I discuss potential research and development directions that may prove relevant within the next ten years. I start with an observation documented by the history of research and product development in information technology, which indicates that it will take approximately ten years of product and market development efforts to transform a well-working laboratory prototype into a successful product. Thus we can estimate what the information technology market will offer ten years from now. Similarly, we can make a prediction about the topics that will be relevant in applied research by reviewing today's long-term research areas.

Keywords
new applications, vision, prediction, broadband, information technology

1 INTRODUCTION

I have been invited and asked to develop my vision of networking in general and, specifically, of broadband communications. While I agree that researchers need to have visions of the not-too-near future - definitely beyond the rather short-term visions that shareholder value oriented organizations tend to have - it is not implied that this requires prophetic capabilities. Visions, in our profession, should be based on evidence - but what evidence should we consider in our fast-living field, where one (Internet) year is said to be worth seven normal years?

To be on the safe side, I will try to generate a vision of the networking scene as it will present itself in the market ten years from now. To this end, let me start with the observation that most of today's established technology was well understood in the research lab roughly ten years before it hit the market, i.e. there appears to be a time constant of ten years for a product to mature from its invention as a basic technology. There are various examples that illustrate this observation:

Broadband Communications   P. Kühn & R. Ulrich (Eds.)
© 1998 IFIP. Published by Chapman & Hall

- Ethernet technology was developed by researchers at the Xerox PARC in the early seventies (see Metcalfe and Boggs, 1976, and the on-line replica of US Patent No. 4063220); however LAN products were available in the market only after 1980, and became an affordable mass product even later.

- The timeline of the Internet is similar. Its key concepts and initial implementations were developed by Cerf and Kahn in 1973 (see Cerf and Kahn, 1974, and Cerf, 1993), and were deployed in a rather restricted academic and military environment in the following years; only between 1985 and 1990, IP routers started to be available in the market from companies like Cisco, 3Com and Wellfleet.

- The history of graphical user interfaces as we know them today again is alike: Developed in the early seventies (again at Xerox PARC), they started to be available for normal users with the arrival of the Apple Macintosh, which was launched in 1984 (see http://www.apple-history.pair.com/128k.html).

It may be argued that all these examples are similar as they pertain to basic technologies developed in the early seventies. However, the deployment of ATM technology shows a similar pattern: Early papers on ATM, a technology initially called fast packet switching or asynchrounous time-division multiplexing, were published as early as 1984 (see Thomas, Coudreuse and Servel, 1984, for a discussion of an experimental system, or Turner, 1986, for an overview of early research), but reasonably priced equipment was not available in the market before 1995.

It is obvious that the interval of ten years is not just needed for readying a technology for the market, but also for the creation of the market itself, and for refining the production processes which will enable the industry to deliver products at prices that are acceptable to the consumers.

2   THE STATE OF THE ART IN NETWORKING

Following the observation given in the introduction, let's consider the technologies under study in the research labs at this time, and we will know what we will use ten years from now. In the sequel, I will also identify steps to be taken in the near future to further each of the technologies considered.

Broadband networking (Gbit/s backbone and Mbit/s access)

In the research labs, the basic technologies for broadband networking have come of age. It will be the challenge of the coming years to deploy Gbit/s technology to provide a universally connected infrastructure at affordable prices. An crucial issue will be to implement broadband access networks (using traditional copper, optical or wireless media) that will allow users to take advantage of the advanced applications possible with broadband networks.

Security technology

Security technology in the form of basic cryptographic mechanisms has been available for some time; however, it was never really deployed due to several reasons: Political measures (e.g. export or usage control applied by various governments) were - and are still - an obstacle preventing their general use. In addition, the market pressure has not been strong enough to create the infrastructure needed for key distribution (e.g., trusted certification authorities). The situation is shifting rapidly with the introduction of electronic commerce, which calls for a broad use of security technology for confidential and committed exchange of information. Besides the deployment of a suitable trust infrastructure, the development of workable and scalable digital payment systems (also for micropayments) will be major challenges. We are still far from being able to do digitally all we can do with traditional forms of money and value representation.

Distributed multimedia applications and integrated services networks

These topics have been very hot ones in the last few years, but despite great efforts the capabilities of computer-based multimedia systems and integrated services networks needed to deploy distributed multimedia are still far from satisfactory. Audio and video quality are often poor, delays due to encoding, transmission and decoding are too long, and signalling systems for providing QoS in networks have become much more complex than anticipated (this applies both to ATM signalling and resource reservation protocols in the Internet). I think that the breakthrough is still ahead of us, and new approaches are dearly needed.

Assuming that these problems will eventually be solved, a new problem awaits a solution (especially in the scope of the Internet): A network providing different service classes will only be workable if their usage is charged to individual users or to groups of users. However, the importance of user-based pricing, charging, accounting and billing in open and cooperatively operated networks have only recently started to draw upon the attention of the research community.

Optical networks

Unlocking the power of optical transmission for communications has been one of the major goals of many research projects. It may be anticipated that the deployment of Wavelength Division Multiplexing (WDM) technology and advances in Code Division Multiple Access (CDMA) will provide the basis for many services requiring broadband access and transmission. Nevertheless, functionally replacing current electronic switching technology with all-optical switching lies far ahead (see section 3).

Mobility

Mobile systems have been one of the fastest growing markets in the past, especially for voice communication. It is obvious that the fascination of mobility will also apply to all kinds of computer based communication, leading to facilities as we know them from Startrek and other science fiction movies. Various problems still need to be solved, however, such as broadband wireless multiaccess transmission, QoS management for multimedia applications, and providing an architecture for a mixed

mobile/terrestrial infrastructure offering a seamless and reliable service to the trav-
elling user. Besides humans, also machines (processes) will be mobile communication partners, and parts of the infrastructure will be mobile as well (just imagine
airborne users accessing the Internet via an airborne subnetwork).

Flexible and easy service creation

The explosive growth of distributed multimedia applications, such as video conferencing, video-on-demand, and collaborative distributed environments, has led to a
variety of networked multimedia services. To support this growth effectively, both
the network and the end systems require flexible service-enabling platforms, also referred to as middleware. These platforms include the functions of accessing and
sharing various types of resources, and they provide generic support for service creation, service delivery and service management. The task of such a platform can be
compared to that of an operating system providing support to applications on a computer.

Service enabling platforms are being made possible by the convergence of several
technologies, specifically, distributed computing, multimedia systems and telecommunications architectures. The challenge will be to develop and build sufficiently
high-utility and high-performance platforms that run on top of advanced networks
with fixed as well as mobile/wireless components[1].

Component-based software

As illustrated with developments in the Internet, there is a trend towards a decom-
position of full-blown applications into smaller components, which can dynamically
be activated and integrated into a given software environment. Java applets and the
(marketing) promise of the network computer are just the beginning of a development which may well create a whole new paradigm of software reusability and for
the way we create applications. If successful, these new approaches will lead to the
dynamic composition of multi-functional applications and pave the way for pay-per-
use software. The main incentive to pay for each use would be that this economic
model requires a minimum investment by the customer, of course offset by a higher
cost for actual usage. Many users may take advantage of such services, or will prefer
a mix of pay-per-use and pay-for-ownership models. It is obvious that component-
based software can only be implemented with the security services that were previously mentioned.

---

1. This text is taken from the Call for Papers of an upcoming issue of the *IEEE
    Journal on Selected Areas in Communications* on the theme "Service Enabling
    Platforms for Networked Multimedia Systems". The author is one of the guest
    editors of this issue.

3 NEW APPLICATIONS AND DIRECTIONS FOR FUTURE RESEARCH

The developments above, which are not speculative but based on work currently carried out in research and development laboratories, will be complemented by the ever increasing complexity and performance of semiconductor chips, accompanied by a commensurate decrease in per-function prices. They will enable new applications, such as:

- Multimedia-based, mobile group communication, whenever and wherever we wish or need to have it, with service guarantees (if at all required by the user).

- New forms of entertainment systems, in which the public may actively participate (e.g. with the use of virtual reality approaches). Traditional TV as we know it today may disappear altogether (however, not within the ten years considered here).

- The technological progress will enable applications needed for distance education: Virtual tele-presence will be possible without restriction in quality or quantity, and instruction material and live simulations will readily be available to the student. By consequence, dramatic changes in our systems for higher and continuing education have to be expected, changes that will probably be relevant for most of the readers that are not yet close to retirement. In line with changes in education, the same technology will foster a shift from the traditional workplaces to electronic home workplaces, which may be a preferred way of working for many people active in information-oriented businesses and professions.

- It is apparent that the long-awaited convergence of computers and communications is now happening, triggered by the Internet wave that swept the world in the past few years. We will see that technologies and applications form the areas of database systems and information retrieval will be joined with methods from telecommunications and computer networking. The former disciplines traditionally used architectures with central control, and gradually introduced aspects of distributed systems, while the latter started out with a distributed approach and attempted to achieve reliability and consistency of information, providing to the user the illusion of a centralized system. It is my strong belief that the gradual transformation of the Internet from a computer network infrastructure (providing connectivity and transport of information as a primary service) to a huge information resource is just a beginning. Whenever we realize how difficult it actually is to find the right information fast, using today's Internet search engines, it becomes obvious that we still have a long way to go.

- New approaches to control systems will evolve, likely to be denoted as "remote xy", where "xy" equates to any member of the set {surgery, factory control, control of transport systems, etc.}.

- Finally, electronic commerce will be a booming domain, especially since in the information society many goods to be traded will be information. Such trade is easily done on networks, and will generate the need for new small or large scale systems for general trade, brokerage, funds management, and clearing, etc.

However, some of the more far-reaching research carried out today will not have its effect within the next ten years. Just consider a few of these areas (without striving for completeness, of course):

All-optical (photonic) switching

While photonic switching has made some progress lately, especially on the component level, system-oriented work is still some years away. However, if available, all-optical networks would turn the wheel of development back in certain respects: Optical networks are analog by nature, tend to be noisy (due to optical amplifiers) and are subject to nonlinearities (see Thylén et.al., 1996), as opposed to the digital representation possible today in electronic switching and computing. This means that we will probably see a combination of photonic and electronic switching systems, as soon as photonic switching components will become available. One specific concern, which is due to the noise problem present in optical switching systems, is their limited scalability, which will dictate the establishment of photonic switching domains interconnected with optoelectronic converters.

Active networks

The concept of active networks has fast become a hot research topic. The incentive to make networks active is to enable users or user groups to tailor a general networking infrastructure to the need of their specific applications, e.g. by embedding application-oriented functions into the processing nodes of a network. This is well connected with the topic of (dynamic) service creation discussed above, as it will allow to implement services to be offered by the network dynamically and on demand. However, much research needs to be done in this area, and it is not obvious that all these efforts will prove beneficial, as considerable difficulties have to be overcome. Security will be a primary issue, as well as the effciency, correctness and reliability of the services to be implanted into an active network.

Quantum computing and quantum cryptography

Quantum computing[1] has the (faint?) potential of liberating us from the sequential nature of computing in which we have been imprisoned since von Neumann's invention (sequential computation is the rule even in the area of massively parallel processing, which just replicates sequential processors by the thousands). Quantum computing, if ever available, would offer mechanisms to solve NP-hard problems in polynomial time, as a quantum computer would be capable of performing an arbitrary number of operations in parallel and link the results in a probabilistic manner.

---

1. see http://aerodec.anu.edu.au/~qc/ for an updated bibliography on quantum computing

As a side-effect, quantum computers would render obsolete public-key cryptosystems as we know them today, since these rely on the impossibility to compute the solutions for certain large problems. But, as a bonus, quantum-based systems could provide new methods for the exchange of key material. Quantum cryptographic systems take advantage of Heisenberg's uncertainty principle, according to which measuring a quantum system in general disturbs it and yields incomplete information about its state before the measurement. Monitoring a quantum communication channel therefore causes an a disturbance which cannot be prevented by the eavesdropper, alerting the legitimate users. This yields a cryptographic system that can be used for the distribution of a secret random cryptographic key between two parties initially sharing no secret information. Research on quantum cryptography was pioneered by Wiesner (Wiesner, 1983); Gilles Brassard provides an excellent overview of the subject (see "A Bibliography of Quantum Cryptography", http://www.iro.umontreal.ca/~crepeau/Biblio-QC.html).

4   CONCLUSIONS

The vision given in this paper certainly is a forward-looking one. It should be obvious that the state of the art, as sketched in the respective section of this paper, still exposes many unsolved problems and opportunities for research. In the past we have achieved progress in all these areas, but to actually make the promises of the technologies currently under study come true, much research is required, primarily application-oriented research in which systems are designed, implemented and evaluated.

One should not under-estimate, however, the implications of economic, political and legal constraints on the deployment of technology. In parallel with technology deployment, a legal framework fitting the needs of the information society has to be established, not only within countries, but also on an international level.

In my deliberations, I have met some assumptions about the economic and political framework:

- The liberalisation of the telecommunications market which was established in Europe at the very time of writing will eventually lead to a distinct decrease of per-unit price for bandwidth. Broadband applications will not be possible if the unit price for bandwidth were still tied to the price of a simple telephone connection, as has been the case in the past of the monopolistic regime in telecommunications.

- Competition in the IT market will not be hampered by emerging de-facto monopolies, such as the Microsoft-Intel complex and the follow-ups of monopolistic telecommunication carriers which can already be identified after the magic date of the 1st January 1998.

Assuming this, and also anticipating that no global catastrophe will hit us in the near future, I think that the field of broadband networking will remain a very interesting and active area, in terms of research, development of products and the impact on our society as a whole.

5   REFERENCES

Cerf, V. and Kahn, R. (1974): A Protocol for Packet Switching Network Interconnection, IEEE Transactions on Communications, vol. COM-22, pp. 637-648, May 1974.

Cerf, V. (1993) How the Internet Came to Be, in "The On-line User's Encyclopedia" by Bernard Aboba, Addison-Wesley, Nov. 1993, ISBN 0-201-62214-9.

Metcalfe, R.M, Boggs, D.R. (1976) Ethernet: Distributed Packet Switching for Local Computer Networks, Commun. of the ACM, vol. 19, pp. 395-404, July 1976.

US Patent No. 4063220,
    http://patent.womplex.ibm.com/details?patent_number=4063220.

Thomas, A., Coudreuse, J.-P., Servel, M. (1984) Asynchronous time-division techniques: An experimental packet network integrating videocommunication, Proc. Int. Switching Symp. (ISS), Florence, May 1984.

Thylén, L., Karlsson, G., Nilsson, O. (1996) Switching Technologies for Future Guided Wave Optical Networks: Potentials and Limitation of Photonics and Electronics, IEEE Communications Mag., Vol. 34, No. 2, February 1996.

Turner, J. (1986) New Directions in Communications (or Which Way to the Information Age?), Proceedings of the Zurich Seminar on Digital Communication, pp. 25-32, 1986.

Wiesner, S. (1983) Conjugate coding, Sigact News, vol. 15, no. 1, 1983, pp. 78 - 88; original manuscript written circa 1970.

2

Fast address lookup for Internet routers

Stefan Nilsson
Helsinki University of Technology
P.O. Box 1100, FIN-02015 HUT, Finland, `sni@cs.hut.fi`

Gunnar Karlsson
Swedish Institute of Computer Science
P.O. Box 1263, SE-164 29 Kista, Sweden, `gk@sics.se`

Abstract

We consider the problem of organizing address tables for internet routers to enable fast searching. Our proposal is to to build an efficient, compact and easily searchable implementation of an IP routing table by using an *LC-trie*, a trie structure with combined path and level compression. The depth of this structure increases very slowly as function of the number of entries in the table. A node can be coded in only four bytes and the size of the main search structure never exceeds 256 kB for the tables in the US core routers. We present a software implementation that can sustain approximately half a million lookups per second on a 133 MHz Pentium personal computer, and two million lookups per second on a more powerful SUN Sparc Ultra II workstation.

Keywords

Address lookup, LC-trie, packet forwarding, internet protocol

1 INTRODUCTION

The high and steadily increasing demand for Internet service has lead to a new version of the internet protocol to avoid the imminent starvation of the present address space. The new version 6 being standardized will replace the current 32-bit addresses with a virtually inexhaustible 128-bit address space. Another consequence of the growth is the need for higher transmission and switching capacities. Links may read-

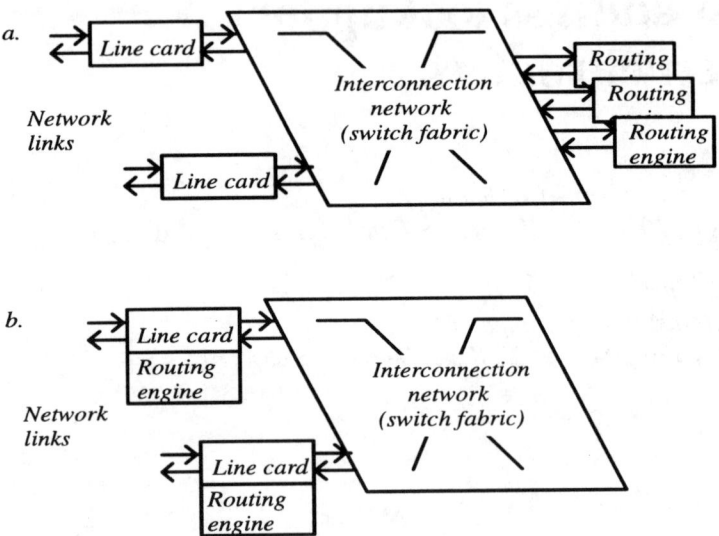

Figure 1 Two router architectures: a. one or more centralized routing engines, b. one routing engine per port.

ily be upgraded in speed and added in number to increase the network's transmission capacity. Equipment for the synchronous digital hierarchy is now being deployed to provide link rates of 155.5 Mb/s, 622 Mb/s and 2.5 Gb/s. Increases in switching capacity are not equally accessible and the bottlenecks in today's Internet are chiefly the routers.

With the upgrade of the transmission infrastructure follows that routers have to interconnect both more and faster links. But few if any of today's routers can provide switching with aggregate throughput of 50 to 100 Gb/s, as needed for a few tens of ports with bit rates of up to 2.5 Gb/s. Yet this can be accomplished by fast packet switches, as proven by the availability of high-capacity ATM switches on the market. It is important to note that there are no fundamental limits as to why the same performance cannot be expected for IP routers. The problems are rather pertaining to design choices and to practical implementation.

In this paper we consider address lookup as one of the basic functions of a router, along with buffering, scheduling and switching of packets. All of these functions have to be studied in order to increase throughput. The variable lengths of IP packets complicate buffer management and scheduling compared to protocol-data units of fixed length, but an average packet length far longer than a cell may compensate by reducing the number of such operations per second for a given link rate. Switching of packets could be made in parallel if routers incorporate space-division interconnection fabrics rather than the customarily used data buses. The remaining function that has been deemed critical for the viability of multi-Gb/s routers is the address lookup [11].

The address lookup has traditionally been performed centrally by a processor that serves all input ports of the router (Figure 1a). An input module sends the header of a packet to the processor (often called a forwarding or routing "engine"). The processor uses the destination address to determine the output port for the packet and its next-hop address (it also modifies the header). When the header has been returned along with the routing information, the input module forwards the packet across the interconnection network to the designated output port. The next-hop address is used at the output port to determine the link address of the packet, in case the link is shared by multiple parties (as an Ethernet, Token Ring or FDDI network), and it is consequently not needed if the output connects to a point-to-point link.

When a single routing engine cannot keep up with the requests from the input ports, the remedy has been to employ multiple engines in parallel. An advantage of having engines serving as one pool is the ease of balancing loads from the ports when they have different speeds and utilization levels. The disadvantage is the round-trip transfer of packet headers across the interconnection network. This load is concentrated to a few outputs and can therefore be problematic for a space-division network. When the network links are upgraded to higher bit rates, the natural modification is to place one routing engine on each input module (Figure 1b) simply because there will not be much idle time in the processing to share with other ports. This means that each engine only needs to offer a rate of address resolutions appropriate for the link, and that the interconnection network is not loaded unnecessarily by the transfer of packet headers (which will become a multiple of 40 bytes long by the introduction of IP version 6). To support a 2.5 Gb/s link means that 1.25 million lookups should be sustained on the average for today's mean packet size of 250 bytes.

In this paper we present an efficient organization of IP routing tables that allows fast address lookup in software. Our implementation can process approximately half a million addresses per second on a standard 133 MHz Pentium personal computer. The performance scales nicely to fully exploit faster memory and processor clock rates, which is illustrated by the fact that a SUN Sparc Ultra II workstation can perform 2 million lookups per second. The advantage with a software solution, such as ours, is that the processor can run separate lookup routines for multicast and flow-based routing (based on source address and a 24-bit long flow label in IP version 6). It can also process various routing options as specified by extension headers. The processor's caching protocol automatically exploit temporal correlations in packet destinations by keeping the most accessed parts of the data structure in the on-chip cache. A low-cost implementation could consist of a field-programmable gate array, instead of a microprocessor, and less than a megabyte of random-access memory.

2 ADDRESS LOOKUP FOR THE INTERNET PROTOCOL

An IP address has traditionally consisted of one to five bits specifying an address class, a network identifier and a host identifier. The network identifier points to a network within an internet, and the host identifier points to a specific computer on that network. Routing is based on the network identifier solely. Each of the classes A through C had a predetermined length of the network identifier which made the address lookup in routers straightforward.

This class-based structure has been abandoned in favor of the classless interdomain routing (CIDR) [8]. An IP address can now be split into network and host identifiers at almost any point. Address lookup is done by matching a given address to bit strings of variable lengths (prefixes) that are stored in the routing table. If more than one valid prefix matches an address in a routing table, the information associated with the longest prefix is used to forward the packet. For instance, the address 222.21.67.68 matches 222.21.64.0 with 18 bits and 222.16.0.0 with 12 bits. The first of these two is longer and should consequently be used for forwarding the packet.

The matching prefix is not necessarily the same in all routers for one and the same IP address. This allows the tables to be smaller in size since a large number of destinations can be represented by a short network identifier. Most routers use a default route which is given by a prefix of length zero. It matches all addresses and is used if no other prefix match. The core routers in the Internet backbone are not allowed to use default routes and their address tables tend to be larger than in other routers. We have used tables from these routers in our evaluation to ensure that we test with realistic worst cases.

The result of the lookup is the port number and next-hop address. A next-hop address is not needed for a point-to-point link and the corresponding routing-table entry would only need to contain the port number. Even when next-hop addresses are needed, there are usually fewer distinct such addresses than there are entries in the routing table. The table can therefore contain a pointer to an array which lists the next-hop addresses in use.

The address structure for IP version 6 is not fully decided even for unicast addresses. It is, however, suggested to keep the variable-length network identifiers (or subnetwork identifiers) [10]. Thus, a subnetwork can be identified by some n bits in a router, while the remaining $128 - n$ bits form the interface identifier (replacing the host identifier of version 4). Our data structure has been designed to handle version 6 addresses when needed.

There has been remarkably little interest in the organization of routing tables both for hardware and software based searches during the last years. Hardware implementation, which we do not consider here, is discussed in [17]. We therefore place our proposal in relation only to the three most recent works from the literature. Our search structure and implementation is akin to and has been inspired by the work of Brodnik et al. [4]. They use a different data structure from ours and are concerned with the size of the trie to ensure that it fits in a processor's on-chip cache memory. As a consequence, it is not immediately clear if the structure will scale to the longer addresses of IP version 6. A main idea of their work is to quantify the prefix lengths to levels of 16, 24 and 32 bits. This exploits the old class-based address structure but may suffer from the ongoing redistribution of addresses that tends to smooth the distribution of the prefix lengths. Rather than expanding the prefixes to a few levels, we use level-compression to reduce the size of the trie. Thus, we obtain similar performance in our simulations with a more general structure and without making any assumptions about the address structure.

Waldvogel et al. [15] use a different technique. Prefixes of fixed lengths are stored in separate hash tables and the search operation is implemented using binary search on the number of bits. Using 16 and 24 bits for the first two hash table lookups this search strategy is very efficient for current routing tables. However, for a routing table with longer prefixes and a smother distribution of prefix lengths this approach may not be as attractive.

The work on prefix matching by Doeringer, Karjoth, and Nassehi [6] uses a trie structure. One of their concerns is to allow fully dynamic updates. This results in a large space overhead and less than optimum performance. The nodes of the trie structure contains five pointers and one index. More general but slightly dated works on Gb/s routers that could be of interest to the reader are presented in [3, 14, 16].

3   LEVEL-COMPRESSED TRIES

The *trie* [7] is a general purpose data structure for storing strings. The idea is very simple: each string is represented by a leaf in a tree structure and the value of the string corresponds to the path from the root of the tree to the leaf. Consider a small example. The binary strings in Figure 2 correspond to the trie in Figure 3a. In particular, the string 010 corresponds to the path starting at the root and ending in leaf number 3: first a left-turn (0), then a right-turn (1), and finally a turn to the left (0). For simplicity, we will assume that the set of strings to be stored in a trie is prefix-free, no string may be a proper prefix of another string. We postpone the discussion of how to represent prefixes to the next section.

This simple structure is not very efficient. The number of nodes may be large and the average depth (the average length of a path from the root to a leaf) may be long. The traditional technique to overcome this problem is to use *path compression*, each internal node with only one child is removed. Of course, we have to somehow record which nodes are missing. A simple technique is to store a number, the *skip value*, in each node that indicates how many bits that have been skipped on the path. A path-compressed binary trie is sometimes referred to as a Patricia tree [9]. The path-compressed version of the trie in Figure 3a is shown in Figure 3b. The total number of nodes in a path-compressed binary trie is exactly $2n - 1$, where n is the number of leaves in the trie. The statistical properties of this trie structure are very well understood [5, 12]. For a large class of distributions path compression does not give an asymptotic reduction of the average depth. Even so, path compression is very important in practice, since it often gives a significant overall size reduction.

One might think of path compression as a way to compress the parts of the trie that are sparsely populated. *Level compression* [1] is a recently introduced technique for compressing parts of the trie that are densely populated. The idea is to replace the i highest complete levels of the binary trie with a single node of degree 2^i; this replacement is performed recursively on each subtrie. The level-compressed version, the *LC-trie*, of the trie in Figure 3b is shown in Figure 3c.

nbr	string
0	0000
1	0001
2	00101
3	010
4	0110
5	0111
6	100
7	101000
8	101001
9	10101
10	10110
11	10111
12	110
13	11101000
14	11101001

Figure 2 Binary strings to be stored in a trie structure.

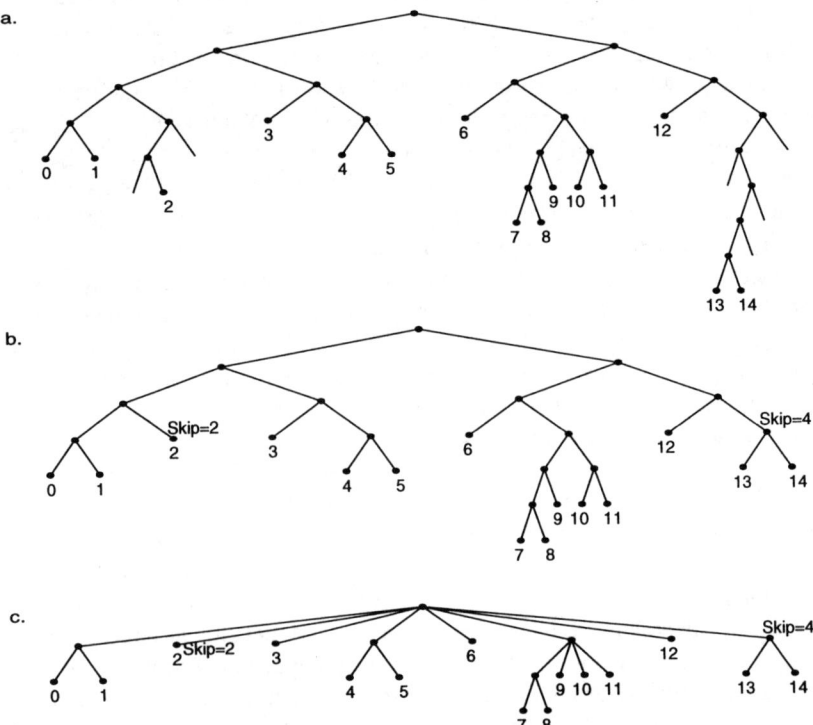

Figure 3 a. Binary trie, b. Path-compressed trie, c. LC-trie.

	branch	skip	pointer
0	3	0	1
1	1	0	9
2	0	2	2
3	0	0	3
4	1	0	11
5	0	0	6
6	2	0	13
7	0	0	12
8	1	4	17
9	0	0	0
10	0	0	1
11	0	0	4
12	0	0	5
13	1	0	19
14	0	0	9
15	0	0	10
16	0	0	11
17	0	0	13
18	0	0	14
19	0	0	7
20	0	0	8

Figure 4 Array representation of the LC-trie in Figure 3c.

For an independent random sample with a density function that is bounded from above and below the expected average depth of an LC-trie is $\Theta(\log^* n)$, where $\log^* n$ is the iterated logarithm function, $\log^* n = 1 + \log^*(\log n)$, if $n > 1$, and $\log^* n = 0$ otherwise. For data from a Bernoulli-type process with character probabilities not all equal, the expected average depth is $\Theta(\log \log n)$ [2]. Uncompressed tries and path-compressed tries both have expected average depth $\Theta(\log n)$ for these distributions.

If we want to achieve the efficiency promised by these theoretical bounds, it is of course important to represent the trie efficiently. The standard implementation of a trie, where a set of children pointers are stored at each internal node is not a good solution, since it has a large space overhead. This may be one explanation why trie structures have traditionally been considered to require much memory.

A space efficient alternative is to store the children of a node in consecutive memory locations. In this way, only a pointer to the leftmost child is needed. In fact, the nodes may be stored in an array and each node can be represented by a single word. In our implementation, the first 5 bits represent the *branching factor*, the number of descendants of the node. This number is always a power of 2 and hence, using 5 bits, the maximum branching factor that can be represented is 2^{31}. The next 7 bits represent the skip value. In this way, we can represent values in the range from 0 to 127, which is sufficient for IP version 6 addresses. This leaves 20 bits for the pointer to the leftmost child and hence, using this very compact 32 bit representation, we can store at least 2^{19} strings. Figure 4 shows the array representation of the LC-trie in Figure 3c, each entry represents a node. The nodes are numbered in breadth-first order starting at the root. The number in the branch column indicates the number of bits used for branching at each node. A value $k \geq 1$ indicates that the node has 2^k children. The value $k = 0$ indicates that the node is a leaf. The next column contains the skip value, the number of bits that can be skipped during a search operation. The value in the pointer column has two different interpretations. For an internal node, it is used as a pointer to the leftmost child; for a leaf it is used as a pointer to a base vector containing the compleat strings.

The search algorithm can be implemented very efficiently. Let s be the string searched for and let EXTRACT(p, b, s) be a function that returns the number given by the b bits starting at position p in the string s. We denote the array representing the tree by T. The root is stored in T[0].

```
node = T[0];
pos = node.skip;
branch = node.branch;
adr = node.adr;
while (branch != 0) {
    node = T[adr + EXTRACT(pos, branch, s)];
    pos = pos + branch + node.skip;
    branch = node.branch;
    adr = node.adr;
}
return adr;
```

Note that the address returned only indicates a possible hit; the bits that have been skipped during the search may not match. Therefore we need to store the values of the strings separately and perform one additional comparison to check whether the search actually was successful.

As an example we search for the string 10110111. We start at the root, node number 0. We see that the branching value is 3 and skip value is 0 and therefore we extract the first three bits from the search string. These 3 bits have the value 5 which is added to the pointer, leading to position 6 in the array. At this node the branching value is 2 and the skip value is 0 and therefore we extract the next two bits. They have the value 2. Adding 2 to the pointer we arrive at position 15. At this node the branching value is 0, which implies that it is a leaf. The pointer value 5 gives the position of the string in the base vector. Observe that it is necessary to check whether this constitutes a true hit. We need to compare the first 5 bits of the search string with the first 5 bits of a value stored in the base vector in the position indicated by the pointer (10) in the leaf. In fact, our table (Figure 2) contains a prefix 10110 matching the string and the search was successful.

4   REPRESENTATION OF A ROUTING TABLE

The routing table consists of four parts. At the heart of the data structure we have an *LC-trie* implemented as discussed in the previous section. The leaves of this trie contain pointers into a *base vector*, where the complete strings are stored. Furthermore we have a *next-hop table*, an array containing all possible next-hops addresses, and a special *prefix vector*, which contains information about strings that are proper prefixes of other strings. This is needed because internal nodes of the LC-trie do not contain pointers to the base vector.

The base vector is typically the largest of these structures. Each entry contains a string. In the current implementation it occupies 32 bits, but it can of course easily be extended to the 128 bits required in IP version 6. Notice that it is not necessary to store the length of the string, since the length will be known to the search routine after the traversal of the LC-trie. Each entry also contains two pointers: one pointer into the next-hop table and one pointer into the prefix table. The search routine follow the next-hop pointer if the search was successful. If not, the search routine tries to match a prefix of the string with the entries in the prefix table. The prefix pointer has the special value -1 if no prefix of the string is present.

The prefix table is also very simple. Each entry contains a number that indicates the length of the prefix. The actual value need not be explicitly stored, since it is always a proper prefix of the corresponding value in the base vector. As in the base vector each entry also contains two pointers: one pointer into the next-hop table and one pointer into the prefix table. The prefix pointer is needed, since it might happen that a path in the trie contains more than one prefix.

The main part of the search is spent within the trie. In our experiments the average depth of the trie is typically close to 6 and one memory lookup is performed for each node traversed. The second step is to access the base vector. This accounts for one additional memory lookup. If the string is found at this point, one final lookup in the next-hop table is made. This memory access will be fast since the next-hop table is typically very small, in our experiments less than 60 entries, and can be expected to reside in cache memory.

Finally, if the string searched for does not match the string in the base vector, an addition lookup in the prefix vector will have to be made. Also this vector is typically very small, in our experiments it contains less than 2000 entries, and it is rarely accessed more than once per lookup. In all the routing tables that we have examined we have found only a few multiple prefixes. Conceptually a prefix corresponds to an exception in the address space. Each entry in the routing table defines a set of addresses that share the same routing table entry. In such an address set, a longer match corresponds to a subset of addresses that should be routed differently. Overlapping prefixes could also be avoided by expanding the shorter prefixes. For instance, instead of having the overlapping prefixes 101 and 10111 in a table, 101 could be expanded to 10100, 10101 and 10110 which all would point to the same routing information. The expansion increases the trie size and we have therefore chosen to use a separate prefix vector. Furthermore, we expect this special case to disappear with

Site	Routing Entries	Next-Hops	Number of Entries			Aver. Depth	Lookups	
			Trie	Base	Prefix		Sparc	PC
Mae East	39 819	56	62 991	38 380	1439	5.92	1.5	0.42
Mae West	14 618	55	24 143	14 291	327	6.08	2.1	0.48
AADS	20 299	19	33 159	19 846	453	6.28	1.6	0.45
Pac Bell	20 611	3	33 591	20 171	440	6.32	1.9	0.45
FUNET	41 578	20	63 623	39 765	1813	8.31	2.3	0.58

Table 1 Experimental data. The speed is measured in million lookups per second.

the introduction of IP version 6 since the new address space will be so large as to make it possible to allocate addresses in a strictly hierarchical fashion.

5   RESULTS

The measurements were performed on two different machines: a SUN Ultra Sparc II with two 296-MHz processors and 512 MB of RAM, and a personal computer with a 133-MHz Pentium processor and 32 MB of RAM. The programs are written in the C programming language and have been compiled with the gcc compiler using optimization level -04. We used routing tables provided by the Internet Performance Measurement and Analysis project (URL http://www.merit.edu/ipma). We have used the routing tables for Mae East, Mae West, AADS, and Pac Bell from the 24th of August, 1997. We did not have access to the actual traffic being routed according to these tables and therefore the traffic is simulated: we simply use a random permutations of all entries in a routing table. The entries were extended to 32 bits numbers by adding zeroes (this should not affect the measurements, since these bits are never inspected by the search routine). We have, however, been able to test our algorithm on a routing table with recorded traces of the actual packet destinations. The router is part of the Finnish University and Research Network (FUNET). Real traffic gives better results than runs of randomly generated destinations and owes to dependencies in the destination addresses. The time measurements have been performed on sequences of lookup operations, where each lookup includes fetching the address from an array, performing the routing table lookup, accessing the nexthop table and assigning the result to a volatile variable.

Some of the entries in the routing tables contain multiple next-hops. In this case, the first one listed was selected as the next-hop address for the routing table, since we only considered one next-hop address per entry in the routing table. There were also a few entries in the routing tables that did not contain a corresponding next-hop address. These entries were routed to a special next-hop address different from the ones found in the routing table.

Table 1 gives a summary of the results. It shows the number of unique entries in the routing table, the number of next-hop addresses, the size of our data structure, the average of the trie, and the number of lookups measured in million lookups per second. In our current implementation an entry in the trie occupies 4 bytes, while the entries in the base vector and the prefix vector occupy 12 bytes each. The largest table, FUNET, still occupies less than 0.8 MB of memory, of which the trie-part is less than 256 kB.

The average throughput corresponding to the number of lookups per second is found by multiplying it with the average packet size, which currently is around 250 bytes. The routing system could be modeled as a G/G/1 queue in order to find the number of pending lookups as a function of the routing system's load. The arrival process would be given by the trimodal packet-length distribution (peaks around 40, 550 and 1500 bytes) divided by the link rate and the service distribution by the lookup times.

6 FINAL REMARKS AND CONCLUSIONS

The Internet has in practice become the long-sought broadband integrated services digital network, in the meaning of a global communication infrastructure for multimedia services [13]. The intention was, however, that the B-ISDN should have been based on the asynchronous transfer mode rather than on IP. Now when that raison d'être for ATM is disappearing there is naturally great interest in salvaging the enormous investment that has been made by finding a role for ATM as carrier of IP packets. Thus, we find IP routing combined with ATM switching. The idea is to establish a virtual circuit for a flow of packets in order to amortize the cost of IP address lookup over several packets [11]. We believe that pure IP routing is a competitive alternative and have shown that software-based address lookup can be performed sufficiently fast for multi-Gb/s systems.

We have demonstrated how IP routing tables can be succinctly represented and efficiently searched by structuring them as level-compressed tries. Our data structure is fully general and does not rely on the old class-based structure of the address format for its efficiency. Even though the data structure does not make explicit assumptions about the distribution of the address, it does adapt gracefully: path compression compacts the sparse parts of the trie and level compression packs the dense parts. The average depth of the trie grows very slowly. This is in accordance with theoretical results. Recall that the average depth of an LC-trie is $\Theta(\log \log n)$ for a large class of distributions. Actually, our experiments show that in some cases the average depth is smaller for a larger table. This can be explained by the fact that a larger table might be more densely populated and hence the level compression will be more efficient. The inner loop of the search algorithm is very tight; it contains only one addressing operation and a few very basic operations, such as shift and addition. Furthermore, the trie can be stored very compactly, using only one 32-bit machine word per node. In fact, the largest trie in our experiments consumed less than 256 kB of memory. The base vector is larger, but is only accessed once per lookup.

The results show that the routinely made statements about possible processing speeds for IP addresses, such as those put forward in [11], are not valid. In many cases, for instance in [4], the trie implementations cited do not reflect the state of the art. Thus, we argue that our new data structure is superior to earlier presented software methods for the organization of IP routing tables.

We thank Erja Kinnunen and Pekka Kytölaakso of the Center for Scientific Computing at Helsinki University of Technology for providing the FUNET routing table and associated packet traces. This research was done when G. Karlsson was visiting professor at the Telecommunication Software and Multimedia Laboratory at the Helsinki University of Technology. This support is gratefully acknowledged.

REFERENCES

[1] A. Andersson and S. Nilsson. Improved behaviour of tries by adaptive branching. *Information Processing Letters*, 46(6):295–300, 1993.

[2] A. Andersson and S. Nilsson. Faster searching in tries and quadtrees – an analysis of level compression. In *Proceedings of the Second Annual European Symposium on Algorithms*, pages 82–93, 1994. LNCS 855.

[3] A. Asthana, C. Delph, H. V. Jagadish, and P. Krzyzanowski. Towards a gigabit IP router. *Journal of High Speed Networks*, 1(4):281–288, 1992.

[4] M. Degermark, A. Brodnik, S. Carlsson, and S. Pink. Small forwarding tables for fast routing lookups. *ACM Computer Communication Review*, 27(4):3–14, October 1997.

[5] L. Devroye. A note on the average depth of tries. *Computing*, 28(4):367–371, 1982.

[6] W. Doeringer, G. Karjoth, and M. Nassehi. Routing on longest-matching prefixes. *IEEE/ACM Transanctions on Networking*, 4(1):86–97, February 1996.

[7] E. Fredkin. Trie memory. *Communications of the ACM*, 3:490–500, 1960.

[8] V. Fuller, T. Li, J. Yu, and K. Varadhan. Classless inter-domain routing (CIDR): an address assignment and aggregation stragegy. Request for Comments: 1519, September 1993.

[9] G. H. Gonnet and R. A. Baeza-Yates. *Handbook of Algorithms and Data Structures*. Addison-Wesley, second edition, 1991.

[10] R. Hinden and S. Deering. IP version 6 addressing architecture. Request for Comments: 1884, December 1995.

[11] P. Newman, G. Minshall, T. Lyon, and L. Huston. IP switching and gigabit routers. *IEEE Communications Magazine*, 35(1):64–69, January 1997.

[12] B. Rais, P. Jacquet, and W. Szpankowski. Limiting distribution for the depth in Patricia tries. *SIAM Journal on Discrete Mathematics*, 6(2):197–213, 1993.

[13] W. D. Sincoskie. Viewpoint: Broadband ISDN is happening – except it's spelled IP. *IEEE Spectrum*, 34(1):32–33, 1997.

[14] A. Tantawy, O. Koufopavlou, and M. Zittertbart. On the design of a multigigabit IP router. *Journal of High Speed Networks*, 3(3), 1994.

[15] M. Waldvogel, G. Varghese, J. Turner, and B. Plattner. Scalable high speed IP routing lookups. *ACM Computer Communication Review*, 27(4):25–36, October 1997.

[16] R. J. Walsh and C. M. Özveren. The gigaswitch control processor. *IEEE Network*, 9(1):36–43, January/February 1995.

[17] C. A. Zukowski and T. Pei. Putting routing tables into silicon. *IEEE Network*, pages 42–50, January 1992.

3

Design and evaluation of an advance reservation protocol on top of RSVP

A. Schill, S. Kühn , F. Breiter
Dresden University of Technology; Chair for Computernetworks
01062 Dresden, Germany, call: 049351/4638261
schill/kuehn/breiter@ibdr.inf.tu-dresden.de

Abstract
Existing reservation protocols such as RSVP implement so-called immediate reservations (ImRe) that are requested and granted just when the resources are actually needed. This paper describes the design, implementation and evaluation of a novel advance reservation protocol. Its major property is that resources are reserved well in advance to the actual usage phase. This way, the probability of resource availability can be increased, and network utilization can improved. Specific aspects covered in this paper are the mapping of ReRA (resource reservation in advance) onto RSVP, the evaluation and selection of an appropriate call admission control strategy, the internal management of advance reservations, the mapping of the approach onto ATM, as well as its implementation and evaluation.

Keywords:
ATM, RSVP, resource reservation in advance, ReRA, quality of service

1 INTRODUCTION

Considering, for instance, the „real world", activities such as a meeting in a conferencing room are scheduled for a specific time. With this example, reservation means that limited resources are reserved a certain time in advance to get an assurance that the resources are available at the requested time. The definition of reservation in current QoS based networks like ATM covers another kind of semantics: reservations are being performed in conjunction with a connection establishment; this means that reservations are done at the time when the network resources are actually needed. This is an element of uncertainty, because in the worst case the required resources are used by other applications and therefore the application request must be rejected. However, using ReRA, a future lack of resources during actual reservations can be avoided.

From our point of view, it shall be possible to schedule applications, e.g. an important videoconference with several partners, for a given time in the future. The system shall then calculate and virtually reserve the required resources for the

Broadband Communications   P. Kühn & R. Ulrich (Eds.)
© 1998 IFIP. Published by Chapman & Hall

specific time, however without immediately blocking them. In order to grant the required QoS, it is necessary to provide mechanisms of resource reservation in advance as part of reservation protocols.

Resource reservation in advance allows a more flexible handling of applications by decoupling the starting time of the service from the time the service request is made. This is implemented by our ReRA mechanisms as an extension of existing reservation protocols (RSVP on top of ATM). As an application example, these mechanisms can support the computation of the earliest time for Scalable Video on Demand (SVoD); so the user can play the movie with the desired QoS at the earliest possible time when the required resources are available. (Hafid, 1997) shows that SVoD achieves high resource utilization and notably decreases the blocking probability of user requests in comparison with typical VoD.

As most current and future networks are not purely based on ATM but consist of a mix of various technologies (ATM, FDDI, Ethernet etc.), higher-level protocols on top of ATM such as IP are necessary to enable heterogeneous interoperability. To achieve QoS guarantees, reservation protocols such as RSVP (Resource ReServation Protocol) are necessary at this level. In order to address advance reservations, we present an implementation of a ReRA signaling mechanism in RSVP. Moreover, an admission control strategy combining ATM and ReRA algorithms are discussed and simulation results are presented.

The paper is organized as follows: In section 2 we shortly discuss differences between typical immediate and advance reservations and the suitability of extending RSVP with ReRA mechanism. Moreover, related work concerning resource reservation in advance is outlined. Section 3 describes the geneneral model and its mapping onto the communication mechanisms in RSVP in detail with references to specific problems and associated solutions. In view of using ReRA on top of a RSVP/ATM protocol stack, an admission control algorithm will be discussed in section 4. In section 5, implementation aspects of ReRA as an RSVP extension are presented as a synergy of the former considerations.

2 FOUNDATIONS

ATM networks (Alles, 1995; ATM1, 1995; ATM2, 1995; Cidon, 1995; Partridge. 1994) with traffic classes such as constant bit rate (CBR), variable bit rate (VBR) and available bit rate (ABR) (Konst, 1995) offer quality of service and traffic parameters in order to specify and guarantee QoS. Resource reservations in terms of bandwidth, buffers and other resources performed accordingly. QoS requirements, however, can not be specified explicitly with conventional IP over ATM, multiprotocol over ATM (MPOA) or LAN emulation services (Jeffries, 1994; Laubach, 1994).

However, emerging protocols such as RSVP (Resource Reservation Protocol) (Braden, 1995; Bordon, 1995) or the older ST-II approach (Delgrossi, 1994) are able to map transport-level QoS requirements onto ATM (Crowcroft, 1995). As opposed to the use of native ATM, this facilitates the integration of heterogeneous networks (Malamud, 1992), for example including Fast Ethernet, Gigabit Ethernet, FDDI and other technologies. Together with the emerging IPv6 protocol (Gilligan,

1995), RSVP will form the basis for the Integrated Services Internet and for Internet-II with its gigabit links. Higher-level quality of service architectures enable a further abstraction of QoS specifications, facilitating the implementation of QoS-aware applications (Campbell, 1994; Vogel, 1995).

However, the reservation approach of RSVP and related protocols attempts to reserve resources just at the point in time when they are actually required. No advance planning or resource scheduling is possible this way. Therefore, the likelihood of receiving a rejection of required reservations by the network increases. As an interesting solution to this problem, several authors have suggested advance reservation techniques. Basic concepts of advance reservation are discussed in (Reinhard, 1994), with a consideration of existing reservation protocols presented in (Reinhard, 1995). A detailed discussion of major issues and trade-offs of advance reservation techniques is also given in (Wolf, 1995). More recently, simulation studies (Degermark, 1997; Ferrari, 1997) have been conducted in order to investigate performance effects concerning the duration of reservations and the advance booking time. (Ferrari, 1997) also emphasizes the coexistence of advance reservations and immediate reservations within the same environment. Moreover, (Degermark, 1997) presents initial considerations concerning the mapping of advance reservation protocols onto RSVP.

Our work is based on the conceptual considerations of these related approaches. However, we particularly emphasize the following specific issues:

Admission control: We investigate several alternative admission control strategies and adapt the selected approach to the requirements of ReRA.

Management of reservations: We analyze different mechanisms for supporting coexistence of advance and immediate reservations. We also propose mechanisms for optimizing resource utilization based on flexible periods of validity of reservation requests. Moreover, in case of rejection of advance reservations, we propose appropriate alternatives to the applications.

Validation by simulation and implementation: In addition to further simulation studies concerning admission control and resource utilization, we are also able to present first results of a running implementation of ReRA in order to validate our concepts. Finally, we attempt to integrate our model with the existing policy control of RSVP (Herzog, 1996; Herzog, 1997).

In the following section, our reservation model and protocol is being presented, followed by details of the call admission control and the implementation.

3 RESERVATION MODEL AND PROTOCOL

3.1 General Model

Advance reservations are characterized by the announcement of events for a given time in the future. So one of the necessary parameters is the starting time, that means the time the resources will be needed. Furthermore, it seems to be meaningful to specify a duration which describes how long the reservation will be alive. Using this information the admission control is able to recognize whether the reservation is overlapping with others or not.

Based on these general ReRA characteristics our model is shown in Figure 1 describing the extended communication to set up a resource reservation in advance. Three different phases can be distinguished which are closely connected with our general model. We use these phases in analogy to (Reinhard, 1995). Moreover, in our detailed view we have introduced some time marks at which messages are sent or actions are performed within the phases.

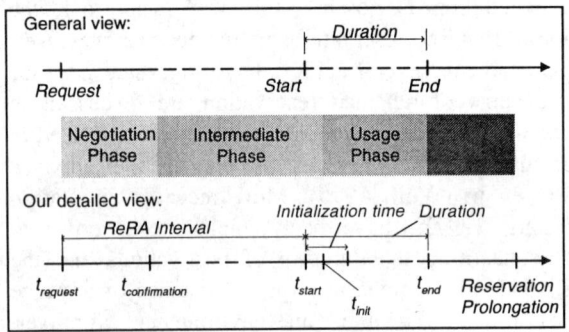

Figure 1: General Model

To be more precise, at the point $t_{request}$ the client issues a request to specify how much of the resource capacities have to be reserved. It also specifies the points in time that define beginning (t_{start}) and duration (d_{res}) of the reservation. After completed negotiation and admission control, the negotiation phase is finished by getting a confirmation message ($t_{confirm}$).

After a successful negotiation, the users might want to re-negotiate the communication resources. This covers the cancellation or an increase or decrease of the requested resources. If the changes are acceptable or the reservation was successfully released, the service provider acknowledges it. Re-negotiation concerns not only the traffic and QoS parameters of the reserved connection but also the starting point and duration. A re-negotiation can be performed until t_{end} is reached.

When the start time is reached, the activation of the requested resources has to be performed within the initialization time. If the client is not sending a demand message by then, the reservation will be aged and all state information corresponding to it will be removed. We prefer this additional communication at t_{start} because last changes should be possible and resources are only hard and physically reserved if they are really needed. In contrast to (Wolf, 1995) we don't need a special preparation mechanism for the resource usage phase, for example for preempting running applications with immediate reservations. We just have to convert a logical reservation into a physical one as the resources are already available.

3.2 RSVP protocol extensions

In the following essential changes and extensions of RSVP to support ReRA will be described. In particular, this includes:

- the negotiation of ReRA information,
- the modification of messages to carry the additional information (ReRA object),
- new RSVP-API functions and
- a special handling of RSVP in case of short and longer termed router failures before the starting time of the advance reservation is reached.

Moreover, to realize a complete extension of the RSVP specification to support

ReRA it also includes an extension of: the state control blocks (Path-, Reservation- and Traffic Control blocks) to support the pseudo hard state concept (see below) and to hold the new reservation parameters, the message processing rules and the specific RSVP interfaces (e.g. traffic control API).

Negotiation of ReRA information

To describe the negotiation of ReRA information, we have introduced service primitives which are derived from the general model:
ResvReq/ ResvInd,
ResvRsp/ ResvConf,
ResvModReq/ ResvModInd,
ResvModRsp/ ResvModConf,
ActivateReq/ ActivateInd.
These service primitives were completed to announce a future data flow:
AnnounceReq/ AnnounceInd,
AnnModReq/ AnnModInd.
Looking at the communication (see Figure 2) we can recognize that it is possible to map the service primitives of the advance reservation to appropriate RSVP messages. So we use PATH messages to announce the future data flow and RESV messages to setup an advance reservation for that flow. In this way we don't have to introduce new RSVP messages. Merely an extension of the currently defined messages is needed to carry the additional parameters.

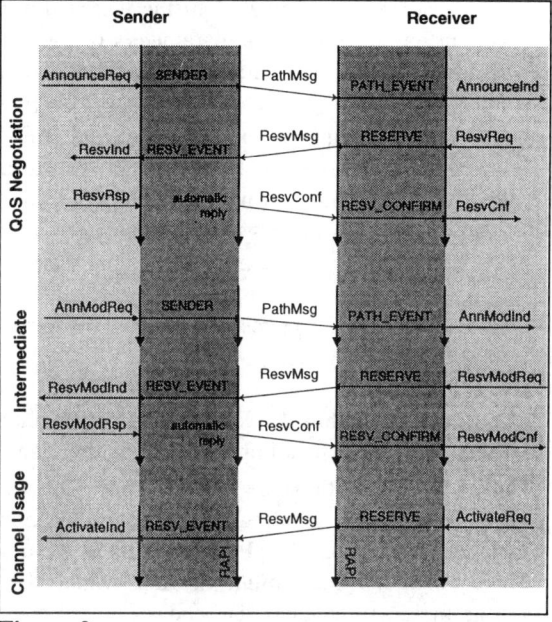

Figure 2: Mapping of ReRA primitives to RSVP messages

ReRA - Object

The negotiation of ReRA information is done by sending PATH and RESV messages carrying an additional instance of a new RSVP (Figure 3). As the ReRA object is an additional and optional object in an RSVP message, a result of this solution is that the mechanisms for sending and receiving RSVP messages need not

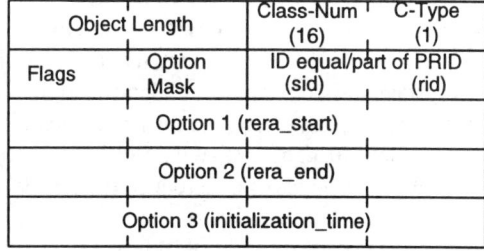

Object Length	Class-Num (16)	C-Type (1)
Flags	Option Mask	ID equal/part of PRID (sid)	(rid)
Option 1 (rera_start)		
Option 2 (rera_end)		
Option 3 (initialization_time)		

Figure 3: ReRA Object

to be changed and the semantics of the messages will be preserved for ImRe.

The definition of the class value (Class-Num 16; C-Type 1) has two reasons: first, to be compatible with other RSVP platforms and secondly to get error information from non ReRA nodes instead of dropping ReRA messages.

A ReRA object consists of a common header, followed by a object specific body. The first byte of the body includes the following *Flags*:

- WILDCARD: If this flag is set and the start and/or end time are not specified then the missing time parameters of the reservation will be filled automatically by the system based on the time parameters of the PATH message. This operation will also be enforced after a modification of the data flow start and end time.
- ACTIVATE (for sender and receiver): This flag is used, for example in case of a receiver, to distinguish between a normal RESV (refresh) message and a RESV message, which is send at the starting point to indicate that the user wishes to use the reserved resources (the latter complies with ActivateReq). If the receiver is not sending a RESV message (with this flag) within the initialization time, a RESV-TEAR message will be sent automatically and all state information corresponding to it will be removed. Automatically means, that we send a RESV-TEAR message instead of a RESV refresh message if the current time is greater than the starting time of the reservation.

The next byte within the ReRA object includes an *Option Mask* which is used to initiate a special handling of the information (option) specified in that class. Currently three options are defined (rera_start, rera_end, initialization_time)[1]. Each option has a corresponding bit within the option mask. For example, in the RESV message, the *Option_1* and *Option_2* bit of the option mask is set to indicate that the object includes the points in time when the reservation starts and finishes. Furthermore, to change the end of an already requested reservation only the *Option_2* bit is set and the options field carries the new end time.

Moreover, an *ID* (identifier) is part of the object to realize a unique identification of a data flow and a reservation as part of a session. As it is possible that a sender can announce more than one data flow and a receiver can request more than one reservation per data flow[2], the sender and receiver have to number it consecutively in *Sid* (sender id) and *Rid* (receiver id).

Hence, a data flow is identified by: *<Session, Sender_Template, Sid>* and
a reservation is identified by: *<Session, Filter_Spec, Sid, Recv_Address, Rid>*

Another possible way is, that the ReRA object consists of two options including rera_start and rera_end (for identification) and two additional options including a new rera_start and rera_end for a modification. Using an ID to identify data flow and reservation, respectively, we don't have to include the rera_start and rera_end time in each message and so we reduce the RSVP overhead.

---

[1] Currently we use a fixed option length. In the future the option field will have a variable length.

[2] In a video conference for example.

Extending RSVP with such a mechanism is supporting and ideally complementing the newly introduced policy control of the IETF (Herzog, 1996; Herzog, 1996; Herzog, 1997). These policies follow the bilateral agreements model. The model assumes that network clouds (providers) contract with their closest point of contract to establish ground rules and arrangements for access control and accounting. The advanced reservation scenario is almost identical to the simple access control. Moreover, it is assumed that each verbal bilateral pre-registration is identified by a PRID (pre-registration ID). The policy data object defined by the IETF has the type of AR (Advanced Reservation) and the form PD (AR, PRID, UID).

Using our mechanism an automatic bilateral pre-registration is realized. Sending the ReRA request the object includes one additional option: the UID. Moreover, the PRID is defined by the identifier in the ReRA object, the session and addresses of sender and receiver. As result of these combined mechanisms the activation and an additional access control for authorized users of an advanced reservation will be performed by sending only the policy object in a RESV message.

New RSVP-API functions

The following functions are introduced to realize a more flexible handling of advanced reservation within the RSVP-API:

- LEAVE: This function allows the user to leave a session without releasing the soft states. It is used to inform the RSVP daemon that the application is not actually available for the next time until the application will make a modification or the starting point is reached. The application should save session information to be able to re-register by calling the session function.

- TEARDOWN: As we allow more than one reservation in one session (which can be distinguished by the start and end time of the reservation) this function can be called to remove a specific reservation.

Short and longer termed router failures

As the advance notice of an advance reservation request can take a long time, the occurrence of special failures and its removal have to be considered. This concerns among other things the actual availability of endsystems and the reaction to short and longer termed router failures (Figure 4) which implies a changed topology of the net. In order to recognize and to handle such kind of failures we have developed a special algorithm. Based on the expired cleanup timer for each reservation, which is the result of missed RSVP refresh messages, longer termed failures are recognized and temporary router failure are ignored by increasing the lifetime of the soft states (see below).

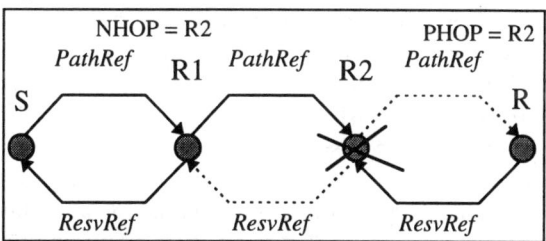

Figure 4: Router failure of router R2

Concerning the expiration of the lifetime there are two reasons. Either there is a router failure (crash, topology changes) or an endsystem is unplugged from the net. In the case of longer termed router failures a new route will be determined using the routing protocol. In this context it should be emphasized that current routing protocols are concerned only with topology information and not network resources. This can result in alternate paths and hence in rejection of advance reservation within the intermediate phase. In order to avoid it, route pinning has been suggested

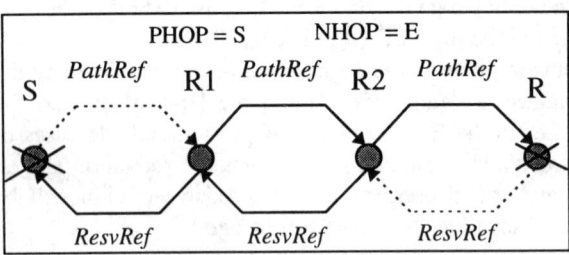

Figure 5: Unplugged endsystems

in (Crowley, 1996): „Route pinning means that an existing route with a reservation will not be replaced by a better route unless the existing one is no longer usable because of a topology change directly related to the existing route".

If an endsystem is unplugged from the network (for instance a PC which was turned off; Figure 5), this results in missed PATH and RESV refresh messages and in a loss of the corresponding soft states of the neighboured router. In order to avoid this, the routers which are situated immediately behind the sender and the receiver, respectively, should react if no new refresh message arrives after the expired lifetime. Because these routers take over the function of the endsystems in sending PATH and RESV refresh messages, they will be called proxy routers. Moreover, each router has to determine if the next host is the receiver downstream or the sender upstream, itself:

- NHOP = receiver address[3] (if RESV refresh message is expected) or
- PHOP = sender address (if PATH refresh message is expected).

If one of these cases is fulfilled, this results in holding the soft states in the router. More precisely, these soft states are kept and only removed if the user sends a teardown message or if the starting time of the reservation is already exceeded. Therefore, the states are called pseudo hard states and so we are able to deal with unplugged endsystems.

To avoid that soft states which hold the ReRA information will be deleted caused by temporary router failure we use a modified refreshing mechanism. Considering the soft state approach of RSVP, the periodic sending of PATH and RESV messages (default: every 30 seconds) produces a considerable overhead. The refreshing of the states is controlled by the refresh time R and the lifetime L. The relationship between R and L is expressed by the following equation: $L \geq (K + 0.5) * 1.5 * R$ (K-1 successive messages may be lost without state information being deleted) and is shown in Figure 6.

---

[3] To provide the receiver address in unicast/multicast sessions the RESV message has to include always a RESV_CONFIRM object.

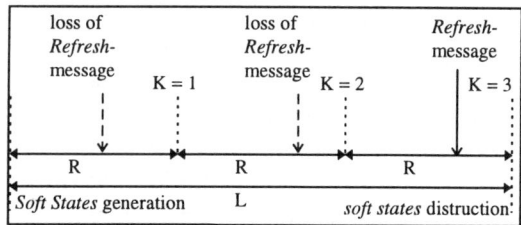

Figure 6: Refreshtime and Lifetime

From our point of view, L should increased in such a way that reservations are insensitive to temporary router failures and that the overhead of RSVP messages is minimized. This can be achieved by an adequate increase of R and K. K is a constant value, which should be configured manually (e.g. as a parameter for the RSVP daemon) or via management functions. R is calculated when the ReRA connection is requested and is reduced step by step until the starting time is reached. At this time, R is set to the default value. Assuming that R is calculated by R=*standard_refresh_time(30s)* + (*starting_time* - *current_time*) / D, R is dependent on the current time and a granularity D. It should be emphasized that concrete values of R and K have to be determined based on future measurements in our experimental environment.

On the other hand concerning receivers which have joined a multicast group during the intermediate phase the increasing of R and K results in delayed receiving of PATH messages. In order to avoid this, RSVP calls the mrouted deamon periodically. Based on the information about new receiver addresses joined to a multicast session, RSVP sends PATH messages to the multicast group immediately.

Moreover, it is possible that a receiver can leave a multicast group during the intermediate phase (e.g. in case of unplugged endsystems). This results in loss of refresh messages following the path to the receiver. In our implementation all multicast information of advanced reservation are held within the soft states as an additional list of outgoing interfaces.

4 CALL ADMISSION CONTROL
4.1 Sharing versus Partitioning

The design of a suitable ATM traffic control is considered as a fundamental challenge for the success of the ReRA mechanism. The primary role of traffic control is to protect the network and the user in order to achieve predefined network performance objectives. Dealing with advance reservation, an additional role of traffic control is to optimize the use of network resources for the purpose of achieving: (1) sufficient network efficiency, that means the effectiveness in utilizing bandwidth, (2) a minimal blocking probability and (3) a fair handling of the advance and the immediate reservations. Obviously, it seems to be difficult to combine these features in such a way that the Call Admission Control (CAC) will achieve all these goals in an optimal way. So in (Schill, 1997) we have considered influence parameters and their effects on the characteristics of the CAC as a first step.

Basically, there are the following policies of resource awarding or administration to handle the two different kinds of reservations: sharing and partitioning. Sharing the resources allows the awarding of the whole bandwidth for example in the case of an aggressive access to ReRA connections. A partitioning of the bandwidth for instance for ImRe and ReRA is heading towards an equitable handling of the single reservations as the bandwidth is dedicated to each reservation type. But, assuming an adverse allocation of resources between ImRe and ReRA, it can result in an inefficient utilization of bandwidth caused by the two boundaries.

A dynamical partitioning (Ferrari1, 1997) seems to be a reasonable trade-off. A partitioning with a moveable boundary means, that each reservation has its own partition, but with the option of using available resources of the other partition if the own partition is completely allocated. But considering this in more detail, it resembles sharing across the whole bandwidth or at least sharing of bandwidth in between exclusive partitions (based on watermarks to protect e.g. the advance partition; see Figure 7), respectively. As an appropriate value of the watermarks for delineating exclusive partitions is not easy to determine, sharing the whole bandwidth seems to be the most common way.

Based on sharing we have developed a basic admission control for an ATM network.

Dealing with the RSVP/ ATM protocol stack the CAC of the QoS based ATM and its specific service classes have to be investigated for a ReRA extension. In view of the growing importance of VBR and ABR services, we consider a suitable ATM VBR algorithm in the next section which takes into account the statistical multiplexing of variable flows.

resource sharing

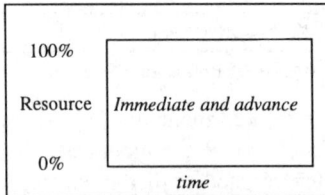

resource partitioning

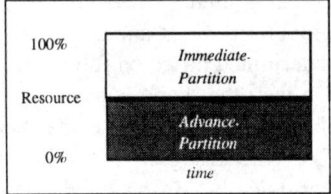

dynamical resource partitioning

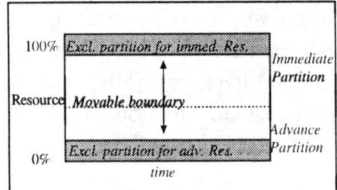

Figure 7: Policies of resource awarding

4.2 Administration of advance reservations

As a basis for introducing our ReRA ATM admission control as part of the next section, we briefly discuss the management or administration of advance reservations.

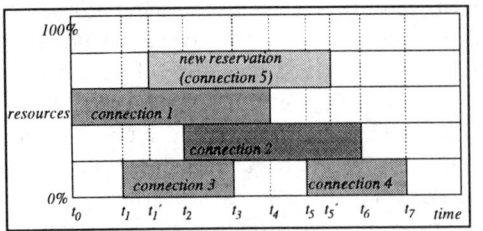

Figure 8: Administration of ReRA information

If a new reservation request arrives, the check for available resources has to be performed over the whole duration of the new connection. Resource changes are only caused by other reservation start or endpoints. So we have considered two methods of administration in interval tables: predefined and variable time intervals. We prefer the latter because no adaptations of start and end points as a multiple of granularity intervals is necessary. Moreover, we assume, that the number of variable intervals will be less than the account of predefined intervals in general. The reservation data base consists of a list of all intervals. Each interval is characterized by a begin, end and a list of overlapping reservations (Figure 8).

4.3 Equivalent Capacity

In general, the congestion control as an important area in ATM networks can be classified into preventive and reactive control. A combination is currently used in ATM: CBR and VBR use preventive schemes and ABR is based on a reactive scheme. Below, we examine the CAC algorithms of the preventive control which can be grouped in non-statical allocation (or peak rate allocation used for CBR) and a statical allocation (which makes economic sense dealing with bursty sources - VBR).

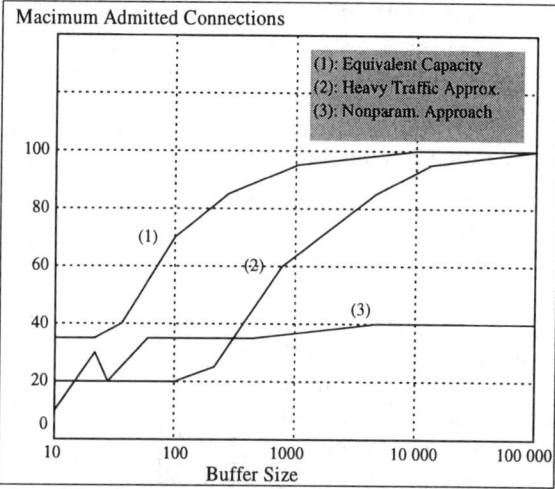

In view of extending one algorithm for advance reservation we have compared the following CAC schemes:

- Equivalent Capacity
- Heavy Traffic Approximation and
- Non-parametric Approach

The simulation comparisons of the performance of the above algorithms has shown, that the Equivalent Capacity achieves the highest statistical gains

Figure 9: Comparison of admission control algorithms

(defined as the maximum number of admitted connections divided by the number of connections that can be accepted using peak rate allocation) depending on buffer size (Figure 9) and the required cell loss probability.

Table 1: Comparison of admission control algorithms

	Equivalent Capacity	Heavy Traffic Approximation	Non-parametric Approach
source description	interrupted fluid process	interrupted bernoulli process	no traffic model
traffic parameters	peak rate, average cell rate, average burst length, standard deviation	peak rate, average cell rate, average burst length,	peak rate, average cell rate
effectivity	high	high (large buffer)	low
computation overhead	low	low	higher

In Table 1 the three algorithms are compared relevant to some characteristics like

their effectivity or traffic parameter description. As a major result of these considerations, we select the Equivalent Capacity (EC) as the basis of a ReRA extended algorithm for bursty (VBR) sources. In case of a new connection the EC of this single source can be obtained by solving the simplified equation (1) for c. The result c has to be less or equal the maximum link cell rate. In the case of N sources, and given that the buffer has the capacity K, the equivalent capacity is again the service rate C (2) which ensures that the cell loss is ε. As a simplification again, C (2) is the combination (minimum) of the Fluid Flow Approximation (which is

---

Equivalent Capacity:
statistical multiplexing of variable flows

with $c = R\dfrac{a - K + \sqrt{(a-K)^2 + 4apK}}{2a}$ (1)

and

$C = \text{MIN} \left(C_{(F)} \approx \sum_{i=1}^{N} c_i, C_{(S)} = m + \alpha\sigma \right)$; (2)

$a = \ln(1/\varepsilon)b(1-\rho)R$

$\alpha \approx \sqrt{2\ln(1/\varepsilon) - \ln 2\pi}$

m : average cell rate   K : buffer
b : average burst length   R : peak rate
σ : standard deviation ε : loss
                                            probability
$\rho = \dfrac{m}{R}$: source utilization

Figure 10: Equivalent Capacity

---

the EC-sum of all running connections including the new one) and the Stationary Approximation which has been observed to follow approximately the Normal distribution supposing a high number of connections. The new connection will be accepted if C (2) is less or equal the maximum link capacity (Figure 10).

Based on own simulation results (Figure 11), the following Table 2 summarizes why it is useful to determine the minimum of the Fluid-Flow approximation and the Stationary approximation.

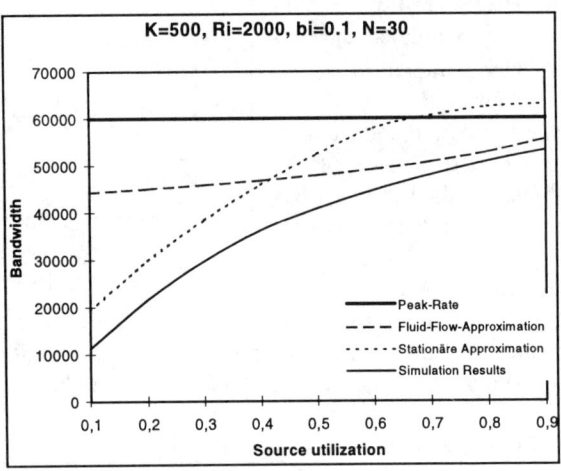

Figure 11: Simulation results

Table 2: Comparison of fluid flow and stationary approximation

Fluid-Flow Approximation is more suitable for:	Stationary Approximation is more suitable for:
small number of sources	higher number of sources
higher utilization of the source	smaller utilization of the source
shorter burst length	longer burst length
smaller peak rates	higher peak rates

As an example, Table 2 presents the suitability of both approximations and the peak rate allocation to draw a parallel with the actual simulation results. While varying the utilization of the 30 sources (namely the average rate) and keeping the buffer size b, the peak rate and burst length constant, the Stationary Approximation is a better approach to the simulation results (basing on a IFP source model) until reaching a source utilization of 0.4. Coming closer to a 100 % utilization (or to R) the Fluid-Flow approximation is the better approach. In principle, it can be said that both reservations are complementary to each other. So the calculated minimum of both registers the highest statistical gain.

4.4 An ATM ReRA-CAC Algorithm based on Sharing

Sharing the resources between ImRe and ReRA, there is a problem with the currently undefined duration of immediate reservation. Retaining these characteristics of ImRe, pre-emption has to be performed to realize that enough resources will be available looking ahead for the connection reserved in advance. As pre-emption is not desirable in most cases, we prefer ImRe with (whenever possible) a given duration. Otherwise ImRe with a wildcard duration is limited by the minimum advance notice. From this, it fellows that an exact separation between ImRe and ReRA is possible.

The following admission control is a combination of the above considerations. That means we use sharing as the resource awarding strategy in combination with the equivalent capacity for variable flows. In general, each new reservation request in ATM is characterized by:

reservation_request := (reservation_time, source_traffic_descriptor)

source_traffic_scriptor := (PCR, SCR, MBS, CLR, MaxCTD)

reservation_time := (start, duration)

Using the equivalent capacity for VBR streams in ATM, an appropriate mapping of the ATM specific parameters onto the algorithm parameters have to performed (with $D^2 = m(R-m)$):

ATM-Parameter	PCR	SCR	MBS	MaxCTD	CLR
CAC-Parameter	R	m, D^2	b	K	CLP

The average burst size b is calculated as follows: $b[s] = \dfrac{MBS[Cell]}{PCR[Cell/s]}$. Moreover, the dimension of the buffer capacity depends on the maximum cell delay variation (Saito, 1994) supposing a FIFO buffer:

$$MaxCTD[s] \geq \frac{K[Cell] * Zellenlänge[bit/Cell]}{Linkkapazität[bit/s]} \text{ and: } CLP = CLR = \varepsilon.$$

The following CAC steps are illustrated in Figure 12:

(A) Assuming, there is a data base with M intervals I_0, I_1, ..., I_M. The first one begins with time 0 (current time), the last one is limited with the maximum advance notice.

(B) Moreover, a new reservation request is given, whereby ImRe is handled as a special case of ReRA with starting time 0 and a duration less or equal the minimum advance notice.

(C) The CAC determines all intervals I_P ... I_Q, which are overlapped by the duration of the new request. Moreover, for all intervals I_P ... I_Q the CAC is performed.

Computation of the equivalent capacity in case of variable bit rates:

```
if res_request == VBR
```

- Assume N is the number of connections in the current interval. The fluid flow approximation $C_F(N+1)$ is calculated for N+1 connections including the new one:

$$C_F(N+1) = C_F(N) + c_{F,N+1}$$

- Moreover, the stationary approximation for N+1 connections is computed:

$$C_S(N+1) = m(N+1) + \alpha * \delta(N+1$$

- Finally, the equivalent capacity is the result of the combination of both approximations:

$$C_E(N+1) = \min\{C_F(N+1), C_S(N+1);$$

```
else //res_request is CBR
    C_S(N+1) = C_F(N) + res_request.PCR;
```

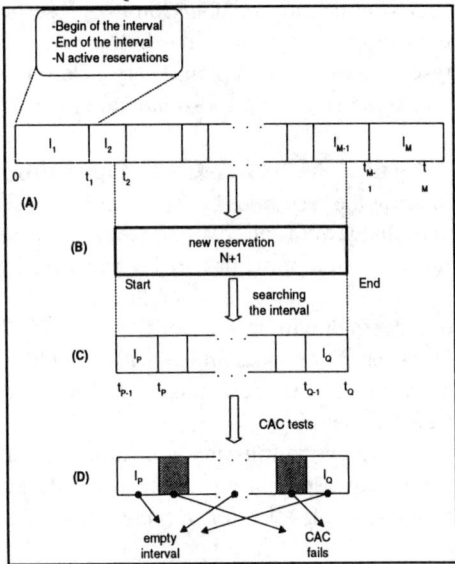

Figure 12: CAC algorithm

(D) In case of a resource shortage the appropriate interval is tagged. If one of the intervals is tagged, the new request is rejected, otherwise accepted.

Accepting the new request causes an updating of the interval table on the appropriate host or intermediate node as described in section 4.2. If a rejection on a subsequent node occurs (Release) all interval tables on the way to the sender have to be changed to the state as it was, before the new request has arrived.

Problems of Resource Fragmentation

The introduction of advance reservation is closely connected with the fragmentation of resources. In order to optimize the utilization of bandwidth for reservation request, two ways are discussed in the following. To cope with the above aim we have introduced an extended time model for ReRA (Figure 13). So each reservation is not only characterized by a starting point and a duration, but a period of validity.

This is useful dealing with Scaleable Video on Demand Systems. The user specifies the time of validity in addition to the duration and the desired beginning of the requested video, for example 5 till 10 p.m. The advantages of this modified ReRA model are an optimized utilization of bandwidth and a lower blocking probability.

To be able to inform the user about alternative reservations (accurate beginnings and ends), an overview about the current reservation situation of all nodes along the route must be gathered. This is done by overlapping the reservation situations for the period of validity of all nodes. Alternatives are determined as follows:

- a list of best matches with the desired bandwidth and duration,
- a bandwidth alternative if the requested resources are not available, and
- a time alternative, if the desired duration cannot be satisfied.

As this model is not part of the current signalling model in ATM, we expect to extend the signaling for sending an ADSPEC like in RSVP. This way, the pessimistic user can gather the desired reservation information before the actual request (SETUP) is sent.

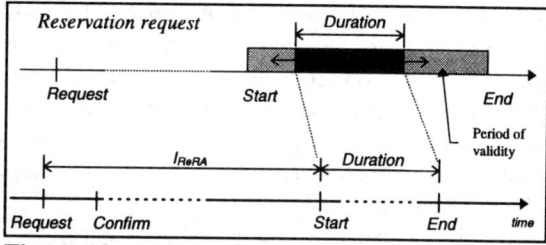

Figure 13: Extended time model for ReRA

In addition to the „ADSPEC" suggestions, a bandwidth and duration alternative is given to the user even after a rejection on a single node. This can be used for calculating and scheduling alternative reservations.

4.5 Deferred Acknowledgements

To optimize the bandwidth utilization and to reduce blocking probability, an investigation of immediate versus deferred acknowledgement of reservations is interesting. Generally considered, there are two models based on the time when scheduling decisions are made. In the case of immediate acknowledgement, scheduling decisions are made as soon as a request arrives. The other model is based on delayed acknowledgements, only applicable to reservations in advance. Every request has an associated decision point which is given by the system. Focusing on a reduced blocking probability, for example, the following simple heuristics can be used. We assume, that the bandwidth requests of all reservations with a synchronous starting point are sorted in an ascending way. Accepting all reservations beginning with the smallest resource request until the whole bandwidth is awarded, results in a reduced blocking probability. Obviously, assuming synchronous starting points of an amount of reservations is an abstract model and only conceivable with relatively coarse-grained time intervals. In the future, we also have to investigate and develop reasonable heuristics to achieve a better utilization of resources by a special handling of requests according to their attributes (e.g. bandwidth requirements).

5 IMPLEMENTATION ARCHITECTURE AND DETAILS

Our implementation environment consists of several workstations of type DEC 3000/300 and DEC 3000/700 which are connected with our DEC GigaSwitch/ATM via fibre optic links.

Within our development work, we use a RSVP (version 1 Release 4.0a7 of the Information Science Institute (ISI)) implementation on top of ATM on the Digital

Unix operating system. As part of the IP/IPng convergence modules (CM), and hence part of the ATM subsystem, the RSVP specific API supports real time handling and transport of IP/IPng flows. The API is able to receive reservation information from a local RSVP daemon. Based on information contained in a given flow specification, a new ATM VC reservation is performed after completing the mapping of service classes and parameters. The classification of the IP/IPng flows belonging to a dedicated reserved reservation (virtual channel) is done in the IP/IPng CM, too. The mapping of RSVP onto ATM specific connection parameters has been checked by using the multi-generator (MGEN) utility. After specifying the RSVP parameters the toolset generates real-time traffic patterns so that the network can be loaded in a variety of ways with constant and poisson distributed streams. The following figures 14 and 15 show one flow per vc measured on the sending and receiving side (part I), and the second stream mapped into another vc (part II).

A comparison of the reserved rate and measured rate on the sending node (always including UDP-IP and ATM overhead) shows that the reservation overhead (inaccurate mapping) is about 389 Bit/s, that means the actual throughput is 99.996% of the reservation. But this is a result of the packet generation, too, which is not actually conforming to the reserved cell stream per second. As expected, considering both reservations at the same time, no influences on the required QoS of each flow have been considered.

At present, our ReRA-RSVP version supports an extended API between the RSVP daemon and the

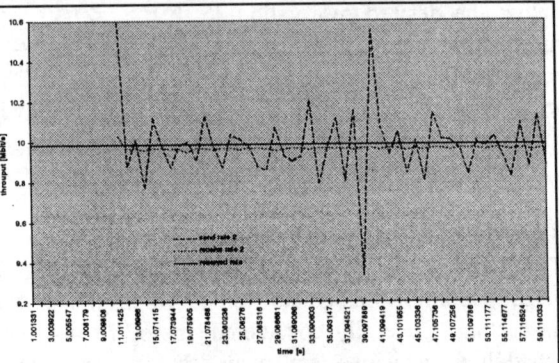

Figure 15: RSVP over ATM measurement - part II

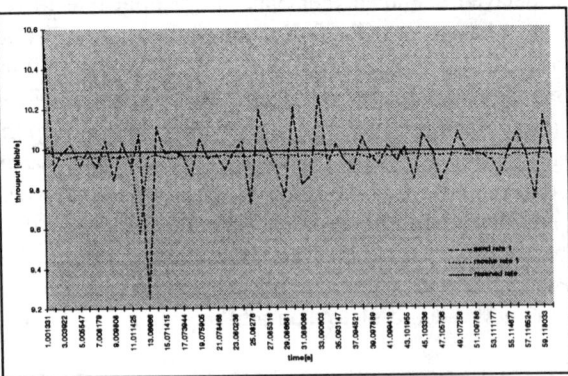

Figure 14: RSVP over ATM measurement - part I

application. In view of the ReRA object, the wildcard and activation flags are supported. The performed tests have shown that it is possible to extend RSVP in such a way described in section 3. With exception of the proxy router model, all extensions have been tested. Solely, the traffic control has not been extended as the implementation of ReRA in ATM is work in progress.

6 CONCLUSIONS

As a main contribution of this paper, a general model of advance reservation was described and an enhancement of RSVP and ATM was presented to support reservation in advance.

Special emphasis was put on the problem of signalling the appropriate ReRA information within the integrated services Internet. Most importantly, the basic structure of RSVP could be maintained by just introducing additional RSVP classes with associated objects. This way, smooth migration of an existing network using RSVP towards ReRA is enabled. Moreover, a simple admission control algorithm for the ReRA algorithm within an ATM environment was designed. Based on the Equivalent Capacity approach and additional internal management mechanisms, a relatively high number of accepted connections and a correspondingly high utilization of resources can be achieved. This was confirmed by initial simulation results. The associated implementation of ReRA via RSVP has also been discussed and has yielded an acceptable performance for data streams using reserved bandwidth.

The presented concepts for resource reservation in advance are also under refinement and we expect to have further implementation experiences soon. To improve the performance of the CAC we are interested in considering neural network structures as connection admission control in ATM environments dealing with ReRA mechanisms. As the implementation of ReRA in RSVP is finished, our ongoing implementation work is aimed at finishing the realization of an extended signaling and admission control in ATM.

7 REFERENCES

Alles, A., „ATM Internetworking", Cisco Systems, Inc., 1995

ATM1, „UNI Signaling 4.0", ATM Forum 94-1019R6, Work in Progress, 1995

ATM2, „Traffic Managment Specification Version 4.0", ATM Forum 95-0013R6, Work in Progress, 1995

Bordon, M., Crawley, E., Davie, B., Batsell, S., „Integration of Real-time Services in an IP-ATM Network Architecture", RFC 1821, August 1995

Braden, R., Zhang, L., Berson, S., Herzog, S., Jamin, S., „Resource ReSerVation Protocol (RSVP) -- Version 1 Functional Specification", RFC 2205, September 1997

Campbell, A., Coulson, G., Hutchison, D., „A Quality of Service Architecture", Computer Communication Review, Vol. 1, No. 2, April 1994, pp. 6-27

Cidon, L., Guerin, R., Khamisy, A., „An Investigation of Application Level Performance in ATM Networks", IEEE Infocom, Boston, 1995, pp. 845-852

Crowcroft, J., Wang, Z., Smith, A., „A Rough Comparison of the IETF and ATM Services Models", 1995

Crowley, Zhang, Sanchez, Salkewicz, „Quality of Service Extension to OSPF", Internet Draft, draft-zhang-qos-qospf-00.txt, 1996

Delgrossi, L., Herrtwich, R.G., Hoffmann, F., „An Implementation of ST-II for the Heidelberg Transport System", Internet-working: Research and Experience, Vol. 5, 1994, pp. 43-69

Degermark, M., Köhler, T., Pink, S., Schelén, O., „Advance Reservation for Predictive Service in the Internet", ACM Multimedia Systems, Vol. 5, 1997, pp. 177-186

Ferrari, D., Gupta A., Ventre, G., „Distributed advance Reservation of real-time connections", ACM Multimedia Systems, Vol. 5, 1997, pp. 187-198

Gilligan, R., Nordmark, E., „Transition Mechanism for IPv6 Hosts and Routers", Work in Progress, 1995

Hafid, A., „Providing a Scalable Video-on-Demand System using Future Reservation of Resources and Multicast Communications", IWQOS '97, New York, May 1997

Herzog, S., „RSVP Extensions for Policy Control", Internet Draft, Apr. 1997, draft-ietf-rsvp-policy-ext-02.ps

Jeffries, R., „ATM LAN Emulation", Data Communications, No. 9, 1994, pp. 95-100

Konstantoulakis, G., Stassinopoulos, G., „Transfer of Data over ATM Networks Using Available Bit Rate (ABR)", IEEE Symposium on Computers and Communications, Alexandria, Egypt, 1995, pp. 2-8

Laubach, M., „Classical IP and ARP over ATM", RFC 1577, January 1994

Malamud, C., „Interoperability in Today's Computer Networks, Prentice-Hall, Englewood Cliffs, N. J., 1992

Partridge, C., „Gigabit Networking", Addison Wesley Publishing Company, 1994

Reinhardt, W., „Advance Reservation of Network Resources for Multimedia Applications", Proceedings of the 2nd IWACA94, Heidelberg, 1994

Reinhardt, W., „Advance Resource Reservation and its Impact on Reservation Protocols", TR, 1995

Saito, H., „Teletraffic Technologies in ATM networks", Artech House 94

Schill A., Kuehn S., Breiter F., „Resource Reservation in Advance in heterogeneous networks with partial ATM infrastructures", INFOCOM 97, Japan, 1997

Vogel, A., Kerherve, B., Bochmann, G., Gecsei, J., „Distributed Multimedia and QoS: A Survey", IEEE Multimedia, Vol. 2, No. 2, 1995, pp. 10-19

Wolf, L.C., Delgrossi, L., Steinmetz, R., Schaller, S., Wittig, H., „Issues of Reserving Resources in Advance", IBM European Network Center Heidelberg, Technical Report 43.9503, 1995

4

Buffer Reservation for TCP over ATM Networks[*]

Clarence S. C. Lee, K. F. Cheung, Danny H. K. Tsang
Department of Electrical and Electronic Engineering
The Hong Kong University of Science and Technology
Clear Water Bay, Hong Kong
Phone: (852) 2358-7045   Fax: (852) 2358-1485
Email: eetsang@ee.ust.hk

Abstract

Transmission Control Protocol (TCP) has already been a widely used protocol in many networks. Recent studies show that the performance of TCP over ATM networks is severely degraded. In this paper, we propose to apply the buffer reservation technique on TCP traffic over ATM networks. While the ATM block transfer (ABT) method (Boyer, 1992) is to reserve bandwidth for bursts of cells, our proposed buffer reservation scheme is to reserve buffer space for TCP packets. We study the performance achieved by the Delayed Transmission (DT) and the Immediate Transmission (IT) buffer reservation schemes. Simulation results show that under the DT buffer reservation scheme, TCP can achieve better throughput and delay performance compared to the plain TCP and the TCP with Early Packet Discard (EPD) except in the WAN environment with short TCP packet length. In the WAN environment with short TCP packet length, the IT scheme can achieve similar throughput performance but smaller delay compared to the plain TCP and the TCP with EPD.

Keywords
Transport control protocol, TCP, ATM, buffer reservation

[*]This work is supported by Hongkong Telecom Institute of Information Technology grant HKTIIT 93/94. EG01

Broadband Communications   P. Kühn & R. Ulrich (Eds.)
© 1998 IFIP. Published by Chapman & Hall

1 INTRODUCTION

Since TCP has already been a widely used protocol in many networks to provide best effort service, the performance of TCP on ATM networks is of major interest. Recent studies show that the TCP performance with UBR service on ATM networks is severely degraded due to the fragmentation of TCP packets into ATM cells as required by ATM networks (Romanow, 1995). Usually a TCP packet is segmented into several 53-byte ATM cells, which will then be transmitted in an ATM network. As a result, even only one of the cells in a TCP packet is dropped due to switch buffer overflow, the whole TCP packet cannot be reconstructed at the destination. Therefore, network resources are wasted to deliver the corrupted packet.

The ATM block transfer (ABT), also known as the fast reservation protocol (Boyer, 1992), is to reserve bandwidth for bursts of cells, which are also referred to as blocks. A preceding RM cell is sent to request the bandwidth required for the block. Another RM cell follows at the end of the block to release the reserved resources. Each block negotiates for a change in the bandwidth allocation by either Delayed Transmission (DT) or Immediate Transmission (IT).

In this paper, we propose a buffer reservation technique for TCP traffic over ATM networks. Buffer space in ATM switch for TCP packets is reserved in our proposed reservation scheme while bandwidth is reserved for the ATM blocks in the ABT method. Our proposed reservation scheme is simple to implement because no target reservation bandwidth is to be determined.

In our proposed reservation scheme, before sending TCP data, a pilot packet is sent to reserve the buffer space in the switches along the connection route. Successful buffer reservation allows the entire TCP packet to be transmitted to the destination. It can dramatically reduce the number of corrupted TCP packets that otherwise can waste network resources. In our investigation, both Immediate Transmission (IT) and Delayed Transmission (DT) buffer reservation schemes are compared to the plain TCP and the TCP with Early Packet Discard (EPD) (Romanow, 1995).

2 TCP OVER ATM WITH BUFFER RESERVATION

In conventional TCP/ATM stack, the TCP packets are given to the AAL layer and are converted to ATM cells. Since the TCP/IP is a standardized protocol, we suggest implementing our proposal by adding a buffer reservation module (see Figure 1) between the TCP/IP and the AAL layer. In this way, users can obtain the buffer reservation improvement without changing the original TCP/IP infrastructure. The major functionality of this buffer reservation module is to handle pilot packet for buffer reservation and pass data between the TCP/IP and the AAL layer. The AAL layer is assumed to be able to convert the pilot packet into the ATM cell format and can loop back the pilot packet to its source.

Two modes of buffer reservation are investigated. In the DT mode of buffer reservation, the data are buffered at the source and the pilot packet is sent first. When the pilot packet reaches the destination, it will be looped back to the source to indicate the buffer reservation status. The data cannot be sent until the buffer space of the switches in the path has been reserved. As a result, TCP/IP data units are stored temporarily at the source. In the IT mode, the data are sent immediately after the pilot packet. The buffer reservation is performed on a hop-by-hop basis. When the network is congested, the pilot packet and the data packet are immediately dropped. The TCP will retransmit the data, and the buffer reservation module will resend the pilot packet along with a reduced number of data packets.

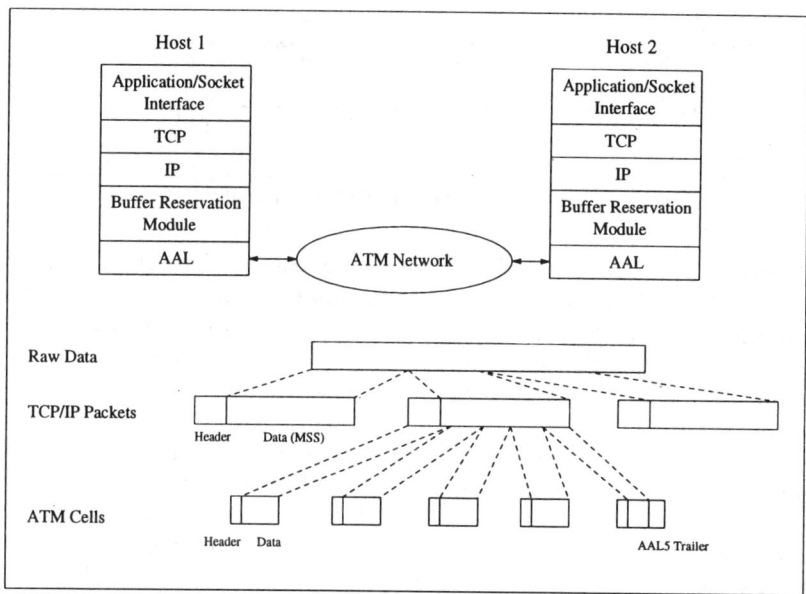

Figure 1 TCP on ATM protocol stack with buffer reservation module.

2.1 Buffer Reservation Module

The buffer reservation module can be implemented at the source as shown in Figure 2. It consists of a sending buffer, a receiving buffer and a pilot packet processing unit (PPPU). In the sending buffer, a sliding window algorithm (Jacobson, 1988) same as the TCP with the slow start and congestion avoidance phases is implemented to make advice on the reservation buffer size. The pilot packet should contain three fields: pilot packet indication bit (PI) is used to indicate a pilot packet, buffer space reserved (BUF_RES) is used to specify the amount of data buffer to be reserved, and reservation successful indication bit (RES_SUCC) is used to indicate whether the reservation is successful.

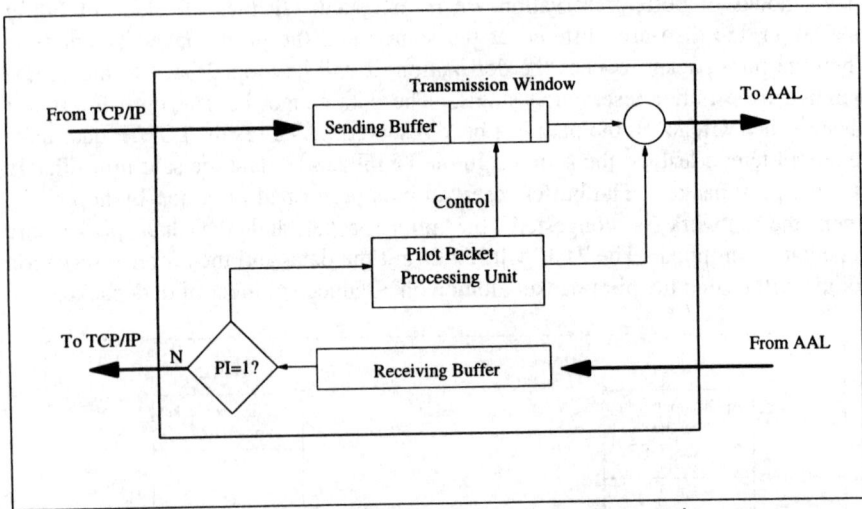

Figure 2 Buffer reservation module implementation.

The PPPU needs to keep a record of the amount of data that have successful buffer reservation. After all the data that have successful buffer reservation are sent and acknowledged, another pilot packet will be sent for the next window of data. If the reservation is unsuccessful, the transmission window will be reduced down to one segment size. After a backoff delay, another pilot packet with the desired buffer space equal to one segment size will be sent.

A time-out timer is also implemented in the buffer reservation module. Whenever a transmission time-out is triggered, the packet is put in the retransmission list. The retransmission packet also requires a pilot packet to reserve its buffer space. When packets wait in the reservation module, the TCP time-out timer in the upper TCP layer may also be triggered for packet retransmission if the packets are not acknowledged. When these retransmission packets are received by the buffer reservation module, the duplicate packets will be dropped if the same packets are still in the sending buffer. On the other hand, if the same packets have already been transmitted but without acknowledgment, these packets will be put into the retransmission list.

The pilot packet is sent with the reservation successful indication bit (RES_SUCC) set to one. When the pilot packet traverses along the connection path, the switches at which the reservation is unsuccessful will reset the RES_SUCC bit. When a return pilot packet is received, the RES_SUCC bit is checked. When the RES_SUCC bit is equal to one, the reserved buffer size in the BYTE_RES field is successfully reserved by all the switches in the connection path. In the DT mode, this indicates that the block of data can be sent. In the IT mode, this indicates that the complete block of data has been successfully transmitted. If RES_SUCC=0, the reservation is unsuccessful. At the destination, non-pilot packets will be passed to the upper TCP/IP layer as normal data packets.

2.2 Buffer Reservation ATM Switch

In Figure 3, the block diagram of the buffer reservation ATM switch is shown. One forward direction output queue is shown in detail while others can be implemented similarly. An unreserved buffer size (UBS) memory is implemented with an initial value equal to the maximum buffer size of the output queue. When an ATM cell enters (leaves) the output queue, the UBS is increased (decreased) by one cell. The switch keeps a record of the buffer size that is not reserved. To ensure efficient buffer reservation, the pilot packet has a higher priority such that it will be put in front of the non-pilot ATM cells in the output queue. Note that in the IT mode, data with unsuccessful reservation are dropped at the switch.

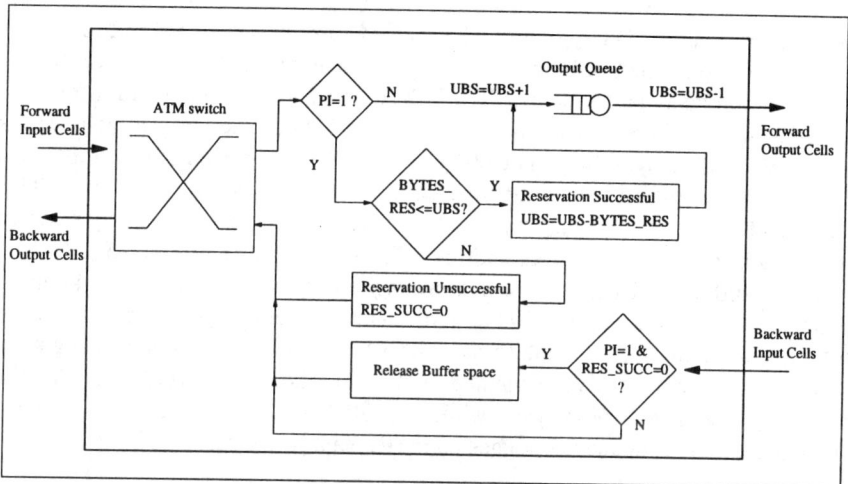

Figure 3 Buffer reservation ATM switch.

When a pilot packet is switched to an output queue, the BYTES_RES field is compared with the UBS memory. If UBS is larger than the desired reservation value, the reservation will be granted and the UBS memory is updated by subtracting the value of BYTES_RES. Otherwise, the RES_SUCC field will be set to zero. In our simulation setup, the pilot packets will only be looped back at the destination. The switch will then check the RES_SUCC field of the backward pilot packets and release the previously reserved buffer space appropriately.

Alternatively, the unsuccessful pilot packets can be looped back at the switch. If the UBS is smaller than the request reservation size, the pilot packet will be looped back to the source in the backward direction. The switch can then check for the RES_SUCC field of the backward pilot packets. In this way, the switch can release the buffer space faster than the destination loop back mechanism. The performance can be increased. However, the switch must be implemented with the functionality to loop back the pilot packet when necessary. In particular, if the pilot packet cannot be switched to the next switch in the forward direction, the pilot packet is looped back to inform the previous switches to release their buffer space.

3 PERFORMANCE EVALUATION IN LAN SCENARIO

The simulation setup in Figure 4 models an ATM network supporting TCP. Acknowledgment and pilot packets are returned to the source in the return path with no background traffic. 20 TCP sources are implemented in the LAN scenario to model a situation with a moderate number of connections. The Maximal Transmission Units (MTUs) of the TCP packets are chosen to be 512 and 9180 bytes to model the effect of long and short packet lengths for the simulations of the throughput and delay performance. The TCP sources are modeled by on-off source sources with mean burst length equal to 1000 and 20000 bytes respectively for the cases of MTU=512 bytes and MTU=9180 bytes. The off period is set to different values to achieve different loading conditions. The transmission speed of the TCP sources and the switch capacity are 150 Mbps. The link propagation delay between the source/destination end system and the ATM switch is taken to be 0.5 μs, which is equal to 100 m in distance. The link propagation delay in the ATM backbone is set to 0.05 ms which is equal to 10 km in distance. The simulation time is set to 2 s, which is reasonably long for the transfer of 30 MB of data. In our investigation, network components are implemented with 4.3-Tahoe BSD TCP version with window size equal to 160 kbytes, while the EPD threshold is set at half of the switch maximum occupancy as suggested in (Romanow, 1995).

3.1 Throughput Performance

As shown in Figure 5, the effective throughput of the simulation model is plotted against different network loading while the switch buffer size is fixed to 160 kbytes. The effective throughput is defined as the number of correct TCP packets that travel through the ATM network in an elapsed time. Retransmitted, loss or duplicate packets are excluded from the calculation. The effective throughput is measured as a percentage of maximum network capacity. An effective throughput of 100% is achieved when the ATM link is never idle and no packet is corrupted and retransmitted. The loading of the network is measured by the amount of data generated by the sources.

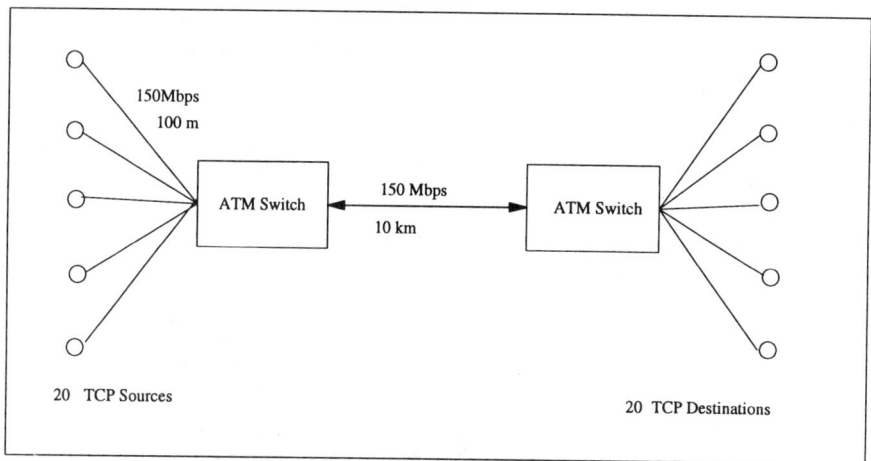

Figure 4 Simulation scenario for ATM LAN.

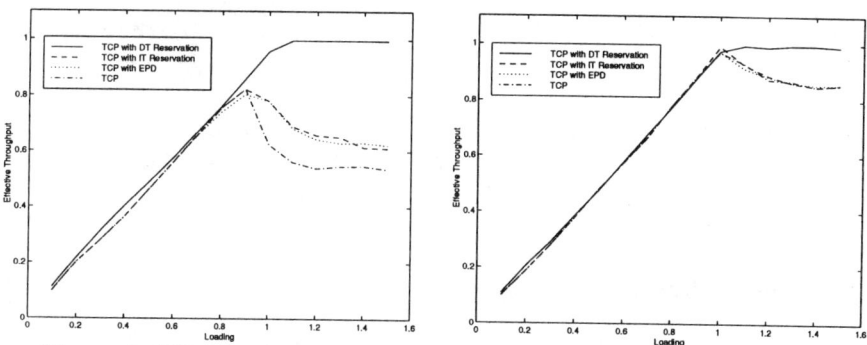

Figure 5 Effective throughput vs. loading (160 kbytes switch buffer size): (left) MTU=9180 Bytes, (right) MTU=512 Bytes.

When the loading becomes larger, throughput degradation is observed for the plain TCP. The major reason for the degradation is the transmission of partially dropped TCP packets that lead to a large number of retransmissions and wastage of bandwidth resource.

The TCP with EPD shows an improvement with respect to the plain TCP; however, throughput degradation is still observed. The number of corrupted packets is reduced since the whole packet is either transmitted or dropped. However, the utilization of buffer is limited due to the EPD dropping threshold.

For the DT buffer reservation scheme, almost no throughput degradation is observed with short and long TCP packet lengths, because every packet has buffer space reserved in advance. The whole TCP packet can be served without being partially dropped and nearly no retransmission is required.

For the IT scheme, the performance is similar to that of TCP with EPD because both schemes will either admit or drop the whole TCP packet at the switch.

With short TCP packet length, both TCP with EPD and plain TCP perform similarly with small throughput degradation. Since the TCP packet length is short, the loss of resources due to corrupted packets is small compared to that of long TCP packet length.

We also investigate the throughput performance with different switch buffer sizes and the results are shown in Figure 6. In all the schemes investigated, larger switch buffer size can always achieve higher effective throughput. With both long and short TCP packet lengths, the DT buffer reservation scheme achieves the highest effective throughput among all the schemes.

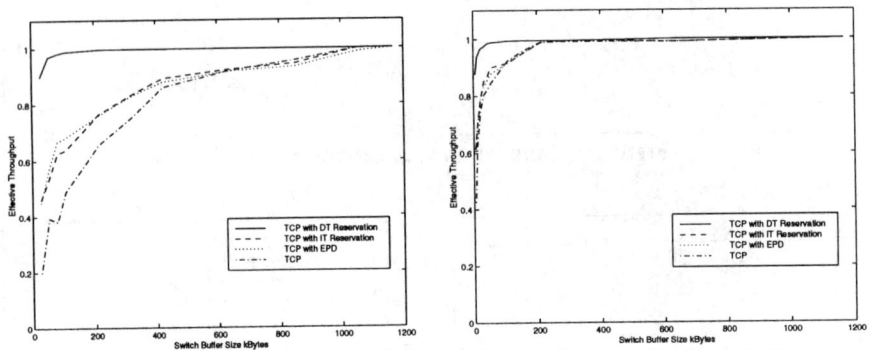

Figure 6 Effective throughput vs. switch buffer size (loading=1.1): (left) MTU=9180 Bytes, (right) MTU=512 Bytes

3.2 Delay Performance

The mean packet transmission delay is investigated. It is defined as the time between the packet generated at the source and the packet correctly received at the destination. The mean packet transmission delay with long and short TCP packet lengths are plotted in Figure 7.

From the results, the mean packet transmission delay becomes longer when the loading increases. It is because the packets are easier to be corrupted and the number of retransmissions is larger at heavy load. With long TCP packet length, the mean packet transmission delays of the DT scheme, the IT scheme and TCP with EPD are smaller than that of plain TCP. The DT buffer reservation scheme can still achieve small mean packet transmission delay because of the short propagation delay time. Packets do not have to wait for a long time to receive the buffer reservation notification.

With short TCP packet length, the loss of resources due to segmentation is reduced. As a result, TCP with EPD and the IT scheme behave similarly compared

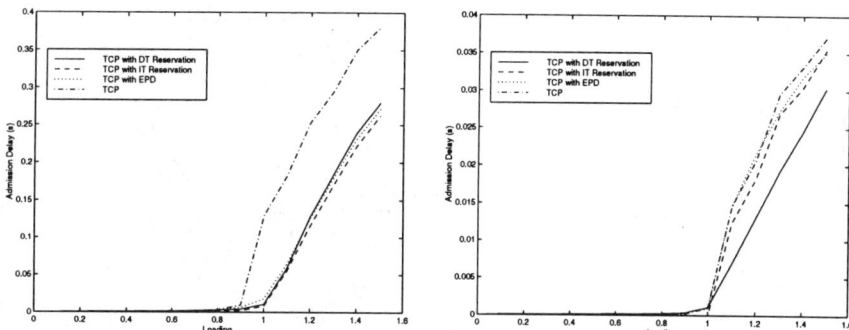

Figure 7 Mean packet transmission delay vs. loading (160 kbytes switch buffer size): (left) MTU=9180 Bytes (right) MTU=512 Bytes

to the plain TCP. At high network loading, the mean packet transmission delay of the DT scheme is smaller than that of the others. It is because few retransmissions are required in the DT scheme. As a result, data packets wait at the source only due to network congestion but not due to frequent retransmission time-out. The network can transmit data more efficiently in the DT scheme.

4 PERFORMANCE EVALUATION IN WAN SCENARIO

To study the buffer reservation schemes over the ATM WAN, we perform similar investigation as in the LAN case. As shown in the WAN simulation scenario (Figure 8), the distance between the ATM switches is set to 1000 km, which is equal to 5 ms propagation delay time. 100 TCP sources are used in this simulation to model a large number of TCP connections in this ATM network.

4.1 Throughput Performance

As shown in Figure 9, the maximum effective throughput is approximately 0.7, which is smaller than the results in the LAN case (effective throughput ≈ 1). The major reason is the relatively long propagation delay time compared to the packet transmission time. After a window of TCP packets are transmitted, the sources can transmit data again when either the acknowledgments return or the retransmission time-out is triggered. Usually, the retransmission time-out is longer than the time for the acknowledgment to return if the data are not lost. As a result, the link idle time is longer in the WAN scenario, which leads to a lower effective throughput. With long TCP packet length, all throughput performance is similar to that of the LAN scenario. The DT scheme does not suffer from throughput degradation while all the other schemes show throughput degradation due to packet corruption and a larger number of retransmissions.

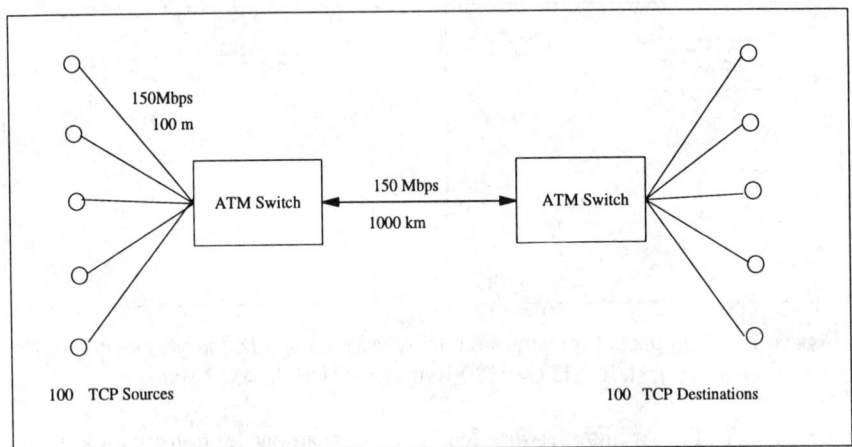

Figure 8 Simulation scenario for ATM WAN.

With short TCP packet length, the DT buffer reservation scheme achieves a smaller effective throughput than the other three schemes. This is because the link idle time in the case of short TCP packet length is longer compared to that in the case of long TCP packet length.

In Figure 10, the effective throughput is plotted against the switch buffer size. In all the schemes, larger switch buffer size always results in higher effective throughput. With long TCP packet length, the DT scheme achieves higher effective throughput than the others except for very small buffer size. For very small buffer size, the DT buffer reservation scheme is not able to reserve the required buffer for data packets effectively. Most of the buffer space is often allocated for the ATM cells in transit. Hence, the incoming pilot packets at the switch cannot easily reserve buffer space successfully. The IT buffer reservation scheme and the TCP with EPD perform similarly with respect to the switch buffer size.

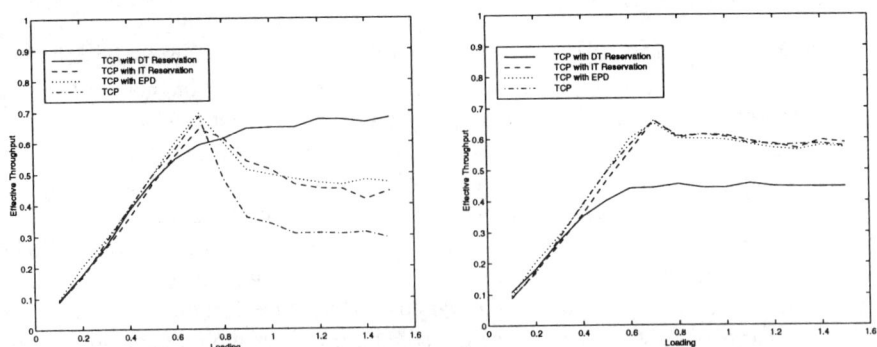

Figure 9 Effective throughput vs. loading (160 kbytes buffer size):
(left) MTU=9180 Bytes, (right) MTU=512 Bytes.

With short TCP packet length, the DT buffer reservation scheme achieves smaller effective throughput than the other schemes. This is due to the longer link idle time when the sources wait for the return of the pilot packets.

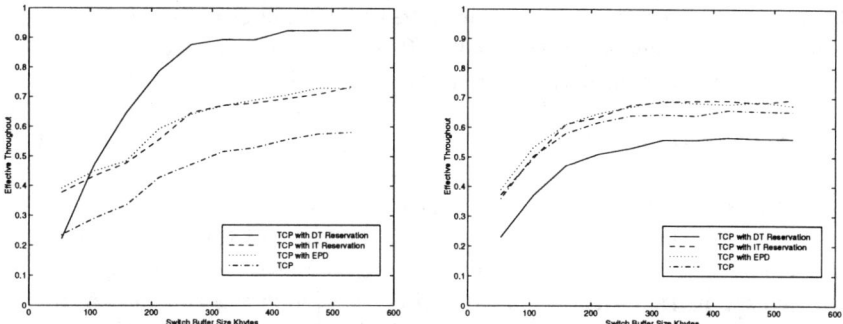

Figure 10 Effective throughput vs. switch buffer size (loading=1.1): (left) MTU=9180 Bytes, (right) MTU=512 Bytes

4.2 Delay Performance

The mean packet transmission delay with long and short TCP packet lengths are plotted in Figure 11. For the DT buffer reservation scheme, TCP packets can be sent after the pilot packet is returned with successful reservation indication. This round-trip delay time is significant compared to the packet transmission time. As a result, at small loading the mean packet transmission delay of the DT buffer reservation scheme is longer than that of the other schemes.

With long TCP packet length, the mean packet transmission delay of the DT scheme increases slower than that of the other schemes as the loading increases. This is because buffer overflow and packet drop rarely happen in the DT scheme.

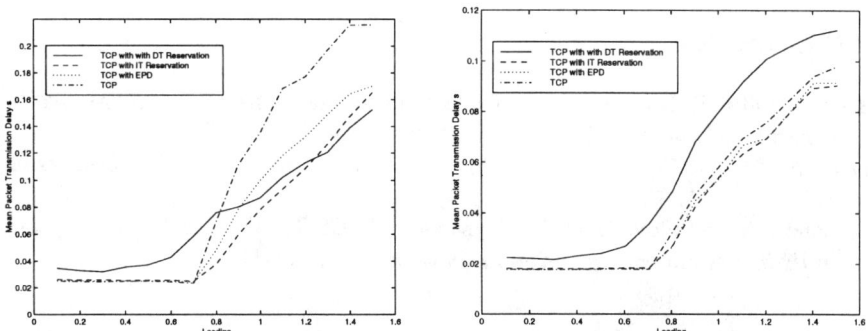

Figure 11 Mean packet admission delay vs. loading (160 kbytes buffer size): (left) MTU=9180 Bytes, (right) MT=512 Bytes

Hence, retransmission that must wait for time-out trigger is very unusual in the DT buffer reservation scheme. Note that at high loading, the DT and the IT schemes perform similarly.

With short TCP packet length, the mean packet transmission delay of the DT scheme is longer than that of the other schemes. Because of shorter TCP packet length, packet corruption in TCP with EPD and plain TCP does not lead to a significant increase in the mean packet transmission delay.

5 CONCLUSIONS

We proposed to apply the buffer reservation technique for TCP on ATM networks. It can support packet-based protocols such as TCP over ATM networks with higher performance. The DT and IT buffer reservation schemes were investigated. We compared the performance of the two buffer reservation schemes to that of plain TCP and TCP with EPD by simulations. In the LAN environment, results show that the DT buffer reservation scheme out-performs the plain TCP and the TCP with EPD. It can achieve higher throughput and similar delay performance compared to the plain TCP and the TCP with EPD. The IT buffer reservation scheme performs similarly compared to the TCP with EPD. In the WAN environment, the DT buffer reservation scheme can achieve better performance with long TCP packet length. However, for short TCP packet length the DT buffer reservation scheme achieves a lower effective throughput due to the longer propagation delay time. On the other hand, the IT buffer reservation can achieve similar throughput performance and smaller delay compared to the other schemes in the WAN environment.

The major disadvantage of the proposed buffer reservation scheme is the modification of TCP over ATM stack, which requires sending a pilot packet before sending the TCP packets. Therefore, we suggest that the current TCP systems can achieve the buffer reservation improvement by simply adding the buffer reservation module and the corresponding ATM adaptation system. In this way, the TCP infrastructure does not have to be rebuilt.

REFERENCES

Boyer P. (1992) Reservation Principle with Applications to the ATM Traffic Control, in *Computer Networks and ISDN Systems,* **24**, 321-34.
Jacobson V. (1988) Congestion Avoidance and Control, in *Proceedings of SIGCOMM 88*, 314-329.
Romanow A. and Floyd S. (1995) Dynamics of TCP Traffic over ATM Networks, in *IEEE Journal on Selected Areas in Comm.*, **13**, 633-41.

Broadband Technologies, Architectures and Network Design

Access Networks

5

Impact of time division duplexing on delay variation in slotted access systems

Sašo STOJANOVSKI and Maurice GAGNAIRE
Ecole Nationale Supérieure des Télécommunications
46 rue Barrault - 75634 - Paris cedex 13 - FRANCE
Email: {sassos, gagnaire} @res.enst.fr

Abstract

Slotted access systems and Time Division Duplexing (TDD) are two techniques well suited to various local loop configurations. Access networks are built on a shared physical medium with point-to-multipoint architecture. A centralised approach with one master entity and several slave entities is adopted. A polling-based MAC protocol enables the slave entities to dynamically share the medium. In this paper we assume that the master entity grants permits to the slaves using either the Virtual Clock or the RCSP service discipline. Both constant- and variable-length TDD frames are considered. The impact of time duplexing on delay variation in constant bitrate connections with variable bitrate background traffic is particularly investigated. For that purpose, several traffic parameters are considered, such as offered load and number of connections.

Keywords

Asynchronous Transfer Mode (ATM), Time Division Duplexing (TDD), access network, Cell Delay Variation (CDV), Rate-Controlled Static Priority (RCSP), Virtual Clock, MAC protocol.

1 INTRODUCTION

Asynchronous Transfer Mode (ATM) has been chosen in the mid '80 as the switching and multiplexing technique for the future B-ISDN networks. Its standardisation reaches the final stage. At the present time the network configuration to be adopted for the last mile of broadband networks remains an open problem. The only solution initially proposed within ITU-T was the Fibre-To-The-Home (FTTH) paradigm, which turned out to be a very expensive solution, since it requires a dedicated fibre from the local broadband exchange to each customer's premises. Recently a multitude of cost-effective solutions for the access network have been proposed, such as: several Digital Subscriber Line techniques (xDSL), Hybrid Fibre-Coax networks (HFC), ATM-based Passive Optical Networks (APON, SuperPON) or Wireless ATM (WATM). Some of these solutions tend to re-use the existing copper and coaxial infrastructures (xDSL, HFC), others require a new infrastructure, at least in the feeder and the distribution part of the access (APON, SuperPON), while WATM only requires a new spectrum band. Another way for classification of these techniques is their architecture. Namely, all the xDSL techniques are point-to-point, while the HFC, APON and WATM techniques are point-to-multipoint. Therefore, the latter techniques require a specific MAC protocol in the upstream direction. This MAC protocol must have a centralised architecture, since that is the only way to offer QoS support to the ATM layer. This is the single common point for all the MAC protocols considered so far in the literature (Acampora. 1996), (Angelopoulos *et al.* 1996), (Angelopoulos SuperPON*et al.* 1996), (Ayanoglu *et al.* 1996), (Charzinski. 1996), (Correia *et al.* 1997), (Miah *et al.* 1997), (Raychaudhuri. 1996), (Walke *et al.* 1996). On the other hand, each solution may differ by the duplexing mechanism (TDD, FDD), the access mechanism (reservation, contention, polling), the introduced MAC overhead, the service discipline for the bandwidth requests etc. In this paper, we consider a slotted access system that uses TDD scheme. Although we do not insist on identifying this access system with one of the existing ATM proposals, we may just note that it is mostly related to WATM for two reasons. First, the majority of recent WATM proposals consider TDD scheme (Raychaudhuri. 1996), (Acampora. 1996). Second, TDD schemes represent an efficient solution only when the propagation delay is small, which again is the case of WATM (cell radius of 100 metres). However, if channel utilisation is not an issue, then other environments may also be considered. In section 2, we describe the architecture of the access system, the details of the polling-based MAC protocol and the service disciplines considered. In section 3 we present the results obtained by simulation and give an intuitive explanation. The conclusions are given in section 4.

2   ACCESS SYSTEM

The ATM technology, as initially devised, is a purely switched technology with no notion for shared medium. However, recent trends in the access portion of broadband networks introduced the notion of shared medium for which a medium access control protocol has to be defined.

Only a MAC protocol with a centralised architecture can offer QoS support to the ATM layer. Therefore, the access system considered here consists of a single master entity and several slave entities, as shown in Figure 1. The master entity is the Optical Line Termination (OLT) in APON systems, the headend in HFC systems or the base station in WATM systems. Each slave entity corresponds to a single virtual connection. Note that this approach is rather different from the majority of proposals available in the literature where

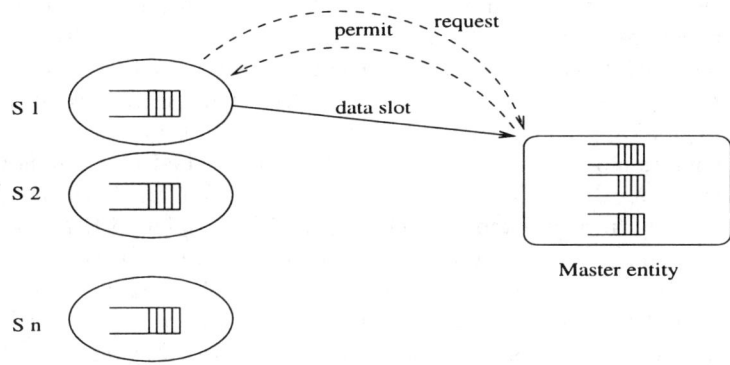

Figure 1 The access system.

the slave entity corresponds to a device which represents an *aggregation* of virtual connections. Such device is the Optical Network Unit (ONU) in APON systems, the cable modem in HFC networks and the mobile stations in WATM. In the following the terms *slave* and *connection* will be used interchangeably.

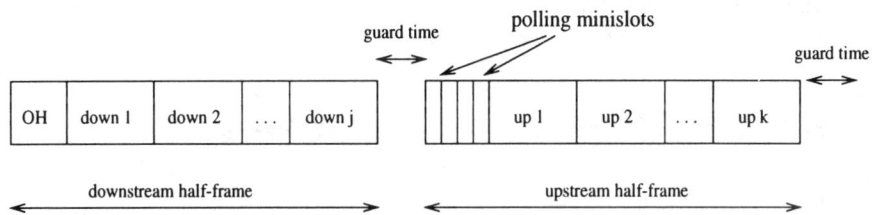

Figure 2 The TDD frame structure.

The TDD frame structure is shown in Figure 2. Note that the system considered here is the one already presented in (Stojanovski *et al.* 1997). As one can see from the figure, the TDD frame structure consists of two parts: a downstream half-frame (master to slaves) and an upstream half-frame (slaves to master). The downstream half-frame consists of j downstream slots (*down 1* to *down j*), whereas the upstream half-frame consists of k upstream slots (*up 1* to *up k*). The variables j and k may take any value up to a predefined maximum value. The downstream slots in general have a different size from the upstream slots. The latter are usually bigger, since they contain a physical layer preamble. The MAC protocol has a mechanism for the slave entities to request bandwidth and a mechanism for the master entity to distribute the transmission permits. The transmission permits are announced at the beginning of each downstream half-frame (in Figure 2 this is the OverHead (OH) slot). Also, polling of a number of slave entities is performed at the beginning of the downstream half-frame within the OH slot. The bandwidth requests of the slave entities may be sent to the master in two ways. First, as a response to the polling in the OH slot, they can send their requests in specific minislots at the beginning of the upstream half-frame. Second, they can also piggyback their requests along with the upstream data slots. The frame size is variable and each of its downstream and upstream parts can vary dynamically and independently from one another. The range of fluctuation of this size may go from zero data slots up to a predefined maximum value. These two parts are separated by a guard interval, whose length is at least equal to the propagation time between the master and the most distant slave. From the MAC layer viewpoint, we consider that all the slave entities experience the same propagation delay from the master. This is either almost true in WATM context (negligible distance disparity), or can be achieved by means of ranging in all other contexts. The master keeps track of a separate counter for each of the connections. These counters reflect more or less accurately the buffer state at the slaves. A service discipline is being applied to these counters. We consider two service disciplines: Virtual Clock and Rate-Controlled Static Priority (RCSP).

Virtual Clock is a *work-conserving* discipline that approximates the very popular WFQ (Weighted Fair Queueing) discipline. A state variable $AuxVC$ is associated to each connection. This variable is updated upon arrival of every new request n according to the formula:

$$AuxVC(n) = max\{TNow, AuxVC(n - 1)\} + 1/Peak. \tag{1}$$

where $Peak$ is the negotiated peak rate for the connection and $TNow$ is the instant of the request arrival. The newly calculated value for $AuxVC$ is used for tagging the corresponding request. When the server wants to send a permit to some of the slaves, it chooses the request with the smallest tag value. Virtual Clock has the interesting property that conforming connections are properly served even in presence of misbehaving connections.

The RCSP is a *non-work conserving* service discipline, which means that the scheduler may not serve an outstanding request even if there is available bandwidth. As a consequence, RCSP increases the average delay when compared to Virtual Clock or other work-conserving disciplines. Unlike Virtual Clock, RSCP has the very interesting property of controlling the delay variation. As shown below, this property is preserved even in the TDD context. From Figure 3 one can see that the RCSP server has two components: a rate-controller and a static priority scheduler. The rate-controller consists of a set of traffic regulators, a separate traffic regulator being associated to each connection. The parameters for the regulators are chosen according to the traffic contract of the underlying connection. Each newly arriving request

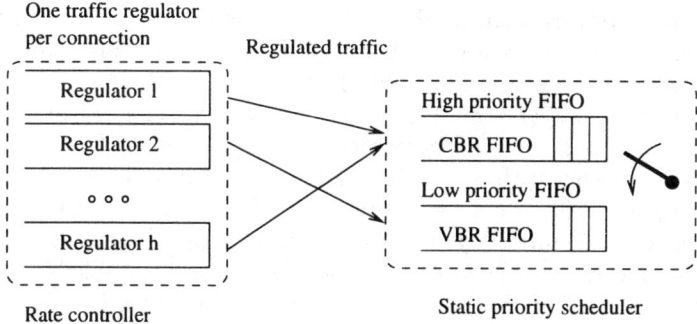

Figure 3 RCSP server.

is being tagged with an *eligibility time ET* which is calculated according to the formula:

$$ET(n) = max\{TNow, ET(n-1)\} + 1/Peak. \qquad (2)$$

This formula is identical to the one for Virtual Clock. The difference is in the way the tags are interpreted. Whereas in Virtual Clock the $AuxVC$ tag is a *virtual* retransmission instant and is used just for sort-out purposes, in RCSP the ET tag corresponds to the *real* retransmission instant in case there are no other competing requests. Only after this instant is reached, is the queued request transferred from the traffic regulator to one of the static priority queues in the scheduler. In other words, the traffic regulators *re-shape* the traffic according to the underlying traffic contract. Another very interesting property of the RCSP algorithm is that it decouples the bandwidth allocation from the delay requirements. Namely, the bandwidth allocation is handled by the traffic regulators, whereas the delay requirements can be met by placing the eligible requests into the appropriate scheduler queue. There are as many scheduler queues as there are different delay classes. The service discipline in the scheduler itself is a non-preemptive one with static priorities: it always serves the request which is at the head of the highest priority non-empty queue. More details about these disciplines can be found in (Zhang. 1995).

3 PERFORMANCE EVALUATION

3.1 Simulation context

Our simulations focus on the constant bitrate traffic. We investigate the impact of the TDD scheme on access delay and access delay variation. The results are expressed as a cumulative probability density function for both the delay and its variation. In all the simulations we consider a constant bitrate traffic in presence of a variable bitrate background traffic. The constant bitrate traffic at the sources is as ideal as can be i.e. the generated packets are equally spaced at $1/Peak$. However, since the system is slotted, we round the generation instant to the beginning of the next slot. After a number of simulations we came to the conclusion that the choice for the background variable bitrate traffic is not very important. Many types of background traffic lead to equal results. However, for the sake of preciseness, we define the traffic type that we use in our simulations. It is a single variable bitrate traffic source characterised by its peak rate $Peak$, sustainable rate $Mean$, maximum and minimum burst size $(MaxBS, MinBS)$. The duration of the bursts is uniformly distributed between $MinBS$ and $MaxBS$. Once the burst is generated, the subsequent period of inactivity is determined in such a way that the overall mean of the generated traffic can reach the declared sustainable rate. This traffic model is conforming to the ATM Forum's VBR traffic model (ATM Forum TM. 1996). A value of 5 is chosen for the $Peak/Mean$ ratio (or *burstiness*) and the simple GCAC formulas (ATM Forum PNNI. 1996) are used to calculate the equivalent bandwidth for the VBR connection. The traffic regulator for VBR traffic is implemented in the following way: for each newly arriving request we calculate two tags, one for the peak cell rate ET_p and another for the sustainable cell rate ET_s, according to the formulas:

$$ET_p(n) = max\{TNow, ET_p(n-1)\} + 1/Peak. \tag{3}$$

$$ET_s(n) = max\{TNow, ET_s(n-1)\} + 1/Mean. \tag{4}$$

The request becomes eligible only when both of the following conditions are met:

$$ET_p(n) <= TNow. \tag{5}$$

$$ET_s(n) <= TNow + BT. \tag{6}$$

where BT is the *Burst Tolerance* and is calculated according to the formula in (ITU-T I.371. 1996), which we reproduce here as:

$$BT = (MaxBS - 1)(1/Mean - 1/Peak). \tag{7}$$

Note that an implicit policing mechanism is integrated in the VBR traffic regulator, since non-conforming VBR bursts will experience intolerable delays. The static priority scheduler distinguishes two traffic delay classes. A high priority and a low priority are associated to the CBR traffic and to the VBR traffic, respectively.

Normally, the slot size would be different for the upstream and downstream directions, as already mentioned above. However, we assume throughout the simulations that the upstream slots have the same size as the downstream ones. The number of connections which are polled at the beginning of the downstream half-frame is fixed to 4. In the Virtual Clock case, the connections are polled in a simple round-robbin fashion. On the contrary, in the RCSP case only those connections which have no outstanding requests at the master are polled. We note that these different choices are well suited to each of the considered disciplines.

3.2   Variable vs constant frames

We consider a scenario with 32 constant bitrate connections with identical peak rate and a single variable bitrate connection. All the connections are symmetrical. The constant bitrate connections are dephased randomely, the phase being uniformly distributed within one period $1/Peak$. All the CBR connections occupy 50% of the link capacity, the other half being allocated to the VBR connection. The latter has a burstiness of 5, and according to the simple GCAC formulas, one can calculate that the sustainable rate ($Mean$) is equal to 34.2% of the allocated capacity. We assume a guard time of one slot between the upstream and downstream frame. Figure 4 and Figure 5 consider the cases of variable frames, for the RCSP and Virtual Clock disciplines, respectively. In these diagrams, as well as in the remainder of this paper, the bold curves represent the cumulative probability density function (pdf) for the access delay, whereas the thin curves represent the 1-point CDV pdf. The abscissa values stand for the number of time slots. On each figure there are three curves corresponding to three different values for the maximum half-frame size: 12, 18 and 24 data slots. Note that these values do not take into account the constant overhead (OH slot, minislots and guard times).

By comparing Figure 4 and Figure 5, one can draw several conclusions. First, in both cases the access delay increases with the maximum half-frame size. Second, the access delay distribution strongly depends on the service discipline: in case of Virtual Clock it is concentrated around much lower values than for RCSP. This is another confirmation of the well known result that work-conserving disciplines imply lower *average* delay. In the RCSP case, it is the traffic regulators that increase the delay by re-shaping the traffic. Nevertheless, the values for the *maximum* access delay are comparable in both cases. Third, the delay variation (measured with the 1-point CDV) increases with the maximum half-frame size, whatever the service discipline. However, while the maximum values for the delay variation are of the order of the access delay for Virtual Clock, these maximum values are *bounded* for RCSP. This is the most interesting result and it will be confirmed on the following figures, as well. Note that the delay variation bound depends on the maximum half-

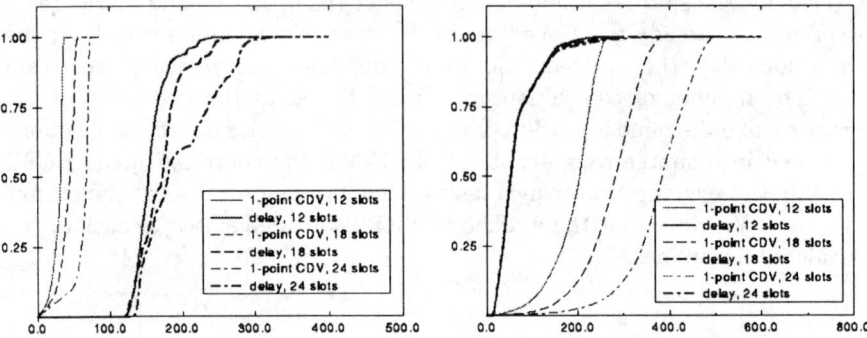

Figure 4 Variable frames, RCSP, max half-frame size of 12, 18 and 24.

Figure 5 Variable frames, VClock, max half-frame size of 12, 18 and 24.

frame size, and its value roughly corresponds to three times the maximum half-frame size. Namely, the maximum values for the delay variation are: 39, 57 and 75 for the maximum half-frame size of 12, 18 and 24, respectively.

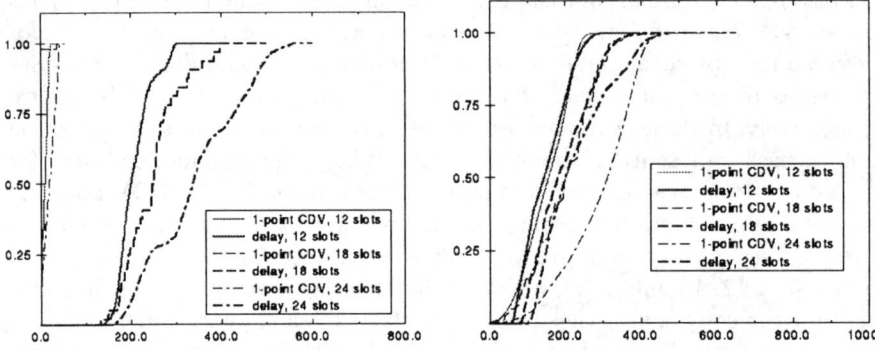

Figure 6 Constant frames, RCSP, max half-frame size of 12, 18 and 24.

Figure 7 Constant frames, VClock, max half-frame size of 12, 18 and 24.

On Figure 6 and Figure 7 we compare the service disciplines in the case of constant frames, with half-frame size equal to: 12, 18 and 24. One can note that access delays increase and the delay variation decreases in the case of fixed-length frames, compared to the case of variable frames. Again, one can note that the delay variation is of the order of the access delay for Virtual Clock, while in case of RCSP the delay variation is bounded. Furthermore, one can see that this bound is even smaller than in the case of variable frames. In spite of the fact that with constant frames one can reduce the delay variation (at the expense of slightly increasing the access delay), this case will not be considered further. The fact that statistical gain in case of asymmetrical traffic is achievable only with variable-sized frames justifies this choice.

3.3 Impact of traffic parameters

Figure 8 and Figure 9 present the variable-frame case in which the parameter is the number of CBR connections: 1, 16 and 64. In all the cases, the generated CBR traffic represents 50% of the link capacity i.e. the cell interarrival time for the CBR connections is equal to 5, 80 and 320 slot times.

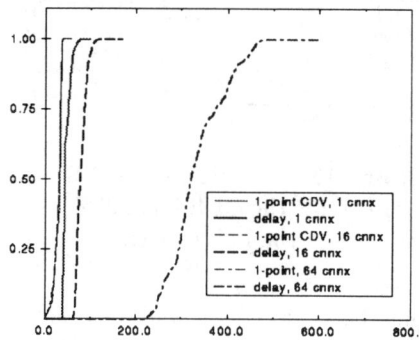 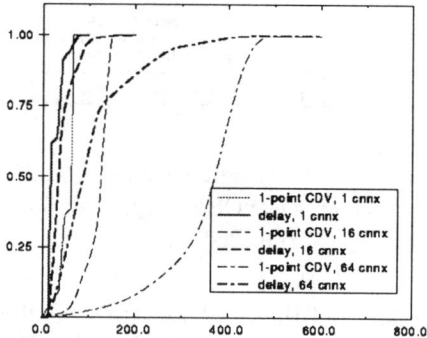

Figure 8 Variable frames, RCSP, parameter is number of CBR connections: 1, 16 and 64.

Figure 9 Variable frames, VClock, parameter is number of CBR connections: 1, 16 and 64.

One can see that delay variation is insensitive to the number of connections with RCSP. In fact, the three curves representing the 1-point CDV cumulative pdf are indistinguishable from one another and appear as a single curve in Figure 8 (the first curve from the left). On the contrary, in the Virtual Clock case the delay variation increases with the number of connections and has maximum values very close to the maximum access delays.

Figure 10 and Figure 11 consider variable-frame cases with 32 CBR connections, the changing parameter being the CBR traffic load. One can see that by increasing the CBR traffic load (25%, 50% and 75% of the link capacity), the delay variation increases with Virtual Clock, whereas it remains unaffected with RCSP.

3.4 Comments about the RCSP discipline

In this section we try to justify the values obtained for the delay variation bound. Let us consider the case with RCSP service discipline and variable frames with maximum half-frame size of 12. In a "steady state" the CBR bandwidth requests arrive more or less periodically into the traffic regulators. After being held for some time, they become eligible and arrive as credits into the scheduler's CBR FIFO. At the beginning of each downstream half-frame, the server announces the permits for the following upstream half-frame. It

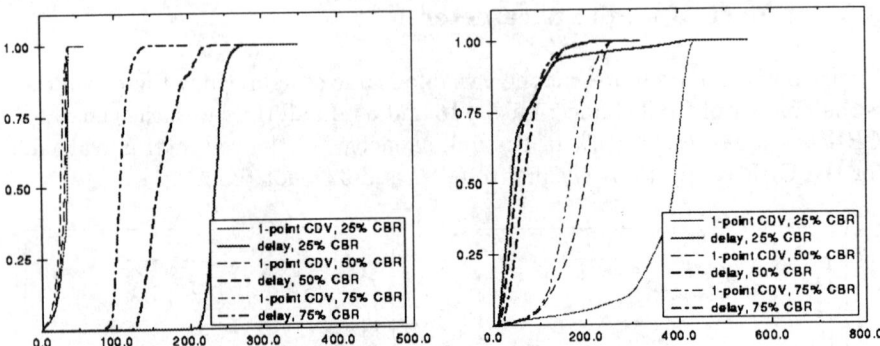

Figure 10 Variable frames, RCSP, parameter is the CBR load: 25%, 50% and 75% of the link capacity.

Figure 11 Variable frames, VClock, parameter is the CBR load: 25%, 50% and 75% of the link capacity.

collects all the available CBR credits from the CBR FIFO, and then looks for the credits in the VBR FIFO. The server collects VBR credits up to the moment when either the VBR FIFO gets empty or the total number of credits reaches the value of 12 (the maximum half-frame size).

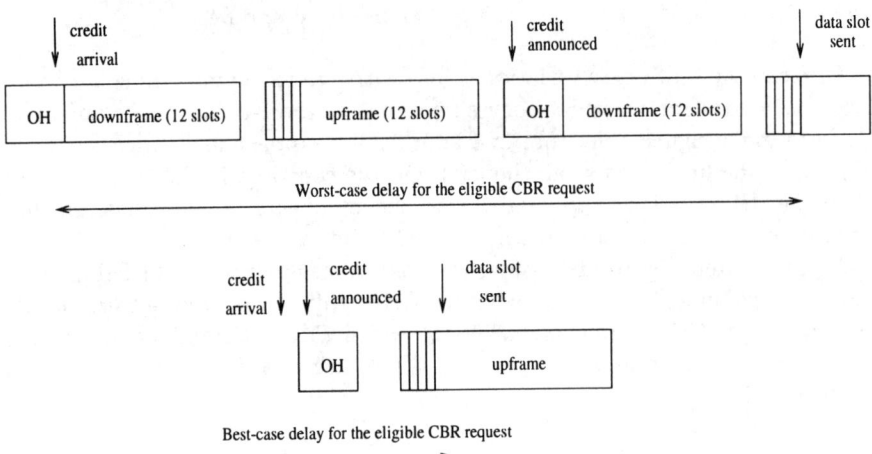

Figure 12 Worst-case and best-case waiting time.

Whenever the VBR connection has simultaneous bursts in both directions, the frame size reaches the value of 12 + 12 data slots. In the worst case, as shown in Figure 12, a new CBR request (or a "credit", as marked in the figure) becomes eligible and arrives in the CBR FIFO just after the OH slot i.e. the instant when the server already announced the 12 credits. Consequently, this request will wait for the duration of a full-sized frame (12 + 12) before being announced. After being announced in the next OH slot, there is again a whole downstream half-frame (12 slots) of waiting time before the upstream half-frame starts. Therefore, the worst-case waiting delay T_w for the credit is equal to:

$$T_w = 3 * 12 DataSlots + 3 * GuardTime + 1 * OHslot + 2 * MiniSlots. \quad (8)$$

On the opposite side, when the VBR connection has no burst for transmission in any of the two directions, the frame size is much shorter. It is even possible to have frames with zero data slots, given that the CBR connections are dephased and the two traffic directions are uncorrelated. In the best case, as shown in Figure 12, a new CBR request becomes eligible and arrives in the CBR FIFO just before the beginning of a new downstream frame. With the assumption that there are no downstream data for the current downframe, the best-case delay T_b experienced by the credit is the delay induced by the constant frame overhead:

$$T_b = 1 * OHslot + 1 * GuardTime + 1 * MiniSlots \quad (9)$$

Hence, the maximum delay variation experienced by the credits is roughly equal to the difference $(T_w - T_b)$, which in this specific case is equal to 39 slot durations. With similar reasoning one can conclude that in the constant-frame case the maximum delay variation is equal to two half-frames (instead of three).

4   CONCLUSION

In this paper, we have investigated a generic slotted system for access to broadband distribution networks. It uses time-division duplexing and a centralised polling-based MAC protocol. Although we have tried to remain as general as possible throughout the paper, it is true that the applicability of the TDD approach is limited to small-range systems, such as WATM. Nonetheless, this approach could be applied to other contexts, too, if bandwidth efficiency is not an issue. We have considered the advantages of using a non-work-conserving discipline for request scheduling over using a work-conserving one. It has been shown through simulations that the interesting properties of the non-work-conserving disciplines are preserved even in the specific TDD context. The comparison has been performed by changing several parameters, such as the traffic load or the number of connections. Simulations with both constant and dynamically-varying frames have been performed. From all the simulations we have drawn several conclusions. First, delay variation is very influenced by traffic parameters in case of Virtual Clock and is almost insensitive to traffic parameters in case of RCSP. Second, in the RCSP case the maximum value for the delay variation is *predictable* and is approximately equal to the duration of two maximum-sized downframes plus the duration of one maximum-sized up-frame. Clearly, our simulation results show that in case of temporary overloads due to the background traffic, it is the maximum frame size that determines the upper bound for the delay variation of constant bitrate connections. This predictability for the maximum delay variation is a nice feature, since it corresponds actually to the CDV Tolerance to be used with regard to the first upstream ATM switch. The values expressed above in slot durations should be divided by two since the slot duration in the context of TDD is half of a non-TDD slot duration. Third, by using frames of constant size one can decrease the delay variation, but increase the average access delay. Fourth, one can see that the average access delay is always bigger with RCSP than it is with Virtual Clock. However, they have comparable values for the maximum access delay.

REFERENCES

Acampora, A. (1996) Wireless ATM: A Perspective on Issues and Prospects *IEEE Personal Comm.*, **August 1996**, 8–17.

Angelopoulos, J.D., Fragoulopoulos E.K. and Protonotarios E.N. (1996) Efficient Support of Best effort Traffic in Passive Tree Local Loops *SPIE Journal*, **Vol. 2609**, 182–191.

Angelopoulos, J.D., Koulouris J. and Fragoulopoulos E. (1996) A MAC Protocol for an ATM-based SuperPON *SPIE Journal*, **Vol. 2919**, 279–287.

ATM Forum (1996) Traffic Management Specification *Version 4.0*, **1996**.

ATM Forum (1996) ATM Forum PNNI Specification *Version 1.0*, **1996**.

Ayanoglu, E., Kai Y.E. and Karol M.J. (1996) Wireless ATM: Limits, Challenges and Proposals *IEEE Personal Comm.*, **Vol.3 No.4**, 18–34.

Charzinski, J. (1996) A New Approach to ATM Access Networks *SPIE Journal*, **Vol.2917**, 108–119.

Correia, L.M. and Prasad, R. (1997) An Overview of Wireless Broadband Communications *IEEE CommMag.*, **Vol.35 No.1**, 28–33.

ITU-T (1996) I.371 Traffic Control and Congestion Control in B-ISDN *Geneva*, **June 1996**.

Miah, B. and Cutbert, L. (1997) An Economic ATM Passive Optical Network *IEEE CommMag.*, **Vol.35 No.3**, 62–68.

Raychaudhuri, D. (1996) Wireless ATM Networks : Architecture, System Design and Prototyping *IEEE Personal Comm.*, **Vol.3 No.4**, 42–49.

Stojanovski, S. and Gagnaire, M. (1997) A New Wireless ATM Access Protocol for the Local Loop *First IFIP/IEEE International Workshop on Mobile and Wireless Communications Networks*, **May 1997**, Paris.

Walke, B., Petras, D. and Plassmann, D. (1996) Wireless ATM: Air Interface and Network Protocols of the Mobile Broadband System *IEEE Personal Comm.*, **Vol.3 No.4**, 50–56.

Zhang, H. (1995) Service Disciplines for Guaranteed Performance in Packet-Switching Networks *Proceedings of the IEEE*, **Vol.83 No.10**.

6

A Lab demonstration of a SuperPON optical access network

*J. Vandewege [a], X.Z. Qiu [a], B. Stubbe [a], C. Coene [a], P. Vaes [a], W. Li[a],
J. Codenie[a], C. Martin [b], H. Slabbinck [b], I. Van de Voorde [b],
P. Solina [c], P. Obino [c]*

[a] *University of Gent, IMEC, Sint Pietersnieuwstraat 41, B 9000 Gent,
Belgium TEL: +32 9 264 33 16 FAX: +32 9 264 35 93*
Email: jan.vandewege@intec.rug.ac.be

[b] *Alcatel Telecom Research Centre, Antwerpen, Belgium*

[c] *Centro Studie Laboratori Telecommunicazioni S.p.A. (CSELT),
Torino, Italy*

Abstract

This paper contains experimental results obtained from the ACTS PLANET
(Photonic Local Access NETwork) Lab demonstrator, which confirm the technical
feasibility of the SuperPON concept. The SuperPON is an optical fibre-based
ATM access network which can support a large number of subscriber Optical
Network Units (ONUs) up to 2048 and cover a long distance of 100 km.
This contribution shows detailed results of upstream experiments carried out on
newly developed fast switching and gain controlled Optical Repeater Units
(ORUs). Performance data on cascaded Semiconductor Optical Amplifiers (SOAs)
include Bit Error Ratio (BER) measurements, and Signal Noise Ratio (SNR)
degradation due to the introduction of cascaded SOAs. Experimental results
confirm that a SuperPON can service 2048 ONUs over a distance of 100 km.

Keywords

High splitting PON, EDFAs, SOAs, Upstream, TDM/TDMA, Burst mode BER.

Broadband Communications   P. Kühn & R. Ulrich (Eds.)
© 1998 IFIP. Published by Chapman & Hall

1   INTRODUCTION

In the continuing endeavour towards large scale broadband network access, SuperPON networks offer a unique combination of a very high bitrate continuous downlink, and a high bitrate burst mode uplink over a shared, tree-like fibre plant. Improvements in network capacity and range obtained within the ACTS PLANET project, show that a SuperPON can be a promising choice for offering a broadband basket with a mixture of distributive and interactive services to a large number of ONUs. The cost reduction offered by the capacity increase is mainly due to the physical concentration of all subscriber upstream information during propagation in the network, and to the distribution of downstream information from a single point of injection. This yields an increased sharing of the fibre plant, ORUs and the Optical Line Termination (OLT), - to be balanced with redundancy requirements - , and a more concentrated management. The range improvement can lead to extensive node consolidation, i.e. the omission of local exchanges, where switching functions are no longer required [1].

2   THE STRUCTURE OF THE LAB DEMONSTRATOR

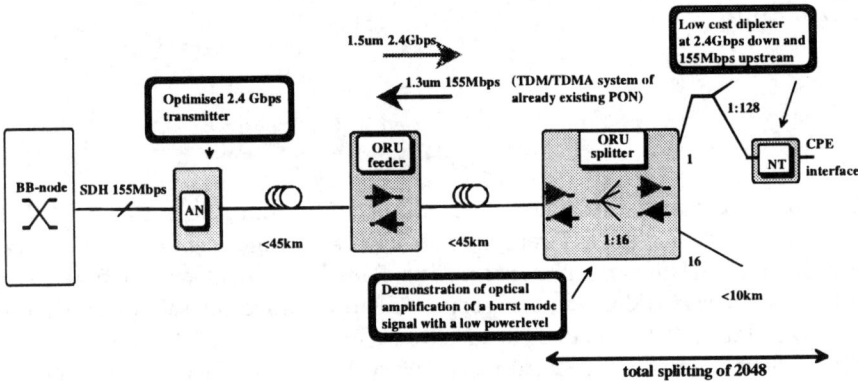

Figure 1 Overview of a SuperPON network.

Figure 1 shows a SuperPON network [2]. Downstream transmission is performed at the wavelength of 1550 nm with 2.5 Gb/s Time Division Multiplexing (TDM), and upstream transmission is operated at 1310 nm with 155 Mbit/s ATM-based Time Division Multiple Access (TDMA). The upstream/downstream Wavelength Devision Multiplxers (WDMs) allow for single fibre operation in the drop section, whereas the ORUs and the feeders use a double fibre approach. The maximum range of the feeder section is 90 km, and of the drop section is 10 km.
From a single Access Node at the OLT, the SuperPON will provide 2048 ONUs with high bit-rate interactive services. The use of Optical Amplifiers (OAs) allows

to support a much larger number of subscribers and to cover a much longer range (up to 100 km) than conventional Passive Optical Networks (PONs).

A Lab demonstrator configuration [3] suitable for the study of the SuperPON concept is shown in Figure 2. It contains 2 optical amplifier branches with a total of 6 ONUs at the drop section. Two cascaded ORUs (ORU1A,B and ORU2A,B) are in parallel branches of the Amplified Splitter.

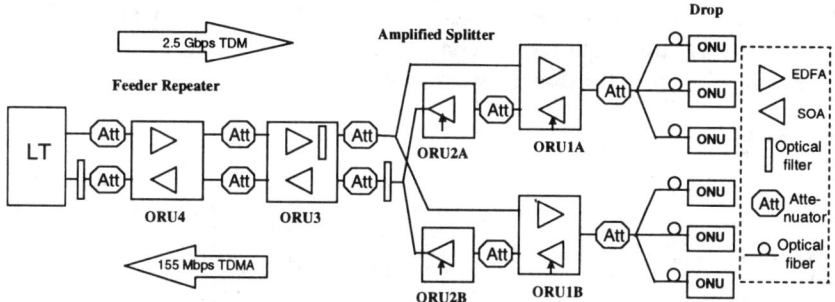

Figure 2 Network configuration of the Lab demonstrator.

2.1 2.488 Gbit/s downlink

The bit rate of the downstream link is 2.48832 Gbit/s, which is a SDH STM-16 bit-rate. This rate is exactly 16 times the 155.52 Mbit/s downstream bit-rate. In the downstream direction, 3 Erbium Doped Fiber Amplifier (EDFA) stages are cascaded [4]. Two branches of the downstream network send data to six 2.488 Gb/s receivers (RXs). As the downstream communication is in a continuous mode, no fast gain setting is required. This allows us to employ EDFAs with high gain, low Noise Figure (NF) as well as high saturation power as ORUs. Only one continuous transmitter using a TDM scheme feeds the downstream with data.

2.2 The 155.52 Mbit/s uplink

For upstream transmission, multiple OAs are placed in parallel branches and the optical outputs are combined at the amplified splitter section. Therefore, upstream ORUs should have nanosecond switching capability to minimise the accumulation of Amplified Spontaneous Emission (ASE) noise. Short carrier lifetime SOAs are used, which have to operate at low signal levels as they are placed behind the passive optical combiners which join the fibres from a large number of ONUs. Design for a high combining factor in the drop section is a major issue for the overall system performance. For dynamic range compression, cell by cell gain control is also required at the upstream ORU1A,B and ORU2A,B. The small signal gain of switchable SOAs is set to 22 ± 2dB, and the gain of gain fixed SOAs at ORU3,4 is set to 24 ± 2dB. Each SOA has maximum 4 dB gain uncertainty due to polarisation, wavelength and ageing effects. The minimum guaranteed small signal gain of the switchable SOAs is 20 dB for the worst case over the whole lifetime.

3 THE UPSTREAM NETWORK TOPOLOGY

Figure 3 illustrates the upstream network topology of the Lab demonstrator. Based on the calculated optical power budget, four cascaded UpStream ORUs (US-ORU1-US-ORU4) are required to compensate a total (Optical Distribution Network) ODN loss of 107 dB in the worst case.

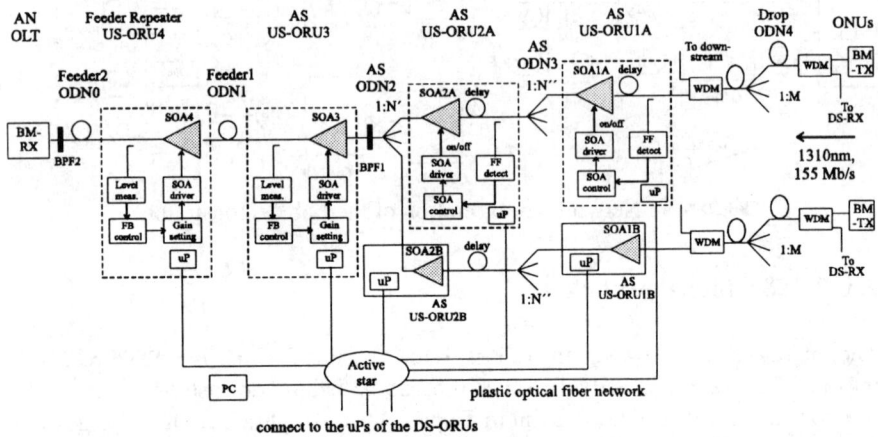

Figure 3 Upstream network topology of the Lab demonstrator.

The amplified splitter repeater uses 3 US-ORUs (ORU3, ORU2 and ORU1) for compensation of the large splitting loss. US-ORU1B and US-ORU2B have the same function as US-ORU1A and US-ORU2A. US-ORU4 is a remote feeder repeater required for extending the feeder length.

In Figure 3, the total splitting factor for upstream is N'×N''×M. According to the simulation and subsystem experiments, the maximum achievable M is 64. For obtaining a total splitting factor of 2048, N'' can be 16 or 8, and depends on the network architecture. It is obvious that N' is equal to 2 or 4.

ODN4 to ODN0 represent the network losses, caused by optical isolators, WDM, splitter, optical fibres, bandpass filters (BPFs) and optical connections. ODN4 represents the optical loss of the drop including the losses of two cascaded WDMs, a fibre length of max. 10km, and a 1:64 splitter. ODN1and ODN0 are the losses of the two feeders.

3.1 Upstream optical power budget

1310nm bi-directional Multiple Quantum Well (MQW) laser amplifiers CQF882/E [5] from Philips are used in the upstream direction, showing a saturated output power of typically 10dBm. Two external isolators are employed to prevent reflection problems.

Table 1 gives calculated upstream ODN losses in the worst case, typical case and best case respectively according to the European Telecommunication Standardisation Institute (ETSI) standard. The high loss at the drop section is typical for a high split upstream architecture. In order to avoid saturation of SOA3 and SOA4 by ASE, a BPF with 10 nm 3 dB bandwidth is placed at the output of SOA2. In front of the RX a second BPF is included to reduce the offset and eventual ASE power.

Table 1 Upstream network loss with 4 ORUs for the Lab demonstrator

	ODN4 (dB)	ODN3 (dB)	ODN2 (dB)	ODN1 (dB)	ODN0 (dB)	Total loss (dB)
max. loss	31.4	14.1	15.2	21.5	24.7	106.9
typ. loss	28.2	12.9	12.8	20.25	22.25	96.4
min. loss	21.5	11.7	10.4	19.0	21.4	84
splitter /length /BPF	1:64 /10km (max.)	1:8	1:4 /BPF1	45km	45km /BPF2	1:2048 /100km /2 BPFs

3.2 Functions of the upstream ORUs

To obtain the timing and signal level information required for switching the SOAs and implementing the level control function, a feed forward detection technique is employed as shown in Figure 3. An activity detect and level measurement is performed on each incoming cell at the input of the switchable SOAs. The control hardware of the US-ORU1,2 contains two major parts, a feed forward loop performing fast real time control [6], and a slow feedback loop monitoring the output power of the SOA. An optical delay line is inserted at the input of the SOA for compensating the required process time of the feed forward control loop. The feedback loop is closed over a microprocessor (μP) and all μPs are linked to an active star via a plastic optical fiber (POF) network.

US-ORU3 and US-ORU4 only require the feedback loop in order to compensate slow variations caused by eventual degradation of the SOA device and temperature dependency of the drive circuitry. Each ORU uses a cell clock that triggers the complete SOA control hardware. As this cell clock is running continuously, we need the activity detect mechanism telling us whether a cell passed by or not.

3.3 Operation of the upstream SOAs in switching mode

To be compatible with the upstream TDMA frame format, the burst-mode gain setting of the SOAs must be stable within 3 bits (19.2ns at 155.52Mb/s). A fixed delay between the control signal, which initiates the switching, and the SOA gain change, can be compensated. The important parameter of the SOA switching behaviour is the rise and fall time. From Figure 4 we can see the SOA switching between the off-state and either a gain of 22 dB or a gain of 12 dB both within a time of 10 ns, which corresponds to less than 2 bits at 155 Mbit/s. The switching time is not limited by the SOA itself but by the driver electronics on the ORU board. Switching off an US-ORU can also be done within 10 ns.

A fixed bias current allows for faster switching, but has a penalty on the isolation of the SOA in the off-state. The measured isolation of the SOA is reduced from 40 dB to 27 dB, when the bias current is increased from 10 mA to 15 mA.

gain = 22 dB, rise time = 6.4 ns   gain=12 dB, rise time = 9.7 ns

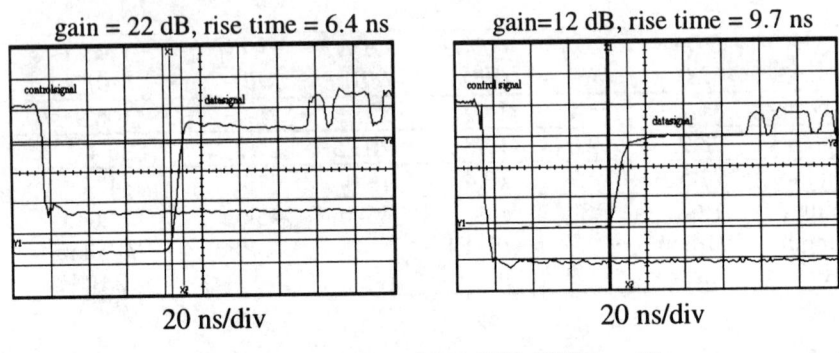

20 ns/div   20 ns/div

Figure 4 Switch on time of US-ORU1: < 10 ns.

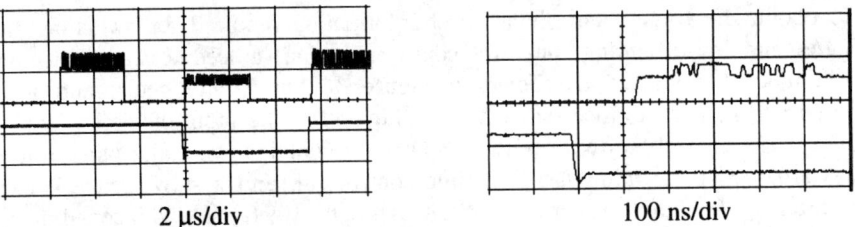

2 µs/div   100 ns/div

Figure 5 gain switching of the US-ORU (gain from 22 dB to 17 dB).

Figure 5 gives a general and detailed view of the fast gain switching of the US-ORU. This experiment shows the gain switching of the SOA on a cell basis under the assumption that no successive cells occur in the same branch. The gain is being switched from 22 dB to 17 dB. For visualisation, the level difference between cells was set to 5 dB. However, fast gain switching on a cell by cell basis and gain control over a range of 12 dB has been achieved. The measured dynamic range was limited by the feed forward detection, which has a dynamic range of 12 dB. It should be noted that this limitation is only presented in the Lab demonstrator, because we use an existing APON system for evaluation of the technical feasibility of the SuperPON network. Within the ACTS PLANET project a field trial is under development, where downstream grants will be extracted by an Operation, Administration & Maintenance ONU (OAM-ONU) located at the amplified splitter. This allows us to do the real time control of the SOAs without activity detection.

4 EXPERIMENTAL RESULTS

4.1 Performance of the operation in Continuous Wave (CW) mode

In the uplink, the main issue is to amplify weak upstream signals in order to obtain a maximum loss of the drop section. A low input power results in a reduced SNR because of the ASE power from the SOA. The signal-spontaneous (sig-sp) beat noise is caused by beating of the signal with spontaneous emission on a square-law photo detector. This is a fundamental limitation on the receiver SNR when the signal is relative strong. The spontaneous-spontaneous (sp-sp) beat noise contribution is generated by heterodyne mixing between frequency adjacent ASE components, which predominates in the total ASE noise when the input signal is getting weak.

For investigating the maximum power budget of the drop, or minimum input level of SOA1, a first subsystem depicted in Figure 6 containing one continuously activated SOA and two ODNs was tested. The two ODNs are emulated with optical variable attenuators. Here the loss of ODN4 and ODN3 represents the drop section loss and the feeder loss respectively. The loss of the feeder ODN3 was set in such a way that the incoming signal power at the RX is about -25 dBm peak, so that the system performance will be not limited by the RX sensitivity. A 9nm BPF is placed in front of a CW- RX.

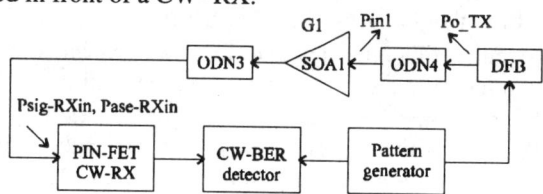

Figure 6 Set-up of 1 SOA in CW mode operation.

Table 2 Parameter settings of the 1-SOA set-up

	Po_TX (peak)	Drop ODN4 1:128/10km	Pin1 (peak) (dB)	G1 (dB)	Feeder ODN3 (with 9 nm BPF1)
Case1	+ 1.5 dBm	34.5 dB	-33 dBm	20	12 dB
Case2	+ 1.5 dBm	34.5 dB	-33 dBm	25	17 dB

Table 2 lists two sets of parameters for this 1-SOA set-up. Here Po_TX is the peak output power of the Transmitter (TX) used in the experiment; Pin1 is the peak input power of SOA1. In Case1, the gain of SOA1 was set to 20 dB (fiber to fiber gain), and ODN3 was set to 12 dB including BPF1. In Case2, the gain of SOA1 was increased to 25 dB (fiber to fiber gain), and ODN3 was increased by 5 dB. The BER was measured versus the input power of SOA1. When Pin1 is equal to - 35 dBm (peak), the measured CW-BER is 1.9×10^{-10} for Case1, and 3×10^{-10} for Case2.

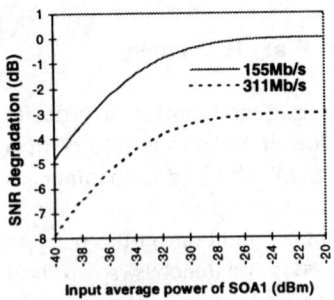

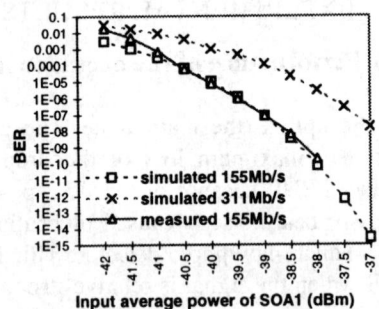

Figure 7 SNR degradation versus input average power of SOA1.

Figure 8 BER versus input average power of SOA1.

Figure 7 shows the calculated SNR degradation as a function of the mean signal power at the SOA input. The graph indicates that the SNR penalty is approximately 1.4 dB at an input average level Pin1 of - 34 dBm, and 3.4 dB at an input level Pin1 of - 38 dBm, both at a bit-rate of 155 Mbit/s. The performance at a bit rate of 311 Mb/s has been also evaluated, as the upstream data rate will be upgraded to 311 Mb/s in the field trial.

Figure 8 compares the simulated and the measured BER. The BER was calculated based on the SNR at the RX. It is clear that the two curves at 155 Mb/s in Figure 8 are merged together, this indicates that the measurement result agrees with the calculation. From the results shown in Figure 8, the minimum input signal power required to obtain a BER less than 10^{-9} is - 38 dBm (avg.) at 155 Mbit/s. However, 3 dB system margin is necessary to be added, since the SNR is degraded quickly when the incoming signal is too weak. Therefore, we can only set the minimum input power to -35 dBm (avg.) taking into account the margin. This value corresponds to a drop loss of 33.5 dB, which contains the loss from a 1:64 splitter, 10 km fiber length and other passive optical components at the drop section (the worst case) and 2 dB margin of the power budget.

From the above analysis, one can see that the maximum achievable optical power budget of the drop section is not dependent on the gain setting of the first SOA if this gain is high enough. The budget is limited by the sp-sp beat noise of SOA1. Therefore, the NF of SOA1 plays an important role in the whole system performance. Moreover, it is clear that ODN3 can be increased by an extra 5 dB if G1 is increased from 20 dB to 25 dB.

To investigate the SNR degradation due to the cascaded SOAs, CW-BER measurements were performed on similar 2-SOA and 3-SOA set-ups. The measurements were based on a maximum length 2^{23} -1 Pseudo Random Binary

Sequence (PRBS) pattern generated at 155.52 Mb/s. The decision threshold level was manually set in the middle of the eye opening.

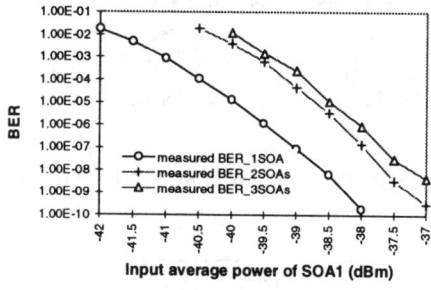

Figure 9 BER as a function of the input average power of SOA1.

Figure 10 Eye diagram of 155 Mbit/s signal having passed through 3 SOAs.

Figure 9 shows the BER as a function of the input average power at SOA1. The BER curves for 1 SOA, 2 SOAs and 3 SOAs are marked with circles, crosses, and triangles respectively. In the configuration with 1 SOA, a minimum input power of -38 dBm (avg.) was found to cause a BER of 1.9×10^{-10}. After the insertion of the second SOA, a BER of 3×10^{-10} was measured for an input power of -37 dBm. For the 3-SOA configuration, a BER less than 1×10^{-9} was found at an input level of -36.5 dBm.

Figure 10 gives an eye diagram measured at the RX. The input mean power at SOA1 was - 36.5 dBm (avg.) and the received average power was - 28 dBm (avg.) We can conclude that there is 1dB degradation on CW-BER from 1-SOA to 2-SOA set-up, and that the total system penalty is 1.5 dB for the worst case scenario, due to the introduction of SOA2 and SOA3. The penalty will be less if the network topology is optimised, whereas the above experiments were performed according to the worst case scenario emulation.

4.2 Performance of the operation in switching mode

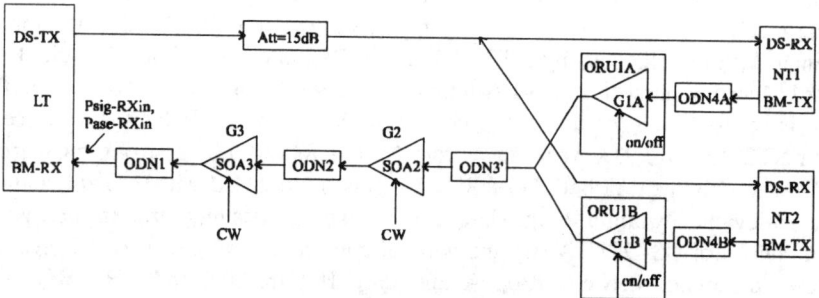

Figure 11 Set-up with 3 cascaded US-ORUs and two switchable branches.

Figure 11 illustrates a set-up with 3 cascaded SOAs and two switchable branches using an Alcatel ATM-PON system [7]. In this experiment, optical power budgets of drop A (ODN4A) and drop B (ODN4B) were measured, and the power levels at SOA1A,B required for a BER of 10^{-9} were also investigated, where US-ORU1A,B were operated in switching mode. Results from two different cases (with and without gain control of US-ORU1) are compared. Table 3 lists the parameter settings for this experiment.

Table 3 Parameter settings for the set-up operating in switching mode with gain control (ctrl.)

spec.ODN4 (split /range)	gain ctrl. G1	ODN3 (split)	gain fixed G2	ODN2 (split)	gain fixed G3	ODN1 (range)
32dB (1:64 /10km)	22 dB (weak cells) 12 dB (strong cells)	16dB (1:8)	22 dB	18dB (1:4)	24 dB	24 dB (45km)

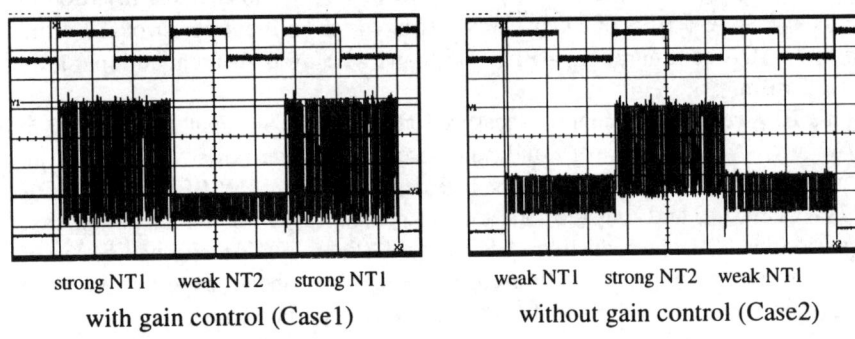

strong NT1   weak NT2   strong NT1   weak NT1   strong NT2   weak NT1

with gain control (Case1)   without gain control (Case2)

Figure 12 Burst data stream in the front of the RX.

Figure 12 shows upstream data streams measured in front of the RX, which contain cells sent by Network Termination 1 (NT1) with 50 % BW allocation from Branch A, and cells sent by NT2 with 25 % BW allocation from Branch B. In Case1, a strong burst from NT1 is followed by a weak burst from NT2, where the gain is set to 12 dB for the strong cells from NT1 and 22 dB for the weak cells from NT2. In Case 2, a weak burst from NT1 is followed by a strong burst from NT2, where the gain of both US-ORU1A,B was set to 22 dB without gain control. The achievable Dynamic Range (DR) of the two drops was measured in the case of gain control of US-ORU1A (B) and without gain control respectively, by creating a level difference between drop A and drop B (DR-AB1 and DR-AB2). The measured dynamic ranges between two drops are listed in Table 4.

From the experimental results, we can see that the dynamic range DR-AB1 (case1) is 16.5 dB with gain control, and 13.5 dB without gain control. The measured DR is limited by the experimental set-up, in which the variable optical attenuator has been set to 0 dB, so DR-AB1 (with gain control) could be higher if the loss of the drop A could go lower. The dynamic range DR-AB2 is 23 dB with gain control, and 15 dB without gain control, which is improved 8 dB by using the gain control.

Table 4 Measured dynamic range between two drops

	Case1,dB (with ctrl.)	Case1,dB (w/o ctrl.)	Case2,dB (with ctrl.)	Case2,dB (w/o ctrl.)
max.Pin1A (dBm)	> - 9.3	- 11.8		
min.Pin1B (dBm)	- 25.8	- 25.3		
DR-AB1 (dB)	> 16.5	13.5		
max.Pin1A (dBm)			- 4.3	- 11.3
min.Pin1B (dBm)			- 27.3	- 26.3
DR-AB2 (dB)			23	15

Moreover, a 5-7 dB degradation on the minimum input level of SOA1A,B for a BER of 1E-9 at the LT was observed when the set-up was changed from CW mode operation to switching mode. This is because the ASE power (or offset) varies a lot from cell to cell, so an optimised BM-RX with a cell based offset compensation is required. However, we use the existing APON LT, which was designed for PON applications where no ASE power variation exists. The actual used BM-RX is AC coupled, and the threshold is set to a fixed level with respect to the peak optical power level, and cannot be optimised on a cell basis for the SuperPON applications. The effect is illustrated in Figure 13.

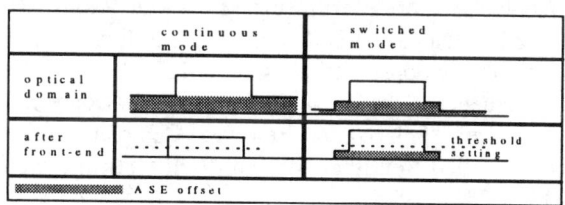

Figure 13 Penalty caused by the current BM_RX design.

By observing the optical signal quality at the input of the BM-RX, we can conclude that the measured minimum input levels of SOA1A,B are limited by the threshold setting of the existing BM-RX. The achievable optical power budget can be increased significantly by using a new receiver optimised for burst mode signals containing strong and variable ASE power, which is currently in development.

5   CONCLUSIONS

Various upstream ORU designs have been completed. Fast on/off switching and gain control of the SOAs results in an important decrease of the accumulation of noise and a substantial reduction of the dynamic range requirements within the system.

Different network topologies have been evaluated and emulated by the Lab demonstrator set-up. The experimental results indicate that even with the worst case scenarios an upstream splitting ratio of 2048 and a range of 100 km are feasible at 155Mb/s if an optimised burst-mode receiver is used.

ACKNOWLEDGMENTS

This work was supported by the European Commission within the ACTS project AC050 "PLANET". We thank our PLANET partners for their co-operation, especially Philips Optoelectronics Centre for the 1310 nm MQW SOAs, and Alcatel Optronics for the EDFAs, STM-16 transmitters and 1.3 DFB lasers used in the Lab demonstrator.

REFERENCES

1. Denis. J.G. Mestdagh and Claire M. Martin, "The SuperPON Concept and its Technical Challenges", Proc. of IFIP/IEEE Broadband Communications 1996.
2. Ingrid Van de Voorde, M.O. van Deventer, P.J.M. Peters, P. Crahay, E. Jaunart, A.J. Philips, J.M. Senior, X.Z. Qiu, J. Vandewege, J.J.M. Binsma, P.J. Vetter, "Network Topologies for SuperPON", Proc. of OFC'97.
3. Claire M. Martin, Hans Slabbinck, Peter Vetter, Xing-Zhi Qiu, Jan Vandewege, Paolo Solina, Denys Haux, Gerlas van den Hoven, Andre Boot, "Realisation of a SuperPON demonstrator", Proc. of NOC '97.
4. Hans Slabbinck, Claire Martin, Ingrid Van de Voorde, Brecht Stubbe, Peter Vaes, Xing-Zhi Qiu, Jan Vandewege, Paolo Solina, Pietro Obino, "Evaluation of a SuperPON Demonstrator", Proc. of SPIE '97.
5. Luuk F. Tiemeijer, Peter J. A. Thijs, Teus van Dongen, J.J.M. Binsma, and Edwin J. Jansen, "Polarization Resolved, Complete Characterization of 1310 nm Fiber Pigtailed Multiple-Quantum-Well Optical Amplifiers", J. LT., vol. 14, pp. 1524-1533, 1996.
6. B. Stubbe , P. Vaes, L. Gouwy, C. Coene, X. Z. Qiu, B. Staelens, J. Vandewege, H. Slabbinck, C. Martin, I. Van de Voorde, "Embedded real time control of optically amplified repeaters in broadband access networks" Proc. of SPIE '97.
7. X. Z. Qiu, C. Martin, J. Zhou, B. Stubbe, C. Coene, B. Staelens, I. Van de Voorde, J. Vandewege, "Upstream Burst-mode Experiments using Semiconductor Optical Amplifiers as Repeaters for a SuperPON", Proc. of Optical Amplifier and their Applications 1997.

7

Request Contention and Request Polling for the Upstream Media Access Control in ATM Access Networks

Joachim Charzinski
Siemens AG, Public Communication Networks Group
Hofmannstr. 51, D-81359 München, Germany
Tel. ++49 89 722 46803, Fax ++40 89 722 26877
e-mail: joachim.charzinski@oen.siemens.de

(The studies reported here were performed while the author was with the Institute of Communication Networks and Computer Engineering, University of Stuttgart, Germany)

Abstract

This paper presents a simulation study comparing different mechanisms used in the upstream channel of ATM access networks. In *request polling*, status information from the access stations is transmitted periodically in minislots dedicated to each access station. In *request contention*, the access stations use a contention based multiple access scheme with a multi-slot stack contention resolution algorithm for random access to the minislots. In addition, in both cases request information is piggy-backed to upstream ATM cells.

In a simulation study, *request polling* and *request contention* are compared using the same system parameters. It is shown that (1) the optimum stack parameter is greater than three if the round trip delay in the system is taken into account and (2) under most conditions request contention produces longer mean transfer delays and delay quantiles than request polling.

Keywords

ATM, Access Network, Media Access Control, Delay Quantile, Comparison, Request Contention, Stack Algorithm, Request Polling

1 INTRODUCTION

In a modern high-speed communication infrastructure, providing cost-effective access to a global broadband network is a crucial issue. Shared media ATM access networks have been under study for some time, mainly using Passive Optical Networks (PON) (Ballance et al. 1990, Glade & Keller 1993, Killat 1996) or the Hybrid Fibre Coax architecture (ATMF 1996). The bandwidth demands of residential users should not be under-estimated as even information

Broadband Communications   P. Kühn & R. Ulrich (Eds.)
© 1998 IFIP. Published by Chapman & Hall

retrieval services like accessing the world wide web can create a high traffic volume for both directions of the network: Except for pure real-time media delivery services, most data transmission employs transport protocols using an acknowledgement mechanism to ensure successful transmission of all data packets. These acknowledgements – in addition to normal user traffic – can pose high bandwidth demands even on the "upstream" channel (from users to the public network) in an access network.

In this paper, two mechanisms employed in the shared media access control for the upstream transmission of cells in ATM access networks are compared in terms of the mean delay and the delay quantiles they introduce into the upstream traffic. In section 2, the investigated system models are introduced, the two media access control mechanisms are described briefly and the traffic scenarios used in the simulation study are defined. The choice of the stack parameter used in contention resolution is discussed in section 3 before the delay results are presented as a function of the ratio of time slots spent to support the MAC mechanisms.

2   SYSTEM MODELS

2.1   Reference Architecture

In the "downstream" direction (from public network to customers), the ATM time slot structure can be maintained on the access network. Each downstrem cell is broadcast to every customer's interface, where a filtering function can substitute all ATM cells not destined for the local interface by idle cells.

In the "upstream" direction (from customer to public network) of PON and HFC access networks, a media access control mechanism is needed to coordinate the times when different stations can access the shared medium. The network elements where the MAC (media access control) protocol is terminated will be called "access station" (AS) and "headend" (HE) to be independent of the actual network architecture used in a specific system. The role of access station can be performed by a network termination or an optical network unit (Killat 1996) and a the headend functions can be found in a line termination (I.327 1993), the ATM Digital Terminal (ATMF 1997) or the DAVIC Access Node (DAVIC 1997).

We assume here that a single return channel is shared between the access stations using TDMA (time division multiple access). The issue of ranging has been discussed elsewhere (van Heijningen et al. 1994), and we can take the access network to be fully ranged, i.e. all access stations show the same reaction time 2τ between the headend issueing a permit and receiving the corresponding upstream ATM cell from the access station. The timing between the headend and one access station is depicted in Figure 1. Note that the sequence of permits issued by the headend is reproduced by the sequence of upstream ATM cells arriving at the headend one round trip delay (2τ) later. The actual transmission time needed to emit an ATM cell on a high speed channel (one time slot of duration t_C) is usually at least an order of magnitude smaller than the round trip delay, so that it is not included in Figure 1 and we assume the transmission times of permits and cells and the associated hardware delays to be included in the round trip delay. For every time slot, the headend can issue a new permit to a different AS, even if the previous permit has not yet resulted in an upstream cell arriving at the head end. This mode of operation is also called "pipeline polling" (Quayle 1996). The destination of a permit is independent of the destination of the ATM cell it is attached to. Note that the whole transfer delay consists of a constant part (3τ) and a variable part $(T_{RW} + T_{PW})$.

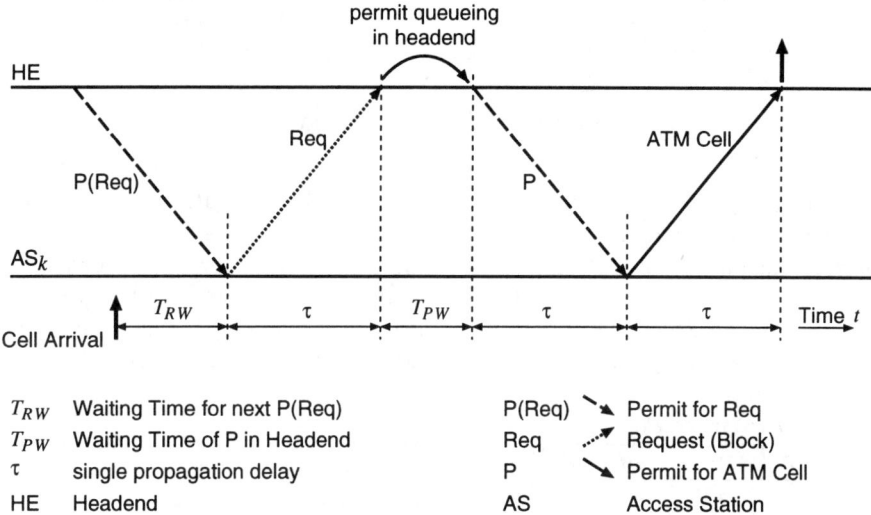

Figure 1 Requests, Permits and Waiting Time Contributions

Figures 1 (timing) and 2 (simulation model) show that an access station, after receiving an ATM cell from one of its sources, stores this cell in its cell queue and transmits a *request* to the headend at the next opportunity. After the request has arrived at the headend, the headend reacts by generating a permit for the respective access station and queueing it in its central permit queue, from which one permit is transmitted downstream in each time slot (except for the request block permits described later). When the access station receives a permit for an ATM cell, the cell waiting at the head of the station's cell queue is transmitted upstream.

Requests to be transmitted to the headend can be piggy-backed on upstream ATM cells from the same access station. However, when an ATM cell arrives at an empty access station queue, there must be an additional way of trans-

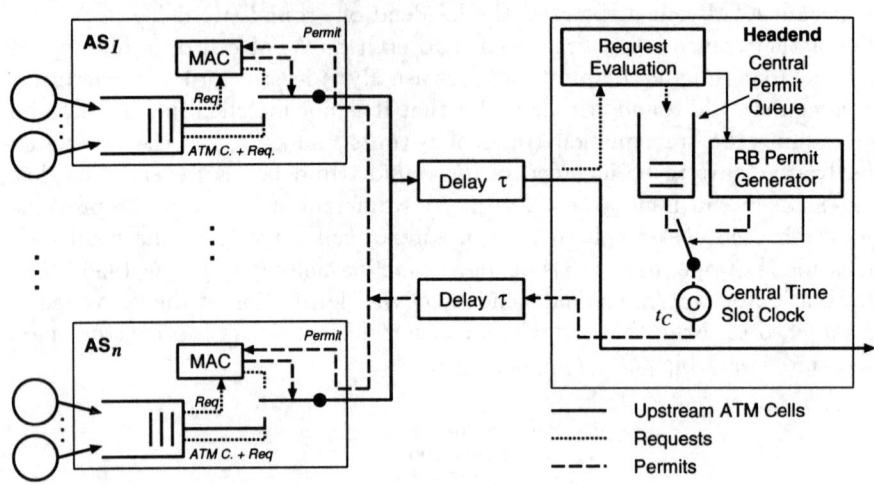

Figure 2 Simulation model

mitting a request. This is where so-called "request blocks" (RBs) are used. An RB has the duration of a normal ATM cell slot and it consists of m minislots, each giving one access station the opportunity to transmit a request. A request is simply the transmission of the value of the current queue length in the access station, which can be transformed into the number of newly arrived ATM cells by a simple algorithm in the headend (Killat 1996, Charzinski 1996).

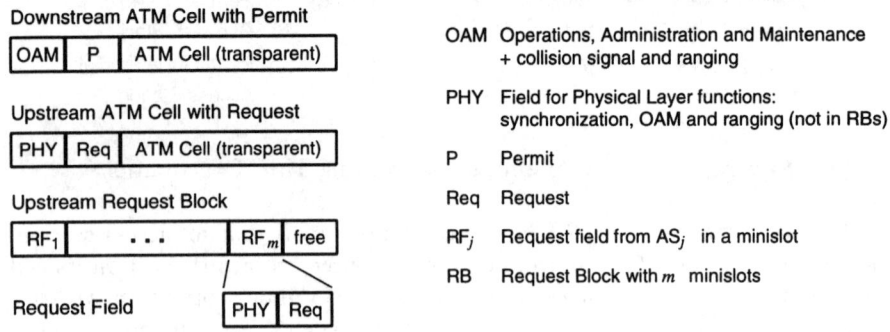

Figure 3 Upstream and Downstream Slot Formats

As illustrated in Figure 3, each access station needs to transmit a synchronization preamble before transmitting an ATM cell or a request upstream in order to keep a minimum gap between successive transmissions and help the burst mode receiver in the headend synchronize to the specific access station's transmitter (Killat 1996). Assuming that the downstream bandwidth is an integer multiple r of the upstream bandwidth, every r^{th} downstream cell has a valid permit attached. For the following studies, we assume that the preamble together with the request field occupies three octets (Killat 1996), so that a slot consists of 56 octets and it can be devided into 18 minislots, of which $m = 16$ will be used for upstream request transmissions.

The following two subsections describe how these minislots are used for request polling or request contention.

2.2 Request Polling

In *request polling*, all access stations connected to the access network are polled periodically for requests. Thus, if there are more than m access stations, each one can transmit in one dedicated minislot of every $\lceil n_0/m \rceil$th request block, with n_0 denoting the number of all access stations. Given the ratio ρ_P of time slots used for request blocks, the time between two request block permits is

$$t_P = t_C/\rho_P \tag{1}$$

and the time between two request block permits for the same access station is

$$t_{PE} = \lceil n_0/m \rceil t_P \quad . \tag{2}$$

The request polling mechanism has been analyzed for Poisson input traffic in (Charzinski 1997) where it was shown that there is an optimum value for the request block ratio ρ_P.

2.3 Request Contention

In *request contention*, only those access stations contend for the request minislots in a request block which have had a new ATM cell arrive to an empty queue. The mechanism used for collision resolution is the Multi-Slot q_W-ary Stack Algorithm described in (Bisdikian 1996) and (Golmie et al. 1996). As illustrated in Figure 4, instead of varying the polling period t_P, the number r_B of consecutive request blocks is varied with ρ_P here to be

$$r_B = \rho_P t_P/t_C \quad . \tag{3}$$

The reason is that in order for an access station to repeat a request in the following request block, it needs to have received information on whether the previous access to a request minislot resulted in a collision (Collision Signal CS in Figure 4). The state transition diagram for the access station is shown in the appendix.

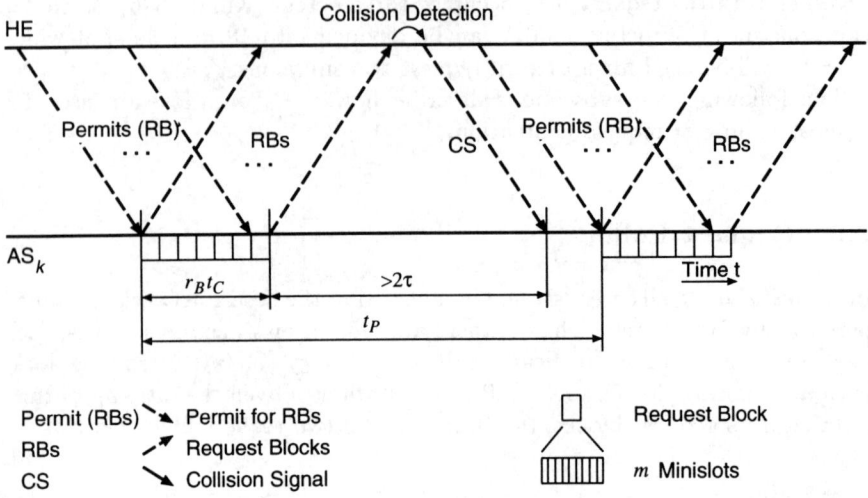

Figure 4 Timing Constraints using Request Contention

2.4   System Parameters and Traffic Scenarios

Table 1 defines four traffic scenarios. The data are divided into three categories: In the upper part of Table 1, the sources are defined. In the middle part, general system parameters are given, and in the lower part, the configuration of sources and access stations is described.

Name		ND100	ND10	TS100	TS10
Source Period	t_S/t_C	100	10	100	10
Burstiness	b	1	1	2	2
Mean Burst Length	c_B	–	–	10	20
Minislots per RB	m	16	16	16	16
Propagation delay	τ/t_C	18	18	18	18
Total Acc. Stations	n_0	128	128	128	128
Active Acc. Stations	n	80	8	70	10
Sources per Station	b_S	1	1	2	1
Offered Load in %	A_R	80	80	70	50
Multiplexing Gain	$b \cdot A_R$	0,8	0,8	1,4	1,0

Table 1 Traffic Scenarios and System Parameters

In two scenarios, the sources have a low cell rate (one cell every 100 slots), so that the mechanism of requests coupled to upstream ATM cells cannot normally be used. In the other two scenarios, the sources have a high cell rate (one cell every 10 slots), so that coupled requests can almost always be used. Two scenarios (ND) use periodic traffic and two use talkspurt-silence (TS) sources with a geometrically distributed number of equidistant cells in a burst and negative exponentially distributed silence periods. The burstiness b denotes the ratio of peak to long-term mean cell rate. A burstiness of $b=2$ is enough to ensure that normally the last ATM cell has left the access station before the next burst arrives, so that the access station becomes idle between bursts.

The single propagation delay is assumed to be 18 cell slots, which roughly corresponds to $50\mu s$ or a distance of $10\,km$ after ranging between access stations and headend at $155.52\,Mbit/s$. The total number of access stations $n_0 = 128$ is not varied, so that the ratio $t_{PE}/t_P = 8$ is fixed for all scenarios.

A target offered load of $A_R = 80\,\%$ was chosen for the ND scenarios. The offered load in the TS scenarios was determined by limiting the burst scale loss probability in a bufferless multiplexer to 10^{-9} using (Roberts 1991, (7.1.5)).

3 RESULTS

Before the results for the ND and TS scenarios are compared in sections 3.2 and 3.3, the influence of the stack parameter q_W (see section 2.3) is investigated in section 3.1. All results have been acquired using discrete event simulation with 10^7 cells. The confidence intervals shown are for a 95 % confidence level. For mean values they have been computed from 10 part test results using a Student-t test. Delay quantiles have been extracted from the empirical delay distribution measured during the whole simulation and their confidence intervals have been computed according to the method given in (Sachs 1992). For simulating the periodic ND scenarios, the phase shift method described e.g. in (Panken 1997) was used.

3.1 Dimensioning the Stack Parameter

For scenario ND100, Figure 5 gives the mean end-to-end delay (EED) and Figure 6 gives the $(1\text{-}10^{-5})$ quantile of EED as a function of the stack parameter q_W. The request block period is $t_P = 50\,t_C$ and there are $r_B = 4$ consecutive request blocks used in each period, leading to a ratio of $\rho_P = 12.5\,\%$ of request blocks per upstream slot. Only integer values of q_W are possible; the discrete points have only been interconnected to increase the readability of the diagram.

For the long round trip delay τ considered here, the optimum value of q_W

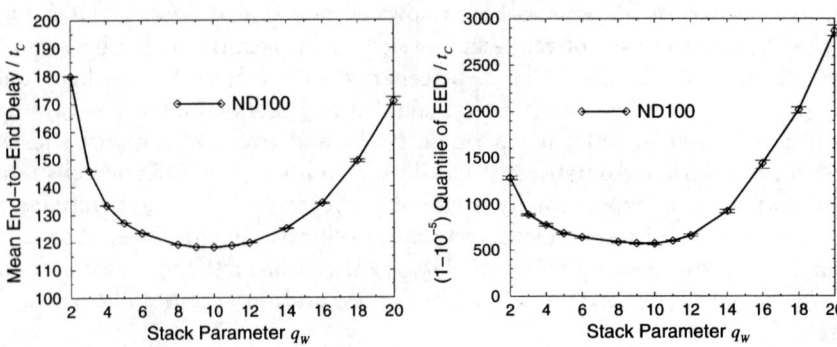

Figure 5 ND100: Mean Delay   **Figure 6** ND100: Delay Quantile

is around 10, which is much more than the value of 3 that has been shown to be the optimum for systems with immediate collision feedback (Bisdikian 1996, Merakos & Bisdikian 1988). For the following investigations, results for both $q_W = 3$ and $q_W = 10$ will be given.

3.2   Constant Cell Rate Traffic

Figure 7 shows the $(1\text{-}10^{-5})$ quantile of end-to-end delay in the ND100 scenario as a function of the ratio ρ_P of upstream slots used for request blocks. As above, the discrete points have been interpolated to increase the readability of this and the following diagrams. Using request contention (RC) on the same number of minislots leads to much longer delay quantiles than request polling (RP). Increasing the stack parameter q_W from 3 to 10 helps in reducing the delay, but leads to instability when ρ_P is less than about 10 %. Due to the inherent stability of an $n^*D/D/1$ periodic service system even at the load limit, the delay quantile reaches a minimum at $\rho_P = 20\,\%$ where the total upstream slot utilization $\rho_P + A_R$ reaches 1.

In scenario ND10, the high source rate ensures that in steady state, all new ATM cell arrivals can be communicated to the headend via requests coupled to upstream ATM cells from the same access station. Therefore the difference between request polling and request contention is much smaller than in the ND100 scenario discussed above. The remaining difference is due to the different distribution of service interruptions for request blocks in the headend permit queue: As shown in Figure 4, multiple request blocks must be blocked together to effectively increase the number of minislots available for contention resolution. This leads to service interruptions in the headend queue lasting for $r_B = \rho_P t_P / t_C$ time slots. On the other hand, for request polling the request blocks are distributed over the whole period t_{PE}, leading to much shorter service interruptions at a time.

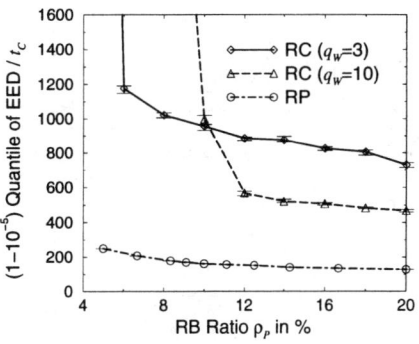

Figure 7 ND100: Delay Quantile

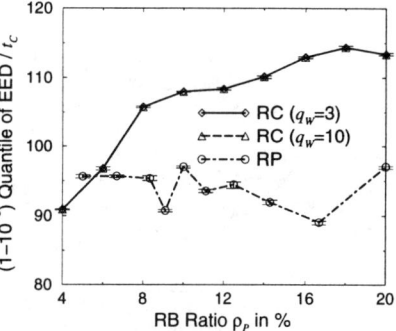

Figure 8 ND10: Delay Quantile

3.3 Variable Cell Rate Traffic

The results for request polling in Figures 9 through 12 show that the optimum RB ratio ρ_P is around 10–15%. For smaller values of ρ_P, the waiting time T_{RW} for a request block dominates whereas for larger ρ_P the headend queueing delay T_{PW} increases as the load limit is approached in the headend queue.

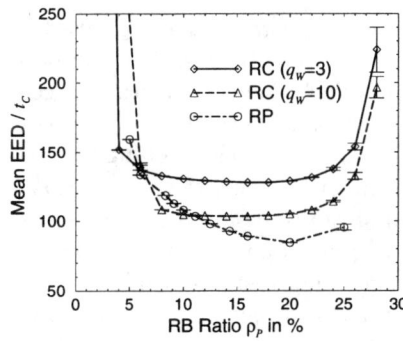

Figure 9 TS100: Mean Delay

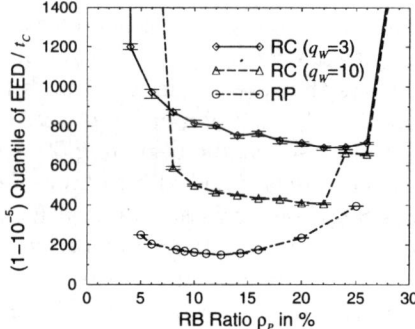

Figure 10 TS100: Delay Quantile

While request contention can lead to lower mean delays than request polling when there is little bandwidth spent for request blocks (Figures 9 and 11), this situation is reversed when ρ_P is increased or when the delay quantiles in Figures 10 and 12 are considered. In scenario TS100 with its many low rate on/off sources and a mean offered load of $A_R = 70\%$, request contention shows instability both at low and high values of ρ_P, regardless of the stack parameter q_W. On the other hand, request polling shows a minimum delay quantile at around $\rho_P \approx 12\%$ in scenario TS100 (Figure 10) and is not so prone to become instable. The results for scenario TS10 shown in Figures 11 and 12 clearly indicate that even if the number of stations contending for access to the request minislots is low (only 10 out of 128 stations are active)

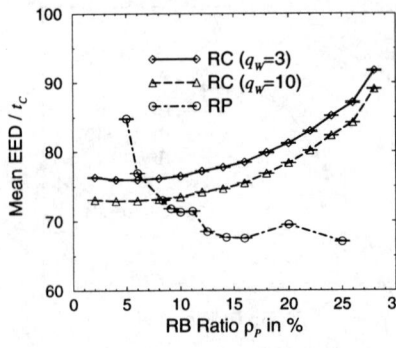

Figure 11 TS10: Mean Delay

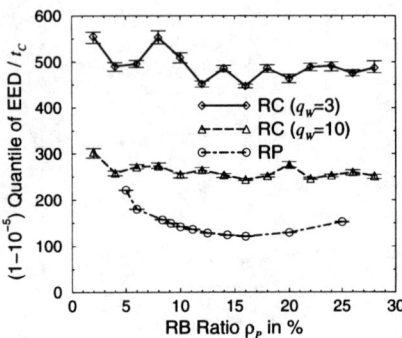

Figure 12 TS10: Delay Quantile

and request contention is only being used at each beginning of a burst, the mean delay and the $(1\text{-}10^{-5})$ delay quantile can be much lower under request polling than under request contention if a sufficient number of minislots is available.

4   CONCLUSIONS

A simulation study has been presented evaluating two mechanisms used in the upstream media access control of ATM access networks. The comparison shows that even under conditions favourable for contention based mechanisms, i.e. a small proportion of active stations and contention based access only being used at the beginning of each data burst, request polling leads to lower mean delays and also to lower delay quantiles when *the same number* of upstream minislots is dedicated for request transmission. In practice, the difference will be even greater because synchronization of a burst mode receiver to an unknown transmitter takes longer than if the transmitter phase and amplitude is known before. Therefore the synchronization preambles for request contention will have to be longer than for request polling and the number of minislots per request block will be smaller, thus further increasing the delay with request contention.

As a side result, it was shown that the optimum stack parameter for request contention is more than 3 if the access stations have to wait for one round trip delay to get feedback information on the success of previous random access to a minislot.

REFERENCES

J.W. BALLANCE, R.F. LEE, P.H. ROGERS, M.F. HALLS, "A B-ISDN Local Distribution System based on a Passive Optical Network." *Proc. GLOBECOM'90*, 1990, pp. 305.4.1–305.4.5

M. GLADE, H. KELLER, "Novel Algorithm for Time Division Multiple Access in Broadband ISDN Passive Optical Networks." *Int. J. of Digital and Analog Comm. Systems* Vol. 6 (1993) pp. 55–62

U. KILLAT (ED.), *Access to B-ISDN via PONs – ATM Communication in Practice*, Wiley/Teubner 1996

THE ATM FORUM, "The UPSTREAMS Protocol for HFC Networks, Revision 1.", ATM Forum/IEEE Doc. 95-1435R1, Jan. 1996

P. VAN HEIJNINGEN, T. MOSCH, A. VAN OOYEN, L. D'ASCOLI, K. DE BLOK, P. SOLINA, "Out-of-Band Ranging Method for ATM over PON Access Systems." *Proc. ECOC'94*, Florence, Italy, 1994

ITU-TSS, ITU-T Recommendation I.327 *B-ISDN Functional Architecture.* March 1993.

THE ATM FORUM TECHNICAL COMMITTEE RBB WG, *RBB Baseline Document Draft.* Doc. BTD-RBB-001.02, April 1997

DIGITAL AUDIO-VISUAL COUNCIL (DAVIC), Specification Part 02 Version 1.2 *System Reference Models and Scenarios.* Rev. 4.0 Geneva, Switzerland, 1997

A. QUAYLE, "Broadband passive optical network media access control protocols." in *Proc. SPIE Photonics East 1996 Conf. on All-Optical Comm. Systems: Architecture, Control and Network Issues II* Vol. 2919, Boston, MA, USA, Nov. 1996, pp. 268–278

J. CHARZINSKI, "A new approach to ATM access networks." in *Proc. of SPIE Photonics East 1996 Conf. on Broadband Access Systems* Vol. 2917, Boston, 1996, pp. 108–119

J. CHARZINSKI, "Performance Analysis of a Multiple Access Mechanism for ATM Access Networks." in *Proc. IEEE ATM'97 Workshop*, Lisboa, Portugal, May 1997, pp. 477–485

C. BISDIKIAN, "Performance Analysis of the Multi-slot n-ary Stack Random Access Algorithm (msSTART)." *IEEE Project 802.14 Working Group* Doc. IEEE802.14-96/117, May 1996

N. GOLMIE, S. MASSON, G. PIERIS, D. SU, "Performance evaluation of MAC protocol components for HFC networks." in *Proc. SPIE Photonics East 1996 Conf. on Broadband Access Systems* Vol. 2917, Boston, 1996, pp. 120–130

J.W. ROBERTS, EDITOR, *Performance evaluation and design of multiservice networks.* Final report of the COST 224 project, Office for Official Publications of the European Communities, Luxembourg, 1991

L. SACHS, *Angewandte Statistik – Anwendung statistischer Methoden.* Springer-Verlag, Berlin, 7. Auflage 1992, S. 338 (in German)

F.J.M. PANKEN, *Design and performance evaluation of multiple-access protocols for ATM-based passive optical networks* Dissertation, Wiskunde en Informatica, Katholieke Universiteit Nijmegen. Thesis Publishers, Amsterdam, NL, 1997

L.F. MERAKOS, C. BISDIKIAN, "Delay Analysis of the n-Ary Stack Random-

Access Algorithm." *IEEE Trans. Information Theory*, Vol. 34, No. 5, Sep. 1988, pp. 931–942

APPENDIX 1   STATE TRANSITION DIAGRAM FOR REQUEST CONTENTION

Figure 13 defines the MAC process in an access station for request contention. The annotation *A[B]/C* along an arrow is to indicate that this transition between two states happens when event *A* occurs and if condition *B* is true. During the transition, action *C* is performed.

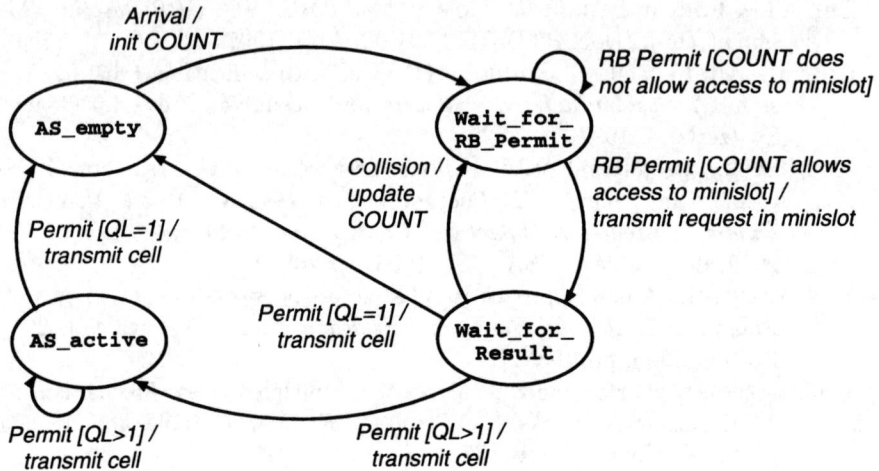

RB: Request Block   QL: Queue length before ATM cell leaves AS
COUNT: Counter for Stack Algorithm

Figure 13 State Transition Diagram for an Access Station with Request Contention

Mobile Access

8

Performance analysis of ATM transmission over a DS-CDMA channel

T. Timotijevic, J. A. Schormans
Queen Mary and Westfield College
University of London
Dept. of Electronic Engineering, Mile End Road,
London E1 4NS, United Kingdom
e-mail: (T.Timotijevic, J.A.Schormans)@qmw.ac.uk
Tel: +44 171 415 3756   Fax: +44 181 981 0259

Abstract

This paper presents numerical results for ATM cell transmission over a DS-CDMA satellite link obtained by mathematical analysis. We show that an increase in capacity can be achieved by using discontinuous transmission detection (DTX) for all ON-OFF sources, thus exploiting the burstiness of ATM traffic. Three scenarios are analysed depending on whether the code is allocated to a user or to a single virtual connection (VC), and whether the transmission rate for different connections is the same or varies across the system. System performance is expressed in terms of link efficiency and relative capacity increase. Initial results show how different transmission rates and loads affect the system performance.

Keywords
ATM, DS-CDMA capacity, traffic analysis, ATM over satellite

Broadband Communications   P. Kühn & R. Ulrich (Eds.)
© 1998 IFIP. Published by Chapman & Hall

1   INTRODUCTION

We describe three architectural scenarios that exploit the bursty nature of ATM traffic to increase the broadband DS-CDMA satellite system capacity, and analyse their cell-level performance. A multirate wireless DS-CDMA system can accommodate connections of varying data rates by adjusting the different data rates to an intermediate transmission rate R prior to spreading by a pseudo-random or other code of a much higher rate (chip rate). We focus on the system performance of the *uplink*, from the *ATM layer* viewpoint, on a satellite link. The results show that the performance on the ATM layer influences the system design on the lower, physical, layer. The bursty nature of ATM is in this case beneficial to the system capacity. The performance is assessed with respect to the system capacity and link efficiency (utilisation). Our aim is to find an optimum scenario with its system design parameters for a given satellite transponder bandwidth.

As CDMA is interference-limited scheme, it can approach and even outperform TDMA in terms of system capacity under conditions that reduce the interference in the system (Elhakeem 1994, Van Nee 1995, Monsen 1995). The capacity increase in a CDMA achievable by DTX has been mainly researched in voice systems (Yang 1994) or multimedia systems with DTX applied only to voice (Fong 1996). Work within RACE 2020 CODIT project, as reported in (Baier 1993, Andermo 1994), applied DTX to other types of traffic. However, no detailed cell-level performance analysis has been presented.

In this study, we investigate the impact of traffic characteristics, as well as the choice of system parameters, on the performance of different scenarios. This serves as a starting point for the development of connection-level control schemes in the satellite IBC networks that use ATM over DS-CDMA. Our mathematical analysis has been validated using a cell level network simulator.

2   SYSTEM DESCRIPTION

We assume DS-CDMA uplink, ideal power control and pure ATM (no framing). Only degradation effects intrinsic to CDMA are taken into account, i.e. CDMA multiple-access interference. Our three configuration scenarios are based on:

- code allocation policy: code per user, and code per virtual channel connection (code per VCC);
- transmission rate R across the system: whether it is constant, or varies among users.

In Code per User allocation policy, VCs are multiplexed prior to spectrum spreading and transmission. In Code per VCC, each VCC gets its own spreading code and multiplexing is done "in the air", in the code domain.

Use of the DTX for CDMA system capacity increase relies on the random nature of sources. In our case, statistical multiplexing is achieved by switching off the carrier when there the source is idle, resulting in the reduction of the overall system self-interference. The drawback is the need for code and carrier re-synchronisation when the source reactivates. In existing CDMA systems transmission during silent periods is continued at reduced power and bit rate in order to preserve carrier synchronisation. The possible increase in capacity in this case may be lower than when the transmission is completely switched off. This trade-off between capacity and efficiency will be considered in future work.

With DTX employed, the empty cells will still be inserted during active periods of the source due to the difference in the transmission and source data rates. For pure ATM, the link efficiency is measured as the proportion of useful cells. Link efficiency is increased when the number of transmitted empty cells is decreased.

2.1 Impact of processing gain on system architecture

The ratio of transmission and chip rates is the CDMA processing gain. It determines the amount of multiple-access interference perceived by a receiver. The approximate formula that binds the processing gain G_p, signal-to-noise-plus-interference ratio (SNR), maximum number of users that could be accepted in a system while maintaining the specified SNR (C_{max}) and Eb/No is (Pursley 1977):

$$SNR = \left(\frac{C_{max} - 1}{3G_p} + \frac{N_0}{2E_b} \right)^{-1}. \tag{1}$$

The first term above effectively defines multiple-access interference from other users. The quality of service of a connection is related to the SNR, and is expressed in terms of the cell loss ratio (CLR). In our model transmitted Eb/No is constant. Depending on whether the transmission rates for different traffic types vary or not, and whether the system allows different QoS for different traffic types, the processing gain G_p can be constant or variable.

2.2 Scenarios

2.2.1 Code per User allocation policy, R =const. (Figure 1)
Multiplexing of VCCs is at the user terminal. Link BER requirement will dictate a limit for CLR at the buffer, which may lead to increase of the queueing delay. The link has to maintain the QoS for the most stringent traffic type. This yields G_p=const. across the system. As R =const. all channels have equal bandwidth

2.2.2 Code per VCC allocation policy, R =const. (Figure 2)
Each connection gets its own spreading code, hence multiplexing at the terminal is avoided. We can expect only very little or no buffering delay. Different connection types can be transmitted in the same channel due to R=const. QoS cannot vary since all processing gains within a channel must be equal.

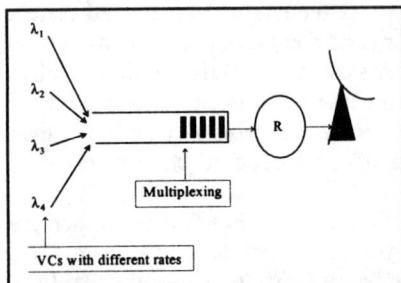 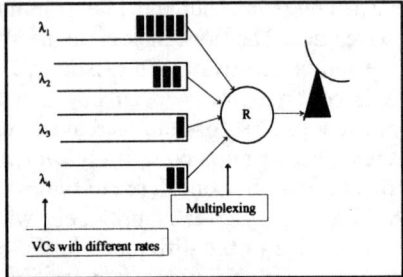

Figure 1   Code per User scenario **Figure 2**   Code per VCC (*R=const*)

2.2.3   *Code per VCC allocation policy, R<>const.*

Each connection has its own spreading code, and different traffic types have different transmission rates. The problem of choosing the optimum R becomes a problem of system channel dimensioning, i.e. determining channel bandwidths for different traffic types and the number of such channels for envisaged corresponding load. Variation of QoS can be achieved by choosing appropriate values of G_p. Within each channel, the increase of capacity is measured as number of connections of that particular type. Thus capacity increase depends on the maximum number of users $C_{max,k}$, QoS for type k, and its activity factor a_k. Illustration of this scenario differs from Figure 2 only in that $R<>const$.

2.2 DTX

The DTX mechanism turns the transmission off after a hangover period if the queue in the buffer is empty. Hangover is equal to the cell inter-arrival time (IAT) of the slowest connection (in case of the Code per User scenario), or just the cell inter-arrival time of the corresponding connection (in Code per VCC scenarios). Signal transmission time is also extended by code re-synchronisation. The effect of acquisition and hangover on busy and idle periods is illustrated on Figure 3.

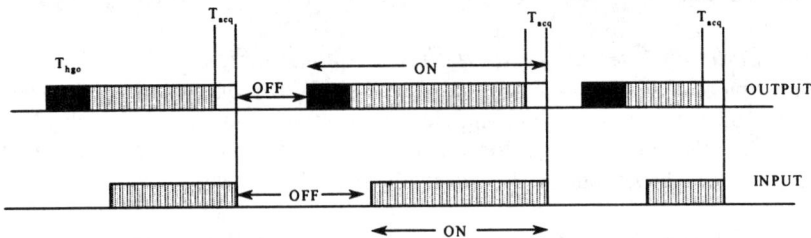

Figure 3 Busy and idle periods of an input stream and a multiplexer output.

3.   ANALYSIS

The analysis is performed on the cell level in discrete time. Sources are modelled as ON-OFF streams with deterministic interarrival times and geometrically distributed state sojourn times. The multiplexer is modelled as an infinite queue with fixed service time that determines the slot rate at its output and geometrically distributed state sojourn times. Each connection type has different probabilities of ON and OFF periods, with different cell rates in ON states. Variables used in derivations of link efficiency (subscript k used in Scenarios 2 and 3 analysis) are:

K	number of traffic types (for numerical analysis $K=3$)
λ_k	mean arrival rate of the kth connection
SR, SR_k	slot rate expressed in [cells/s]
PCR, PCR_k	peak cell rate of a source during busy (ON) period
T_B, T_P	mean duration of a busy period of a multiplexer in Scenario 1
$T_{hgo}, T_{hgo.k}$	mean hangover period
T_{acq}	mean acquisition period
$P_{ON.k}, P_{OFF.k}$	probability that the connection k is ON/OFF,
$T_{ON.k}, T_{OFF.k}$	mean ON and OFF periods of the kth connection,
a_k	activity factor of connection (traffic) type k
$N, N_k, N_k(i)$	number of connections (of traffic type k in Scenario 2; connections of traffic type k in ith channel in Scenario 3)
$C_{max}, C_{max.k}$	maximum theoretical number of signals in a DS-CDMA channel
m_k	number of channels for carrying traffic type k
M	total number of channels
W, W_k	total bandwidth, bandwidth of channel carrying type k

3.1 Code per User scenario

We are interested in efficiency of one user (link) and some equivalent 'aggregate' efficiency of the whole channel. Since in this scenario all users and channels are identical, efficiency of the channel will be the same for all users, and equal to the efficiency of the system. Efficiency with and without DTX (continuous transmission) is:

$$Eff_{DTX} = \frac{(T_B + T_I) \cdot \sum_{k=1}^{K} \lambda_k}{SR \cdot (T_B + T_{hgo} + T_{acq})}. \qquad \dots\dots\dots(2)$$

$$Eff_{CONT} = \frac{\sum_{k=1}^{K} \lambda_k}{SR} \equiv \rho. \qquad (3)$$

It is assumed that all users have the same mean busy and idle periods at the output of the multiplexer, although in reality it may be possible that the connections of a user will not all be active at all times.

3.1.1 Available capacity and CAC
Available unused capacity can be expressed as:

$$\frac{\Delta C}{C_{max}} = 1 - \frac{U(\rho)}{C_{max}} \cdot \frac{T_B + T_{acq} + T_{hgo}}{T_B + T_I}. \tag{4}$$

$U(\rho)$ is a maximum number of users required to achieve the total load on the satellite of ρ. Depending on activity factors a_k of traffic types k, available capacity can be fully utilised by statistical multiplexing of extra users in the code domain. Assuming the number of users (producing load ρ) of type k is $U_k(\rho)$, then the number of extra users of traffic type j that can be accepted above the theoretical value C_{max}, $\Delta U_{j,max}(\rho)$ is:

$$\Delta U_{j,max} = \frac{C_{max} - \sum_k a_k \cdot U_k(\rho)}{a_j}. \tag{5}$$

Expression (5) treats a general case. In Code per User scenario, some "equivalent" activity has to be found before the capacity increase can be assessed. To do this, it is necessary to determine the number of users $U(\rho)$ that produces the load ρ, which is easily done in queueing theory. Expression (5) forms the basis for a connection admission control (CAC) in a satellite IBC system. The controller can assess the available capacity using the known activity factors, which can be measured on-line or provided as traffic descriptors. When a new connection requests access, the CAC algorithm accepts it if $\Delta U_{j,max}(\rho) > 1$.

3.2 Code per VCC, R uniform across the system

In this scenario the server mean idle and busy periods differ from source ON and OFF periods by the hangover and mean code acquisition time. The cells are delayed by the mean acquisition time, which is the same across the system. Efficiency with and without DTX are given by Equations 6 and 7:

$$Eff_{DTX} = \left[1 + \frac{\sum_{k=1}^{K} N_k \left(T_{acq} + T_{hgo,k} \right) \cdot SR \frac{\Delta t}{T_{ON,k} + T_{OFF,k}} + \sum_{k=1}^{K} N_k \left(SR - PCR_k \right) p_{ON,k} \cdot \Delta t}{\sum_{k=1}^{K} N_k \lambda_k \Delta t} \right]^{-1}$$

$$= \left\{ 1 + \frac{\sum_{k=1}^{K} N_k \cdot p_{ON,k} \cdot \left[SR \cdot \left(1 + \frac{T_{acq} + T_{hgo,k}}{T_{ON,k}} \right) - PCR_k \right]}{\sum_{k=1}^{K} N_k \lambda_k} \right\}^{-1}. \tag{6}$$

$$Eff_{CONT} = \frac{\sum\limits_{k=1}^{K} N_k \cdot \lambda_k}{C_{\max} \cdot SR} . \tag{7}$$

3.3 Code per VCC, R varies across the system

For each connection type k of total K types, there are m_k channels of corresponding bandwidth W_k. Channel bandwidths are determined by the QoS requirements of different connection types and their transmission rates. Each channel can have a theoretical maximum of $C_{max,k}$ connections, and a number of active users in the channel is $N_k(i)$, $i=1,...,m_k$. The acquisition time of a code varies according to the QoS requirements and transmission rate (since it depends on the probability of false alarm and probability of detection (Holmes, 1982)). Therefore, each connection type k has a different code mean acquisition time $T_{acq,k}$. We have:

$$Eff_{DTX} = \left[1 + \frac{\sum\limits_{k=1}^{K}\sum\limits_{i=1}^{m_k} N_k(i)\left(T_{acq}+T_{hgqk}\right) \cdot SR_k \cdot \dfrac{\Delta t}{T_{ON,k}+T_{OFF,k}} + \sum\limits_{k=1}^{K}\sum\limits_{i=1}^{m_k} N_k(i)\left(SR_k - PCR_k\right) p_{ON,k} \cdot \Delta t}{\sum\limits_{k=1}^{K}\sum\limits_{i=1}^{m_k} N_k(i)\lambda_k(i)} \right]^{-1}$$

$$= \left[1 + \frac{\sum\limits_{k=1}^{K}\sum\limits_{i=1}^{m_k} N_k(i) p_{ON,k}\left[SR_k \cdot \left(1 + \dfrac{T_{acq,k}+T_{hgo,k}}{T_{ON,k}}\right) - PCR_k \right]}{\sum\limits_{k=1}^{K}\sum\limits_{i=1}^{m_k} N_k(i)\lambda_k} \right]^{-1} . \tag{8}$$

$$Eff_{CONT} = \frac{\sum\limits_{k=1}^{K}\sum\limits_{i=1}^{m_k} N_k(i) \cdot \lambda_k}{\sum\limits_{k=1}^{K} m_k \cdot C_{\max,k} \cdot SR_k} \tag{9}$$

4.   RESULTS

To obtain numerical results we used approximations of $K=3$ types of traffic: voice, video and data, with the parameters:

- voice: 64 kb/s source, with T(on) = 1.54 s and T(off) = 2.75 s, a_1=0.36;
- video: 384 kb/s source with T(on) = 20 s and T(off) = 6.5 s, a_2=0.75;
- data: 4 Mb/s source with T(on) = 1.79 ms and T(off) = 26.7 s, a_3=0.09.

For any traffic source type, when in the ON state, cells are generated at a constant rate; no cells are generated in the OFF state. A number of different transmission rates are being considered: 64 kb/s, 384 kb/s, 512 Kb/s, 1 Mb/s, 1,5 Mb/s, 1,8Mb/s, 2Mb/s and 4 Mb/s. We assumed $SNR = 4.5$ dB and $Eb/No=12$ dB for all types. At the moment only results for the first two scenarios are available. For them, 64 kb/s rate is below the mean rates of video and data traffic, causing buffer overflows. Since the loads on different connections depend on the mean arrival and slot rates, and the mean arrival rates in our model are fixed, then for higher transmission (slot) rates only low loads can be achieved. This is because the maximum number of users or connections C_{max} is fixed.

4.1   Efficiency

Numerical results obtained by analysis are plotted on Figures 4-7. In all graphs, efficiency without DTX is essentially the load of the system (ratio of cell arrival and cell service rates).

4.1.1   Code per User scenario
In the Code per User scenario efficiency of the system with DTX decreases with the increase of the transmission rate (from 0.82 at 512 Kb/s and 70% load to 0.10 at 4Mb/s and 10% load, see Figure 4). The efficiency improvement is constant (17%) for all transmission rates greater than 384 Kb/s. Hence for a more efficient system it is necessary that the transmission rate is as low as possible (so that fewer empty cells are transmitted). However, for low transmission rates (64 Kb/s and 384 Kb/s on Figure 4) the system with DTX is overloaded (efficiency with DTX becomes greater than 1, which is impossible). This is caused by the generation of additional cell rate due to the overheads, and results in the 'aggregate' load being greater than 1. This means that a traffic mix is important in determining the transmission rate for this scenario, and it has to be low enough to keep the system efficient, while maintaining the aggregate load below 1.

4.1.2   Code per VCC, R=const.
Efficiency of this scenario has much higher values than in Code per User scenario (up to 0.99 for 70% loads and 384 Kb/s transmission rate, Figure 5), for all loads and transmission rates. Moreover, the efficiency improvement is much higher than that in Code per User: values of 40% are obtained for 70% loads. The effect of connection activity factor is predictable: for a constant number of connections of a given traffic type, the efficiency improvement will increase with the increase of the number of connections with lower activity factor. However, if the activity factor has very low values, the number of overhead cells transmitted in comparison to the useful cells starts to increase, and the efficiency decreases again. For example, efficiency decreases more quickly if video connections are kept constant and the number of voice connections is increased, than when the voice connections are kept constant and video are increased. This is due to the proportion of overhead present in the connections with lower activity factors: in less active sources, the proportion of DTX overhead is higher (the duration of active states is closer to the

duration of overhead periods). Hence, the capacity gained through having connections of lower activity factors has its cost in the comparatively lower efficiency (with DTX). Overall, Code per VCC scenario outperforms Code per User scenario with respect to efficiency.

4.2 Capacity

The relative capacity increase in Code per VCC scenarios is measured as number of connections of one particular type. We compare the values of relative capacity increase for different connection types. For the same loads, transmission rates and channel bandwidths, Code per VCC outperforms the Code per User scenario in terms of voice and data connections (Figures 8-11). However, in terms of video users, the relative capacity increase in Code per VCC scenario is comparable to or lower than the capacity increase in Code per User scenario. This is due to the fact that the activity factor of video sources (0.75) is close to the activity factor of multiplexed connections in Code per User scenario (0.86), while the available capacity for a particular traffic load is higher in the Code per User than in Code per VCC scenario, because fewer users are necessary in Code per User scenario to achieve the same loads.

4.2.1   Code per User
The capacity improvement decreases with the load and transmission rate. Total system capacity is highest for the 384 Kb/s transmission rate and the 25 MHz channel bandwidth.

4.2.2   Code per VCC
Capacity increase is lower at higher transmission rates for the same loads and channel bandwidths (Figures 9-10). As expected, for a given channel bandwidth and transmission rate, the highest capacity increase results from the connections of lowest value of activity factor (data traffic), and all curves drop with the traffic load. This is because higher loads have less unused capacity available for further statistical multiplexing of the connections.

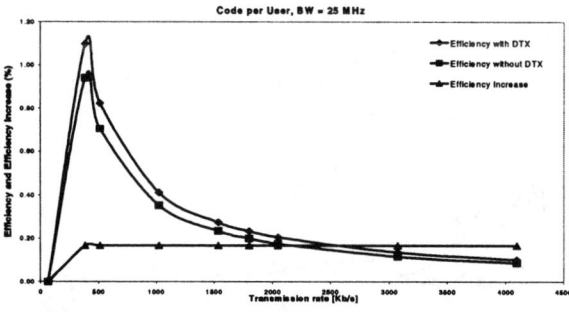

Figure 4.

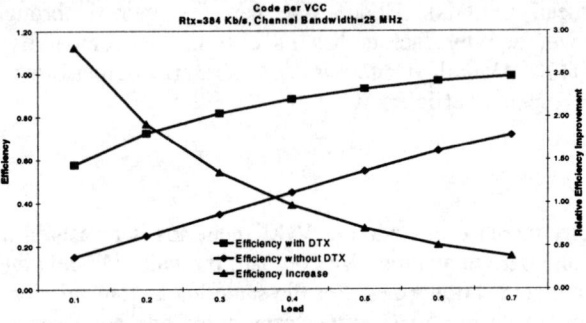

Figure 5

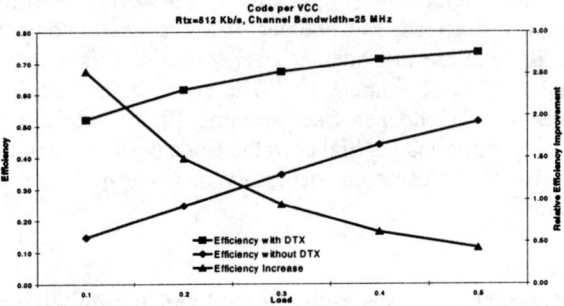

Figure 6

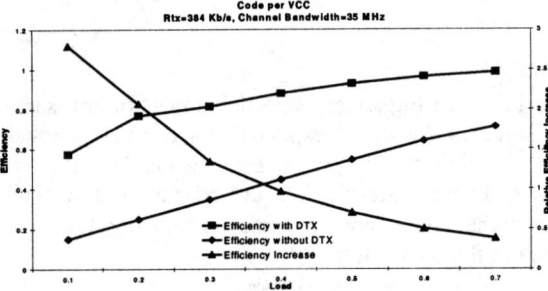

Figure 7.

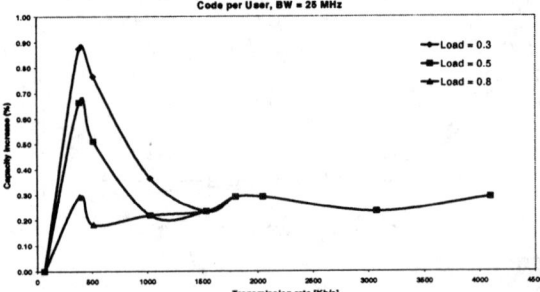

Figure 8.

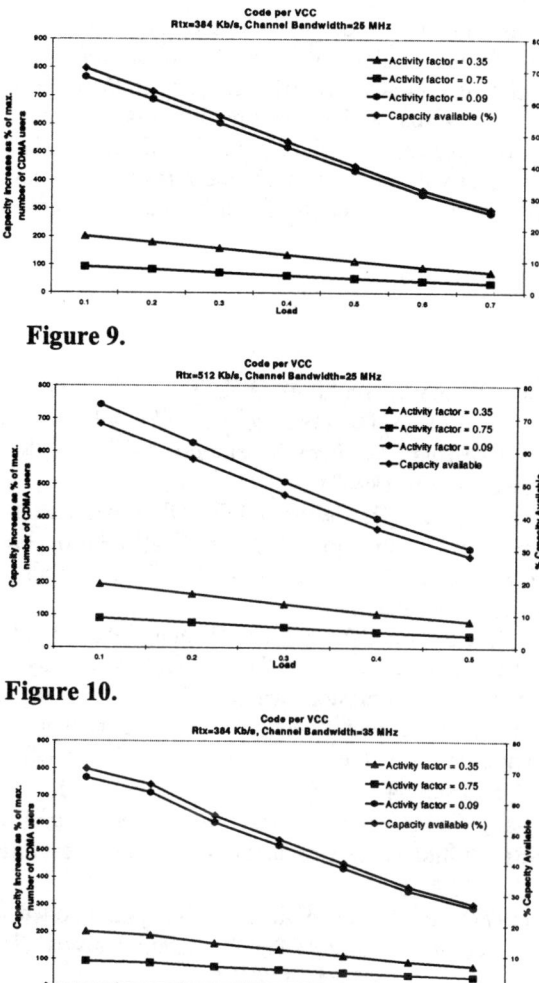

Figure 9.

Figure 10.

Figure 11.

5. SUMMARY AND CONCLUSIONS

Three configuration scenarios are described that use DTX and exploit bursty nature of ATM broadband traffic to increase the CDMA system capacity. We presented numerical results that establish performance gains for two of the three described scenarios. The results show how the system relative capacity increase changes with traffic load, transmission rates and the traffic activity factor, and how efficiency changes with load and transmission rate. The idea for a simple CAC algorithm in a CDMA broadband system that exploits sources' activity factor has been presented.

Preliminary conclusion is that Code per VCC scenario outperforms Code per User scenario. Choice of transmission rate directly affects the system capacity and efficiency, but it also determines the buffering delay in the multiplexer, crucial for the delay-sensitive traffic particularly in satellite systems. The future work will investigate imposed delays and the third scenario - Code per VCC when R varies across the system - and will aim to identify the rules for the choice of the system parameters based on the expected traffic mix, traffic types, and the optimal scenario.

REFERENCES

Andermo, P.G. and Brismark, G. (1994). CODIT, a testbed project evaluating DS-CDMA for UMTS/FPLMTS. *Proc. of the IEEE 44th Vehicular Technology Conference "Creating tomorrow's mobile systems"*, 8-10 June 1994, Stockholm, Sweden, Ch. 389, 21-5

Baier, A. and Panzer, H. (1993). Multi-rate DS-CDMA radio interference for third-generation cellular systems. *Proc. of the 7th IEE European Conference on Mobile and Personal Communications*, 13-15 December 1993, Brighton, England, Ch. 46, Conf. Pub. Np. 387, 255-9

Elhakeem, A.K., Di Girolamo, R., Bdira, I.B. and Talla, M. (1994). Delay and throughput characteristics of TH, CDMA, TDMA and hybrid networks for multipath faded data transmission channels. *IEEE JSAC* **Vol. 12/4**, 622-37

Fong, M.H., Bhargava, V.K. and Wang, Q. (1996). Concatenated orthogonal/PN spreading sequences and their application to cellular DS-CDMA systems. *IEEE JSAC,* **Vol. 14/3**, 547-58

Holmes, J. K. (1982). *Coherent Spread Spectrum Systems.* John Wiley & Sons Inc.

Monsen, P. (1995). Multiple-access capacity in mobile user satellite systems. *IEEE JSAC*, **Vol. 13/2,** 222-31

Pursley, M.B. (1977). Performance analysis for phase-coded spread-spectrum multiple-access communications - Part I: System analysis. *IEEE Transactions on Communications*, **Vol. COM-25**, 795-9

Van Nee, R., Van Wolfwinkel, R.N., and Prasad, R. (1995). Slotted ALOHA and code division multiple access techniques for land-mobile satellite personal communications. *IEEE JSAC*, **Vol.13/2**, 382-8

Yang, W.B. and Geraniotis, E. (1994). Admission policies for integrated voice and data traffic in CDMA packet radio networks. *IEEE JSAC,* **Vol. 12/4**, 654-64

9

An adaptive random-reservation MAC protocol to guarantee QoS for ATM over satellite

T. Örs, Z. Sun and B.G. Evans
Centre for Communication Systems Research,
University of Surrey, Guildford, Surrey GU2 5XH, UK
Phone: +44 1483 259844, Fax: +44 1483 259504
e-mail: {T.Ors,Z.Sun,B.Evans}@ee.surrey.ac.uk

Abstract

In this paper we analyse the performance of an Adaptive Random-Reservation Medium Access Control (MAC) protocol which can support all ATM service classes while providing the required Quality of Service (QoS). Our study focuses on parameter optimisation of the multiple access schemes for ATM over a GEO satellite with on-board processing capabilities, considering various traffic mixes of Constant Bit Rate (CBR), real-time Variable Bit Rate (rt VBR), non-real-time VBR (nrt VBR) and Unspecified Bit Rate (UBR). It is shown that maximum throughput can be achieved by using this access scheme. A TDMA access protocol combining both Random Access and Demand Assignment Multiple Access (DAMA) is particularly suited for a scenario with a high number of terminals and very bursty UBR traffic (e.g. web browsing). UBR sources with short burst length access the slots remaining after the reservation procedure by random access which drastically reduces the slot access delay, at the expense of lower utilisation.

It is shown that the potential user population which can be served is considerably increased by statistically multiplexing bursty traffic over the air interface.

Keywords
ATM, DAMA, Random Access, TDMA, Quality of Service, Satellite, Performance

Broadband Communications P. Kühn & R. Ulrich (Eds.)
© 1998 IFIP. Published by Chapman & Hall

1   INTRODUCTION

In the recent years, significant progress has been made in the research and standardisation of ATM over terrestrial networks. Whilst optical fibre is the preferred carrier for high-bandwidth terrestrial communication services, satellite systems can play an important role in the B-ISDN. The main strengths of satellites are their fast deployment, global coverage and flexible bandwidth-on-demand capabilities.

After the terrestrial broadband infrastructure will have reached some degree of maturity, satellites are expected to provide broadcast service and also cost-effective links to rural areas complementing the terrestrial network. In this phase satellite networks will provide broadband links to a large number of end users through a User Network Interface (UNI) for accessing the ATM B-ISDN. Portable user terminals are expected to have relatively low average and peak bit rates (up to 2Mbit/s) and the traffic is expected to show large fluctuations. Therefore the access scheme will considerably effect the performance of the system. Furthermore the cost and size of the terminal will have a large impact on the suitability of the satellite solution.

Recent proposals for broadband multimedia satellite systems (Fitzpatrick, 1996), (Fernandez, 1997) are examples for this scenario and increased the attention paid to ATM access over satellite links. The methodolgy and standards developed for wired ATM networks are ported to the wireless environment, where possible. This will allow seamless integration of terrestrial and satellite networks.

The use of standard ATM protocols to support seamless wired and wireless networking is possible by incorporating a new radio specific protocol sublayer into the ATM protocol model (ATM-Forum, 1996) as shown in Figure 1.

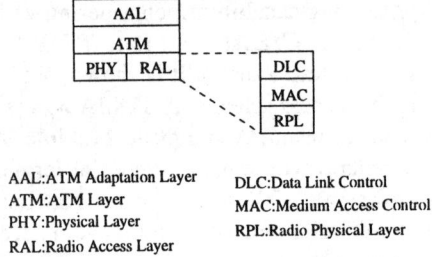

AAL:ATM Adaptation Layer   DLC:Data Link Control
ATM:ATM Layer   MAC:Medium Access Control
PHY:Physical Layer   RPL:Radio Physical Layer
RAL:Radio Access Layer

Figure 1   ATM Protocol Model including Radio Access Layer

Considering that satellite communications uses multiple access on a shared medium, a MAC layer, which is not present in traditional ATM networks is needed. The MAC protocol plays a central role as means of accessing the RPL from the ATM layer. The access scheme refers to the physical layer multiplexing technique to a share a common channel among multiple users of possibly multi-services. The problem of statistical multiplexing at the air interface is slightly different to that in the fixed network. In the fixed network the problem is associated with control of bandwidth on an outgoing link from some multiplexing point after buffering has occurred. It is implicitly assumed that the access links from the source are dimensioned in such a way that they do not impose any constraints on the traffic (e.g. sources can transmit at their peak bit rate). In the air interface the constraint is on the bandwidth available in <u>total</u> to all sources before the buffering/multiplexing point.

2   SATELLITE NETWORK ARCHITECTURE

The satellite network architecture for the provision of ATM services to portable terminals employs a satellite with cell switching capabilities (Ors, 1997). Network requirements are for a full meshed point-to-point and point-to-multipoint system. Suitable satellite architectures for meshed terminal networks are expected to employ a spot beam coverage pattern to achieve the high uplink and downlink gain required for mesh connectivity between portable terminals. On-Board Processing (OBP) functions such as switching, channel set-up and multiplexing result in increased complexity on-board the satellite but offer added flexibility and improvement in link performance. The reasons for using OBP functions in this scenario are:

- To maximise the bandwidth utilisation of the satellite.
- To reduce the reservation delay.
- To improve interconnectivity.
- To reduce the ground terminal RF cost.

The 155 and 622 Mbit/s transmission rates conventionally associated with ATM are well above the maximum rates possible with today's portable terminal technology. However, in practice, most individual users will usually require significantly lower traffic rates, especially if there are only a few data or voice terminals located at a remote location. This large number of users with bursty traffic will need a cost-efficient way to communicate between each other and access the ATM/B-ISDN network.

3   FRAMEWORK FOR PROPOSED MAC PROTOCOL

3.1 Design Objectives

The MAC protocol has to be designed to allow statistical multiplexing of ATM traffic over the air interface, especially in the uplink for the independent and spatially distributed terminals. The following design objectives are taken into consideration:

- Maximise the slot utilisation, especially for bursty traffic.
- Guarantee the QoS requirements for all service classes.
- Maximise frame efficiency by minimising overheads.

The minimisation of overheads is not an easy task, especially for ATM which was designed for channels with very good error characteristics (Bit Error Rates around 10^{-10}). To minimise cell loss over the satellite link, channel coding has to be used to make the transmission more robust. A Logical Link Control (LLC) header to facilitate error recovery mechanisms is optional and not in scope of this study. Finally a satellite specific header with satellite routing and wireless resource management fields is added to form a MAC packet.

3.2   Access Schemes

MAC layer access schemes can be typically categorised into four classes: *Fixed Access, Random Access, Demand Assignment Multiple Access (DAMA)* and *Adaptive Access*. The first three techniques have evolved to meet the needs of constant high traffic with long duration's, sporadic traffic with short to medium duration's, and sporadic traffic with long duration's, respectively (Bohm, 1991). Finally *adaptive access* is used to meet the needs of multiple media which consists of traffic which has all possible traffic patterns mentioned above. Thus to meet the design objectives an Adaptive Access mechanism seems to be the best choice.

3.3   Mapping of ATM Service Classes onto MAC Service Classes

To simplify the conceptual design of the MAC protocol, ATM service classes (ATM-Forum, 1996) çan be mapped onto MAC service classes.

Fixed-Rate DAMA is ideal for connections with a constant bit rate such as the CBR service class in ATM networks. Before a connection is set-up, the terminal and satellite negotiate the Quality of Service (QoS) parameters. These QoS parameters determine the characteristics of the connection. Since the parameters will not be modified during the connection, the amount of bandwidth allocated for that connection will not be changed until the connection is terminated. For ATM CBR connections the Peak Cell Rate (PCR) is allocated to the terminal.

Real-time Variable Bit Rate (rt-VBR) services can also be supported with fixed-rate DAMA. For real-time services, the amount of bandwidth assigned to the connection should be close or equal to the PCR to avoid cell delay. The major drawback of this scheme is that a major portion of the bandwidth is wasted when the cell transfer rate is lower than the assigned bandwidth. The major difficulty to employ variable-rate DAMA in ATM over GEO satellite systems is the effect of the large propagation delay (135ms). The computing and negotiation process between the satellite and the terminal may be too long for real-time VBR services and result in unacceptable QoS. The use of variable-rate DAMA for rt-VBR is only possible if the arriving traffic can be predicted one hop delay in advance. Since this is not possible except in some special cases fixed-rate DAMA will be used for rt-VBR.

A scenario where fixed-rate DAMA is efficient for rt-VBR services is when the terminal can multiplex traffic from multiple services. In this case the aggregate traffic can be approximated as a constant cell flow by using a small amount of shaping.

VBR services which are not time sensitive can use burst reservation. In this case cells can be buffered in the terminal till the required bandwidth is reserved and in case the queue exceeds a certain threshold more bandwidth can be requested. Thus using variable-rate DAMA the bandwidth of a connection can be adjusted according to the change of the data transfer rate.

No numerical commitments are made for the UBR service class and this service category is intended for non-real time applications. UBR services could be supported by variable-rate DAMA. However the fact that this service class has the lowest priority (because no commitments to cell loss or cell delay are made) has to be considered. We propose that UBR could transmit data directly to the unoccupied MAC slots without reservation. The unreserved slots are broadcasted on the downlink to be accessed by random access. This is particularly appealing for bursty interactive services with short duration, for which the long slot reservation delay is unacceptable.

4   THE RANDOM-RESERVATION ADAPTIVE ASSIGNMENT PROTOCOL

The TDMA frame of the adaptive assignment protocol is divided into Request slots, Control slots, Reservation slots and Random Access slots, as shown in Figure 2. The protocol is based on the proposals by (Bohm, 1993), (Celandroni, 1991), (Zein, 1991) with modifications to achieve the design objectives for multi-service networks. Prioritised queuing of requests on-board the satellite is the most important difference from previous proposals.

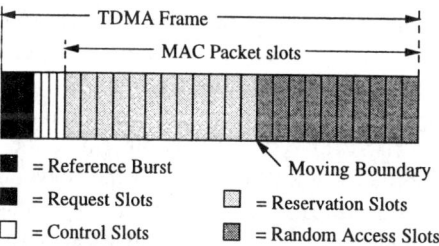

Figure 2 Uplink Frame Structure

A Request slot is that period of time in which terminals report their requests to the on-board Radio Resource Management (RRM). There are only a few Request slots available and a terminal selects one at random without knowing whether another station is using the same slot. If more than one terminal selects the same request slot, a collision occurs and terminals have to retransmit after waiting for a mean retransmit waiting time determined by the collision resolution algorithm. The MAC protocol ensures that the collision probability stays low. The reason for using request slots is because ATM networks support different services which have different loss and delay requirements.

If a single request was received for a request slot (successful request), the on-board RRM module tries to allocate the necessary Reservation slots. If no Reservation slots are available, the request can either be blocked (called blocking probability) or queued. We propose to queue successful requests in a prioritised queue so that the terminal does not need to compete with other terminals for a request slot again. By queuing successful requests, Reservation slots can be allocated according to their priorities.

Once the Reservation slots are reserved (successful reservation), an acknowledgement is transmitted to the terminal in Time Division Multiplex (TDM) mode on the downlink frame. Reservation slots represent the part of the frame in which a terminal can transmit its message after a successful reservation. In every frame there are many reservation slots and the on-board RRM module will assign reservation slots to a particular successful request according to its priority. A reservation slot is assigned to only one terminal and therefore there is no possibility of collision.

On the other hand Random Access (RA) slots represent the part of the frame in which terminals can transmit without the need of making a reservation. The slots available for RA are broadcasted in the downlink frame. This part is for services which don't want to wait for the lengthy reservation procedure. In RA mode it is not possible to guarantee a certain QoS to users although the protocol will try to minimise the number of collisions to maximise throughput by using an adaptive collision resolution algorithm. RA should only be used by UBR sources with relatively small burst length since RA traffic is allowed to reserve slots and have to content for each MAC packet. An improvement in performance for RA is possible by using the UNR+ service class, where a minimum bandwidth is guaranteed for UBR services (Goya, 1996)

Unless the number of request slots per frame is carefully adjusted the result would be either low capacity utilisation and long delays (too many request slots, less capacity available for information transmission) or network backlog (too few request slots resulting in successive collisions and high delay). The number of request slots should be fixed for system behaviour where the number of collisions can be controlled by broadcasting a message in the downlink that services with lower priority should not send/resend requests till the collisions have been resolved. Our analysis has shown that two request slots provide adequate performance for an average call holding time of longer than six seconds. However when the number of collisions can't be controlled (very short average call holding time) new request slots can be added by reducing the number of control slots.

The requests for dynamic slot allocation are done using the Control Slots which are assigned, on a round-robin basis to all terminals which request the variable-rate DAMA MAC class. The number of control slots is set to eight to minimise the frame overhead.

The satellite frame introduces a constant delay equal to the frame length, on the cells of a stream connection. Therefore the selection of the frame size should be small enough to satisfy the delay limit of real-time services (400ms) (ITU-T,1996) taking into account the satellite propagation and processing delays and the delay introduced by the terrestrial B-ISDN.

The MAC packet slot period has been chosen to support a 32 kbit/s CBR stream and corresponds to one frame unit of 384 un-coded information bits every uplink frame. This results in a frame period of 11.9 ms to transmit 84 ATM cells per second using AAL5. There are 64 MAC packet slots to support 2.048 Mbit/s of traffic per spot-beam on the uplink. The actual uplink transmission rate is higher due to ATM and MAC layer overheads.

5 ANALYSIS OF ADAPTIVE RANDOM-RESERVATION MAC PROTOCOLS

5.1 CBR and nrt-VBR Analysis

By queuing requests when no reservation slots are available (buffer size=100) and by using an adaptive collision resolution algorithm the blocking and collision probabilities can be minimised for a traffic mix of CBR and nrt-VBR. Thus this section will mainly focus on the performance of the system in terms of throughput and reservation delay. One of the main evaluating factor of various access protocols is the normalised information throughput defined as:

$$\frac{throughput}{capacity} = \frac{the\ number\ of\ occupied\ MAC\ slots\ in\ a\ frame}{the\ number\ of\ MAC\ slots\ in\ a\ frame}$$

Nrt-VBR connections first have to reserve the initial capacity by using a Request slot and can then use a Control Slot assigned to them in round-robin fashion for other requests. Thus there are two different reservation delay values. The delay for the initial capacity request using the request slot is called 'Call Reservation Delay' (or only Reservation Delay) and the delay for the dynamic bandwidth reservations using the control slot is called 'Burst Reservation Delay'. The slot reservation is relinquished during the silence periods of the source. In this way the air interface bandwidth is shared between multiple sources achieving statistical multiplexing on the air interface.

The offered traffic load of the system can be calculated by multiplying the average number of calls originating per unit time (λ) with the mean call holding time (h). Furthermore a source may occupy more than one slot according to it's PCR which has to be taken into account:

$$Offered\ traffic\ load\ (\text{Erlang}) = (\lambda \cdot h \cdot PCR) / (32 \cdot 10^3) \qquad (1)$$

For VBR source the call holding time has to be divided by the burstiness (β) to calculate the offered load. The average call holding time for all services is assumed to be one minute.

The normalised load is defined as:

Normalised Load = Offered Traffic Load / Number of MAC slots   (2)

Each nrt-VBR source represented by an on-off source model shown in Figure 3 where a^{-1} is the mean burst period and b^{-1} is the mean silence period which are both exponentially distributed. The burstiness is defined as:

$$\beta = \frac{Peak\ Rate}{Mean\ Rate} = \frac{a^{-1} + b^{-1}}{a^{-1}}$$ (3)

The VBR mean burst period is 100ms and the PCR during the burst is 64 kbit/s.

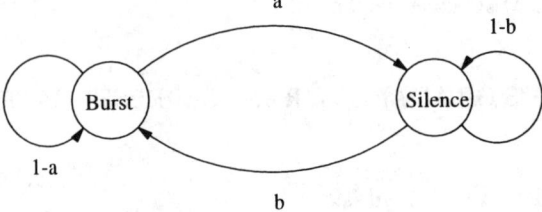

Figure 3 The on-off source model

First the results for only nrt-VBR are presented to show the advantage of multiplexing bursty traffic over the air interface. Simulations are carried out for traffic with burstiness of 5,10 and 20. Note that for all the simulations the number of active terminals is equal to the call arrival rate. The confidence interval for all simulations is 95% and not shown on the graphs for the neatness of the results.

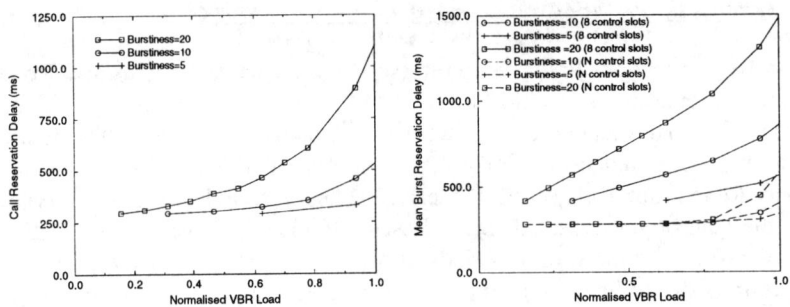

Figure 4 (a) Call Reservation (b) Burst Reservation Delay for nrt-VBR services with various burstiness values.

The simulation results for this scenario show that a throughput of up to 0.95 can be achieved. The reservation delay remains reasonably low, for low burstiness values. As the burstiness is increased so does the mean reservation delay (Figure 4) as more sources need to be multiplexed on the air interface to achieve high utilisation.

Choosing a low number of control slots to minimise frame overheads results in increased burst reservation delay as can be seen in Figure 4 (b). The increase in burst reservation delay become more visible as the number of active terminals increases. If each source is assigned an individual control slot then there will be large frame overhead for N sources while achieving very low burst reservation delays. Note that N (number of terminals) is equal to the call arrival rate which can be calculated from equations (1)-(3). Changing the burstiness for a fixed load changes the mean rate and hence the number of terminals. Since the terminal is expected to buffer bursts at the PCR for the burst reservation delay period, a reduced burst reservation delay results in smaller buffer requirement for the terminal, to avoid cell loss due to buffer overflow.

Next the effects of CBR traffic on nrt-VBR traffic is investigated by fixing the nrt-VBR load at 0.6 and varying the CBR load. The required CBR PCR is assumed 32kbit/s. The increase in the call and burst reservation delay as a function of CBR load is shown in Figure 5. The increased delay is due to the fact that as the total offered load is increased, it exceeds the system capacity and hence request have to be buffered which results in longer delays till bandwidth allocation can be made. Again the burst reservation delay is minimised by allocating a control slot for each VBR source as shown in Figure 5 (b). For eight control slots the burst reservation delay increases since control slots are assigned in round-robin fashion to terminals which can only request bandwidth when the control slot is assigned to them.

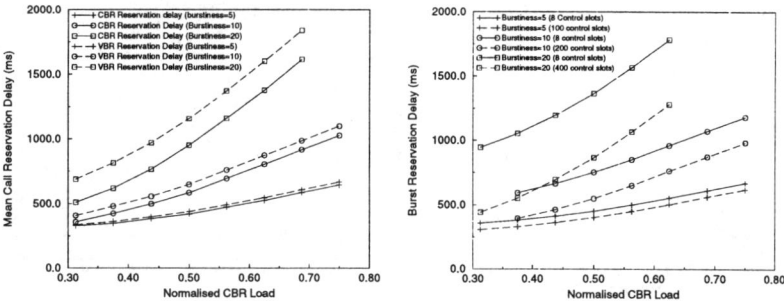

Figure 5 (a) Mean Call Reservation Delay (b) Mean Burst Reservation Delay

The achieved throughput is similar for various burstiness values as shown in Figure 6, due to the low number of collisions and blockings.

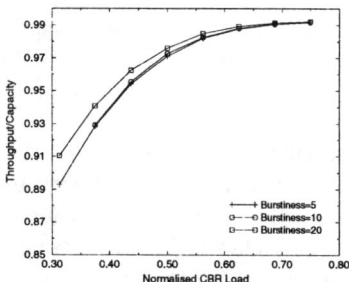

Figure 6 Total System Throughput vs Normalised CBR Load

5.2 CBR, VBR and UBR Analysis

The novelty of the adaptive MAC protocol is it's ability to support lower priority UBR traffic in both reservation and random access mode according to the terminals burst duration.

The UBR load is calculated in the same way as for VBR traffic and the throughput of UBR for a pure-reservation system, (simulation parameters shown in Table-1) in a scenario with a high number of UBR sources with very bursty traffic (β=200-5000) is shown in Figure 7(a). The required PCR for UBR sources is assumed 32 kbit/s and the UBR burst period is varied from 13ms (1 cell) to 416 ms. As it can be seen the amount of carried traffic remains unacceptably low, even for low to medium load due to the high number of collisions, in particular for high burstiness values where the burst period is short and a high number of terminals are admitted to achieve high utilisation. The pure-reservation MAC throughput increases for traffic with longer burst duration (or lower burstiness). Only for burstiness values lower than 200 can a throughput higher than 25% of the frame capacity be achieved.

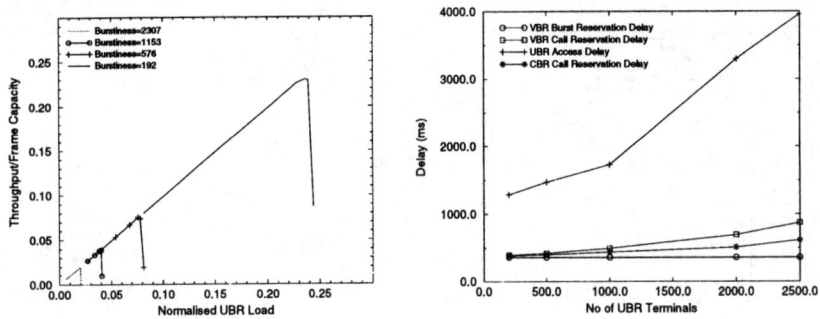

Figure 7 (a) Throughput vs normalised UBR load (b) Reservation Delay vs Arrival Rate (per minute)

Since the offered UBR traffic load is dependent on the burstiness of the traffic Figure 7 (b) shows the delay as a function of the UBR terminal numbers (which is also the number of arrivals per minute). As it can be seen, UBR traffic is increasing the call reservation delay of higher priority CBR and VBR. Since all the terminals content for the same reservation slots, UBR traffic increases the collision probability and the access delay.

To improve the UBR throughput and the access delay we propose that UBR sources access the MAC slots remaining after the reservation procedure by Random Access (RA). This way the lengthy reservation procedure is avoided an the number of collisions reduced. The reservation delay of terminals using reservation (CBR and VBR terminals) is unaffected by RA terminals.

The UBR throughput using RA is limited by 36.8 % of the available RA capacity (theoretical Slotted Aloha limit). This throughput is higher than the pure-reservation throughput for very bursty traffic (β=500-5000). Our simulations showed that in order to minimise RA collisions and access delays, the UBR load has to be kept around 25% of the available RA capacity. The throughput of UBR vs access delay is shown in Figure 8(a). Note that the normalised UBR load is found by dividing the offered load by the number of available RA slots. Since ATM is connection-oriented, the system can refuse new UBR connections to keep the UBR load at acceptable levels. However as the number of available RA slots changes every frame there may be instants of decreased throughput till the collision resolution algorithm optimises system performance by reducing the RA collisions. Figure 8(b) shows the increased collision probability vs UBR load when the number of available RA slots is reduced by 1/3 on the next frame.

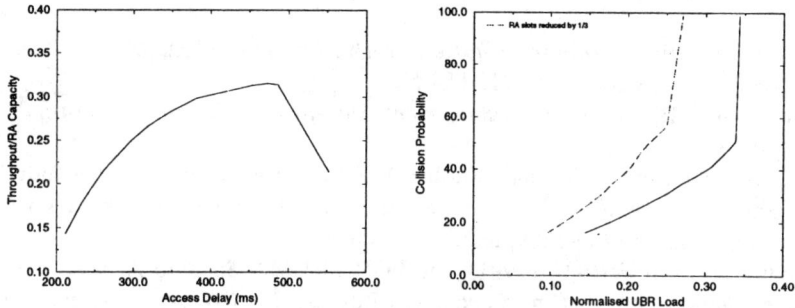

Figure 8 (a) Throughput vs Access Delay (b) Collision Probability vs Normalised UBR Load

The maximum number of UBR terminals which can be supported can be found as:

$$Maximum\ number\ of\ UBR\ sources = Normalised\ Load \cdot \beta \cdot number\ of\ RA\ slots$$

For a normalised UBR load of 0.25 a burstiness of 4500 and 10 RA slots, 11250 UBR terminals can be supported. For pure-reservation this number was around 3000 because of the increase in collisions.

6   CONCLUSIONS

As user demands become more complex, satellite networks are expected to support a much wider range of services. As satellites will play an important role in the deployment of ATM networks, we addressed the optimisation of the capacity allocation scheme, using performance results for an adaptive MAC scheme.

It has been shown that considerable improvements in delay performance and satellite bandwidth utilisation are possible if next-generation satellite technology (OBP) and an adaptive MAC is used. The traditional demand-assignment scheme using a ground terminal as control station has two major drawbacks: long set-up and reservation time and limited channel utility. Both are due to the long propagation delay of the satellite link. Both disadvantages can be improved by processing channel requests in the satellite to allocate frame slots.

The mapping of ATM service classes to MAC classes and the use of a prioritised request queue provides the QoS differentiation required by ATM networks. Furthermore the advantages of a Random-Reservation MAC scheme for a scenario with a high number of very bursty users was illustrated. The user population which can be supported was shown to be much higher than with a pure-reservation scheme while also achieving higher throughput and lower access delay for low to medium UBR traffic load.

7   REFERENCES

ATM-Forum (1996) *Proposed Charter, Work Plan and Schedule for a Wireless ATM Working Group*, 96-0712/PLEN.

ATM Forum (1996) *Traffic Management Specification Version 4*, af-tm-0056, April.

Bohm, S. Elhakeem, A.K. and Murthy, V.K.M. (1993) Analysis of a movable boundary random/DAMA accessing technique for future integrated services satellites, pages 1283-1289, IEEE Globecom

Celandroni,N. and Ferro,E. (1991) The FODA-TDMA Satellite Access Scheme: Presentation, Study of the System and Results., *IEEE Trans. Comms.*, COM-39(12), pages 1823-1831.

Fernandez, R. (1997) The Ka-band Quest Continues, Via Satellite, March.

Fitzpatrick, E.J. (1996) Hughes Spaceway: Wireless Interactive Broadband Service, Second Ka-Band Utilization Conference, September 24-26.

Guerin R. and Heinanen J. (1996), UBR+ Service Category Definition, ATM-Forum 96-1598

ITU-T (1996) *B-ISDN ATM Layer Cell Transfer Performance*, Draft 4R Rec.I.356.

Ors, T. Sun, Z. and Evans, B.G. (1997) A Meshed VSAT Satellite Network Architecture using an On-Board ATM Switch, IEEE IPCCC, pages208-215.

Zein,T. Maral,G., Tondriaux,M. and Seret,D. (1991) A Dynamic Allocation Protocol for a Satellite Network Integrated with B-ISDN., *Proc. 2nd ECSC*, pages 15-20.

10

A collision resolution algorithm for ad-hoc wireless LANs

D. V. Cortizo*, J. García**
*Electronic and Systems Department, University of A Coruña
Campus de Elviña s/n 15071 A Coruña, Spain, phone: +34.81.167150,
fax: +34.81.167160, e-mail:dcortizo@des.fi.udc.es

**Computer Architecture Dep., Polytechnic University of Catalonia
Jordi Girona 1-3, Campus Nord Mòdul D6 08034 Barcelona, Spain,
phone: +34.3.401.6798, fax: +34.3.401.7055, e-mail:jorge@ac.upc.es

Abstract

In this paper* we propose and analyze a Medium Access Control (MAC) protocol for high-speed ad-hoc Wireless Local Area Networks. The MAC protocol is a collision resolution algorithm and aims to approximate the delay characteristics of an ideal multiplexer with FIFO discipline. The protocol groups the packets arriving in given time intervals. Packets belonging to the same set will be transmitted via a CSMA algorithm whereas the different sets will be served following a FIFO discipline. The results obtained by means of simulation and analytical models show that the protocol can support real-time services even when the network load is moderately high.

Keywords

Wireless LANs, Ad-hoc networks, Medium Access Control algorithms, Collision Resolution algorithms, QBD processes

1 INTRODUCTION

The deployment of high speed wireless LANs has become a topic of increasing interest over the past few years. The wireless environment raises some problems (e.g. multipath, interference, hidden terminal node, minimization of power consumption at the mobile terminals, handoff) that either are not present or have marginal importance in wired networks (Acampora 1986).

Wireless LANs usually have two types of realizations: *infrastructured WLANs* and *ad-hoc WLANs*. In infrastructured WLANs Time Division Multiple Ac-

---

*This work was supported by projects XUGA 10503A96 and TIC95-0982-C02-01. Fundacion Caixa Galicia has supported D.V. Cortizo's stay at HUT, where part of the work was carried out.

Broadband Communications   P. Kühn & R. Ulrich (Eds.)
© 1998 IFIP. Published by Chapman & Hall

cess (TDMA) access control protocols are often preferred (Bantz *et al.* 1994). In ad-hoc WLANs mobile terminals establish peer-to-peer communication without the help of base stations and Carrier Sense Medium Access (CSMA) schemes may be advantageous because of their simplicity and because no station is compelled to assume special functions as in TDMA-based protocols. Unfortunately, random access protocols suffer from large access delay quantiles which make them unsuitable for delay-sensitive traffic support.

Modifications to the CSMA basic scheme have been proposed (Panwar *et al.* 1993) (Polyzos *et al.* 1987) to solve the instabilities of CSMA mechanisms and to improve the throughput-delay characteristics. The resemblance to the optimal FIFO discipline performance will be a trade-off with the overhead introduced and the protocol complexity.

The MAC protocol that we present in this paper is a collision resolution algorithm (Georgiadis *et al.* 1987) (Huang *et al.* 1986) aiming at the delay characteristics of a FIFO multiplexer. The protocol builds sets that contain the packets arriving (a packet arrival is defined to occur when a new packet is generated at any user site) within fixed-length grouping intervals. Packets belonging to the same set will be transmitted via CSMA but the different sets will be served following a FIFO discipline. The shorter the grouping time the closer to FIFO performance, but the signalling overhead will increase.

In the remainder of this section we present our assumptions on the system. In section II we describe the MAC protocol. In section III we describe both the simulation model and the analytical model developed to study the performance of the protocol. The results obtained are shown in section IV.

1.1   Assumptions and description of the system

- The network is supposed to be an ad-hoc high-speed wireless local area network. The possibility of handover has not been addressed.
- Either full connectivity among the stations or the existence of a higher level protocol to implement forwarding among nodes.
- Use of either radio or infrared channel providing a bit rate as seen by the MAC sublayer of 25 Mbps (or higher).
- Every succesfully transmitted packet is positively acknowledged by the receiver via a short ACK packet. Multicast traffic has not been considered.
- We assume a constant packet length of aproximately 500 bits. This includes payload, overhead and the time needed to generate, transmit and process an ACK. Therefore all users in the system transmit packets of constant duration $T_s \approx 20$ μs.

- The range of the network is supposed to be $R \approx 30\ m$, so the maximum one-way propagation delay between any pair of users is $a \approx 0.1\ \mu s$.
- We choose T_s as the time *slot* length, and we further define *mini-slots* with length $t_s = 3a \approx 0.3\ \mu s$. This way $T_s = N t_s$, where $N \approx 66$.
- The physical level will identify the different states of the channel: idle (no carrier), pilot tone transmission and packet transmission.

2   DESCRIPTION OF THE PROTOCOL

The basic medium access mechanism is CSMA with random backoff (an integer number of minislots) and adaptive backoff window size. However, this access mechanism is not applied to the whole set of stations having packets to transmit but only to a subset of them. The CSMA application to just some of the first arrived packets pending for transmission will allow to bring the protocol performance closer to the FIFO discipline behavior. Let us recall here the meaning of packet arrival: a packet arrival takes place whenever a packet is generated for transmission at any user site.

Our protocol will generate a collection of (possibly empty) Transmission Sets (TS) $\{S_k\}$, each of them containing the packets arrived during successive disjoint grouping intervals of constant length $W t_s$. These Transmission Sets are served (i.e. the associated packets are transmitted using CSMA) with a FIFO discipline, so that S_k is served before S_j if $k < j$. If every S_k had no more than one packet to be transmitted we would be facing a FIFO-MAC algorithm, with packets being sent in order of arrival.

The number of packets per TS is closely related to the integer W, which determines the length of the grouping interval, and to the network load ρ. The smaller W or ρ are the more similar the behavior of the protocol will be to a FIFO system. On the contrary, the protocol will perform as pure CSMA if W is made big enough. The value of W will be a trade-off between the desired properties of a FIFO system and the protocol overhead.

Every active station keeps track of two sequence numbers SN associated to the TSs: the sequence number $N_s(t)$ of the TS in service and the sequence number of the TS that any packet arrival at time t must join, $N_g(t)$.
The terminals shall include in the packet headers the sequence numbers $(N_g(t_a), N_g(t_{tx}))$, where

- t_a is the packet arrival time, $N_g(t_a)$ is the sequence number of the TS the packet belongs to, and
- assuming the packet starts being transmitted at t_{tx}, $N_g(t_{tx})$ is the sequence number of the TS being generated by the time the packet attempts transmission.

In the next subsections we will address the questions of how to generate, process and eliminate the transmission sets. For this sake the mobile stations will emit pilot tones of length one minislot, which also contribute to the protocol signalling overhead.

2.1   Generation of the transmission sets

Let us assume that the network has already been initialized and that no packets have been generated in the system for a long period of time. The system will be in idle state, characterized by all terminals in the network having $N_s(t) = N_g(t)$. When in this situation a packet is generated at any mobile node, it is immediately transmitted. Once the first transmission attempt takes place a busy period starts. During busy periods a new transmission set (either empty or not) is generated every W minislots. TSs will be empty whenever there are no packet arrivals during the associated grouping intervals, but it is important to note that only during the busy periods empty TSs may be generated. Any packet arriving at t_1 will defer transmission until a time t_2 when its associated TS $N_g(t_1)$ begins to be served $((N_s(t_2) = N_g(t_1))$. The service of a non empty TS consists of the successful transmission of all its packets via a scheduling which will be described in section 2.2.

If the network is stable the system must eventually get empty again, when $N_s(t) = N_g(t)$. At this point the generation of TSs is halted and the system enters an idle period. Packets finding at arrival time the system in idle state will directly attempt transmission, starting a new busy period, the generation of new TSs, and reinitiating the process described above.

2.2   Service of the transmission sets

The service of a Transmission Set consists of the successful transmission of all the packets that entered the system (at any station) during the grouping interval of length W associated to the TS. The active stations must be able to identify not only the beginning and end of service of a non empty TS but also the empty TSs.

At the beginning of a TS service all the stations with packets belonging to that TS will emit a pilot tone during one minislot. This mechanism allows the stations to identify empty TSs (when there is no pilot tone present in the channel at the beginning of their service), discard them and proceed after only one minislot delay to the processing of the next TS. If the TS is non-empty after emitting the pilot tone the stations wait a random number of minislots before sensing the channel and, if idle, start transmission (CSMA with ran-

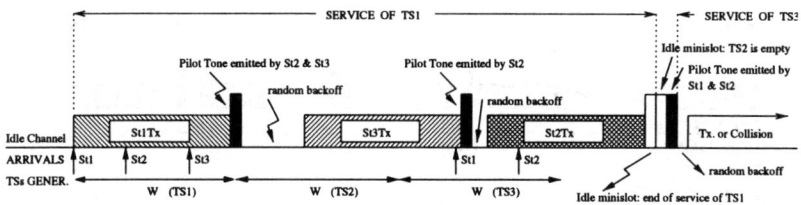

Figure 1 View of the operation of the protocol.

dom backoff window). At the end of the transmission there will have been either a collision or successful transmission, and the random backoff window value will be adapted accordingly.

After the successful transmission of a packet we may have two different scenarios: either there are still more packets in the TS that have not been transmitted (i.e. the service of the TS has not finished yet) or there are no more packets to transmit associated to the currently served TS, and the MAC should start processing the next TS.

These different situations can be easily identified by the active stations if we force all the terminals still having packets to transmit in the currently served TS to emit a pilot tone after every transmission. In the former case the stations waiting for the TS service to end will sense a pilot tone and keep on waiting. In the latter case, the terminals will sense an absence of carrier during one minislot, realize that the service of the i-TS has finished and start processing the $(i + 1)$-TS.

In Figure 1 we have illustrated the operation of the protocol, assuming three Mobile Stations in the system St1, St2 and St3. When the system has been idle for a long enough period the first packet (St1 in the Figure) is immediately transmitted and it also starts the generation of TSs (TS1 in this case). After the transmission of the message (St1Tx) St2 and St3 emit a pilot tone, as they both have packets to transmit belonging to TS1. St3 gains access and after transmission of St3Tx, St2 emits a pilot tone, generates a random backoff and transmits its packet. As no mobile station has any other packet to transmit belonging to TS1 the next minislot remains idle, and the service of the next TS (TS2) starts. However, as there has not been any packet arrival during the interval associated to TS2, there will be another idle minislot indicating that TS2 is empty. In the next minislot the service of TS3 starts with a pilot tone generation by the stations having packets belonging to TS3 (St1 and St2).

3   THE MODELS

Our main objective is to evaluate the density function of the delay suffered by a packet arriving in the system at an arbitrary point of time. This delay

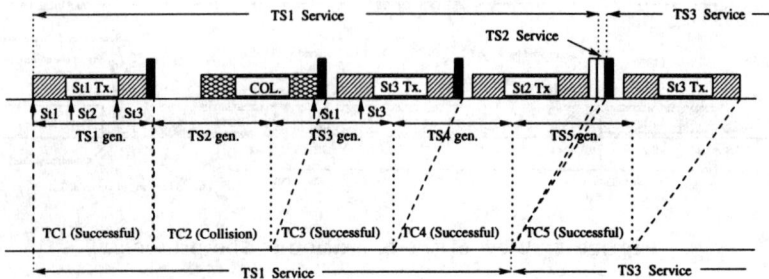

Figure 2 Synchronization achieved with the simplifications of the model.

density function will allow us to assess delay percentiles and therefore estimate the goodness of the protocol to support time-bounded services. We have
developed both a simulation model and an approximate analytical model. For
both models we have assumed that there are m stations in the network and
that the joint packet arrival process is Poisson with rate λ (arrivals per slot
T_s).

3.1   Analytical model

For the analytical model we have made the following simplifying assumptions:

- The minislot used for the pilot tones following every packet transmission
  is regarded as packet overhead.
- We will only consider the case where $W = N$, that is, the length of the
  grouping intervals is made equal to the length of the packets transmitted.
- The signalling minislot used at the beginning of every TSs (empty or not)
  service is neglected.
- The random backoff with adaptive window CSMA is replaced by a p-
  persistent CSMA. Adaptive mechanisms for the probability p of transmission are not included in the analytical model.
- We will neglect the random backoff delay introduced by the stations before
  packet transmission.
- The maximum number of packet arrivals per Transmission Set has been
  bounded to a certain value q_{max}.

In the simplified analytical model the system behaves as a TS-processing
queue. During the busy periods we obtain a sequence of Transmission Cycles (TC) of length W. In every TC a packet transmission occurs (successful
or not) and a new Transmission Set is generated (even if empty). When the
system is idle neither TCs nor TSs are generated. It is the first next packet
arrival who will reinitiate a busy period and the generation of both TSs and
TCs. The simplifications synchronize the instants when packets get success-

fully transmitted and the instants when new TSs are generated (see Figure 2).

The state of the system is defined by (G, q), being G the number of non-empty Transmission Sets pending to be served and q the number of packets awaiting transmission in the TS in service. The pair (G, q) can ve viewed as a Markov renewal stochastic process, and observing the state of the system at the end of every TC we get an embedded Markov Chain. The time elapsed between transitions will have a density function dependent on the state of the system (idle or busy). Being $u(t)$ the step function and $\delta(\cdot)$ the dirac delta function, we have

$$f_{busy}(t) = \delta(t - W \cdot t_s)$$

$$f_{idle}(t) = \frac{\lambda}{W} e^{-\frac{\lambda}{W}(t - W \cdot t_s)} u(t - W \cdot t_s).$$

(1)

(a) System state probabilities and vector solution of the embedded markov chain

The packet delay density function associated to the MAC protocol will be evaluated by conditioning on certain state of the system $(G, q)^*$. The values $\pi(G, q)^*$ represent the probabilities that a packet arrival finds that the last completed Transmission Cycle moved the system into state (G, q). Note that $\pi(G, q)^*$ is not the probability that an arbitrary packet arrives being the state of the system (G, q), but the probability that the packet finds that the *last* state transition of the embedded Markov chain *was* into the value (G, q).

Previous to the assessment of the solution vector $\mathbf{v}$ of the embedded Markov chain (with one-step transition matrix $\mathbf{P}$) we will truncate the maximum value of q to a certain value q_{max}. If this bound is chosen large enough its influence on the accuracy of the results will be negligible.

Let us define the $\mathbf{M}_{q_{max} \times q_{max}}$ matrices $\mathbf{C}$, $\mathbf{D}$ and $\mathbf{E}$, as the matrices $\mathbf{F} \in \mathbf{M}_{(q_{max}+1) \times (q_{max}+1)}$, $\mathbf{E_1} \in \mathbf{M}_{q_{max} \times (q_{max}+1)}$ and $\mathbf{D_1} \in \mathbf{M}_{(q_{max}+1) \times q_{max}}$

$$
\begin{aligned}
\mathbf{C}(q, r) &= P[(G, q) \to (G, r)] & G \geq 1 \quad q, r = 1, ..., q_{max} & \quad (2)\\
\mathbf{D}(q, r) &= P[(G-1, q) \to (G, r)] & G \geq 1 \quad q, r = 1, ..., q_{max} \\
\mathbf{E}(q, r) &= P[(G, q) \to (G-1, r)] & G > 1 \quad\quad q, r = 1, ..., q_{max} \\
\mathbf{F}(q, r) &= P[(0, q) \to (0, r)] & q, r = 0, ..., q_{max} \\
\mathbf{D_1}(q, r) &= P[(0, r) \to (1, q)] & q = 1, ..., q_{max} \quad r = 0, ..., q_{max} \\
\mathbf{E_1}(q, r) &= P[(1, q) \to (0, r)] & q = 1, ..., q_{max} \quad r = 0, ..., q_{max}.
\end{aligned}
$$

The expressions for these transition probabilities are (only the nonzero probabilities are indicated)

$$\mathbf{F}^{0,i} = \begin{cases} e^{-\lambda} & i = 0 \\ x(0,A)x(i,B) + \sum_{j=1}^{i-1} x(j,A)x(i-1-j,B) & i = 1, ..., q_{max} \end{cases}$$

$$\mathbf{F}^{i,i} = e^{-\lambda}c(i,p) \qquad i = 1, ..., q_{max}$$

$$\mathbf{F}^{i,i-1} = e^{-\lambda}t(i,p) \quad i = 1, ..., q_{max}.$$

(3)

$$\left. \begin{aligned} &\mathbf{D}^{q,q} = \mathbf{D_1}^{q,q} = (1 - e^{-\lambda})c(q,p) \\ &\mathbf{C}^{q,q} = e^{-\lambda}c(q,p) \\ &\mathbf{D}^{q,q-1} = \mathbf{D_1}^{q,q-1} = (1-e^{-\lambda})t(q,p) \\ &\mathbf{C}^{q,q-1} = e^{-\lambda}t(q,p) \end{aligned} \right\} q = 2, ..., q_{max}$$

$$\left. \begin{aligned} &\mathbf{C}^{1,q} = x(q,\lambda) \\ &\mathbf{E}^{1,q} = \mathbf{E_1}^{1,q} = e^{-\lambda}x(q,\lambda)/(1-e^{-\lambda}) \end{aligned} \right\} q = 1, ..., q_{max}.$$

(4)

Being p the transmission probability in one minislot of the p-persistent CSMA, $A = \lambda/W$, $B = \lambda - A$ and

$$x(i,\Delta) = \begin{cases} e^{-\Delta}\frac{\Delta^i}{i!} & i = 0, ..., q_{max} - 1 \\ 1 - \sum_{j=0}^{j=q_{max}-1} x(j,\Delta) & i = q_{max} \end{cases}$$

$$t(i,p) = \frac{ip(1-p)^{i-1}}{1-(1-p)^i} \qquad i = 1, ..., q_{max}$$

$$c(i,p) = 1 - t(i,p) \quad i = 1, ..., q_{max}.$$

(5)

With the above definitions matrix $\mathbf{P}$ can be written as:

$$\mathbf{P} = \begin{pmatrix} \mathbf{F} & \mathbf{D_1} & 0 & 0 & 0 & \cdots \\ \mathbf{E_1} & \mathbf{C} & \mathbf{D} & 0 & 0 & \cdots \\ 0 & \mathbf{E} & \mathbf{C} & \mathbf{D} & 0 & \cdots \\ 0 & 0 & \mathbf{E} & \mathbf{C} & \mathbf{D} & \\ \vdots & \vdots & & \ddots & \ddots & \ddots \end{pmatrix}.$$

(6)

From the structure of $\mathbf{P}$ we see that it corresponds to a Quasy-Birth-and-Death (QBD) proccess. Therefore we will assess the vector solution $\mathbf{v}$ applying any of the existing algorithms (Neuts 1981), (Stewart 1994) to solve QBD markov chains. Once we have found the vector solution $\mathbf{v}$ for the embedded markov chain we make use of both the PASTA* property and the key Markov renewal theorem (Cinlar 1975) to calculate the state probabilities $\pi(G, q)^*$

$$\pi(G,q)^* = \begin{cases} \frac{v(G,q)}{1+v(0,0)/\lambda} & (G,q) \neq (0,0) \\ \frac{v(0,0)}{1+v(0,0)/\lambda}(1+\frac{1}{\lambda}) & (G,q) = (0,0) \end{cases} \tag{7}$$

These values will be needed to assess the transmission delay density functions of the the data packets. The derivation of these density functions is rather involved, and we have not included it in this paper (Cortizo *et al.* 1997).

3.2 Simulation model

Most of the simplifying assumptions of the analytical model are overcomed by the simulation model, whose main features are:

- The delay distribution function assessed includes the effect of both the pilot tones (signalling overhead) and the idle minislots associated to the random backoff in the CSMA algorithm.
- As in the analytical model, the random backoff with adaptive window CSMA is replaced by a *p*-persistent CSMA.
- The simulation model does not impose any constraint on the grouping interval length, allowing the study of the CDV percentiles vs. *W*.

4 SIMULATIONS AND NUMERICAL RESULTS

In this section we will analyze the behavior of the fixed-size random back-off window protocol, thus limiting the study to the non adaptive *p*-persistent CSMA case.

According to the assumptions of section 1.1, we have taken minislot lengths $t_s = 0.3\mu s$ and packet lengths $N = 64$ minislots. The graphics obtained by simulation present the average of five experiments of length $64\,000\,000$ minislots and the delay distribution functions $F_D(t)$ are depicted as $log_{10}(1 - F_D(t))$.

Figure 3 shows, for a fixed load $\rho = 0.6$, the protocol performance dependence on the transmission probability p, which is optimal for a value $p = 0.1/0.2$. This value might seem surprisingly low but it must be noticed that the delays due to idle minislots (with $p = 0.2$ a packet will be transmitted, in average, after a backoff of 5 minislots) are well paid off by the drastic re-

---

*Poisson arrivals See Temporal Averages

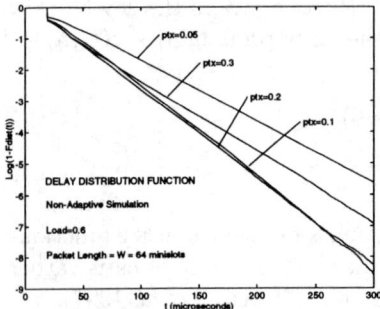

Figure 3 Determination of the optimal p.

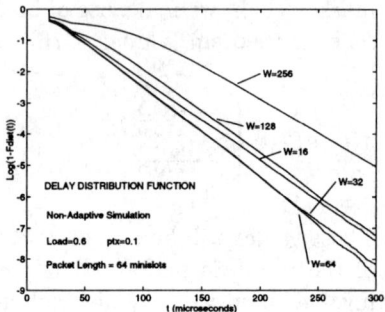

Figure 4 Determination of the optimal W.

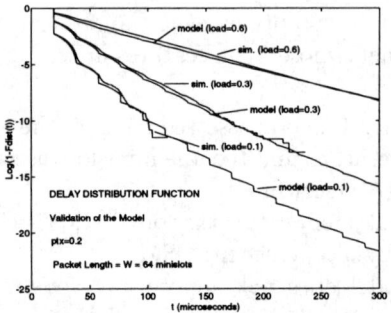

Figure 5 Validation of the model.

duction of the collision probability (each collision adds a delay of 64 minislots).

Figure 4 depicts the influence of the grouping interval length W on the protocol performance. The value $W \approx N$ seems to be the best choice for load $\rho = 0.6$. This result enhances the value of the analytical models, which only apply for this case. Moreover, the results corresponding to low-loaded systems ($\rho = 0.1$) lead to the same choices $p_{opt} = 0.2$ and $W_{opt} \approx N$ (Cortizo *et al.* 1997). For the remainder of the section we will therefore assume $W = N$.

Figure 5 validates the analytical model by comparison with the results obtained by simulation. The model gives always similar results to those obtained by simulation. For lower values of p (less than 0.1) we expect an impairment in the behavior of the analytical approach, as it neglects the idle minislots introduced by the random backoff windows.

Figures 6 and 7 show the 10^{-3} and 10^{-6} packet delay quantiles for different network loads. The quantiles are defined as the delays exceeded with less probability mass than 10^{-3} and 10^{-6} respectively. From these quantiles one can easily find estimations of the Cell Delay Variation introduced by the access protocol (Roberts *et al.* 1996). The graphics show the good behavior of the MAC protocol, allowing the support of real-time services with loads close to 0.8.

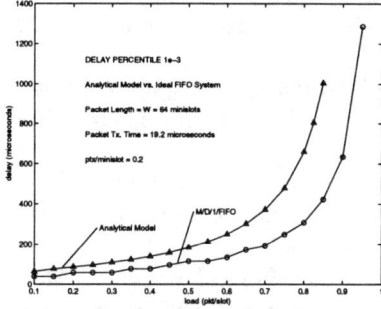

Figure 6 10^{-3} quantile for the packet delay.

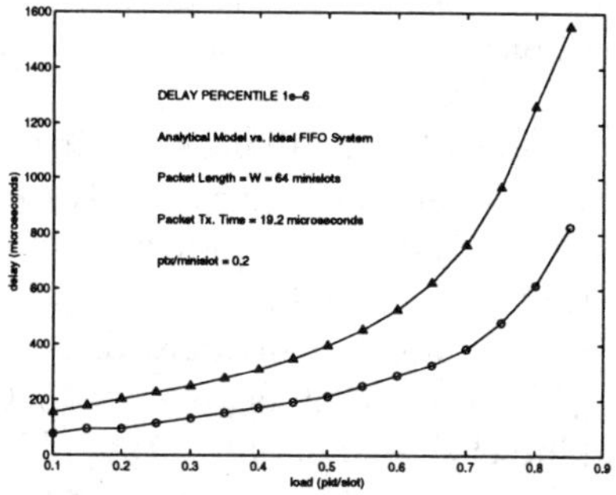

Figure 7 10^{-6} quantile for the packet delay.

5 CONCLUSIONS

In this paper we have presented a MAC protocol oriented to ad-hoc Wireless LANs. This protocol is a Group Resolution Algorithm, as it gathers into Transmission Sets the packet arrivals taking place in grouping intervals of fixed length. Every Transmission Set is served via an adaptive-window random backoff CSMA, and the different TSs are handled with FIFO discipline.

The protocol performance lies between that of CSMA and FIFO algorithms, depending on the length of the grouping intervals. The shorter these are, there will be less packets in average per TS and the MAC will behave as a distributed FIFO scheduling. However, the signalling overhead introduced by the protocol grows when the grouping interval length decreases.

We have developed a simplified analytical model which only considers the situation where the packet lengths equal the size of the grouping intervals. The results obtained show that the protocol can support real-time services even when the network load is moderately high (loads up to 0.8).

REFERENCES

Acampora, A. (1996) Wireless ATM: A Perspective on Issues and Prospects. IEEE Personal Communications, Vol. 3, No. 4, August 1996, 8–17.

Bantz, D.F. and Bauchot, F.J. (1994) Wireless LAN Design Alternatives. IEEE Network, March/April 1994, 43–53.

Roberts, R., Mocci, U. and Virtamo, J. (1996) Broadband Network Traffic, Performance Evaluation and Design of Broadband Multiservice Networks Final Report of Action COST 242. Springer-Verlag, Berlin.

Cinlar, E. (1975) Introduction to Stochastic Processes. Ed. Englewood Cliffs, New Jersey.

Georgiadis, L., Papantoni-Kazakos, P. (1987) A 0.487 Throughput Limited Sensing Algorithm. IEEE Transactions on Information Theory, Vol. IT-33, No. 2, March 1987, 233–237.

Huang, J.C. and Berger, T. (1986) Delay Analysis of 0.487 Contention Resolution Algorithms. IEEE Transactions on Communications, Vol. COM-34, No. 9, September 1986, 916–926.

Marcel F. Neuts, M.F. (1981) Matrix-Geometric Solutions in Stochastic Models, An Algorithmic Approach. The John Hopkins University Press, Baltimore.

Panwar, S.S., Towsley, D. and Armoni, Y. (1993) Collision Resolution Algorithms for a Time-Constrained Multiaccess Channel. IEEE Transactions on Communications, Vol. 41, No. 7, July 1993, 1023–1026.

Polyzos, G.C., Molle, M.L. and Venetsanopoulos, A.N. (1987) Performance Analysis of Finite Nonhomogeneous Population Tree Conflict Resolution Algorithms Using Constant Size Window Access. IEEE Transactions on Communications, Vol. COM-35, No. 11, November 1987, 1124–1138.

Stewart, W.J. (1994) Introduction to the Numerical Solution of Markov Chains. Princeton University Press.

Cortizo, D.V. and García, J. (1997) A MAC Protocol for Ad-Hoc Wireless ATM LANs. Departamento Electrónica e Sistemas, UDC, Internal report, Aug. 1997.

ATM Mobile

11

Air Interface of an ATM Radio Access Network

Dietmar Petras
Bosch Telecom GmbH
Backnang, Germany
tel/fax: +49 7191 13-3905/-4603, e-mail: dietmar.petras@pcm.bosch.de

Abstract

This paper describes system and protocol aspects of the air interface of a fixed ATM Radio Access Network with directional antennas at terminals and sectored antennas at base stations (also called Point-to-Multipoint system). A requirement on the system is that it has to fit transparently into a fixed ATM network. A radio sector of the system is interpreted as a distributed ATM multiplexer. The protocol stack of the air interface implements the statistical multiplexing of ATM cells with a quality of service as in fixed ATM multiplexers with the same link data rate. The multiplexing is controlled by a service strategy that optimizes the resource allocation based on short-term demands of virtual channels and their negotiated quality of service. The MAC protocol realizes the transmission order of ATM cells given by the service strategy. By this, the protocol stack is able to efficiently support all ATM service categories.

Keywords

Broadband Radio Access Network (BRAN), Point-to-Multipoint System, wireless ATM, quality of service, ATM cell scheduler, MAC protocol, random access

1 INTRODUCTION

The development in modern telecommunications is currently determined by the on-going migration of the worldwide infrastructure to a broadband multimedia network. Multimedia applications require circuit-switched services with guaranteed capacity and transfer delays for interactive voice and video services as well as packet switched services with on demand capacity for the transmission of text, pictures and data. Therefore, multimedia networks are based on the Asynchronous Transfer Mode (ATM), since its connection-oriented packet switched transport allows the connection specific guarantee of capacity and delays. The operator of an ATM network has to install, operate and maintain only one common service integrating network technique instead of the previous multiple service specific networks. Beside this saving in costs, ATM networks are a future proof investment, since their service independent transport in general allows the support of future services.

Another trend in modern telecommunications is given by the success of cellular

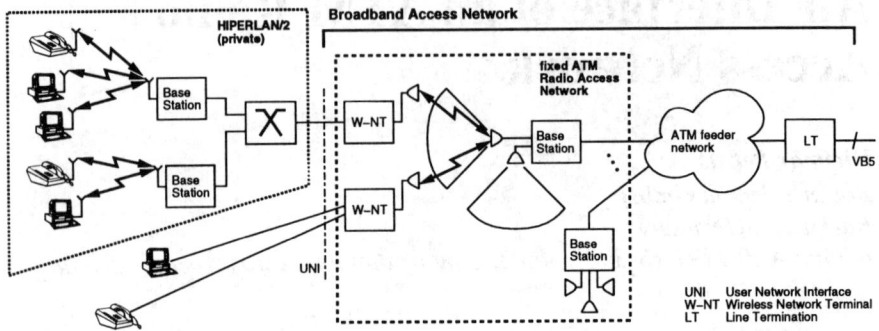

Figure 1 Different applications for wireless ATM systems and architecture of fixed ATM Radio Access Network

mobile radio networks, which serve the users demand for universal accessibility. The current cellular network technique allows the global availability of voice and narrowband data services up to several 10kbit/s. However, the demand for much higher bandwidths is growing rapidly. But making available data rates of several 1Mbit/s, as usual at the user access of cable-based multimedia networks, would require a much wider frequency spectrum. This is only available at frequencies higher than 2GHz where line-of-sight operation is necessary. Thus, a full coverage mobile multimedia network would force an enormous number of base station so that its realization is neither technically nor economically feasible. But there are still two market segments, where ATM radio networks seam to be realistic (Figure 1):

1. ATM-based cordless system or local area radio networks (wireless ATM local area network, W-ATM-LAN) in an unlicensed frequency band with limited mobility in a restricted area (e.g., office building, exhibition area)
2. Broadband fixed radio access (B-FRA) in licensed frequency bands without mobility as replacement of cable-based infrastructure for subscriber access

The second segment is especially stimulated by the current world-wide deregulation in telecommunications, since radio systems allow new operators, which usually don't own a cable-based access network, to gather broad coverage fast and efficiently.

There are a number of research activities in science and industry investigating W-ATM systems (Walke et al., 1996; Radimirsch, 1997; Dinis et al., 1997; Ala-Laurila and Awater, 1997; Eng et al., 1995; Raychaudhuri et al., 1997; Porter and Hopper, 1995; Agrawal et al., 1996; Ciotti, 1996). Also international standardization bodies have started W-ATM activities and are planing first technical standards for end of 1998 (BRAN Project, 1997; Wireless ATM Group, 1997; DAVIC, 1997). The ETSI project *Broadband Radio Access Networks* (BRAN) is dealing with the air interface of two W-ATM systems according to the above two market segments,

Table 1 Parameters of the ATM Radio Access Network

frequency band	3.5, 10, 24/26, 28GHz and higher
channel bandwidth	6 × 14MHz duplex @ 3.5GHz 5 × 30MHz duplex @ 10GHz 18 × 28MHz duplex @ 24/26GHz block allocation @ 28GHz
range	10-15km @ 3.5GHz 7-10km @ 10GHz 3-5km @ 24/26, 28GHz
multiplex	frequency division multiplex (FDM)
duplex	frequency division duplex (FDD)
downstream data rate	8 - 51 Mbit/s
upstream data rate	2 - 25 Mbit/s
medium access	asynchronous multiplex of ATM cells

which are called "HIPERLAN type 2" and "HIPERACCESS" (HIPER = HIgh PERformance)*.

The world-wide release of an adequate unlicensed spectrum for personal communication systems at 5GHz (much more bandwidth is available at higher frequencies bands around 40 and 60GHz) is an import prerequisite for the realization of local area multimedia radio networks. For broadband radio access networks appropriate frequency bands are available region-dependent at 3.5, 10, 24/26, 28GHz, and further bandwidth above 30GHz is under discussion.

The paper presents the protocol stack for the air interface of an ATM fixed radio access network with directional antennas at terminals and sectored antennas at base stations (also called Point-to-Multipoint system). The system is intended mainly for the access of small and medium business users. The basic parameters of the system are summarized in Table 1. A radio access network usually contains multiple base stations each with several radio sectors (Figure 1). A feeder network connects the base stations to the line termination with the VB5 interface. It is assumed that interference between radio channels of neighbouring sectors is avoided by using orthogonal radio resources which are divided by directional antennas, polarization and frequency. Aspects of the physical layer (modem, antennas, etc.) as well as the networks behind the air interface are outside the scope of the paper.

The W-ATM system has to fit transparently into a fixed ATM network. Therefore, the protocol stack at the air interface has to execute statistical multiplexing of ATM

---

*Furthermore, a "HIPERLINK" standard for short distance ATM radio relay systems in an unlicensed frequency band is planed.

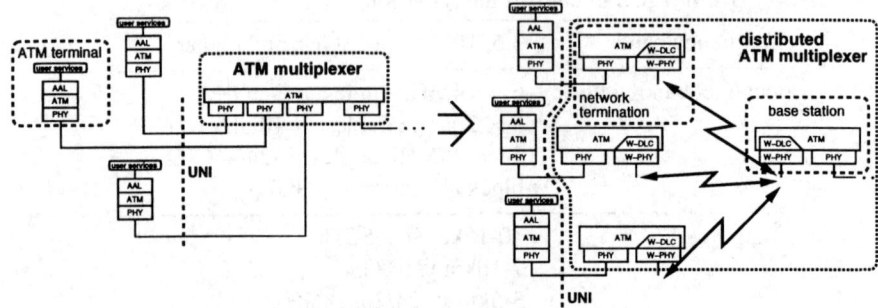

Figure 2 Correspondence between radio sector and ATM multiplexer

cells as in fixed ATM multiplexers. A medium access control (MAC) protocol is required to coordinate the competition of terminals for the shared radio channel.

The approach for the MAC protocol presented here takes into account that the performance of ATM networks is mainly influenced by the *intelligence* of the ATM cell multiplexing in ATM network nodes. Modern ATM multiplexers apply service strategies that optimize the resource allocation based on short-term demands of virtual channels and their negotiated quality of service (QoS). The ATM air interface can be interpreted as a distributed ATM multiplexer. The MAC protocol is centrally controlled by the base station and realizes the transmission order of ATM cells given by the service strategy.

The paper has the following structure. Section 2 introduces the architecture and protocol stack of the ATM air interface. In section 3 the service strategy for the air interface is specified. Section 4 describes how the distributed ATM cell scheduler is realized. The MAC protocol is described in section 5. Section 6 presents a random access protocol for the fast and efficient signaling of capacity requests over the uplink. The paper winds up with a conclusion.

2   ARCHITECTURE OF AN ATM AIR INTERFACE

In general, the users at a W-ATM access network request the same functionality and QoS as users of wired ATM access networks. Figure 2 illustrates how these user requirements can be transformed into the demand on building a distributed ATM multiplexer *around* the air interface which is characterized by a radio channel *inside*. The figure only considers the user plane of the multiplexer. The wireless terminal is equivalent to the network termination (W-NT) of the access network and offers one or multiple User-Network interfaces (UNI) to connect customer premises equipment (see also Figure 1).

At the air interface additional layers must be introduced in the protocol stack. The resulting stack contains a wireless physical layer with the radio modems below the ATM layer and a data link control (DLC) layer that contains the ATM cell scheduler.

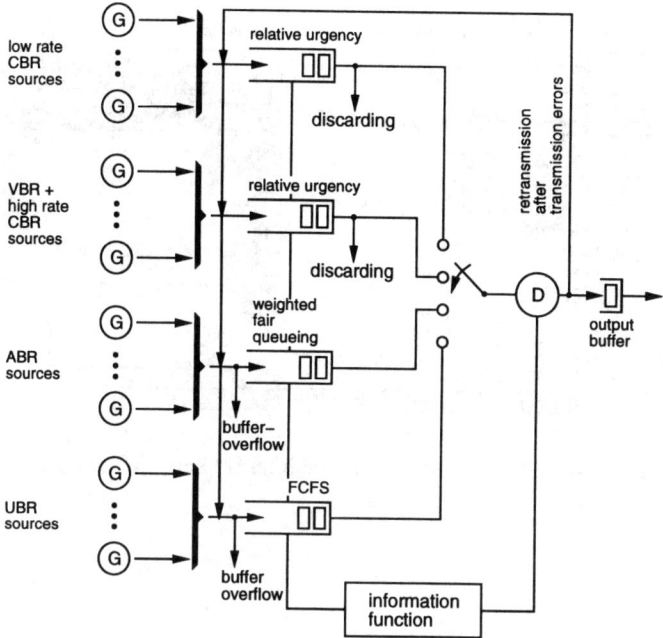

Figure 3 Queueing model for the upstream of the distributed ATM multiplexer

The DLC layer is furthermore divided into a MAC sublayer and a Logical Link Control (LLC) sublayer.

3 SERVICE STRATEGY AT THE AIR INTERFACE

Network nodes (switches and multiplexers) have to determine the transmission order of cells to be sent over a link. An ATM cell scheduler is introduced for each output port/link that controls the statistical multiplexing of cells by employing an appropriate service strategy. Its goal is to optimize the resource allocation based on short-term demands of virtual channels and their negotiated quality of service. Appropriate service strategies usually focus on two key targets: avoiding overflow of buffers and controlling delays of ATM cells. Which issue the strategy has to focus on strongly depends on the transfer rate of the link. While a fast transfer rate of 155 Mbit/s and more in wired ATM networks causes buffer overflows to be the more critical aspect, with slow transfer rates, e.g. 8 Mbit/s ($\approx$ 20.000 cells/s) on the upstream in W-ATM systems, cell delay guarantees become more difficult to fulfill and the service strategy plays a major role in providing QoS.

A service policy has to distinguish between the real-time oriented CBR/VBR services and non-real-time ABR/UBR services. As in most fixed ATM multiplexers, static priorities are introduced between service categories (CBR/VBR > ABR > UBR). For the upstream another high priority level is introduced for low rate CBR

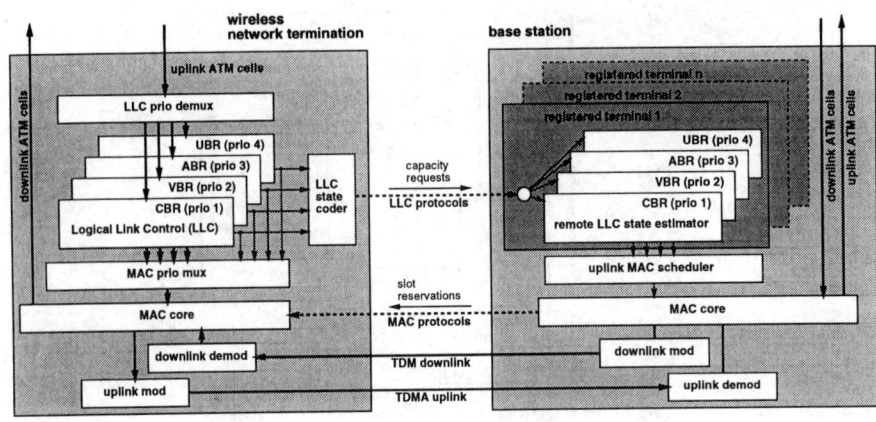

Figure 4 Dividing the upstream scheduler into an LLC and MAC part

connections ($\leq$ 128kbit/s) in order to prevent voice services from congestion events caused by bursty VBR connections (Figure 3). For ABR services the algorithms applied for fixed ATM multiplexers are used (e.g., weighted fair queueing (COST 242 Management Committee, 1996)), while the UBR service category requires fair resource sharing.

For real-time oriented CBR and VBR services a specific service strategy has to controls delays. The connection specific performance parameter *maximum cell transfer delay* (*maxCTD*) is specified by the traffic contract and is related to the end-to-end delay. The air interface as one hop of a virtual channel connection is only allowed to consume a fraction of *maxCTD*. A maximum delay τ_{dmax} on the air interface is not specified. The assumption of $\tau_{dmax} = 0.1\ maxCTD$ seems to be useful. A detailed analysis of appropriate service strategies for real-time services can be found in (Kist and Petras, 1997). The reference recommends dynamic priorities where the priorities of ATM cells depend on their waiting time and their connection specific QoS requirements. An arriving cell gets a due date by adding τ_{dmax} to its arrival time. Accourding to the strategy the cell with earliest due date (EDD) is transmitted first. If all connections are considered to have the same maximum delay, the EDD strategy is equivalent to the First Come First Serve (FCFS) strategy. The more complex Relative Urgency (RU) discipline furthermore considers the maximum cell loss ratio and minimizes the probability for cells being late (exceeding their due dates). ATM cells which exceed their maximum delay will usually be discarded by the receiving application. Thus, discarding delayed cells (due date expired) at the air interface contributes to avoid and resolve congestion events, since the delay of the following cells can be shortened and the probability to exceed further due dates is reduced.

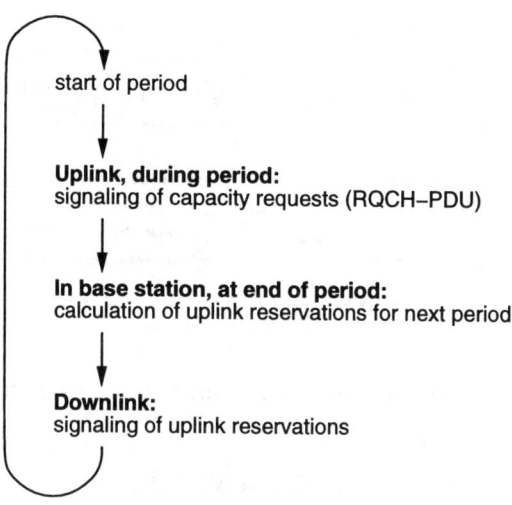

Figure 5 Cycle of signaling periods in DSA++ MAC protocol

4 REALIZING THE DISTRIBUTED ATM CELL SCHEDULER

The main difference between the distributed ATM multiplexer and a fixed ATM multiplexer is the distribution of the ATM cell scheduler for upstream cells between W-NT and the base station. For the downstream, the scheduler is completely contained in the DLC layer of the base station.

For the uplink the ATM cell scheduler is separated into two parts. The lower part in the MAC sublayer selects the W-NT which is allowed to send in a time slot (determination of slot reservations). The upper part of the scheduler belonging to the LLC sublayer contains the send buffers and determines a virtual channel for delivering ATM cells which are transmitted in a reserved time slot.

The LLC sublayer in W-NTs contains an entity for each priority level of the service strategy with the send buffers of the corresponding service category. The base station contains mirrors of the W-NT entities each estimating the state of the send buffers in the corresponding W-NT entity. A signaling protocol is executed for notifying the mirror entities about the states of the W-NT entities. This is done by generating (in the W-NT) and interpreting (in the base station) capacity request messages, which are transmitted as RQCH-PDUs (ReQuest CHannel Protocol Data Unit) by the RQCH service of the MAC sublayer (cf. section 6).

The MAC scheduler uses the capacity requests of W-NTs, which are estimated by the mirror instances, in order to determine the reservation of upstream slots. The MAC core performs a signaling protocol (cf. section 5) that notifies the W-NTs about the slot reservations.

Furthermore, the MAC core is responsible for the transmission of RQCH-PDUs.

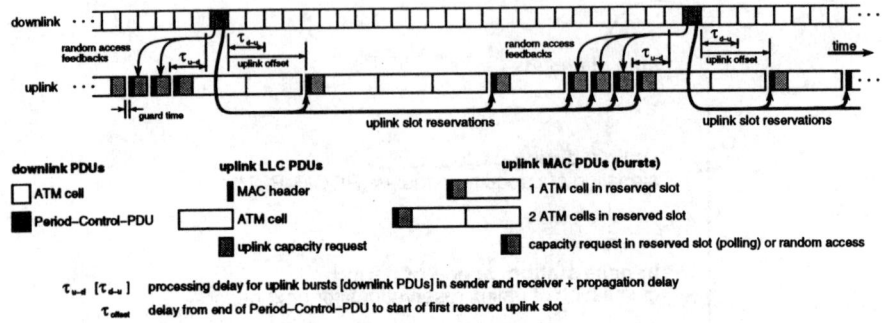

Figure 6 Downlink signaling scheme of DSA++ protocol

5   MEDIUM ACCESS CONTROL PROTOCOL

Active W-NT have to register at the base station. A temporary MAC identifier (MAC-Id, 8 bit) is assigned to each registered W-NT, which is used as a short address within the sector for addressing uplink slot reservations.

On the downlink ATM cells are broadcasted and received by all W-NTs (Figure 6). The virtual path identifier (VPI) is used to detect the target W-NT. MAC-internal signaling messages are carried by ATM cells with a specific mark, e.g., VPI=0 or OAM F3. On the uplink the VPI may also be used to indicate the destination W-NT of ATM cells. For safety reasons, each uplink burst is extended by a MAC header containing the MAC-Id and a short identification of the sector to enable the detection of faulty insertions of bursts received from neighbouring sectors.

The MAC protocol executed by the MAC core is called Dynamic Slot Reservation (DSA++) protocol and has originally been developed for HIPERLAN type 2 systems on a time division duplex (TDD) link (Petras and Krämling, 1997b). For broadband radio access networks it has been modified in order to work on a frequency division duplex (FDD) link with asymmetric link rates.

Uplink slots allow the transmission of one physical burst that contains an information field of variable length together with the necessary overhead of the physical layer (synchronization of terminals to balance propagation delays is assumed). The information field carries one RQCH-PDU and up to four ATM cells. Slots which allow the transmission of a burst with only an RQCH-PDU and no ATM cell are called RQCH slots or short slots.

In order to co-ordinate the channel access, the DSA++ MAC protocol groups uplink slots in so-called signaling periods. The periodic signaling procedure is illustrated in Figure 5. A signaling period consists of a variable number of uplink slots, the number and order of which is determined by the MAC scheduler.

During such a signaling period new capacity requests arrive at the base station and are saved in the mirror entities. At the end of a period the slot reservations of the next period are calculated and transmitted to the W-NTs. This is done with a Period-Control-PDU on the downlink that signals the number of slots in the next signaling period and for each slot its length (number of ATM cells) and the associated MAC-

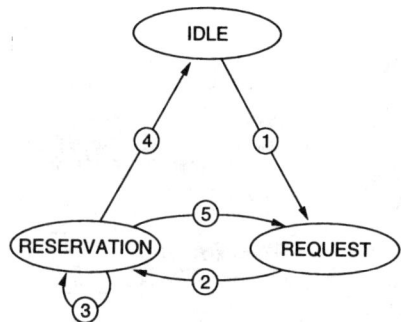

(1) arrival of burst of ATM cells
(2) successful transmission of capacity request message (RQCH-PDU)
(3) piggyback transmission of RQCH-PDU
(4) transmission of last ATM cell from burst, empty send buffer
(5) arrival of ATM cell of service category with higher priority

Figure 7 Reservation states of W-NT

Id. A signaling period may contain a variable number of short slots for RQCH-PDUs (Figure 6). The access to the RQCH slots is performed with random access under control of the RQCH protocol explained in section 6.

The Period-Control-PDU furthermore contains a feedback (MAC acknowledgment) for each random access slots of the last signaling period which consists of the MAC-Id of the received bursts. The feedback is used by the RQCH access protocol to resolve collisions. By also sending a feedback for each ATM cell, erroneous transmission of ATM cells can be detected and retransmissions be triggered. Such an error control protocol is for further study.

The maximum length of a signaling period is given by the limited size of the Period-Control-PDU carrying the slot reservations. Furthermore, too long signaling periods reduce the dynamics of the protocol and cause too long delays for real-time services. Experience has shown that a signaling period should not exceed 0.5ms (Petras and Krämling, 1997b).

6 SIGNALING OF CAPACITY REQUESTS OVER THE UPLINK

Capacity requests (RQCH-PDU) are transmitted depending on the reservation state of a W-NT. The diagram in Figure 7 shows the corresponding state transition diagram. State IDLE corresponds to an empty send buffer so that no capacity request message is to be sent. After the arrival of an ATM cell the W-NT switches to state REQUEST (transition (1)) and tries to transmit its capacity request in a short slot. After a successful transmission the W-NT enters state RESERVATION (transition (2)) and will be served by the scheduler according the urgency of its ATM cells. With the transmission of an ATM cell in a reserved long slot the scheduler is informed about the newest capacity requirements by means of the piggybacked RQCH-PDU (transition (3)). If no further capacity is required, the base station recognizes that the W-NT has returned to state IDLE (transition (4)).

A special case is the parallel existence of virtual channels of different service categories. If a W-NT did request capacity for a low priority service category (e.g. UBR) and thus is in state RESERVATION, than the arrival of an ATM cell of a high

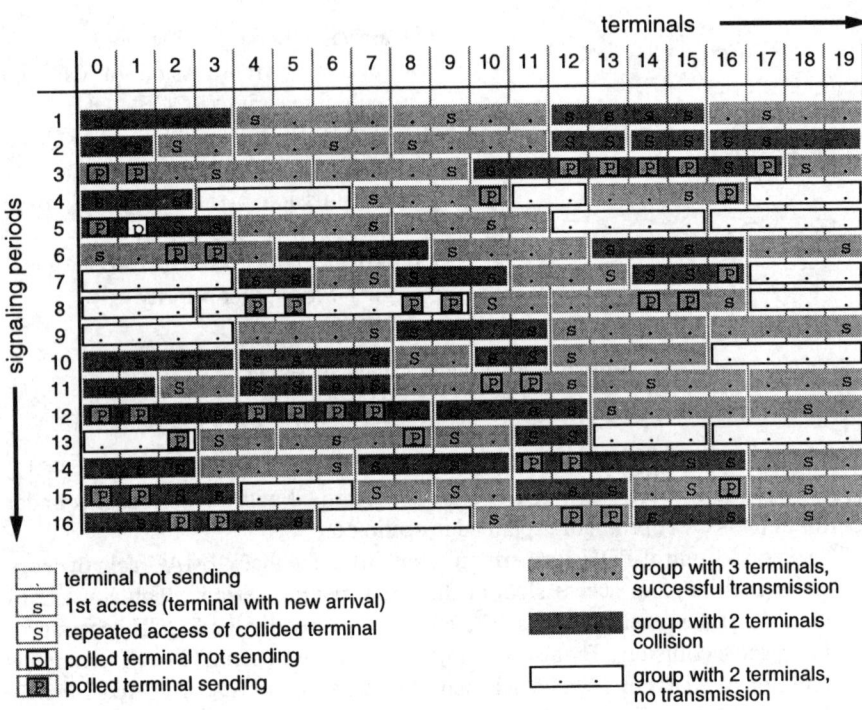

Figure 8 Example protocol sequence of probing algorithm over 16 signaling periods with 20 registered W-NT

priority service category (e.g. VBR) may modify the capacity requirements in such a way, that the W-NT is not able to wait for the next reserved slot in order to piggyback the newest capacity request on ATM cells. Instead it goes back to state REQUEST (transition (5)) and forces the retransmission of its capacity request over the RQCH.

Due to the urgency of the transmission of capacity requests, the random access of the RQCH protocol must not be optimized for throughput but for short delays. A critical item are the delayed feedbacks, because a second random access of the same W-NT is only useful, if the feedback of the first access has been evaluated before. Therefore, dedicated shortened signaling periods are useful to enable fast transmission of feedbacks.

The proposed collision resolution algorithm called probing algorithm (Petras and Krämling, 1997a; Petras and Krämling, 1997b) is a non blocking adaptive identifier splitting algorithm that takes advantage of the known number of stations in contention mode, since only registered W-NTs are allowed to request capacity. The algorithm combines the advantages of random access and polling. Thus, it is able to keep access delays below a predefined W-NT specific limit, as it is required for the support of real-time multimedia services in ATM networks.

At the beginning of each signaling period the probing algorithm divides the identifier space in a variable number t of consecutive intervals and assigns one short slot to each interval. The Period-Control-PDU signals the interval division to the W-NTs by transmitting the start identifier of each interval. Furthermore, it is possible to poll specific W-NTs in dedicated short slots. Such W-NTs are not allowed to send in random access slots.

The width of each interval is determined by considering the probability $p_{send,i}$ that W-NT i will send in a random access slot. After each access $p_{send,i}$ is corrected taking into account the knowledge gathered from the results of the last access. The example in Figure 8 illustrates how intervals with collided terminals are split into smaller intervals in order to resolve the collision (according to identifier splitting). Details about the algorithms for estimating and correcting $p_{send,i}$ are outside the scope of this paper but can be found in (Petras and Krämling, 1997a).

7 CONCLUSIONS

The protocol stack for a fixed ATM radio access network has been presented. The development of the protocol stack was based on the objective to implement an distributed ATM multiplexer with an ATM cell scheduler as in modern fixed ATM network nodes. Therefore, it is able to transparently extend ATM over a Point-to-Multipoint radio channel. Although wireless ATM systems with mobility like HIPERLAN type 2 are still under research (e.g., dynamic channel allocation in unlicensed frequency bands, handover due to mobility, error control on a mobile radio channel), the protocols for fixed W-ATM systems are well investigated. Bosch Telecom currently executes development activities to extend its narrowband Point-to-Multipoint product (Digital Multipoint System, DMS) by broadband ATM services.

REFERENCES

Agrawal, P., Hyden, E., Krzyzanowski, P., Mishra, P., Srivastava, M. and Trotter, J. (1996). SWAN: A Mobile Multimedia Wireless Network, *IEEE Personal Communications Magazine* 3(2): 18 – 33.

Ala-Laurila, J. and Awater, G. (1997). The Magic WAND - Wireless ATM Network Demonstrator System, *ACTS Mobile Summit '97*, Aalborg, Denmark.

BRAN Project (1997). Broadband Radio Access Networks (BRAN): Terms of References, *Technical report*, ETSI.

Ciotti, C. (1996). The AC006 MEDIAN Project, an Overview and State of the Art, *ACTS Mobile Communications Summit.*

COST 242 Management Committee (1996). Methods for the performance evaluation and design of broadband multiservice networks, Part I Traffic control, *Final Report Seminar version*, COST.

DAVIC (1997). DAVIC 1.3 Specification Part 8: Lower Layer Protocols and Physical Interfaces, *Technical Report Rev. 6.2*, DAVIC.

Dinis, M., Lagarto, V., Prögler, M. and Zubrzycki, J. (1997). SAMBA - A Step to bring MBS to the People, *ACTS Mobile Summit '97*, pp. 495–500.

Eng, K., Karol, M., Veeraraghavan, M., Ayanoglu, E., Woodworth, C., Pancha, P. and Valenzuela, R. (1995). A wireless broadband ad-hoc ATM local area network, *Wireless Networks* (1): 161–174.

Kist, H. and Petras, D. (1997). Service Strategy for VBR Services at an ATM Air Interface, *EPMCC'97*, Bonn, Germany.

Petras, D. and Krämling, A. (1997a). Fast Collision Resolution in Wireless ATM Networks, *2nd MATHMOD*, Vienna, Austria.

Petras, D. and Krämling, A. (1997b). Wireless ATM: Performance Evaluation of a DSA++ MAC Protocol with Fast Collision Resolution by a Probing Algorithm, *Int. J. of Wireless Information Networks* 4(4).

Porter, J. and Hopper, A. (1995). An Overview of the ORL Wireless ATM System, *IEEE ATM Workshop*, Washington, DC, pp. 18 – 33.

Radimirsch, M. (1997). ATMmobil Overview, *ACTS Wireless-ATM Workshop 1997*, Brussels, Belgium.

Raychaudhuri, D., French, L., Siracusa, R., Biswas, S., Yuan, R., Narasimhan, P. and Johnston, C. (1997). WATMnet: A Prototype Wireless ATM System for Multimedia Personal Communication, *IEEE J. Selected Areas in Communications* 15(1): 83–95.

Walke, B., Petras, D. and Plassmann, D. (1996). Wireless ATM: Air Interface and Network Protocols of the Mobile Broadband System, *IEEE Personal Communications Magazine* 3(4): 50–56.

Wireless ATM Group (1997). Baseline Text for Wireless ATM Specifications, *Technical report*, ATM Forum.

12

Signalling for Handover in a Broadband Cellular ATM System

K. Keil, H. Bakker, W. Schödl, M. Litzenburger

Alcatel, Corporate Research Center, Stuttgart, Germany
e-mail: {kkeil\hbakker\wschoedl\mlitzenb}@rcs.sel.de

Abstract
This paper presents handover procedures for the Broadband Cellular ATM Access (C-ATM) system which shall provide wireless ATM access for mobile users. Beside the description of the architecture and protocol stack of the system this paper describes message sequence charts for two different handover types: radio and network handover. This differentiation allows most of the handovers to be performed seamlessly, i.e., without loss of ATM cells. This feature is especially important for real time applications such as video.

Keywords
Wireless ATM, Radio Handover, Network Handover, Message Sequence Charts, Connection Admission Control

1   INTRODUCTION

In April 1996 the German Federal Ministry of Education, Science, Research and Technology has launched a project called 'Broadband Mobile Communication for Multimedia based on ATM (ATMmobil)'. Within this ATMmobil project several German companies and universities cooperate in the area of a wireless ATM cell transport to end users. Four different system concepts are explored and prototyped within the frame of ATMmobil, one of these is called "Broadband Cellular ATM Access - C-ATM" which aims at providing wireless ATM access in selected areas enabling mobility features within the coverage of the radio cells. After a short system description of the system architecture the handover procedures are described in detail.

Broadband Communications   P. Kühn & R. Ulrich (Eds.)
© 1998 IFIP. Published by Chapman & Hall

2   SYSTEM ARCHITECTURE

The system architecture of the C-ATM system is depicted in Figure 1. It is divided in three parts: mobile terminals, basestation subsystem and fixed network part.

The C-ATM system distinguishes between two types of mobile terminals: a standard ATM terminal connected to a Mobile Terminal Adapter (MTA) and a terminal with radio transceiver, i.e, the MTA functionality is integrated in the ATM terminal. The MTA comprises the radio part and terminates all mobility related protocols, i.e., it hides the mobility aspects from the standard ATM terminal. Both terminal types support the UNI signalling for set up of transparent ATM connections over the air interface.

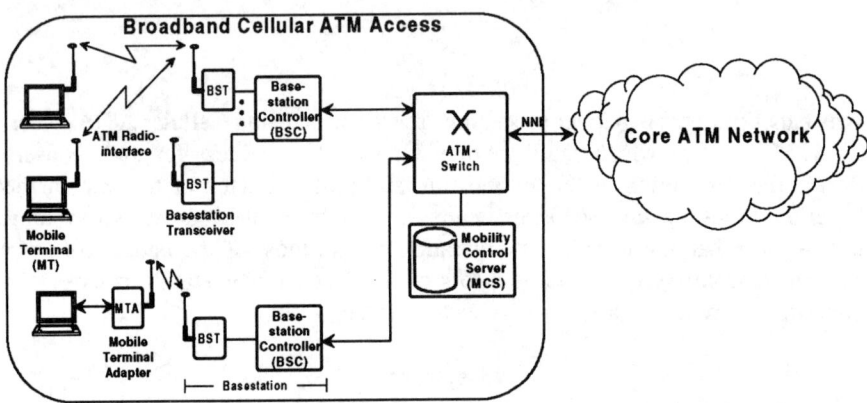

Figure 1: Architecture of the Broadband Cellular ATM access system.

The basestation subsystem consists of Basestation Transceivers (BST) and Basestation Controllers (BSC). A radio cell is set up by a BST which includes baseband processing and the radio frequency part. Several BSTs are controlled by one BSC, i.e. the BSC acts as a multiplexer in the upstream direction towards the fixed network and as a demulitplexer towards the terminal in the downstream direction. As the BST is only working on the physical layer, all mobility related protocols are implemented in the BSC.

The C-ATM system will operate in the 5 GHz frequency range with a bandwidth of about 25 MHz per radio cell in accordance with the (initial) HIPERLAN Type 2 specifications [1]. The maximum peak EIRP (Effective Isotropic Radiated Power) will be 1 W to cover radio cells with a radius of up to 50 m, which leads to a picocellular environment. Furthermore, it is intended to use OFDM (Orthogonal Frequency Division Multiplexing) as modulation scheme.

In our envisaged C-ATM system the fixed network part comprises as central part an ATM switch which connects the basestation subsystem via an NNI interface to the core ATM network. Connections to other C-ATM islands are not considered here. The mobility functions such as Location Management, extended Connection Admission Control (CAC), and Handover management are not handled by the ATM switch, thus a Mobility Control Server (MCS), which is connected to the switch via an ATM interface, is used to support the ATM switch. The introduction of a MCS requires only minor modifications within the Call Control (CC) unit of a standard ATM switch. Both the ATM switch and the MCS are acting towards the core ATM network and towards the radio access system as a mobility enhanced ATM switch. Within a demonstrator platform which will be prototyped by the C-ATM project the CC unit will also be implemented within the MCS as a second task. Following this approach the modifications of the CC which become necessary to support mobility functions (e.g. handover), can easily be implemented. Thus, all UNI messages of the CC protocol (setup, connect, release) are terminated in the MCS and not in the ATM cell site switch.

Figure 2 shows the user plane protocol stack of the C-ATM system. Providing wireless access the user of the C-ATM system should not be aware of the transmission of ATM cells over the air interface, i.e., the transmission should be fully transparent without any degradation compared to fixed access.

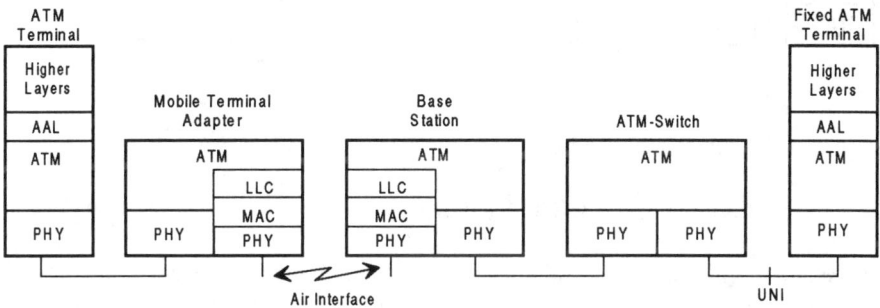

Figure 2: Protocol stack user plane.

To hide the specific air interface characteristics from the ATM layer, an additional Data Link Control (DLC) layer has been introduced. The DLC layer consists of a Media Access Control (MAC) and a Logical Link Control (LLC). While the MAC protocol enables the statistical multiplexing of multiple mobile terminals on the air interface, the LLC protocol improves the ATM cell error rate by a controlled retransmission of corrupted ATM cells. The retransmission will become necessary as soon as the Forward Error Correction - (FEC-) mechanism implemented in the physical layers fails. All other layers are unmodified, enabling a transparent end-to-end ATM Adaptation Layer (AAL) connection between a wireless and a second fixed/wireless ATM terminal. The protocol stack for the control plane is depicted in Figure 3. The functions for Mobility Control

(MC) and Call Control (CC) are handled by the CC/MC layer of the MTA, the BSC and the ATM switch / Mobility Control Server. Within the ATM terminal and the fixed ATM terminal representing the fixed ATM network part, no MC is implemented. Mobility Control messages are exchanges between the MTA / BSC and the Mobility Control Server, which controls the mobility functions mentioned above.

Below the CC/MC protocol layer in a new network layer protocol is introduced, the so-called Signalling Network Layer (SNL) [2]. The main function of the SNL protocol is to deliver signalling messages to the corresponding signalling end point, even to a mobile end point. The SNL hides the fact that the mobile terminals move within the access network from the application layer, i.e. CC/MC.

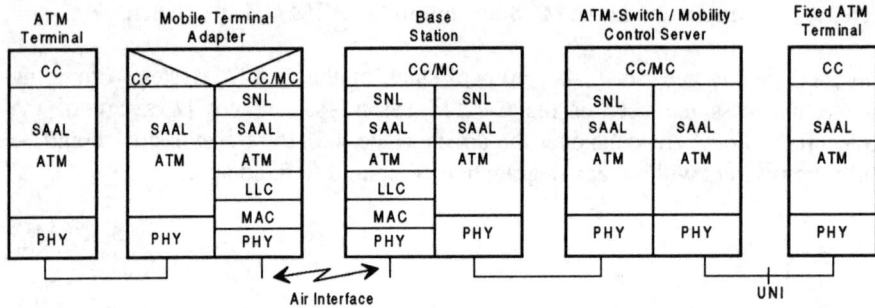

Figure 3: Protocol stack control plane.

2.1   Handover in a C - ATM System

A key function within any mobile cellular communication system is handover, which allows to guarantee a certain requested quality of service for the connections of a mobile user while moving. A handover may also become necessary for reasons of traffic load redistribution among radio cells. This section sketches the mechanisms and signalling procedures for handover in a wireless cellular ATM network.

Most of these procedures are included in the baseline text for wireless ATM specifications of the Wireless ATM working group of the ATM Forum [6].

2.2   Definitions and Functional Description

Based on the architecture of the C-ATM system (Figure 1), we distinguish between two types of handover which are depicted in Figure 4 [3].

- Radio handover: The mobile terminal moves from the coverage area of one BST to the coverage area of another BST, but stays within the area of one BSC.
- Network handover: The mobile terminal changes its location from the area of one BSC to the area of another BSC.

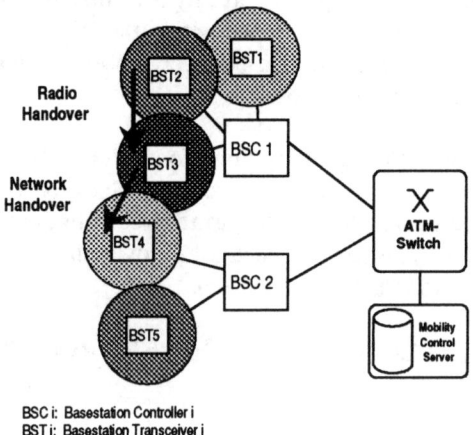

BSC i: Basestation Controller i
BST j: Basestation Transceiver j

Figure 4: C-ATM handover types.

Since radio handovers might occur quite frequently in a pico-cellular environment, it is important that a radio handover can be performed seamlessly, i.e., without loss of data and with minimum additional delay. The ATM switch is not involved in the handover execution. The synchronisation of the ATM cells is performed within the DLC layer and is described in [3].

For a network handover, on the other hand, the output port of the ATM switch will be changed in order to handover a mobile terminal's current connections to another BSC, i.e., the path has to be modified. In contrast to a path extension scheme (cf., e.g., [4]), path rerouting requires no switching functionality within the BSCs. But with the ATM switches currently available, ATM cells might be lost, because it is not possible to synchronise ATM cell streams arriving at different switch ports. Therefore, the network handover will not be seamless.

Concepts for a seamless network handover [7] are currently not discussed in detail within the Wireless ATM working group of the ATM Forum.

The C-ATM handover procedures which combine principles of DECT and GSM have the following properties:

- The handover will be mobile-initiated, i.e., the mobile terminal decides when to request a handover. Together with the request, it shall provide a sorted list of BSC/BST identifiers and radio-link quality measurements of the neighbouring radio-cells. This request is forwarded to the MCS.

- The decision, whether to accept the handover request and which target radio cell to choose, will be performed in the MCS, taking into account the link qualities of the possible target cells and the current traffic load distribution. Therefore, the MCS has to monitor the traffic load in all attached radio cells. An extended Connection Admission Control (CAC) has to be implemented which checks whether the required capacity is available in the new radio cell.

The radio handover execution takes place within the BSC of the two radio-cells involved, while for the execution of a network handover additionally the ATM switch is required for the rerouting of cells

In the following sections the signalling flows for both C-ATM handover types are described. The algorithms to obtain measurement values for the radio-link quality of the current connection and the neighbouring radio cells are not included. These algorithms are necessary to decide, when a handover should be requested which have to be implemented in the mobile terminal.

3   SIGNALLING FLOW FOR HANDOVER PROCEDURES

3.1   Definitions and Assumption

In the following we assume the Mobile Terminal is composed of a standard terminal and a MTA for termination of the radio link - ranging from the physical layer up to the MC function (see Figure 3). So the MTA is the endpoint for all the mobility functions.

Within our demonstration platform we will use fixed separate mobility control channels established between MTA and BSC, MTA and MCS and furthermore between BSC and MSC for the exchange of mobility control message between those entities. Within the C-ATM system concept a more advanced solution based on SNL principles is envisaged [5].

The mobility control messages for any handover procedure to and from the MTA-MC instance are exchanged via the current active radio link (backward handover).

The MTA-MC instance is informed about the session (i.e. number of virtual connections, VPI/VCI and capacity per virtual connection) currently active on the terminal either by the CC resident within the terminal or by the mobility control (MC) instance of the MCS.

Furthermore, as addressed above, two types of handover have been defined. The decision which handover type shall be executed will be made by the MCS, taking into account the availability of capacity within the proposed target radio cells. In cases, where both types are possible, priority is given to radio handover versus network handover, because it performs a seamless transition to another radio cell. The decision to initiate a network handover is based on two criteria:

• The list received from the mobile station does not include an adjacent radio cell belonging to the same BSC, or
• there is no capacity available in an adjacent radio cell belonging to the same BSC.

Concerning the availability of capacity, the C -ATM project has classified handover procedures as follows:

• A normal handover procedure, e.g. normal radio / network handover procedure if enough capacity for the requested handover is available in one or more target cells.
• A restricted handover procedure, e.g. restricted radio / network handover procedure if in none of the proposed target cells the required capacity is available for the handover of a session with multiple active virtual connections. Then the MCS checks if a release of one or more virtual connection(s) of the current session (i.e. reduction of capacity requirement) makes a handover possible. If this is the case the MCS then initiates via the CC Function and based on a standard signalling procedure the release of the defined virtual connection(s) before starting the handover procedure.

In case the required capacity is not available within the proposed target radio cells (neither for the requested handover with only one active virtual connection, nor in case of restriction for a session with multiple active virtual connection) the MCS rejects the handover request from the MTA.

3.2 Handover Procedures

A software process within the MTA responsible for monitoring the transmission quality over the radio link will initiate a handover procedure by informing the MTA-MC instance if the received signal strength falls short of a defined threshold limit. The MTA-MC instance will inform its counterpart in the Mobility Control Server (MCS) by sending a HO_Request_Indication message (Figure 5), which includes among others the following parameters: a sorted list with the BSC/BST combinations. Furthermore the MSC-MC starts the timer T1, which is introduced to monitor the reaction of the MCS-MC instance and the stability of the radio link. In case of time-out of T1, the MTA-MC instance shall start a forward handover procedure.

On receipt of the HO_Request_Indication message the MCS-MC instance assigns a process identity (HO_ID) and checks, based on the BSC/BST combinations received, whether enough capacity for a handover is available in the proposed target cells. Depending on the result of this extended Connection Admission Control (CAC), the MSC-MC instance will define the handover type and the condition for the handover procedure (i.e. normal or restricted) or will reject the handover request if the required bandwidth capacity is not available.

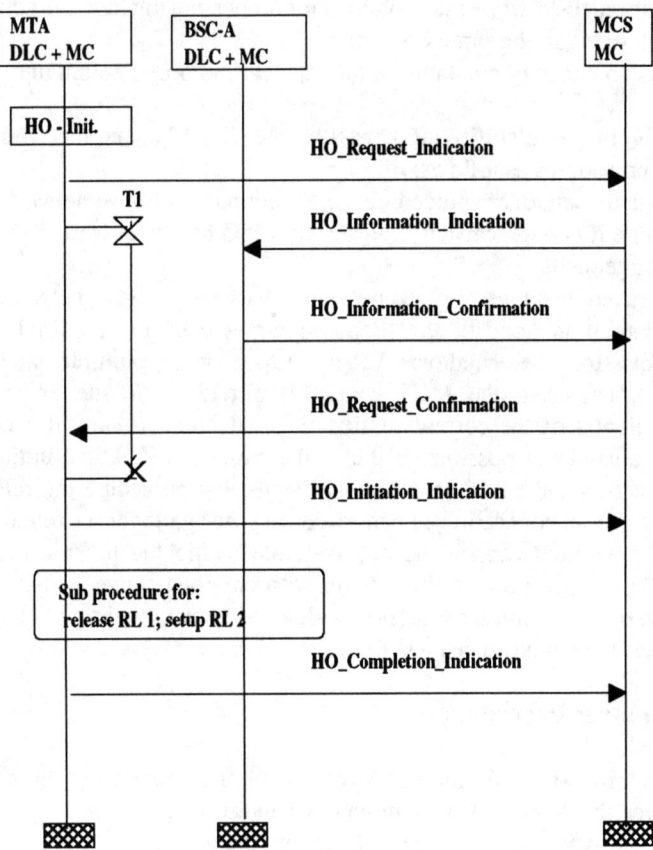

Figure 5: Message Sequence Chart for a normal radio handover procedure.

In the section below, the message sequence flow for each particular case is described in detail.

3.2.1   Normal Radio Handover Procedure

In this case the adjacent radio cell belonging to the same BSC provides enough capacity and the MCS-MC instance defines the corresponding BSC_ID / BST_ID combination for the radio handover and sends this combination with a HO_Information_Indication message to the MC instance of the BSC.

On receipt of the HO_Information_Indication message the BSC-MC instance informs the DLC process to prepare itself for a handover to a transceiver defined by the received parameters. After completion of the preparation phase the BSC-MC instance sends the HO_Information_Confirmation message to the MSC-MC instance to indicate that the BSC-MC and BSC-DLC instances are prepared for the handover.

To inform the MTA-MC about the target cell and the handover type, the MSC-MC instance sends the HO_Request_Confirmation message to the MTA-MC, which includes, among others, an identifier for the handover type (HO-Flag) and the identifier of the new selected radio cell.

On receipt of the confirmation the MTA-MC instance clears timer T1 and checks the selected handover type. For radio handover, the MTA-MC instance sends a HO_Initiation_Indication message to the MSC-MC instance to indicate that the change of the radio links will be started. Then the MTA stops the transmission of ATM cells over the current radio link and establishes a new radio link to the new BST.

If the new radio link is set up, both DLC instances restart the transmission of ATM cells send a confirmation about the completion to their respective MC instances.

On receipt of the confirmation message from the DLC instance, the MTA-MC instance sends a HO_Completion_Indication message to the MCS-MC instance and terminates the handover procedure within the MTA.

On receipt of the HO_Completion_Indication message the MCS-MC instance will release the assigned HO_ID, and will update the database of the MCS, e.g., mobile terminal located in a new location area.

3.3.2 Normal Network Handover Procedure

In cases, where only BSC/BST combinations from a neighbour BSC are proposed, or a radio handover is not possible due to capacity limitations but enough capacity is available in radio cells from a neighbour BSC, the MCS-MC instance decides to perform a network handover.

Compared to the radio handover procedure, additional messages are necessary for a network handover due to the fact that the involved radio cells belong to two different BSCs. In order to inform the MC instances of both BSCs about the planned handover, the MCS-MC instance sends a HO_Release_Indication message to BSC-A and a HO_Information_Indication message to BSC-B (see Figure 6).

As described in the message sequence flow for the radio handover procedure (Figure 5), the MC instances of BCS-A and BSC-B inform and request their respective DLC instances to get prepared for the handover procedure. Then both MC instances sends a relevant acknowlegde message to the MCS-MC instance.

On receipt of the HO_Request_Confirmation message the MTA-MC checks the HO_Flag parameter and handover type, the MTA-MC instance requests from its MTA_DLC to stop transmission of ATM cells.

The MTA-MC then sends the HO_Initiation_Request message to the MCS-MC instance indicating to be ready for the rerouting procedure.

On receipt of the HO_Initiation _Request message the MCS-MC instance requests from the Bearer Control (BC) instance inside the MCS, to reroute the path through the ATM switch for the involved virtual connection(s) and the associated signalling channel by sending a R+S_Request message.

The BC instance modifies the routing tables of the ATM switch so that the ATM cells of the defined virtual connection(s) will be routed to the port where the new BSC is connected. The new BSC, already prepared by the HO_Informaton_Indication message, will store the ATM cells until the new radio link to the MTA has been established.

After termination of the rerouting procedure the BC instance sends the R+S_Confirmation message to the MSC-MC instance as an indication.

On receipt of the R+S_Confirmation message the MSC-MC instance informs the MTA-MC instance by means of the HO_Initiation_Confirmation message that the path through the switch for user data and associated signalling channel has been changed and the change for the radio link can be started. The MTA-MC instance then request from the MTA-DLC instance to change the radio links. The release of the radio link L1 and the setup of the new radio link L2 is achieved via exchange of DLC protocol messages between the MTA-DLC and the DLC instances of BSC-A and BSC-B.

After the change of the radio links is completed the MTA-DLC instance sends a confirm message to the MTA-MC instance.

To inform the MSC-MC instance about the end of the handover procedure within the MTA, the MTA-MC instance then sends the HO_Completion_Indication message. This message is transferred in upstream direction over the new established radio link using the Basestation proprietary mobility control channel, while for the mobility channel in downstream direction to the MTA a rerouting is still required.

Therefore, on receipt of the HO_Completion_Indication message the MCS-MC instance contacts again the BC instance and request by the R+S_request message to change for the downstream mobility channel the path through the ATM switch. The BC instance informs the MSC-MC instance about the termination of the rerouting procedure by sending the R+S_Confirmation message. Using the new downstream path, the MSC-MC instance sends the HO_Completion_Confirmation message to the MTA-MC instance to indicate the termination of the handover procedure on the MSC side.

The MSC-MC instance then releases the assigned HO_ID and updates its own database and other databases relevant for mobility management inside the MCS.

3.2.3   Restricted Handover Procedure

If in none of the proposed target cells the required capacity is available for the handover request from a session with multiple virtual connections, the MSC-MC instance checks and decides which virtual connections shall be released, before a handover procedure can be performed. Thus, for both handover types, the MSC-MC instance will contact the CC instance and request the release of the defined virtual connections, before the handover procedure can proceed.

Based on a standard signalling procedure the CC instance performs the release of the defined virtual connection(s) and confirms the release to the MSC -MC instance.

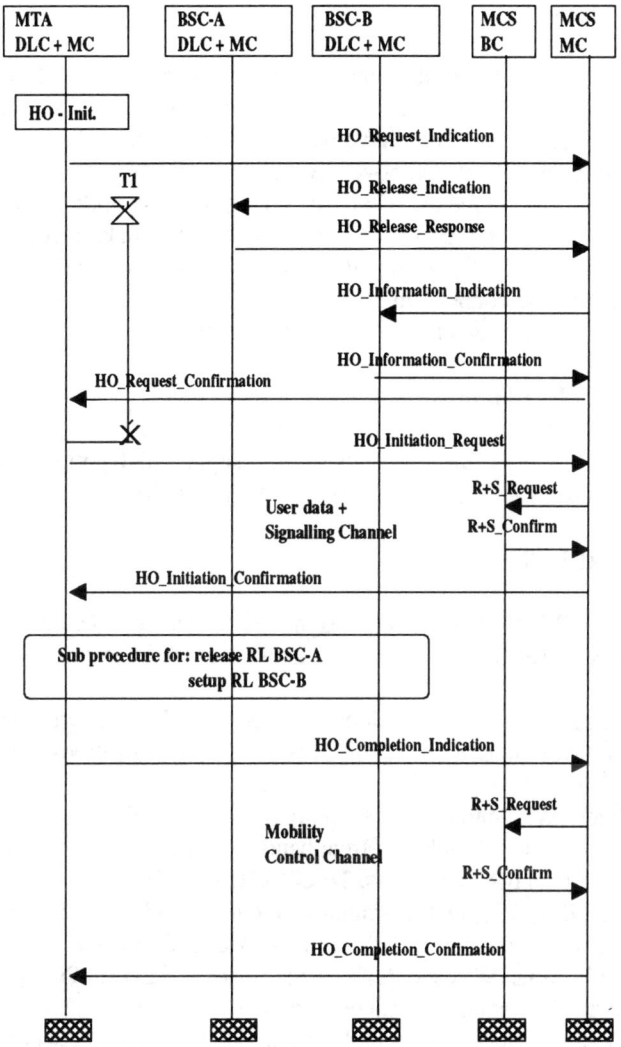

Figure 6: Message Sequence Chart for a normal network handover procedure.

On receipt of the confirmation message, the MCS-MC instance defines the BSC/BST combination for the target cell and the handover type, which determines the further operations for the handover procedure.

3.2.4 *Aborted Handover procedure*

In cases where not enough capacity is available in any of the proposed target radio cells, the MSC-MC instance rejects the HO_Request_Indication message by means of the HO_Request_Rejection message which leads the MTA-MC instance to abort the handover procedure.

4   SUMMARY

This contribution describes the handover procedures which will be used in the Broadband Cellular ATM Access project. After outlining the system architecture two different types of handover have been introduced: the radio handover and the network handover. These handover types are reflected in the corresponding simplified message sequence charts showing the network entities involved. Further work has to be performed to extend the standard ATM Connection Admission Control (CAC). While in fixed ATM networks the CAC function is only used during the connection setup, in C-ATM systems the CAC function has also to be used during each handover to a new radio cell.

5   ACKNOWLEDGEMENT

The work presented in this paper has been supported by the German Federal Ministry of Education, Science, Research, and Technology.

6   REFERENCES

[1] ETSI RES10 Technical Report "High Performance Radio Local Area Networks (HIPERLANs), Requirements and Architecture," Draft, Sophia Antipolis, 1996.
[2] Mitts, H.; Luijten, G.; Korinthios, J.A.; Nelson, J.: Connectionless signalling network layer in UMTS, IEEE Personal Communications Magazine, June 1996.
[3] H. Bakker, W. Schödl: ATM up to the Mobile Terminal- Impact on the Design of a Cellular Broadband System, ISS'97: XVI World Telecommunications Congress, Toronto, 21.09. - 26.09.97, pp. 379 - 385
[4] B. Walke, D. Petras, D. Plassmann: "Wireless ATM: Air Interface and Network Protocols of the Mobile Broadband System," IEEE Personal Communications: Wireless ATM, Vol. 3, No. 4, August 1996, pp. 50 - 56.
[5] H. Bakker, W. Schödl, M. Litzenburger, K. Degenhard: A Channel Concept for Signalling Messages in a Wireless ATM Access Network with Server Based Control, ACTS Mobile Summit 1997, Aalborg, Denmark, October 1997, pp. 441 - 446
[6] ATM Forum: Wireless ATM working group , Baseline Text for Wireless ATM specifications, ATM Forum BTD-WATM-01.05, December 1997
[7] H. Mitts, H. Hansén, J. Immonen: Lossless handover for wireless ATM, Mobicom'96, Proceedings, pp. 85-96, New York, U.S., 10 -12 Nov 96

13

A New Channel Reservation and Prediction Concept in Cellular Communication Systems

Selma Boumerdassi
Université de Versailles Saint-Quentin, Laboratoire PRiSM
45, avenue des Etats-Unis. 78035 Versailles Cedex France
Phone: +33 1 39 25 40 76   Fax: +33 1 39 25 40 57
Email: Selma.Boumerdassi@prism.uvsq.fr

Abstract

In future wireless networks, the determination of the amount of resources that a base station (cell) must reserve to maintain a certain call dropping probability is likely to become a very important issue. We present in this paper a new algorithm for channel assignment in wireless networks called Predictive Reservation Policy (PRP) based on FCA concept. This new dynamic channel reservation scheme, adapts the number of guard channels in each cell to: 1) the current number of ongoing calls in neighbouring cells and 2) on the anticipation of the future localisation of users.

The proposed scheme for radio channel allocation reduces handoff call blocking probability substantially at the expense of slight increases in new call blocking probability by giving resource access priority to handoff calls over new calls. PRP improves classical FCA and FCA with 2 or 3 reserved channel for Handoff exclusively.

Keywords

Adaptive bandwidth resource management, prediction, channel reservation, transition probability

Broadband Communications   P. Kühn & R. Ulrich (Eds.)
© 1998 IFIP. Published by Chapman & Hall

1. INTRODUCTION

The bandwidth in a wireless network is perhaps the most precious and scare resource of the entire communication system. This resource should be used in the most efficient manner. A base station sometimes may need to reserve resources, even if this means denying access to a mobile terminal requesting admission to the network, in order to keep enough resources to support active users currently outside of its coverage area, but who may soon emigrate to its cell. Base stations must maintain a balance between the two conflicting requirements: 1) maintaining maximum resource (bandwidth) utilisation and 2) reserving enough bandwidth resources so that the maximum rate of unsuccessful incoming handoffs (due to insufficient resources) is kept below an acceptable level. The probability of unsuccessful handoff can be established in terms of quality-of-service (QoS) metric, e.g. call dropping probability, that the network agrees to maintain [Lev 97].

In this article, we propose a new approach (Predictive Reservation Policy - PRP [Bou 97]) based on FCA [Cho 82]. This scheme is based on the assumption that a mobile does not move randomly. This is due to the existence of roads, highways, dead-ends, shops, etc. On the road, a car is going along in a determined direction. The presence of a park implies that there are only pedestrians.

This paper shows that if some cell to cell handoff probabilities are known even approximately, it is possible to use them in order to manage the bandwidth efficiently. Indeed, if destinations of mobiles are approximately known, it may be interesting to reserve resources which will be freed when the flow will decrease (end of communication for example).

This paper is divided into the following sections. Section 2 describes the new scheme PRP. Simulation results are presented and discussed in Section 3. The conclusion is given in section 4.

2. DESCRIPTION

Future personal communication networks (PCN's) will employ microcells and picocells to support higher capacity [Yu 97], thus increasing the frequency of call handoffs, and so the network call processor becomes a bottleneck. A possible solution is the virtual connection tree [Aca 94] [Nag 94], for which the geographical areas are divided into zones named "cell clusters". Each cluster is managed by an ATM switch (see Fig. 1). The treatment of handoff within a cluster is local. This approach decentralises the management and prevents a bottleneck at the network call processor. To implement PRP we can use the virtual tree concept.

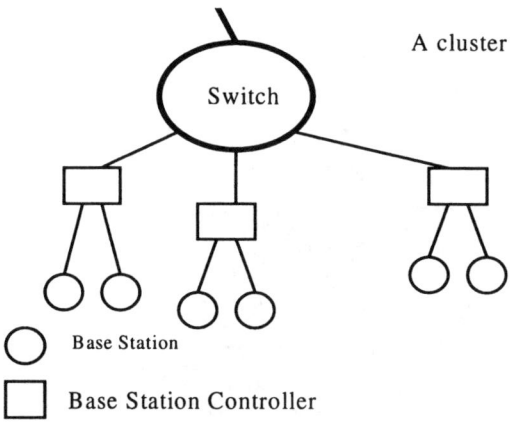

Figure 1 The adaptive architecture for PRP scheme.

PRP dynamically reserves radio resources under certain conditions of traffic, taking into account the actual situation in the covered region (by the base stations). The policy is based on two fundamental and linked elements: reservation and prediction. In this concept, communications in progress have a higher priority than new calls.

2.1 Reservation scheme

We consider a wireless network where $i \in I$ denotes a base station (or a cell), and I is the set of the all base stations in the network. $N(i,j)$ denotes the neighbour number j of cell i ($j \in [1..6]$ in the case of an hexagonal cell). Let $M_i(t)$ denote the number of occupied channel in cell i at time t.

In a base station, when the number $M_i(t)$ of occupied channel reaches a threshold K or a multiple of K (see Fig. 2), cell i reserves one channel in all its neighbours $N(i,j)$. If $N(i,j)$ has an available channel the reservation takes place immediately; else, the reservation has to wait for a free channel. A reserved channel represents a potential arrival of communication. In this case the cell considers that the channel is occupied by a virtual (or a potential) mobile.

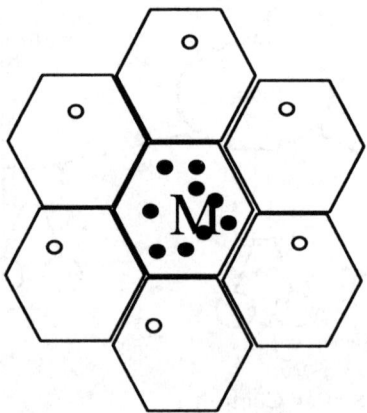

Example : K = 10

$M_i(t) = 10 =>$ 1st reservation (or ask for reservation) in the neighbours of i.
$M_i(t) = 20 =>$ 2nd reservation (or ask for reservation) in the neighbours of i.

Also, the reservation is cancelled when the number of occupied channels leaves threshold.

Example : K = 10

When $M_i(t)$ gets down from 20 to 19 => the 2nd reservation (or ask of reservation) is cancelled.

Figure 2 Principle of reservation.

The channels which are blocked as mentioned in Fig. 2 are reserved for handoffs. It is important to notice that a channel is not only reserved for the cell which reserves; it can also be used by the handoffs from all the neighbouring cells (e.g., a handoff can come from any neighbour). In case of overload (or a state close to overload), the system does not accept incoming calls, it only processes calls in progress.

2.2 Prediction

In a wireless network, call arrival, call departure, and call handoff rates depend on the presence of roads, houses, and other topological features, ... Thus, mobiles do not randomly move: most of the time their trajectories are foreseeable. For example, cars run along roads and are not usually present in dead-ends. In such a context, it would be very interesting to reserve resources to optimise the global performance of the system (i.e. call in progress should have a higher access priority to resources). In the PRP approach, each cell of the system are associated probabilities of transition to its neighbours. Measurements could be performed in order to estimate these probabilities. The reservation threshold must take into account the traffic characteristics.

2.3 Algorithms

PRP can be resumed by the followings algorithms :

Reservation (Uniform traffic case)
If a multiple of the threshold is reached
        Then   A reservation is set in each neighbouring cell

Reservation (Non uniform traffic case)
For each neighbour j
        If a multiple of the threshold is reached
                Then   A reservation is set in neighbour $N(i,j)i$

Getting in cell
If the number of free channels > 0
        Then   The communication is established
                Reservation
        Else   If New Call
                Then   The communication is blocked
                Else If the number of reserved channels > 0
                        Then   The communication is established
                                Reservation
                      Else   The Hand Off is dropping

Unreservation (Uniform traffic case)
If a multiple of the threshold is lost
        Then   A reservation is cancelled in every neighbouring cell

Unreservation (Non uniform traffic case)
For every neighbour j
If a multiple of the threshold is lost
        Then   A reservation is cancelled in neighbour $N(i,j)$

Getting out cell
Unreservation
End of the communication

3. SIMULATION AND ANALYSIS

Two cases were studied:.
3.1 Uniform case

The local topology is not taken into account. It is considered that mobiles enter all the neighbour cells with the same probability (see Fig. 3). In this case, when a threshold is reached in a given cell the base station asks for reservation in all the surrounding cells.

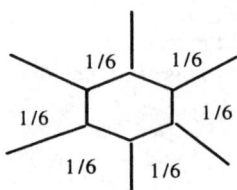

Figure 3 Probability for the uniform case.

3.2 Non uniform case:

In this more realistic case, the transition probabilities depend on the local topology (Fig. 4).Thus, the reservation threshold varies inversely with the probability of visit to the neighbouring cells. The interval [0,1] is divided into three intervals: at each one a reservation threshold (K_0, K_1, K_2) is associated to each one of them.

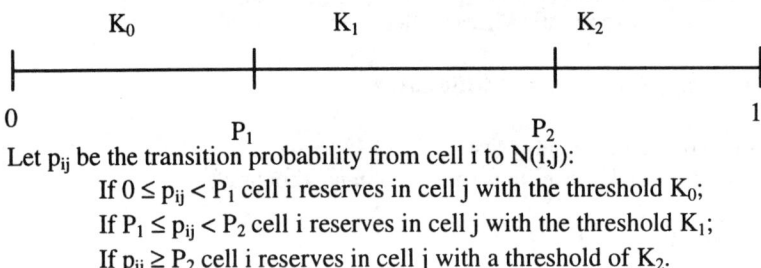

Let p_{ij} be the transition probability from cell i to $N(i,j)$:
If $0 \leq p_{ij} < P_1$ cell i reserves in cell j with the threshold K_0;
If $P_1 \leq p_{ij} < P_2$ cell i reserves in cell j with the threshold K_1;
If $p_{ij} \geq P_2$ cell i reserves in cell j with a threshold of K_2.

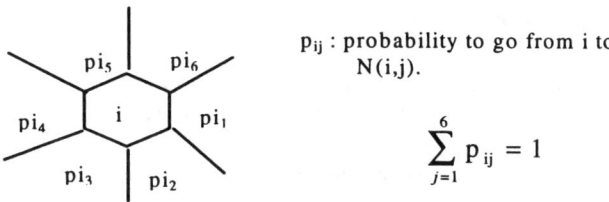

p_{ij} : probability to go from i to N(i,j).

$$\sum_{j=1}^{6} p_{ij} = 1$$

Figure 4 Probability for the non uniform case.

3.3 Simulation model

To simulate the system a cell cluster with 30 hexagonal has been considered. As shown in Fig. 5 edge effects on handoffs at the cluster boundary are handed by wrapping them around thereby assuming that arrival rates and handoff departure rates from cluster to cluster are equivalent. The call duration is exponentially distributed with a mean value of 120 seconds and the time a mobile spends in a cell is also assumed to be exponentially distributed with a mean value of 50 seconds. In our model, the new calls arrival distribution is assumed to be Poisson with parameter λ_i.(for cell i). To obtain a non uniform traffic on the cell cluster different values of the parameters λ_i were chosen. Let λ be the global cell cluster rate we choose:

$\lambda = \sum \lambda_i$ with $\lambda_i = \lambda_j$ $\Rightarrow$ $i = j$

- Each base station manages C=24 channels

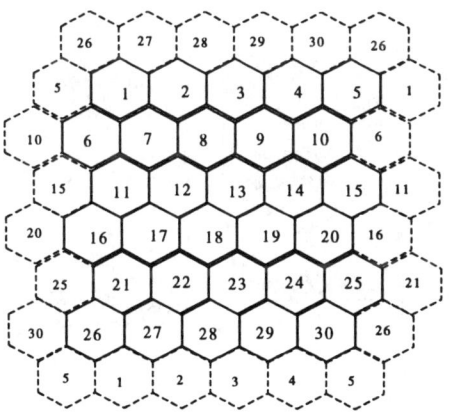

Figure 5 Cluster of 30 cells.

3.4 Results and analysis

In each simulation run, we generate about 2 million calls for the cluster. The result (grade of service and blocking probabilities) concerns all the clusters. The relative confidence intervals for the blocking probabilities were calculated, they are around 10%. The simulation has been written in QNAP2.

The grade of service used is :

GoS = 10* *Hand off failure probability* + *New call block probability.*

3.4.1 Results: uniform case

Several thresholds where tested and Figure 6 shows that the lower the reservation threshold the lower the hand off failure probability. Moreover, when the load is low enough, the reservation has no effect on performance. Inversely, with a reservation under a high load, the number of handoff droppings decreases significantly. Figure 7 shows the influence of reservation on new call blocks: the more reservations are supported, the larger the number of refused incoming calls. Table 1 represents the Grade of Service for different thresholds as a function of the traffic load (Erlangs). PRP results are compared with classical FCA, and this leads to good performance from threshold 14. The best results are obtained with threshold 16 (see Fig. 8, 9 and 10). When the input load is quite low, the reservation algorithm has no effect on the Grade of Service, because few cells are overloaded. When the input load increases, the thresholds are often reached and reservation are consequently efficient.

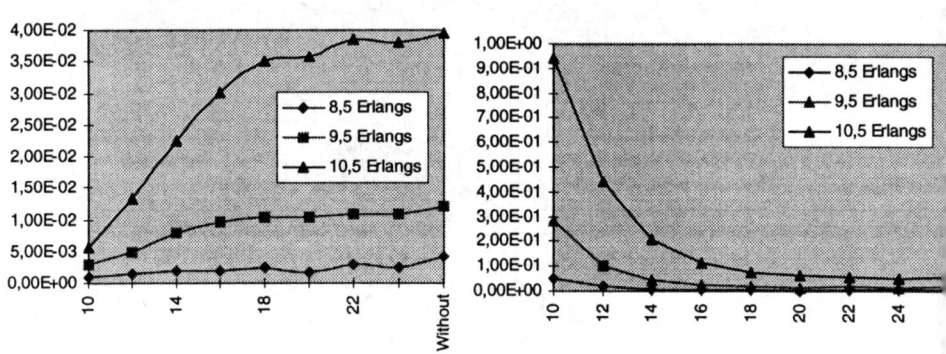

Figure 6 Handoff failure probability for different thresholds.

Figure 7 New call block probability for different thresholds.

Erlangs	Threshold 10	12	14	16	18	20	22	24	Without
7,5	7,87E-03	3,14E-03	4,06E-03	5,41E-03	5,15E-03	5,28E-03	5,52E-03	6,02E-03	4,28E-03
8,5	6,07E-02	3,13E-02	2,71E-02	2,41E-02	2,97E-02	2,04E-02	3,43E-02	2,94E-02	4,49E-02
9,5	3,14E-01	1,49E-01	1,23E-01	1,22E-01	1,26E-01	1,18E-01	1,27E-01	1,25E-01	1,39E-01
10,5	9,99E-01	5,76E-01	4,31E-01	4,17E-01	4,25E-01	4,18E-01	4,41E-01	4,32E-01	4,51E-01
11,5	2,44E+0	1,56E+0	1,16E+0	1,08E+0	1,11E+0	1,18E+0	1,18E+0	1,19E+0	1,22E+0

Table 1 Grade of Service in a uniform traffic case.

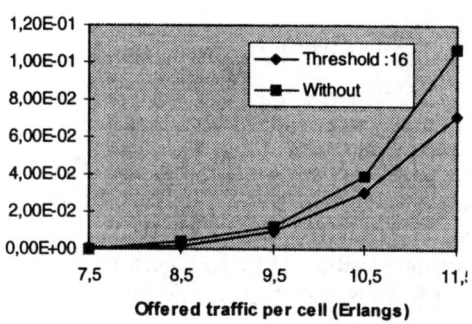

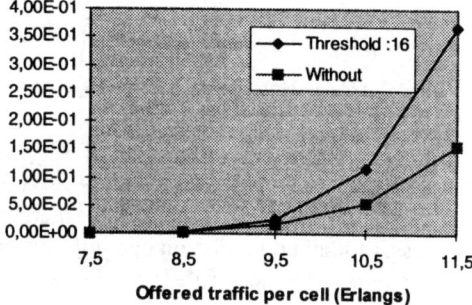

Figure 8 Handoff failure probability with a threshold of 16

Figure 9 New call block probability with a threshold of 16

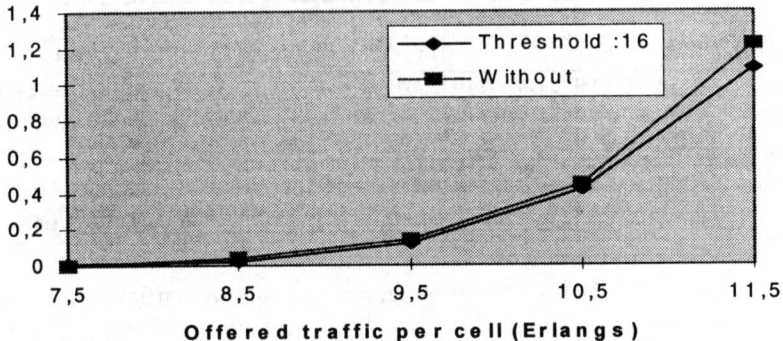

Figure 10 Grade of Service with a threshold of 16

3.4.2 Results: non uniform case

Figure 11 represents the Grade of Service as a function of the traffic load (expressed in Erlangs). PRP reservation results are compared to those obtained using FCA (which corresponds to the case where there is no reservation for handoff traffic). PRP reservation and prediction can improve significantly traditional FCA. Figure 12 and 13 show the influence of PRP on HandOff probability and New Call probability. PRP has to be tested with real data.

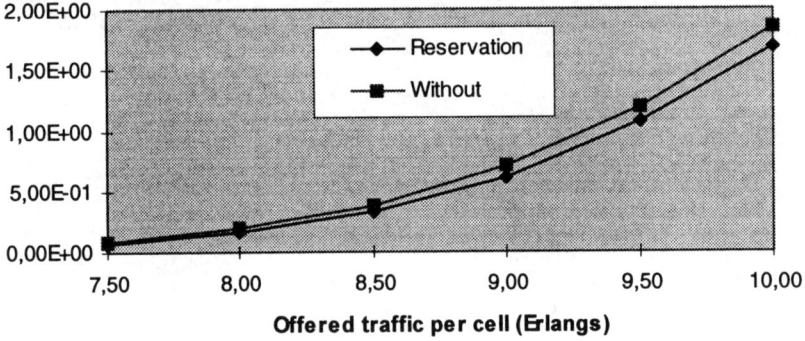

$K_0 = 25$, $K_1 = 14$, $K_2 = 10$ and $P_1 = 0.35$, $P_2 = 0.5$
Figure 11 Grade of Service in non uniform case

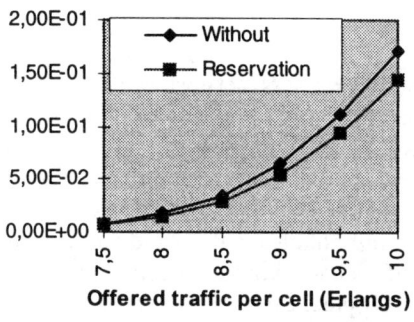

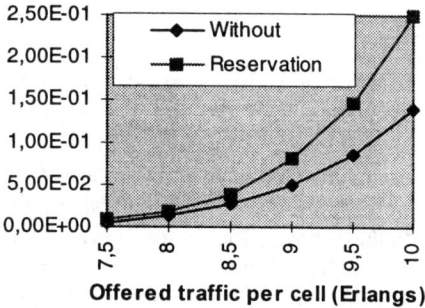

Figure 12 Handoff failure probability   **Figure 13** New call block probability

3.5 Comparison between PRP and FCA with fixed reservation

Figures 14 and 15 represent the Grade of Service as a function of the traffic load (expressed in Erlangs). PRP results are compared to those obtained using a fix number of reservation (2 and 3 in each cell) for HandOff traffic. It is shown that PRP leads better performance results. This is due to the fact that PRP reservations are adapted to the traffic and to the local topology. We obtain the same results with uniform case (with a threshold of 16) and non uniform case.

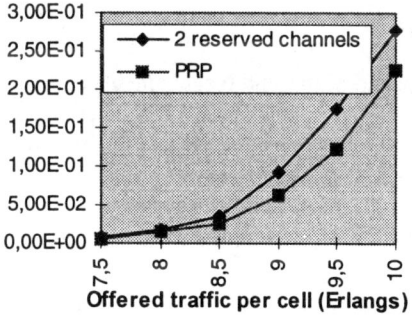

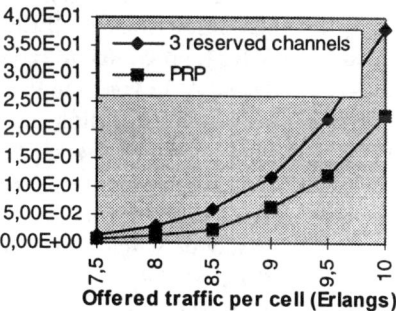

Figure 14 Grade of Service   **Figure 15** Grade of Service

4. CONCLUSION

PRP concept is likely to be very useful in any wireless network with small cells, irregular topology and time-varying traffic load (hot spot for example). We showed in this article the applicability and usefulness of PRP mechanism with simulation experiments. We have shown that with this new approach the bandwidth is better used if we take into account the characteristics of the traffic and the local topology. Indeed, with these simulations, a load corresponds to a threshold and a threshold corresponds to a probability of visit. If real data on the traffic characteristics become available we could find optimal values that result in PRP that is practical and manageable.

For future researches, we are thinking to make base stations "more intelligent" to calculate the thresholds automatically using the characteristics of the environment. Indeed, it is possible to find a daily (or weekly) regularity on the conditions of traffic, and - in the same way - a regularity on the movement of mobiles.

REFERENCES

[Aca 94]   A.S. Acampora and M. Naghshineh - An Architecture and Methodology for Mobile-Executed Handoff in Cellular ATM Networks - IEEE Journal On Selected Areas In Communications, VOL. 12, NO.8, October 1994.

[Bou 97]   S Boumerdassi, A New Concept to Optimize Channel Management in Cellular Communication Systems, MWCN'97, Paris, May 97.

[Cho 82]   G.L. Choudury, S.S. Rappaport - Cellular Communication Schemes Using Generalized Fixed Channel Assignment and Collision Type Request Channels - IEEE Transactions on Vehicular Technology, Vol VT-31, pp. 53-65, May 1982.

[Lev 97]   David A. Levine, Ian F. Akyildiz and Mahmoud Naghshineh - A Resource Estimation and Call Admission Algorithm for Wireless Multimedia Networks Using the Shadow Cluster Concept - IEEE/ACM Transactions on Networking, Vol 5, NO 1, February 97

[Nag 94]   A.S. Acampora and M. Naghshineh - Distributed call admission control in mobile/wireless networks - IEEE Personal Communications, Second Quarter 1994.

[Ray 92]   D. Raychoudhuri, A.D. Wilson - Control and Quality-of-Service Provisioning in High-Speed Microcellular Networks - IEEE Journal on Selected Areas in Communications, Vol SAC-12, n°8, October 1992.

[Yu 97]   O.T.W Yu and V.C.M Leung - Adaptative Resource Allocation for Priotized Call Admission over an ATM-Based Wireless PCN - IEEE Journal on Selected Areas in Communications, Vol SAC-15, n°7, pp 1208-1225, September 1997.

Service Control

14

Signalling/IN Server architecture for a broadband SSP

D.Blaiotta[1], S.Daneluzzi[1], D.Fava[1], D.Lento[2], M.Varisco[1]
[1] Italtel, [2] CSELT
Contact Person: Danilo Fava, Central R&D Labs (C02),
20019-Settimo Milanese (MI) -Italy, Phone +39.2.43889116,
Fax +39.2.43887989, E-mail: Danilo.Fava@italtel.it

Abstract

This paper describes a Signalling/IN Server taking into account the adopted architectural and implementation solutions. The Signalling/IN Server takes part in a Broadband Service Switching Point realised in the framework of the European research project INSIGNIA (IN and B-ISDN Signalling Integration on ATM Platforms). The aim of the INSIGNIA project is to define and implement an advanced IN and B-ISDN signalling integration for satisfying users' requirements in terms of multimedia and broadband applications. The paper focuses on the B-SSP network element that integrates an evolved version of both the Service Switching Function and the Call Control Function. The core of the adopted functional model is the Switching State Model, which presents an abstract view of the network resources involved in the service provisioning.

Keywords
ATM, Broadband Signalling , Intelligent Network, Service Switching Point

INTRODUCTION

The ITU standardisation bodies have been defining a signalling system able to manage the network resources in order to support wideband services. The services

Broadband Communications   P. Kühn & R. Ulrich (Eds.)
© 1998 IFIP. Published by Chapman & Hall

development increases in accordance with the signalling system, but in order to optimally realise such services, the network has to provide a higher level of intelligence than just the establishment of point-to-point connections. This paper reports on results of an European ACTS research project that implements and demonstrates an advanced architecture integrating IN and B-ISDN signalling. The project INSIGNIA, using available end systems and taking a Broadband ISDN network as its starting point, develops prototypes of *Broadband Service Switching Point* (B-SSPs), *Broadband Service Control Point* (B-SCPs) and *Broadband Intelligent Peripheral* (B-IP) in different National Hosts. In order to carry out trials with real end users, attractive multimedia application services have been made available that make use of the advanced network services. The three selected kinds of services are: *Video-on-Demand, Broadband Video Conference* and *Broadband Virtual Private Network.* The INSIGNIA project experiments broadband multimedia services handled by IN in five different sites interconnected by European ATM facilities : Madrid, Milan, Munich, Turin and Berlin (remote user for the Munich site).

In the next section, an introduction inside the functional architecture of INSIGNIA is given. Special emphasis is put throughout this paper on the central network element that provides the switching functionality (B-SSP). The following section describes the B-SSP external Server entity, and the next one gives an overview of the object oriented model solution adopted for the Italian B-SSP software implementation. The last section focuses on the foreseen developments for the Server architecture.

COMMON DESIGN OF THE B-SSP

INSIGNIA network elements

The INSIGNIA architecture of a Broadband IN resumes and widens the classical IN. The main equipment involved in the INSIGNIA base structure are listed below.

- *SERVICE CONTROL POINT* (SCP). It is an intelligent server computer, as in classical IN, on which the Service Logic programs run in order to guarantee the call/connection configurations. Compared to the classical SCP concept the INSIGNIA SCP has got an advanced communication level with B-SSP, based on an object oriented model of Service Logic instead of simple call state dialogue.
- *SERVICE SWITCHING POINT* (SSP). It is the switching network element able to support UNI/NNI signalling dialogue and with service independent IN functionality.
- *END SYSTEMS AND SPECIAL DEVICES.* They change in accordance with the involved service, but since for multimedia applications PC terminals are commonly used as end system rather than telephone equipment, a service is supported in a distributed manner where program parts reside on the host

system. This increased users' capability drives to an SCP overload concerning its user interaction. For this aim a special network element has been introduced: the B-IP. It acts as a server that frees the SCP from heavy service computations, and carries out multimedia communication with the user.

B-SSP Functional Model

The design of the components, which have been developed for the INSIGNIA B-SSPs, covers the functional entities of a *Service Switching Function (SSF)* and a *Call Control Function (CCF)*.

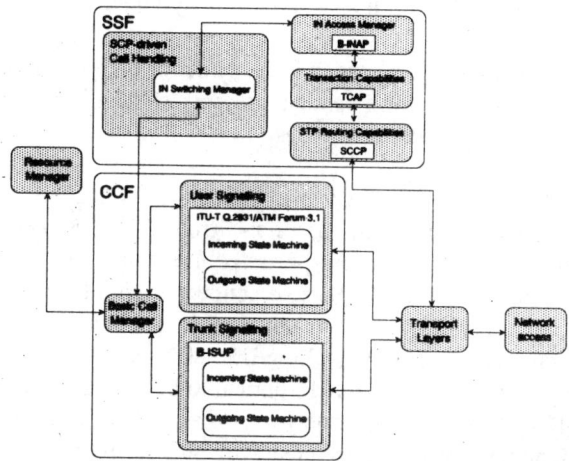

Figure 1   The Italtel/CSELT B-SSP functional Model.

In more detail, the design of the SSF comprises aspects of the SCP access handling, the processing of B-INAP messages, the realisation of TCAP, SCCP and MTP3 protocol layers, the *Switching State Model (SSM)*, the handling of *Detection Points (DP)* and the interaction with the CCF. The design of the CCF has to consider the *Basic Call State Model (BCSM)*, the handling of the NNI/UNI signalling and the establishment of ATM VC connections.

In Figure 1 the functional model used for the INSIGNIA B-SSP is depicted. It is based on the models defined in ITU-T Recommendations for IN Capability Set 1 (CS-1) [ITU-T, Rec. Q.1214] and the Draft Recommendations for IN Capability Set 2 (CS-2) [ITU-T, Draft Rec. Q.1224].

Both the INSIGNIA SSF and CCF are different from the corresponding IN CS-1 functional blocks. In fact CCF is split in two levels, which are the Call Control and the Bearer Connection Control in order to make the IN model adequate for the support of advanced B-ISDN calls. The SSF achieves the capability of handling the session concept introduced in a complex service configuration to solve the problem of the call/connection correlation. Inside the SSF, the IN Switching

Manager provides an object oriented session model, and then offers an abstract view of the whole call configuration to SCP. The BCSM model becomes valid only for the Basic Call Manager and its interaction with the IN Switching Manager. The design of the IN functional entities is developed according to the *Object Modeling Technique (OMT)* object oriented methodology [Blama, 1991].

THE ITALTEL/CSELT B-SSP ARCHITECTURE

The Italtel/CSELT B-SSP is built around two subsystems.

- The Signalling and IN server, which is the same for both B-SSPs.
- The ATM switching platform, which presents some differences according to equipment availability in both premises.

The Signalling/IN server hosts the protocols and the applications which enable the handling of calls coming from the users and from the ATM network. The ATM switching platform offers the access ports for user and network connections and the capabilities needed in order to interconnect them. In addition, commands performing the establishment and tear down of connections are accepted from the signalling/IN subsystem. The Signalling/IN subsystem communicates with the ATM switching platform via an STM1 optical link. The ATM switching platform collects signalling from each user/network interface and presents it to the link towards the signalling/IN server by means of previously built cross-connections. After elaborating signalling, the server instructs the switching platform in order to establish/tear down connections between the users. This process takes place via the Ethernet connection between the two subsystems making up the Italtel/CSELT B-SSP.

ATM Switching platform
The Italtel switching platform used for the first trial is an Italtel prototype (UT-ASM). The system accepts ATM cells on three different underlying supports, namely, optical STM1, electrical E3 and optical TAXI on both user and network ports. The combination of UT-ASM and the signalling server provides switched ATM connections to the Milan site. The CSELT platform is based on the FORE Systems local ATM switch ASX200E. As the control software is installed on an external workstation, the FORE switch provides only the capability to realise connections between the requested end points. This functionality is accessed, for both the two ATM switching platform, by means of SNMP hidden in an API, that translates the commands issued by the signalling process in the correspondent SNMP dialogue. The adoption of this API allows the use of different switching fabric controlled by signalling.

Signalling/IN Server

The Signalling/IN Server consists of a Sun SPARC20 workstation equipped with the Solaris 2.5 operating system, a SunATM155 adaptation board and signalling/IN protocol stacks compliant with the INSIGNIA specifications.

The Signalling/IN Server for the first trial of INSIGNIA offers the following functionality (Figure 2).

- UNI signalling ([ATM Forum 3.1], [ITU-T, Rec. Q.2931]).
- NNI signalling (ITU-T B-ISUP, [ITU-T, Rec. Q.2761-Q.2764]).
- Call Control (interworking between UNI/NNI, UNI/IN, NNI/IN, UNI/UNI).
- Switching Matrix Control (it is able to send commands for ports initialisation, setup and release of ATM bearer).
- Switching State Model (SSM) developed according to INSIGNIA object oriented model. It is able to translate the separated calls view seen by the signalling system into a unified service (or session). In this way it reports to each B-SCP session the service resources status.
- NNI/IN signalling based on INSIGNIA B-INAP protocol.

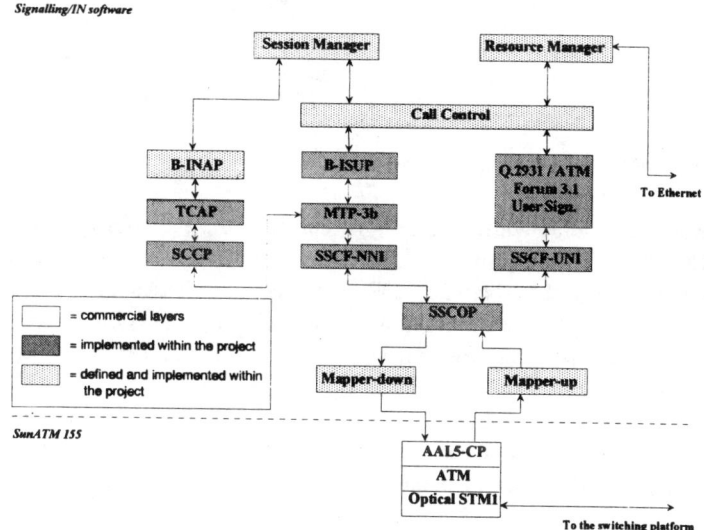

Figure 2   Signalling/IN Server architecture.

SIGNALLING SERVER IMPLEMENTATION ISSUES

Object Oriented model solution

The aim of this section is to describe the object oriented approach adopted by Italtel and CSELT to develop both SSF functionality, which are split in *IN*

Switching Manager (SM) and SCF *Access Manager (AM)*, and CCF functionality. In Figure 3 the CCF/SSF and SSF/SCF interface classes are shown.

A BCSM contains several DPs and it is assigned to each Call Control object that represents one basic call. On reaching an armed Trigger DP during the call processing, a Session object is created. A specific DP can trigger only one IN Service, represented by a Session, because no interaction between different IN services is provided for the first trial. Each instance of the Session class is in one-to-one correspondence with an instance of the SCF AM class. Communication between SCF and SSF is represented by the one-to-one association between SSF AM and SCF AM.

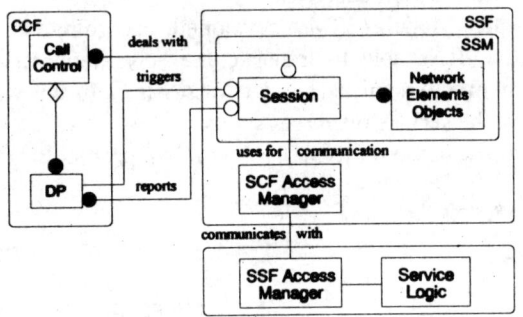

Figure 3   INSIGNIA IN object model.

Basic call handled within the CCF is represented by the objects inside the SSM. The CCF/SSF relationship is represented by two associations: the already mentioned relationship between DP and Session classes and the relationship between Call Control and Session classes. Since a session is an association between calls/connections for the deployment of a single IN service, many Call Control objects can deal with a Session object. When a SCP-initiated call is established, the Session object creates, inside the SSM, the objects that represent the corresponding network elements involved in the call. At the same time a new corresponding Call Control instance is created within the CCF.

CCF description
The CCF implements all the functionality needed to handle calls and, in the INSIGNIA framework, the dialogue with the IN entities.

The core section of this model is represented by the Call View that contains the information related to a particular call instance. The Call View model was derived by the one specified and implemented during the RACE MAGIC project.

In more details, this model combines the abstract representation of the requested service with its real implementation (see figure 4). The classes *TCS (TeleCommunication Service)*, Party, *USM (User Service Module)* and *PartyLevel* give a first view of the service, highlighting the Parties involved in the service, which resources (USM) they are sharing and the relationship (PartyLevel) between the Parties and the resources. The other classes provide a more detailed description of the resources needed in the provision of the service. The ASM (Abstract Service Module) links the USM with the connections' description. The SM (Service Module) stores a description of the user information transported in the related connection while the characteristics (i.e., in terms of bandwidth) are stored in the CE (Connection Element) class. In the INSIGNIA framework, the TCS class contains also the Detection Point class. This class represents the core part of the BCSM and is used in the handling of IN calls to synchronise the interaction with the IN entities.

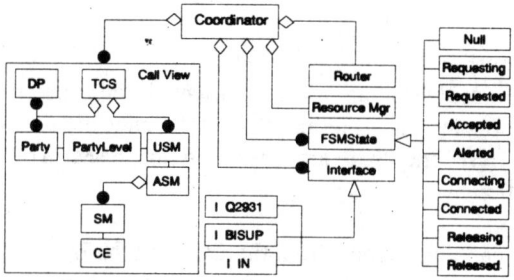

Figure 4   CCF object model.

At this point suppose that the requested service is the TV Distribution where a user wants to be connected to a TV station and suppose that the audio and video information is transported in different connections. From the user point of view, this is express in a TV channel request. The Call View will represent this service as in the following:

TCS: TV Distribution   USM X: TV channel
ASM: TV channel is composed by two connections
Party A: User   SM1: this connection carries the video information
PartyLevel A: User using USM X   CE 1: bandwidth and other informations about connection
Party B: TV Station   1
PartyLevel B: TV Station using USM   SM 2: this connection carries the audio information
X   CE 2: bandwidth and other informations about connection
                                              2

The *Interface class* provide the functionality to translate the signalling primitives coming from the protocol stacks in modifications in the Call View and vice-versa. As shown in the model, the Interface class is a superclass and there is a derived class for each protocol stack.

- InterfaceQ2931: handles the dialogue with the UNI protocol stack.
- InterfaceBISUP: handles the dialogue with the NNI protocol stack.
- InterfaceIN: handles the dialogue with the SSM entity.

In general, the aim of this class is to offer a unique interface to the CCF core in order to preserve the core functions from the underlying protocols. If in the future a new protocol will be specified, the only adoption to the whole CCF model is a new derived Interface call specialised to dialogue with this new protocol.

The ***FSMState class*** implements the Finite State Machine of the CCF. There is a subclass for each state of the CCF and each class implements the operations that have to be performed on the Call View when the call is in that state. The ***Router class*** manages the routing tables and provides routing information to the other classes of the CCF. The ***Resource Manager class*** handles the API used to control the switching fabric. The ***Coordinator class***, as suggested by its name, coordinates the functions performed by all the other classes.

SSM description

The IN SSM provides an object oriented description of SSF/CCF IN call/connection processing in terms of IN call/connection states. The information flows between SSF and SCF are based on the SSM. The SSM performs the following functions:

- reception of incoming signals from the CCF and the IN AM;
- translation of external signals into appropriate actions on the SSM objects;
- transmission of signals towards the CCF and IN AM;
- maintenance of the SSM data structures related to the Session and to the Session objects.

It therefore contains objects that are abstractions of switching and transmission resources. For an integration of IN with B-ISDN, the SSM becomes the main concept since it provides a model for the complex call/connection configurations that may appear in B-ISDN. The complex call connection configurations are represented in the SSM with the Session object. The IN Switching Manager structure is composed of the following object classes and relationships (figure 5).

- ***SMcoordinator*** class: it is instantiated only once in the SSF, and coordinates the dispatching of the B-INAP operations and CCF/SSF commands to the addressed session or call. It is able to coordinate all session and interface instances.
- ***Session class***: it has two different meanings. From an IN point of view the Session is an instance of the provided service, (i.e. the VOD service). In a B-ISDN perspective the session is the composition of resources needed to support the service.
- ***Party Class***: it is the representation of a calling user, a called user or a network element, (i.e. the B-SCP).
- ***Bearer Class***: it is the abstract view of an end-to-end connection between two Parties.

- *Leg Class*: it is the representation of the path communication between the B-SSP and the Party.
- *Interface Class*: it allows to have an abstract view of the communication between the IN Switching Manager and the outside world. It is specialised in two subclasses: InterfaceToBCM (that handles the link towards the Basic Call Manager) and InterfaceToAM (that allows the link towards the SCF Access Manager).

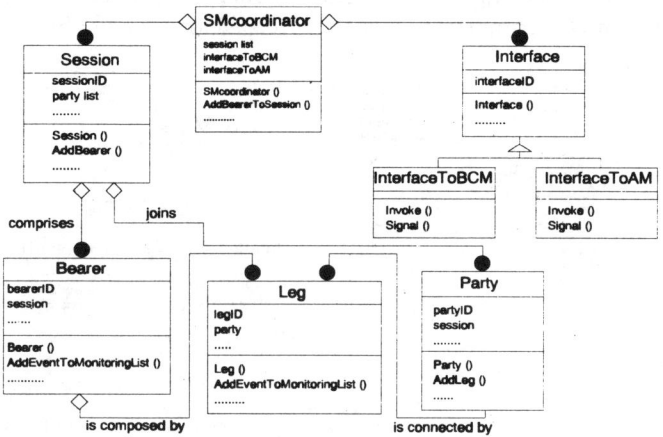

Figure 5 OMT-based Session Manager design.

The main attributes of the SSM objects are described in more details in Table 1.

Table 1 Objects and identifiers

OBJECT NAME	OBJECT IDENTIFIER	ATTRIBUTE	ATTRIBUTE IDENTIFIER	ATTRIBUTE IDENTIFIER VALUES
Session	Session ID	None		
Party	Party ID	Virtual Party	Is_Virtual	FALSE / TRUE
Bearer Connection	BC ID	Bearer Status	Status	BEING SETUP / SETUP / BEING RELEASED
Leg	Leg ID	Leg Status	Status	PENDING / DESTINED / JOINED / ABANDONED / REFUSED

The attribute "Is_Virtual" indicates whether the respective party object is representing a network element (i.e., the SCP for the INSIGNIA first trial) or a true party associated with an end system.

The "Status" attributes of BearerConnection and Leg represent the status of the respective object with respect to call processing. There is a relationship between the status attributes of the bearer connection and the legs. The status of the bearer connection is BEING SETUP, as long as the status attributes of all corresponding legs are not JOINED. Only if all corresponding legs have status JOINED, the bearer connection will have status SETUP.

The IN-SSM view, including relationships between objects, is communicated from B-SSP to B-SCP (or vice versa) through information elements in the Information Flows. The two ownership relations of the SSM class diagram (between Party and Session as well as between Party and BearerConnection), however, are considered to be local at the SSP for the first trial. This means that the ownership relationships are not visible to the SCF, and the Information Flows for the first trial do not transport parameters related to session or bearer ownership.

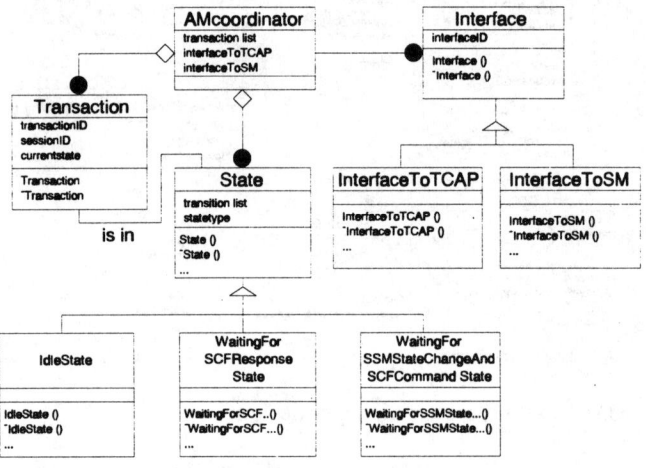

Figure 6   OMT-based Access Manager design.

AM description

The SCF AM handles the communication between the SSF and SCF. It locates the required IN Service in ATM Network and sends messages to/receives messages from the B-SCF. This communication is based on the SSM and therefore on the physical plane is performed through a Broadband Intelligent Network Application Protocol (B-INAP), completely defined in INSIGNIA project. In fact from a methodological point of view the B-INAP specification is structured in a way similar to IN CS-1 INAP (such as use of ASN.1 notation language as abstract syntax and *BER (Basic Encoding Rules)* as encoding rules), but the content of Information Flows exchanged between B-SSF and B-SCF reflects the abstract object view of SSM. Since the B-SCF handles only the objects included in the model (session, party, bearer connection, leg), the SCF/SSF interaction is in terms of IN SSM events (SSM state changes) and not in terms of Call Model events (BCSM state changes).The OMT approach has been adopted also for the design of the SCF Access Manager.

The structure of this functional block shown in figure 6 foresees the following object classes and relationships.

- The *AMcoordinator* class is instantiated only once in the SSF to coordinate the handling of the B-INAP operations from/to the SM and of the messages from/to TCAP, addressing the appropriate instance. It is able to coordinate all transactions, states and interface instances.
- The *Interface* class allows to abstract the communication between the SCF AM and the outside world; it is specialised in two subclasses: InterfaceToTCAP and InterfaceToSM.
- The *InterfaceToTCAP* class represents the link towards the TCAP ASE.
- The *InterfaceToSM* class represents the link towards the IN SM.
- The *Transaction* class instance represents the current state of the communication between the SSF and SCF for each active session instance. Each Transaction instance has a "is in" association with only one State subclass instance and this association changes during the transaction lifetime according to the AM dynamics.
- The *State* is the class that allows to describe all allowed transitions from the specific state to the destination states, depending on the received operation.

Server software architecture

The Signalling/IN Server software architecture is composed by software modules that may execute concurrently and message queues provide the mechanism that allows the communication between them.

The architecture of some software modules in B-SSP reflects their SDL description. In this case, the process evolves like a Finite State Machine, that is, performs specific operations only on the basis of its initial state and the input message. A module structured in such a way, takes advantage from facilities offered by ad-hoc developed tools and libraries. A code generator, starting from a description of the expected events (in terms of messages) and the associated state transition, performs the application framework in C language. This code contains the matrix that binds initial state, input message, function to be performed and ending state. This matrix is scanned by library routines allowing the operation flow control. The library routines allow the generation of several instances of the same process; instances may be created either at startup time or run-time and they may evolve independently (but not concurrently during state transitions). The adopted solution does not overload the host, since a new instance only needs allocation of little amount of memory. This memory is used to store information regarding state, timers and protocol dependent values. The software architecture allows the linkage of several modules into a single UNIX process. The communication between modules is always performed via library functions that are based on UNIX IPC queues or on process internal queues, depending on whether the modules are parts of the same process or not. The library routines post the output messages into the destination queues and, when messages are received, schedule the execution of the associated routines. The binding between queues and software modules is statically determined, and the routing of messages is done by analysing the message header that contains information about source and destination module. Libraries also employ facilities to handle timers by sending messages to the involved instances when the requested timers expire. It has to be noticed that software modules that do not have a Finite State Machine structure, may use the library routines for interprocess communication without the need for modifications.

CONCLUSIONS AND FUTURE DEVELOPMENTS

This paper describes the external Signalling/IN Server characteristic of a Broadband SSP system realised with implementation solutions conceived according to the OMT approach. The B-SSP described in this paper has been used in the first INSIGNIA trail in order to support the B-VC and VOD services. The experimentation allows to confirm new IN based solutions for multimedia applications. In particular it will be possible to analyse, in a wide area network, the IN solution efficiency concerning some services otherwise provided with less advanced network structures. An example of these is the VOD service, which can also be supported according to other kinds of architectures. The IN solution provides a simplified service handling method and assures a fast way to enrich services, without necessarily modifying the end systems.

For the second trial (summer 1998) an upgrading towards a more sophisticated Server architecture is planned. First of all, the Server will be supplied with advanced B-ISDN Signalling aspects, compliant with the CS-2 study group standard, and also considerable for multimedia services. Porting of the software architecture on a real-time platform is also planned.

REFERENCES

ITU-T. (1993) Recommendation Q.1214. Study Group XI, Helsinki.
ITU-T. (1996) Draft Recommendation Q.1224. Study Group XI, Nice.
ITU-T. (1994) Recommendation Q.2931. Study Group XI, Geneva.
ITU-T. (1994) Recommendation Q.2761-Q.2764. Study Group XI, Geneva.
ATMU Forum. (1994) UNI Specification 3.1.
Blama, M. and Rumbaugh, J. (1991) *Object-Oriented Modelling and Design.* Prentice Hall, New Jersey.

15

A signaling based approach to broadband service control. Application to access signaling

P. Martins
Ecole Nationale Supérieure des Télécommunications, Paris, France.
e-mail: martins@res.enst.fr

C. Rigault
Ecole Nationale Supérieure des Télécommunications, Paris, France.
e-mail: rigault@res.enst.fr

N. Raguideau[1]
Telecommunication systems Business Unit,
Hewlett Packard, Grenoble, France.
e-mail : Nicolas_Raguideau@grenoble.hp.com

Abstract
This paper describes a new signaling based approach for the implementation of a control plane for the B-ISDN network. A control architecture is developed, with a separation between service session, access session and call. This architecture provides the guidelines for the definition of signaling protocols. The differences between this architecture and other existing control plane proposals are reviewed. The emphasis is put mostly on the access session signaling and the physical architecture. However some indications on the service control signaling are given.

Keywords
B-ISDN signaling, Control plane, Call Model, Session concept

---

[1] The ideas presented in this paper represent the author's point of view and not necessarily the point of view of their respective institutions.

Broadband Communications   P. Kühn & R. Ulrich (Eds.)
© 1998 IFIP. Published by Chapman & Hall

1   INTRODUCTION

The purpose of this paper is to define suitable signaling schemes for the control plane of the broadband network. We start from a control architecture compatible with already pending proposals such as B-ISDN Capability Set 3 (B-ISDN CS3), IN Capability Set 3 (IN-CS3), and TINA-C (Darmois, 1996) (Barr, Boyd, Inoue, 1993). However our scheme differs from the above mentioned architectures by the importance given to signaling, both for the service session and the call session. Differences with other approaches will be underlined as we explain the architecture. The model will then be used to develop the access signaling and to introduce service session signaling on a service example.

2   THE PROPOSED ARCHITECTURE COMPARED TO PREVIOUS APPROACHES

The proposed architecture has been described in details (Rigault, Kovacikova 1997). It is based on the TINA-C view of the telecommunication business model (Figure 1). This model is characterized by five business roles: the "consumer" who asks for services and gets billed for it, the "retailer" providing access to the services and reselling to the consumer telecom usage bought in bulk at bargain prices to a "connectivity provider", the "third party service" providers using the broadband network to provide an infinite variety of services, and the "broker" used to find out which services are provided and their identity.

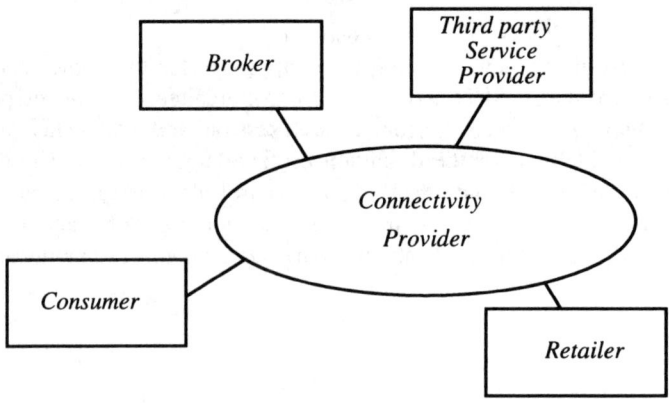

Figure 1   Business roles in the B-ISDN network.

Before B-ISDN CS3, the concept of a call was equivalent to the negotiation and allocation of communication resources between users, independently of service specific functionalities, and user roles in the service. Such a call concept was mixing in the same process the logical establishment of the association between parties, the search for network resources and the actual connection of the bearer facilities between these resources and parties. In order to solve the problems of the variety of connectivity requirements both in topology and set up sequences for the expected new services, we adopt the clear separation introduced by B-ISDN CS3 between the session concept and the call concept (Figure 2).

The Service Session in B-ISDN CS3 is a single instance of service use with global significance. It includes the end to end view between user applications, while the call includes the end to end view between network access points. The Service Session takes place in the service provider platform (eventually distributed over several servers linked by networks). The Service Session takes care of service parameters like coding schemes, but does not take care of network aspects like QOS parameters. On the other hand the call takes place in the network (or the networks) involved by the service. In its new meaning, the call represents an association of one or more parties, using a telecommunication service to communicate through the network. The correspondence between session and calls may be of the "one to many" type, since a given Service Session may involve several calls or no calls at all. A session may be kept set up, while the corresponding call is released for some time. Upon resumption of the session activity a new call will be set up.

The Service-Call separation is a necessity for services to be provided by other stake holders than the network operators.

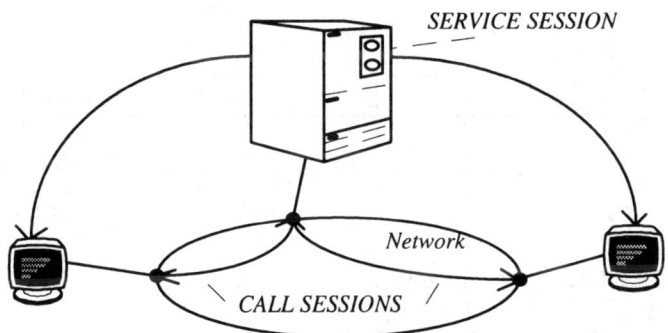

Figure 2   Service Session and calls in the B-ISDN network.

Due to mobility, it will be very frequent that the user of a given terminal will not be it's owner. However each user will have unique characteristics such as environment preferences, personal subscriptions and so on. It is therefore advisable, following again the TINA approach, to provide an Access Session. The access session brings the user profile down to the actual access switch of the terminal presently used by the consumer. More generally, it takes care of customer related functionalities, like the Home Location Register (HLR) of mobile telephony does it today, and presents a generic interface to the Service Session. The Access Session is executed on the platform of the retailer providing telecommunication services to this consumer.

All above considerations finally lead to the control plane functional architecture shown in Figure 3. This architecture entails the following functional entities.

• Terminal Manager Function (TMF) : terminal related, provided by the operating company supplying connection services to the terminal in use. The TMF is a generalization of the role played by the Visitor's Location Register (VLR) of mobile telephone networks. It gets from the access function the user profile of the consumers using its related terminals.

• Access Session Function (ASF) : customer related, provided by a retailer type of business which may also be an operating company. The ASF is a generalization of the role played by the Home Location Register (HLR) of mobile telephone networks. It keeps information on the exact status of all consumers associated with this access provider.

• Service Session Control Function (SSCF) : third party service supplier related or operating company related. The switches will normally contain the SSCF for basic services

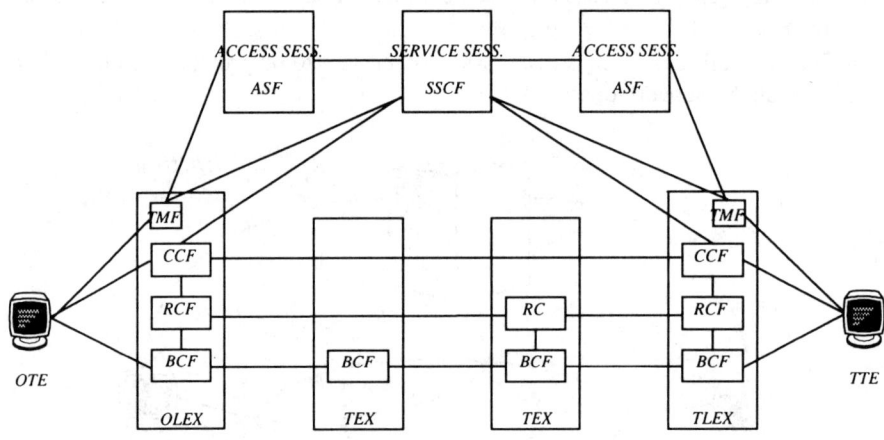

Figure 3   Control plane Functional Architecture.

To underline the differences between this architecture and Intelligent Network (IN) solutions we remark that IN uses the call as a device to trigger a service instance, giving therefore precedence of the call over the Service Session. The main issue then, is of the feasibility of an "universal, any service call model". Could any type of service be triggered from a call ? This idea is very much challenged and some services are known today as "call unrelated" services (such as updating the location of a mobile user, a short message service, etc.). Actually services constitute an open set. It is likely that new services that cannot be triggered from the current call model will always be proposed. We therefore suggest the **service precedence**

principle according to which the service should be triggered first, and the call, if any call is required, should be triggered afterwards. Only in this manner generic call procedures may be designed. This service precedence is not found in IN architectures or in B-ISDN architectures. It is however found in the TINA-C proposal.

The first key point of our control architecture lies therefore in the provision of a direct signaling path between the Consumer and the Service Provider platform to initiate the service session first. If required, the call or the calls (some services may require several calls in different networks) may then be originated, either by an instruction of the service instance to the originating customer equipment, or as a direct third party connection request to the network.

A second key point in this architecture concerns the nature of the communication paths between the functional entities. These paths are direct signaling paths and do not require the use of a middleware like the Distributed Programming Environment (DPE) of TINA-C, making there an important difference between our architecture and the TINA-C proposal. While we use the object oriented technology for design, we prefer not to resort to the DPE for the implementation.

Indeed, the important objectives of our architecture is compatibility with legacy system, sufficient control of network parameters from the service application and performance regarding service and call setup times. Parties should not experience significant delays after joining agreement, mostly if these parties are information servers in the case of information retrieval services.

Due to these performance considerations, a DPE should be considered as a fast real time system. In particular, one of the issues to be considered for a DPE is the implementation of efficient naming / trading services for telecommunications. An other issue is to be found when taking in account high availability aspects in distributed object oriented architectures where the semantic attached to certain information elements like object references needs to be revisited for keeping performance. Regarding such issues, and others, it is generally agreed that the achievement of a performant real time DPE, able to interwork with a sizable amount of distributed objects is not to be available in a near future. As an example of such concerns, the CORBA environment presently considered as the best candidate for the DPE does not provide mechanisms for end to end QOS definition, does not provide a priority system for requests and does not provide a blocking protection when serving the requests (Schmidt 1997).

We therefore propose an architecture that is compatible with the service objectives of TINA-C and which is implementable as an evolution of legacy systems without having to wait for the availability of an adapted DPE environment. Furthermore our architecture, by its compatibility, does not prevent the development of distributed middleware when the technology will become available. In the meanwhile, our proposed architecture is adapted to favor fast service setup times.

An other point to be underlined is the compatibility of this architecture with Connectionless and Connection oriented Services. The SSCF has two different ways to establish a call. In a first method, the called party address is loaded into the user's Originating Terminal Equipment (OTE). A connectionless call may then be initiated from this OTE.

A connection oriented call may also be originated in this manner via signaling from the OTE. However, it seems more advantageous to initiate connection oriented calls via a third party connection request sent by the service session to the call control function of the OLEX.

The call may now proceed by a Call Control instance, including look ahead procedure to know if the terminals are able to accept the call and to negotiate the QOS, resource control is activated to locate suitable network resources, and bearer control is finally performed to set up all the bearer connections.

3 ACCESS NETWORKS PHYSICAL ARCHITECTURE

The access network configuration in the new telecommunication infrastructures will be based on SDH loops connecting remote Broadband concentrators to the connectivity provider switches (Figure 4). Some cooperation has to take place between connectivity providers to share access to given SDH loops, and broadband concentrators must have enough switching abilities to direct calls to one operator switch or to an other since it is expected in the future that the user will be allowed to select his connectivity provider.

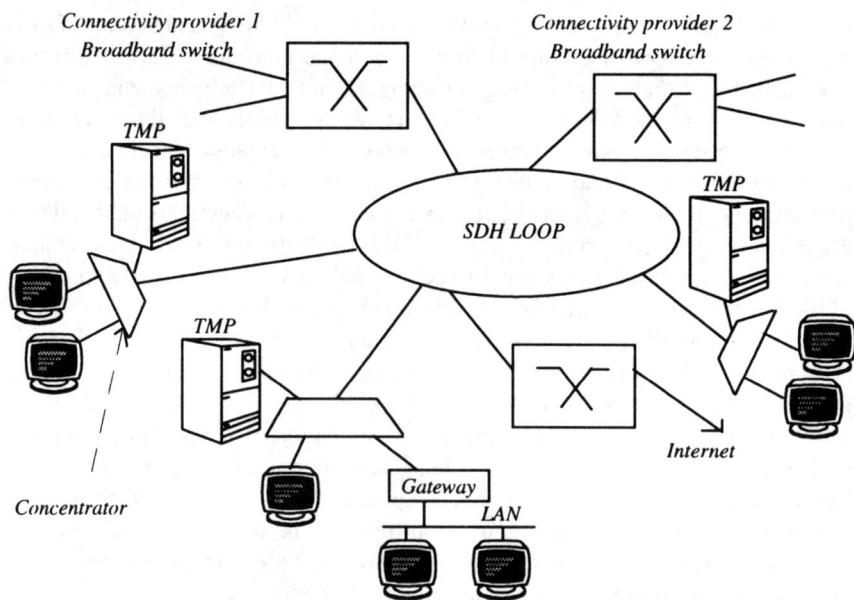

Figure 4 TMP and the access networks.

The TMP is the control interface between the user and the access and service session. Concentrators and TMPs are operator equipment. The TMP should be placed behind the concentrators so that each terminal gets a full availability access to it. In addition such a location at some operator site is required given its role in critical processes like billing and user's authentication. Whenever a user initiates a service session, the service provider, via service session signaling messages (between SSCP and TMP), informs the user of which connectivity providers he can access. The user selects on his screen, a connectivity provider, and this choice is sent in a signaling message. The TMP will redirect the chosen connectivity provider identity to the call control function of the concentrator. For terminals directly connected in ATM to the concentrators, a connectionless signaling path is established between the terminal and the TMP.

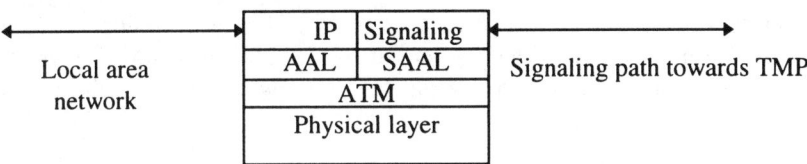

Figure 5 A gateway.

However terminals may be connected via some Local Area Networks not necessarily based on ATM technology. In such a case, a new gateway equipment (Figure 5) providing interworking functionalities should be provided to interface with the ATM network. Such an approach for a gateway in a local area network based on TCP/IP stack is used in the ITU-T H.323 gateway recommendation (H.323). In this case, a connectionless signaling path is established between the gateway and the TMP.

4 ACCESS SESSION SIGNALING

In order to describe how a service session is activated and calls are set up in our architecture, we must first define the concepts of *"user profile"* and *"login file"*.

User Profile : We call "user profile" a set of information characterizing a particular consumer that are necessary to be known in order to establish telecommunication services for this consumer. Examples of such information are :
- his authentication information
- his preferred graphic environment (including wallpaper, colors, screen saver...)
- pointers for his various mail services (email, voice mails...)
- his bookmarks
- his applications
- his billing records

- his subscriptions in the case of non generally available services
- his list of terminals associated to his eventual "one number" service. (a list of terminals where he is likely to be joined)

It is the responsibility of the retailer or access provider to keep the user profile available and updated in the Access Session Function. A copy of this user profile is sent to the Terminal Manager Function of the switch hosting the terminal that is currently used by this consumer.

Login file : We call "login file" an information element sent by means of signaling by the terminal to Terminal Manager Function which contains among others :

- the identity of the calling consumer (unrelated to the terminal).
- the identity of the retailer or access session provider for this consumer

This login file may be memorized on various media, such as a SIM card or any other portable memory device, that would be generally accepted by all terminal equipment.

The Economy Principle : We shall use consistently this principle according to which no signaling process should be made systematic if not strictly required. As an example, if the user profile is already cached in the Terminal Manager Function (like it will be in most cases), no signaling should take place to the Access Session Function.

Having defined important datas regarding the consumer, we will now describe the exchange of signaling messages by which the access session may develop by using these data.

Originating access session sequences

The user initiates the access session by placing his Login file storage medium in a terminal. The access session sequence entails the authentication of the user followed by the downloading of the user profile into the TMP (in case it is not already cached there).

Signaling steps (Figure 6) are the following :

- A login message is sent to the TMP. It contains the International Terminal Equipment Identity ITEI of the Originating Terminal Equipment OTE and the International Subscriber identity ISI of the person presently using the OTE. The ISI points to the ASP and to the name of the subscriber in the ASP. The TMP checks if the user's profile is in its cache. If it is the case, an authentication procedure similar to the MAP authentication, is performed and a Login_ack message, containing the calling user's profile (UPF) is returned to the OTE. Otherwise, a send_authentication_info request is sent to the ASP. After completion of the authentication procedure the TMP sends a Send_user_profile message to the ASP to request the user's profile.

- The ASP returns the user profile to the TMP in a Send_user_profile_ack message. The user profile is then loaded in the terminal. The information elements received are the mail pointers (M_P), the service pointers (SS_P for subscribed service and US_P for unsubscribed services), the user's environment (ENV) and the broker pointers (B_P). The user may now invoke a service.

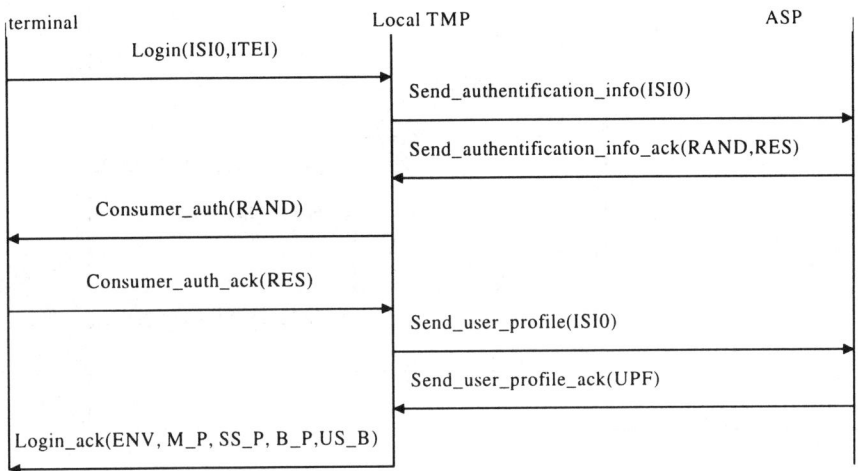

Figure 6 Authentication and user's profile retrieval procedures

Terminating access session sequences

We now assume that a service session has been initiated in an SSCP, and that a called terminal has to be invited to join in the service session. The SSCP contacts the called party's ASP. This ASP responds with an Terminal_info_type message (Figure 7) containing the called user's network location and the features of the terminal he is presently using. If the type of terminal is compatible with the service logic, the SSCP sends a Terminal_info_type_ack. Then the called ASP contacts the TMP where the called user is located with a terminal_notification message. The TMP notifies the terminal of an incoming service session with a Consumer_notification. The terminal acknowledges the service session notification via a Consumer_notification_ack message. If the called user accepts to join the service session, an Accept_service_session message is sent to the TMP. Then the TMP initiates an authentication procedure such as described previously. After that, the TMP retrieves the user profile from the ASP and acknowledges the request for service session with an Accept_service_session_ack. Finally, the TMP sends a Terminal_notification message to the ASP, and the ASP acknowledges the join service session message.

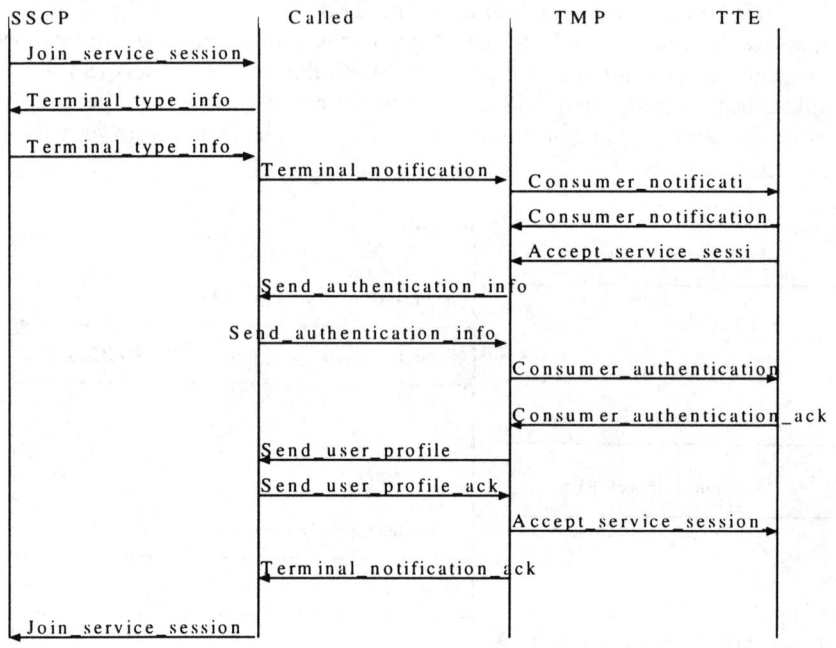

Figure 7   Terminating access session procedures

5   GENERAL OUTLINE OF SERVICE SESSION SETUP

The service invocation may proceed according to the following steps (Figure 8). The user clicks on a subscribed service bookmark and a subscribe_service_invoke is directed to the TMP. This message contains the Subscribed service Pointer SS_P which makes it possible to retrieve the network address of the service provider SSCP. Then a Send_user_status is sent to the SSCP containing the user identity and the access provider identity.The SSCP uses these identities to verify the subscription status of the invoking user and access provider. If the status allows, a service session may be instanced and a Send_user_status_ack is returned to the TMP containing service availability, cost, terminal requirements, eventually identity of other parties already involved in the service. These information are forwarded to the user by a Subscribed_service_invoke_ack message. The user may then confirm his service request with a service_confirm message. This message is relayed by the TMP to the SSCP. The SSCP initiates the service session and interacts with the invoking user to acquire all necessary service instance data such as Identities of other parties to be joined in the service, user languages... When all parties have agreed to join in the service, a Service_confirm_ack message is returned to the TMP indicating that the service session is setup. However in most cases calls will have to be setup. The Service_confirm_ack message will therefore indicate the calls that have to be initiated, the called parties identities along with the connectivity providers available for each of these parties (CP_P or connectivity providers pointers), and the suitable connecting mode for this service i.e. connection oriented or connectionless. This message is forwarded to the user. The user makes his choices for the current service session, and these options are sent in a Call_Request message to the TMP. The connectionless calls are directly initiated from the TMP. Connection oriented call_request messages are relayed to the SCCP. The SSCP can now contact the connectivity provider to request call setup.

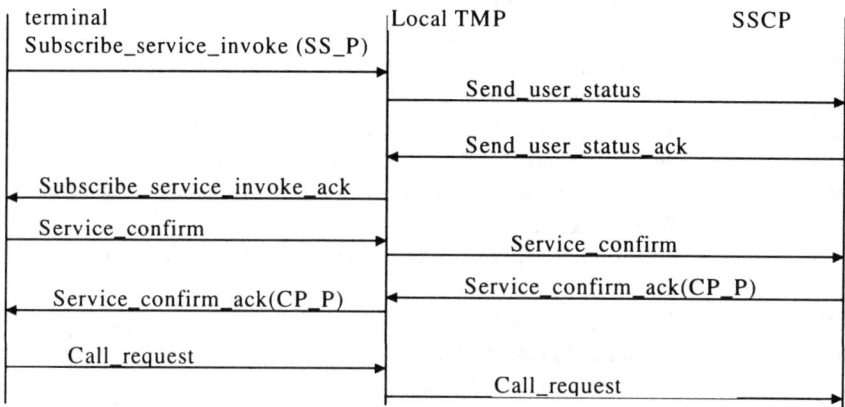

Figure 9 Service invocation

6 CONCLUSION

We may conclude from the above considerations that it is possible to develop signaling mechanisms allowing the implementation of a service control architecture that is compatible with the service objectives of TINA-C, which is implementable as an evolution of legacy systems and which takes into account the work of B-ISDN standardization without having to wait for the availability of an adapted and performant Real Time DPE environment. However we have mostly emphasized in this paper the service control aspect of the Broadband Telecommunication. We expect that the signaling mechanisms underlined here, provided that they are complemented with an efficient Call Control signaling will contribute to an efficient Broadband network.

This paper emphasized the signaling requirements at the access network. This allowed us to identify (or make assumptions on) the user and network data that the network infrastructure will have to maintain. In a future paper we will present our current work on how to organize, distribute and maintain the network information model in large scale broadband networks.

Glossary

ASF	Access Session Function
ASP	Access Session Point
B-ISDN	Broadband Integrated Services Digital Network
CORBA	Common Object Request Broker Architecture
DPE	Distributed Processing Environment
HLR	Home Location Register
IE	Information Element
IN	Intelligent Network
MAP	Mobile Application Part
OTE	Originating Terminal Equipment
OLEX	Originating Local Exchange
SIM	Subscriber Identification Module
SSCF	Service Session Control Function
SSCP	Service Session Control Point
TEX	Transit Exchange
TINA-C	Telecommunication Information Network Architecture Consortium
TLEX	Terminating Local Exchange
TMF	Terminal Manager Function
TMP	Terminal Manager Point
TTE	Terminating Terminal Equipment
VLR	Visitor Location Register

References

C. Rigault, T. Kovacikova: A signaling scheme for B-ISDN and IN integration. *IEEE ATM 97 workshop,* Lisboa, Portugal

B-ISDN CS3, Signaling requirements. Feb 96 *ITU-T Recommendations*
IN CS3, *ITU-T Recommendations*

E. Darmois: TINA: from concept to reality.
Proceedings ICIN 1996, Bordeaux
W.J. Barr, T. Boyd, Y.Inoue: The TINA Initiative
IEEE Communications Magazine, March 1993

H.323, visual telephones systems and equipment for local area networks which provide a non guaranteed quality of service.
Feb 96ITU-T Draft Recommendations

D. C. Schmidt et al: A high performance end system architecture for real time CORBA.
IEEE Communications Magazine, Feb 97

16

Communication Support for Knowledge-intensive Services

J. Berghoff, J. Schuhmann, M. Matthes, O. Drobnik
Department of Computer Science, University of Frankfurt/Main
Robert-Mayer-Str. 11-15, D-60054 Frankfurt (Germany), email:
{juergen, josefine, sascha, drobnik}@tm.informatik.uni-frankfurt.de

Abstract
Knowledge-intensive services gain significant importance for the development of innovative products. Up to now they depend mainly on face-to-face communication. Technological progress in artificial intelligence, groupware, and broadband systems favour computerized support of such services. In this field we focus on the computerized communication support for human interactions. We try to enrich human communication through 3D visualization of problems and research artefacts. A prototype system is under development to assess the required network capacity. It consists of tools for knowledge modeling and group interaction supplemented by a tailor-made multicast protocol.

Keywords
Network, CSCW, Graphics, Multicast, Collaborative Learning

1 INTRODUCTION

The advent of multimedia applications combined with networked desktop computers has shown a gap between communication needs of modern applications and current communication protocols (services). With respect to the requirements of real-time multimedia applications a variety of protocols and QoS models have been developed (for example: RTP, RSVP etc.) to bridge the gap. Besides the types of data which determine the communication needs, the kind of interaction has to be taken into account: one-to-one, one-to-many, many-to-many. The concurrent interaction of more than two entities which is typical for many CSCW-applications require adequate multicast protocols.

Driven by the World Wide Web and its needs for visualization of more and more complex objects and environments, new 3D visualization and animation techniques (VRML) enter the scene. This leads to new communication requirements. Again the gap between application and communication layer demands an integrating solution (e.g. Rhyne (1997), Brutzman (1997)).

On the application level, we have focused our research on groupware to support knowledge-intensive services. Such services are characterized by non-standardized human to human communication. We try to support this kind of communication by means of computerised 3D-visualization techniques and teleconferencing tools.

2   KNOWLEDGE-INTENSIVE SERVICES

Knowledge-intensive services gain significant importance for product development, eg., in the industry or financial enterprises to increase innovation. They aim in particular at the development of new products starting with rather rough ideas about the final product. Therefore, their degree of standardization is low and the individual processes to solve the problem are highly specialized. Often the product is developed and detailed in tight interaction between customer and service provider. Thus, solutions and innovations are the result of a collaborative learning process of both parties. The support of knowledge-intensive services is not restricted to the interaction between costumer and provider but also considers the production process itself. The spectrum of cooperation scenarios between firms (joint venture) or between developers (teamwork) is rather broad.

Non-standardized communication in all phases of a business relation is typical for such services. The complexity of a knowledge-intensive service comes from the service description on one side, and from the development and mediation of the solution on the other side.

Because of the highly interactive character of knowledge-intensive services their workflows include several face-to-face meetings. Face-to-face communication does not consist only exchanging information, which could be done by simple messaging solutions like email, but is often necessary for the process of finding general agreements, making decisions, getting a common understanding of problems and developing solutions.

At a first glance, face-to-face meetings could be supported by multimedia conferencing systems. But such systems must be enriched by tools to illustrate problems and points of discussion. These tools should support the collaborative construction of concepts for problem solving and the description of research artefacts. Since the kind of communication process immanent to knowledge-intensive services could be considered as an act of collaborative learning, solutions from this research area could be adopted. An important subject in collaborative learning is the modeling and representation of knowledge.

A widespread form of knowledge representation are so called "concept maps". A concept map is a diagrammatic representation which shows concepts (as nodes) and meaningful relationships (as links) between the concepts. Facts, evidences etc. are represented as nodes and are connected to each other via links. The link type reflects the kind of relation between the connected nodes.

Important topics are the process of creating knowledge and the representation of knowledge. Especially in collaborative knowledge construction the aggregation of individual views (from individual persons) of the problem to a commonly accepted representation is an important task. A system that addresses both problems is CLARE (Wan 1994), which stands for "Collaborative Learning and Research Environment". CLARE is a distributed learning environment and incorporates two semi-formal methods: RESRA and SECAI.

RESRA ("Representational Schema of Research Artifacts") is a semi-structured knowledge representation language designed specifically to facilitate collaborative learning from scientific text. RESRA can be seen as a special class of concept maps, where each node and link is of a certain type of a predetermined set of node types and link types. An example of an abstract RESRA-representation is shown in figure 1.

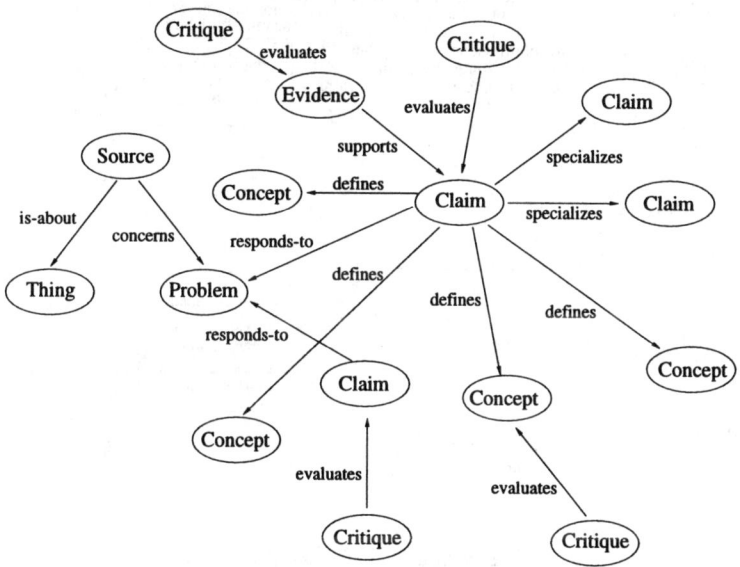

Figure 1 *RESRA example.*

SECAI ("Summarization, Evaluation, Comparison, Argumentation, and Integration") defines two phases for collaborative learning from scientific text: *exploration phase, consolidation phase* (cf. 3.1).

The concepts of our supporting tools are based on the ideas of CLARE. We are developing a system for collaborative modeling and visualization of problems and research artefacts. It will be integrated into a multimedia conferencing system to supplement or replace conventional face-to-face meetings. The overall concept of our approach is presented in more detail in the following section.

3   AN INFRASTRUCTURE FOR COLLABORATIVE LEARNING

Our infrastructure consists of a distributed meeting system for concurrent interaction of humans and incorporates video and audio capabilities for video conferencing and presentation tools for 3D-visualization of knowledge. Figure 2 shows an overview of the infrastructure. A central component is the control unit, which runs on the workstation of each participant (group member). The control unit manages the interaction between group members and controls the supporting applications. The video and audio components exist on each workstation and exchange data directly via multicast. In a first prototype, the visualization components for knowledge modeling are connected to a central server (a distributed version is planned). All actual representations of knowledge are stored in this server.

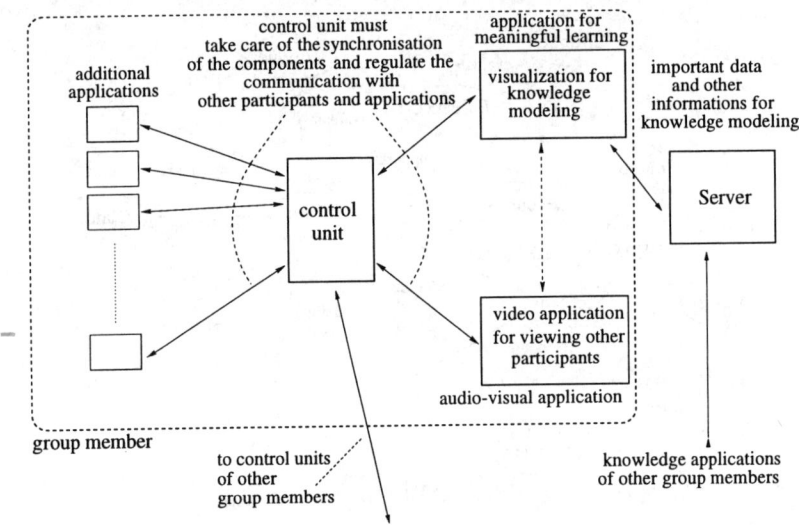

Figure 2 *Infrastructure overview.*

Several additional applications can be connected to the control unit. Most of these applications are collaboration-aware and consist of a number of replicated components (one on each local system of the participants) which communicate and interchange information via their own communication facilities. Examples of such applications are Whiteboards, Web-Browsers and other presentation components. The control unit is able to start applications and controls their behaviour in dependence of a role scheme. A participant has a certain role in a group. An example of roles and a role scheme is given in the virtual seminar section. Often data streams of different application components have to be synchronized. Synchronization information is distributed by the control units among the application components. One of the application components functions as a synchronization master and sends synchronization events to the control unit. Furthermore the control unit has the following functions: *role management and application control, member management, distribution of synchronization information.*

In the following section we concentrate on those components developed especially to support knowledge intensive services.

3.1   Knowledge modeling and representation

In conventional systems, concept maps have a 2-dimensional graphical representation. For complex problems or artifacts a map may become large and meshed and is difficult to understand. We propose a 3-dimensional interactive map representation to overcome these difficulties. Our implementation is based on VRML 2.0 (e.g. VRML96 (1996), Hartmann (1996)). VRML 2.0 enables the description of dynamic and animated worlds which responds to the actions of the user (e.g. if the user clicks on a door bell he could hear the corresponding sound). The possible actions of the user are registered by so called sensors which dispatch the events to the right nodes (In the example a *touch sensor* registeres the mouse klick and dispatches a corresponding event to a *sound node*). Another possibility to respond to a user action is to start a (small) program via *script nodes.* Script nodes contain either a URL at which the program could be found, or the source code itself. The programs can be written in *JavaScript* or *Java* (Descartes 1996). We use a VRML 2.0 viewer with support for *Java* script nodes.

(a)   Exploration phase:
In this first phase all group members work on their own to build a concept map. Nodes and links can be built, modified and deleted via a graphical user interface. For each node or link a textual annotation of any length can be stored. The input of the graphical user interface is transmitted to a so called *file generator* which modifies and updates the VRML-file. The file generator stores all informations in a database. After updating the VRML-file the file generator signals the viewer to update its display. To view the information associated with a node or link, the user simply clicks on the appropiate element on the screen, and the associated information is displayed in a separate window. Figure 3 shows the architecture of the knowledge modeling system. A canonical form of a concept map is displayed to help the user in building a representation of the problem.

(b)   Consolidation phase:
In a second phase all group members work together to aggregate the individual concept maps to a common representation of the problem. Differences and conflicts are discussed using multimedia conferencing capabilities. The knowledge modeling systems of the participants are connected to a central server

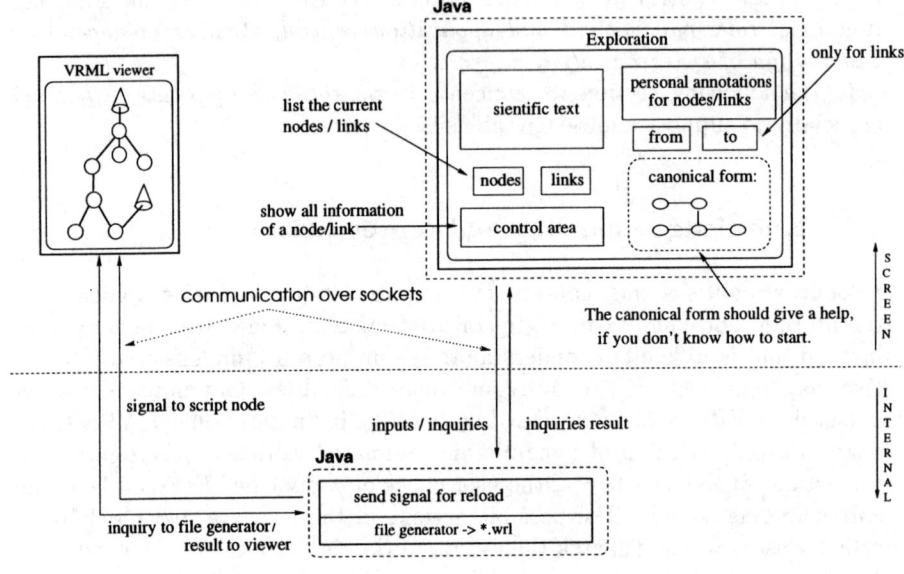

Exploration

Figure 3 *Architecture of the knowledge modeling system.*

which contains a common database. The access to the various representations is synchronized by the server. Updates and modifiactions of the representations are distributed by the server to each participant (point-to-multipoint communication).

The consolidation phase consists of many interactions between group members. Besides the exchange of audio and video information through the conferencing system the concurrent manipulation of the representations requires multiple and timely updates of the displays of all involved persons. A reliable multicast protocol is necessary to ensure the consistency of all displayed representations.

4 VIRTUAL SEMINAR

We implemented the *Virtual Seminar System* to evaluate the general infrastructure for collaborative learning. It will also be used to estimate the QoS-requirements. Each participant is provided with a replicated software environment which enables him to participate in a seminar from his home terminal. The communication mechanisms required for the data transfer are based on the services provided by the Internet.

The system encompasses several components which support audio-visual communication among the participants as well as the presentation of papers and charts. The role of the participants can change in the course of the seminar and determines which possibilities of interaction a participant has at any given time during the seminar. The structure of the Virtual Seminar System is shown in figure 4.

The control unit is responsible for the administration of the participants as well as the allocation of the roles. Each component has an interface which distributes information about the current roles of the participants.

Audio data synchronize the media information flow. As the human brain is most sensitive to disruptions in the perception of audio data, it is essential that speech sequences are replicated as clearly as possible. The audio component distributes local synchronization events for synchronizing the replication of audio data with the remaining media information flow.

The cross-system data exchange between the components is realized by using the IP-Multicast protocol which reduces the bandwidth required for distributed real-time group applications.

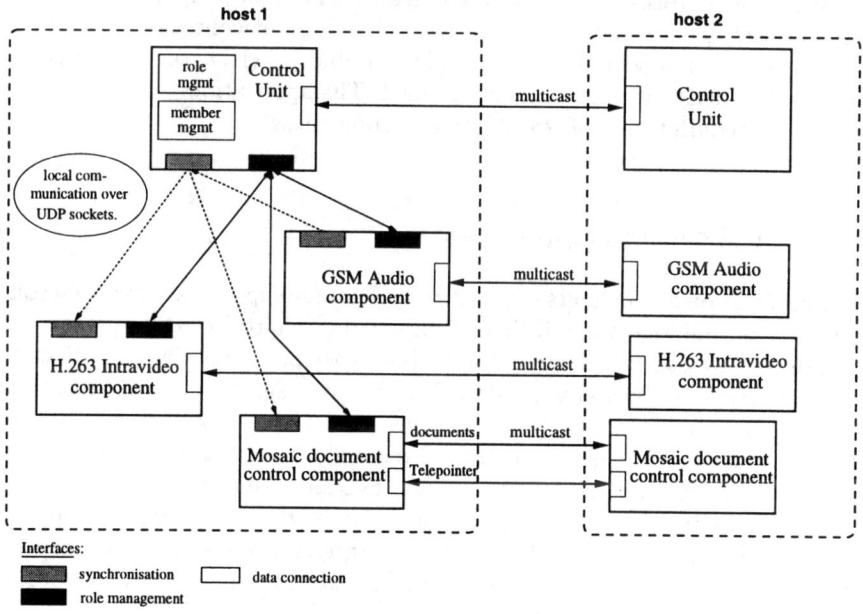

Figure 4 *Virtual Seminar Architecture.*

4.1 Role Administration

The role scheme reflects the hierarchical allocation of tasks in normal seminars. There are three active roles which can be held by a participant: *initiator, lecturer, talker.*

The initiator starts the seminar with an introduction into the subject of the respective seminar and allocates the role of the lecturer.

The lecturer presents a paper on a certain subject. He can allocate the role of the talker temporarily to another participant who has a question or comment.

The role of the talker is held by the initiator at the start of the seminar and by the lecturer during the presentation of the paper. It can be allocated to other participants to ask questions.

Participants who do not hold any active role are listeners.

The *Control Unit* administers and allocates the three active roles and is therefore responsible for the consistency of the distributed group application. At any given time during the seminar, each of the three active roles can be held by one participant only. This requires special control mechanisms such as the use of a protocol for the allocation of active roles. This protocol has been specifically adjusted for the Virtual Seminar. It uses the IP-Multicast protocol for addressing different participants simultaneously and uses other reliable Unicast protocols where direct point-to-point connections would be sufficient. That way some of the typical problems which tend to occur in reliable multicast protocols can be avoided. The application-specific protocol meets the requirements of scalability and robustness.

4.2   Application components

The application components of the *Virtual Seminar* support audio-visual communication and the presentation of papers and charts. Data sequences are transferred by using the unreliable IP-Multicast protocol. The loss of single data packets in audio and video data flows can be tolerated as the information of the individual data packets are only needed for playback at a certain time. Delayed data can be discarded. Due to the real-time character of the *Virtual Seminar* it is most likely that data sequences which are transferred repeatedly are received after they are needed for playback. As this would imply additional load to the network, it is sufficient to use unreliable data transfer methods. In order to reduce the required network capacities the media data sequences are compressed before transfer.

(a)   Video component

The video component realizes the transfer of video images of the lecturer or talker. The picture formats sub-QCIF (128x96) and QCIF (176x144) are supported. The compression is based on the intra compression method for video images which includes the Discrete Cosinus Transformation (DCT), quantisation and subsequent variable length encoding according to Huffman. As the compression is a pure software solution, there is no compensation of movement by calculating motion-vectors. For a higher compression rate a mechanism has been integrated which recognizes motion and only transmits those parts of the image which have changed. Moreover, there is a successive refresh procedure which updates the entire image periodically. Thus the refresh procedure prevents errors in the transmitted image.

(b)   Audio component

For the transfer of speech sequences the data are compressed according to the European GSM 6.10 standard used for digital mobile telephone communication in Europe which is particularly suitable for the compression of speech. The audio component recognizes speech activities so that only active speech sequences are transferred and pauses do not take up unnecessary network capacities.

(c)   Document component

The WWW-Browser Mosaic is used to present papers and charts in the *Virtual Seminar*. The browser can be accesed via its CCI interface. A telepointer feature has been integrated in the browser to provide higher comfort for the lecturer. The telepointer uses its current position in the document as orientation and moves the displayed part of the document in the receiving browsers accordingly. In their browsers the participants can therefore see the part of the document which is currently presented.

The procedures and components of the *Virtual Seminar* support an efficient collaboration of group members via the Internet. The specific needs of the consolidation phase have to be considered for the allocation of roles to group members.

5   MULTICAST PROTOCOL

An efficient infrastructure for groupware should make use of the multicasting capabilities of the network. Designing universal multicast protocols for a broad range of services is difficult. Diot (1997) gives a survey of multicast protocols and functions. We decided to design an application-specific multicast protocol based on IP-Multicast (Deering 1989).

The IP-Multicast protocol provides a service which is unreliable. There is no guarantee for a correct and complete data transfer. The IP-Multicast protocol can be extended by adding suitable control mechanisms to provide reliable data transfer.

When implementing reliable multicast services, the mechanisms used by IP-Unicast protocols (TCP) cannot be used as the scalability of the protocol would be poor (the protocol overhead would be too high in lage multicast groups). If every member of a group has to acknowledge the receipt of every single data packet to the sender, the result will be an acknowledgement implosion (Figure: 5).

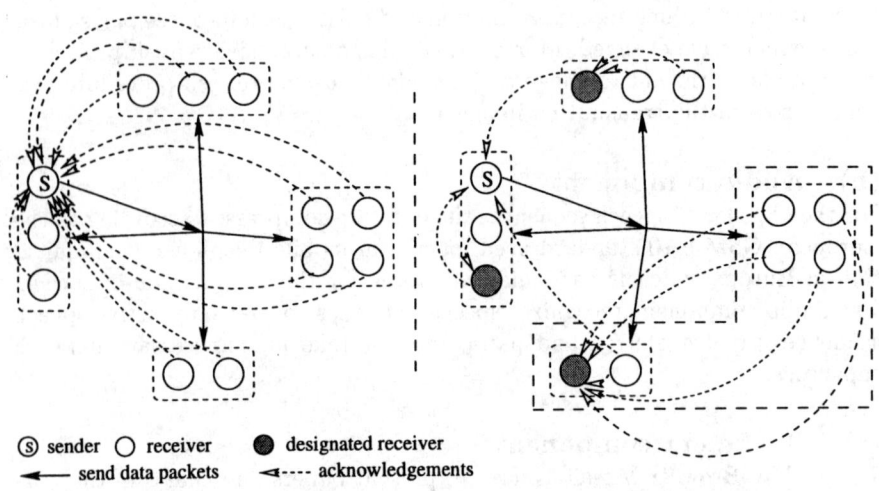

Ⓢ sender ◯ receiver   ● designated receiver
◀—— send data packets ◀---- acknowledgements

Figure 5 *Acknowledgement Implosion and sub-groups.*

One way of avoiding such situations is to define sub-groups. In this case several group members acknowledge the receipt of data packets together. Another solution is to confirm data transfers after individual delays. As a result the sender does not have to process all data transfer acknowledgements at the same time.

In addition to the extension of mechanisms for reliable data transfer, special processes need to be designed to guarantee that the transmitted data sequences are received in the correct order. This could be ensured by using sliding windows methods with numbered data packets to obtain the so-called source order. If the objective is to secure the correct order of all data sequences transmitted by different senders to the same group, i.e. the so-called total order special control structures need to be developed. This could be achieved by giving data sequences to be transferred global sequence numbers (e.g. Armstrong (1992), Bormann (1994)) or by giving already transmitted data sequences relative time marks (Birman 1991).

In the following we outline our multicast protocol, which was designed to provide high scalability and robustness. A protocol is considered robust when its performance is not affected by temporary breakdowns of systems involved in the data transfer.

Scalability: The allocation of the three active roles according to the role scheme has to be carefully synchronized. This requires a reliable data transfer.

Before allocating an active role to a person it has to be taken from the person who is presently holding it. This is realized by using a reliable point-to-point connection with the current role holder. All participants have to be informed about this change so that the components can adjust to the new allocation of roles. The allocation is therefore done by using the unreliable

IP-Multicast protocol. Only the affirmative acknowledgement of the new role holder is necessary to secure the correct allocation of the active roles.

Using the unreliable IP-Multicast protocol for the allocation of roles implies that not all participants are informed about the role changes if the data packets about the new roles are lost. However, the components are capable of recognizing that a role change has taken place as soon as they start receiving data sequences from a new sender. In this case a status request about the current role allocation is issued. This request is sent via the IP-Multicast protocol, thus all participants are informed and do not have to issue the same status requests. A status request is delayed briefly to avoid negative acknowledgement implosions just in case another participant has already sent a request.

Robustness: When the data transfer connections between the participants in a seminar are temporarily down, the protocol should reestablish the connections once the systems are back on line. The problem is the consistency of the overall status of the distributed group application, i.e., the allocation of the current roles.

If a participant who does not hold an active role is disconnected, the course of the seminar is not affected. The participant can join the seminar again as soon as the connection has been reestablished. If required, a status request can be issued to inform the participant about the current allocation of roles.

However, if a participant is disconnected who is holding an active role the seminar cannot be continued. The active role which has been lost cannot be allocated to a new person, as it has not been taken from the previous person. In that case the holder of the hierarchically higher role can give the lost role to another participant without acknowledgement so that the seminar can continue. As long as the connection to the previous holder of the role is down, the consistency of the group application is maintained. As soon as the previous holder is back on line, however, the active role is allocated twice. If the previous holder of the active role receives data sequences from the present holder or if the other participants receive data sequences from the previous holder, status requests are issued. As a result a status report on the present allocation of roles is transmitted and a consistent status is reestablished.

The application-specific protocol developed for the *Virtual Seminar* is therefore robust in case of system breakdowns and supports the continuing of the seminar in spite of breakdowns. Participants who have been disconnected can be reintegrated in the seminar. Thus the protocol does not only tolerate errors but addresses errors efficiently as well.

6 CONCLUSION

In this paper we introduce an infrastructure to support knowledge-intensive services. We use modern 3D-visualization techniques to enrich the human-to-human interface and integrate the knowledge modeling tools into a multimedia conferencing system. On the communication side we propose customized multicast protocols for efficient data exchange. A prototype system is under development; the virtual seminar and the exploration phase already exist. Experiments with the prototype are planned to assess the required network capacity. Part of this work (especially those focused on knowledge intensive services) is done in the context of the research program "Competitive Advantage by Networking" at the University Frankfurt (for more information about the research program see http://www.vernetzung.de). Further research will be done to advance the process of collaborative knowledge modeling and to support heterogeneous system environments.

REFERENCES

Descartes, Alligator (1996) Interfacing Java and VRML. *UK Java Developer's Conference Paper,* Nov. 1996.

Hartmann, J. and Wernecke, J. (1996) The VRML 2.0 Handbook. Addison Wesley Developers Press, 1996.

The Virtual Reality Modeling Language Specification (1996). Version 2.0; ISO/IEC CD 14772; August 1996.

Wan, D. Johnson, P.M. (1994) Computer Supported Collaborative Learning Using CLARE: *the Approach and Experimental Findings.* Proceedings of CSCW 94; New York 1994.

Rhyne, T-M. Brutzman, D. Macedonia, M. (1997) Internetworked Graphics and the Web; Computer, August 1997.

Bormann, C. Ott, J. *et al* (1980) MTP-2: Towards Achieving the S.E.R.O. Properties for Multicast Transport; ICCCN 94, San Francisco, Sept. 1994.

Armstrong, S. Freier, A. Marzullo, K. (1992) Multicast Transport Protocol; RFC 1301, 1992.

Birman, K. Schiper, A. (1991) Lightweight Causal and Atomic Group Multicast; ACM Transactions on Computer Systems, Vol. 9, No. 3, Aug. 1991.

Brutzman, D. Zyda, M. Watsen, K. Macedonia, M. (1997) virtual reality transfer protocol (vrtp) Design Rationale. Workshop on Enabling Technology (WET ICE), Cambridge, June 1997.

Diot, C. Dabbous, W. Crowcroft, J. (1997) Multipoint Communication: A Survey of Protocols, Functions and Mechanisms. IEEE Journal on Selected Areas in Communications, Vol. 15, No. 3, April 1997.

Deering, S. (1989) Host Extensions for IP Multicasting. RFC 1112, August 1989.

QoS and Charging Schemes

17

A study of simple usage-based charging schemes for broadband networks*

C. Courcoubetis[1], F. P. Kelly[2], V. A. Siris[1], and R. Weber[2]

[1]ICS-FORTH and Dept. of Computer Science, University of Crete,
P.O. Box 1385 GR 711 10 Heraklion, Greece. ˙
email: {courcou,vsiris}@ics.forth.gr

[2]University of Cambridge, Statistical Laboratory,
16 Mill Lane, Cambridge CB2 1SB, UK.
email: {fpk,rrw1}@statslab.cam.ac.uk

Abstract

Operators of high-speed networks are interested in implementing simple charging schemes with which they can fairly recover costs from their customers and effectively allocate network resources. This paper describes an approach for computing such charges from simple measurements (the duration and transferred volume of a connection), and relating these to bounds of the effective bandwidth. A requirement for usage-based charging schemes is that they capture the relative amount of resources used by connections. Based on this criteria, we evaluate our approach for Internet Wide Area Network traffic. Furthermore, its incentive compatibility is displayed with an example involving deterministic multiplexing, and the effect of pricing on a network's equilibrium is investigated for deterministic and statistical multiplexing.

Keywords

Usage-based charging, effective bandwidths, incentive compatibility, ATM, Internet

1 INTRODUCTION

A method for charging and pricing is an essential requirement in operating a high-speed network. Pricing is not only needed for recovering costs. There are compelling reasons that pricing is needed as a method of control. The congestion that has plagued the Internet, where pricing is based largely on *flat rate* pricing, highlights the fact that without usage-based pricing it is difficult to control congestion or divide network resources amongst customers in a workable and stable way (Mackie-Mason and Varian 1995, Mackie-Mason

---

*This work was supported in part by the EC under ACTS Project CASHMAN (AC-039).

Broadband Communications   P. Kühn & R. Ulrich (Eds.)
© 1998 IFIP. Published by Chapman & Hall

and Varian 1995, Gupta *et al.* 1994). Furthermore, in a competitive environment, besides offering sophisticated service disciplines, providers will need to price services in a manner which takes some account of network resource usage (Parris *et al.* 1992, Cocchi *et al.* 1993).

There are many considerations that influence the price of network services, such as marketing and regulation. However, these considerations are not particular to the operation of a communications network which is closely related to technological constraints (e.g., the quantities of services that it can support with a given network installation). A special consideration arises from the fact that a broadband communications network is intended to simultaneously carry a wide variety of traffic types and to provide certain performance guarantees. For example, in ATM networks a traffic contract is agreed among the customer and the operator. The customer agrees that his traffic will conform to certain parameters (e.g., which bound his peak rate and the size of his bursts), while the operator guarantees to carry this traffic with a particular quality of service (expressed, e.g., in terms of delay and cell loss ratio). The traffic contract gives the operator information by which he can bound the network resources that will be required to carry the call.

This paper is concerned with just one important part of the charging activity: that part which aims to assess a connection's resource usage. To avoid repeatedly having to qualify our remarks with a reminder that this is the focus, we shall henceforth simply refer to this component as *"charging"* and of computing a *"charge"*.

Some desired properties of tariffs

The role of tariffs is not only to generate income for the provider, but to introduce feedback and control. This happens via the mechanism that is automatically in effect as each individual customer reacts to tariffs and seeks to minimize his charges. For example, tariffs may be set which make it economical for some customers to shape their traffic, and by their doing so the overall network performance may be enhanced. This is the key idea of *incentive compatibility*. Tariffs should guide the population of cost-minimizing customers to select contracts and use the network in ways that are good for overall network performance (e.g., to maximize social welfare (Low and Varaiya 1993)). Tariffs which are not incentive compatible give the wrong signals and lead customers to use the network in very inefficient ways.

Well-designed tariffs should also have what we call the *fairness* property.[†] By this we mean that charges should reflect a customer's *relative* network usage. This raises the interesting question of when one charging scheme is more accurate than another, where accuracy is measured not in terms of the absolute value of the charges, but in terms of their correspondence to true network usage.

The above remarks naturally lead one to ask whether it possible to design tariffs that are sound, both in terms of incentive compatibility and fairness, but which are also not too complex, and whose implementation does not require the network operator to make overly sophisticated or unrealistic measurements. Incentive compatibility will be hard to achieve if tariffs are too complex, since customers will find it difficult to determine what effect the decisions under their control, such as whether or not to shape their traffic, might have on the charges they incur.

[†]In the case of differential pricing and/or time-of-day pricing, the fairness property is considered for customers of the same "class" which use network services at the same time period.

Contribution of the paper

In this paper we provide the framework for constructing incentive compatible charges that reflect effective usage. Our approach is based on the notion of effective bandwidth as a proxy for resource usage. In this sense our work differs from (Low and Varaiya 1993, Sairamesh *et al.* 1995) which investigate optimal pricing strategies assuming that network resources (buffer and capacity) are charged separately, and (Wang *et al.* 1996) which also deals with optimal pricing, but does not address the issue of measuring the amount of resources used by connections.

Our charging schemes are simple and can be cast in the same formats that are used today, namely the charge depends on *static* contract parameters (access line speed, policing parameters, anticipated average rate) and on *dynamic* parameters of the connection (actual average rate). Our approach is quite general, and can also be used to design that part of a tariff which prices the network usage of large customers connected to an Internet service provider. Furthermore, it can be complemented with other pricing mechanisms such as time-of-day pricing (Shenker *et al.* 1996).

The novelty of the approach lies in the following two points. First, we provide an interpretation of effective bandwidths that is right for our purposes. In (Courcoubetis, Kelly and Weber 1997) we provide the mathematical foundation of our charging framework where we show that the effective bandwidth of a connection depends on the actual state (composition of the traffic mix) of the links in a network, hence can not be defined in isolation. Furthermore, this dependence is only through a pair of parameters (the s, t parameters discussed in Section 2.1). The same connection will potentially exhibit different effective bandwidths at different times of the day. An important consequence of the approach is that it treats deterministic and statistical multiplexing in a unifying way.

The second contribution is in the way we transform simple tariffs of the form $a_0 T + a_1 V$, where T is the duration and V is the volume of a connection, into sound approximations of the effective bandwidth of the connection, by casting all the information from the static contract parameters and the operating point of the network into the coefficients a_0, a_1.[‡] Based on experimentation, we believe that our simple tariffs can serve their purpose well and can provide the right incentives for efficient and stable network operation.

The rest of the paper is organized as follows. In Section 2 we briefly explain our charging methodology by reviewing some key notions and results for the simpler case of a network consisting of a single shared link. In Section 3 we discuss issues related to the fairness of charging schemes, based on which we evaluate our approach for Internet Wide Area Network traffic. In Section 4 we discuss the incentive compatibility of the approach and work through a complete example in the simpler, but illuminating, case of deterministic multiplexing. Our conclusions and some open issues are discussed in Section 5.

[‡]The theory developed in (Courcoubetis, Kelly and Weber 1997) allows for the construction of more elaborate charging schemes where the network measurements can be arbitrarily complex.

2 A THEORY FOR USAGE-BASED CHARGING

2.1 Effective bandwidths as a measure of resource usage

Suppose the arrival process at a broadband link is the superposition of independent sources of J types: let n_j be the number of connections of type j, and let $n = (n_1, \ldots, n_J)$. We suppose that after taking into account all economic factors (such as demand and competition) the proportions of traffic of each of the J types remains close to that given by the vector n, and we seek to understand the relative usage of network resources that should be attributed to each traffic type.

Consider a discrete time model and let $X_j[0, t]$ be the total load produced by a source of type j in epochs $0, \ldots, t$. We assume that the increments of $\{X_j[0, t], t \geq 0\}$ are stationary. Then, the *effective bandwidth* of a source of type j is defined as

$$\alpha_j(s, t) = \frac{1}{st} \log E \left[e^{sX_j[0,t]} \right], \tag{1}$$

where s, t are *system defined* parameters which depend on the characteristics of the multiplexed traffic and the link resources (capacity and buffer). Specifically, the *time* parameter t (measured in, e.g., msec) corresponds to the most probable duration of the buffer busy period prior to overflow. The *space* parameter s (measured in, e.g., kb^{-1}) corresponds to the degree of multiplexing and depends, among others, on the size of the peak rate of the multiplexed sources relative to the link capacity. In particular, for links with capacity much larger than the peak rate of the multiplexed sources, s tends to zero and $\alpha_j(s, t)$ approaches the mean rate of the source, while for links with capacity not much larger than the peak rate of the sources, s is large and $\alpha_j(s, t)$ approaches the maximum value of $X_j[0, t]/t$.

Let $L(C, B, n)$ be the proportion of workload lost, through overflow of a buffer of size $B > 0$, when the server has rate C and $n = (n_1, n_2, \ldots, n_J)$. Assume that the constraint on the proportion of workload lost is $e^{-\gamma}$ (we will assume that the Quality of Service -QoS- is expressed solely through this quantity). The *acceptance region* $A(\gamma, C, B)$ is the subset of $\mathbb{Z}_+^J$ such that $n \in A(\gamma, C, B)$ implies $\log L(C, B, n) \leq -\gamma$, i.e., the QoS constraint is satisfied.

If n is on the boundary of the region $A(\gamma, C, B)$, and the boundary is differentiable at that point, then the tangent plane determines a half-space which is well approximated, when C, B, and n are large, by (Kelly 1996)

$$\sum_j n_j \alpha_j(s, t) \leq C + \frac{1}{t} \left(B - \frac{\gamma}{s} \right), \tag{2}$$

where (s, t) is an extremizing pair in the equation (called the *many sources asymptotic*; see (Courcoubetis and Weber 1996))

$$\lim_{N \to \infty} \frac{1}{N} \log L(NC, NB, nN) = \sup_t \inf_s \left[st \sum_{j=1}^{J} n_j \alpha_j(s, t) - s(Ct + B) \right]. \tag{3}$$

The asymptotics behind this approximation assumes only stationarity of sources, and illustrative examples discussed in (Kelly 1996) include periodic streams, fractional Brownian input, policed and shaped sources, and deterministic multiplexing. Note that the QoS guarantees are encoded in the effective bandwidth definition through the value of γ which influences the form of the acceptance region.

We must stress the network engineering implications of the above results. For any given traffic stream, the effective bandwidth definition (1) is nothing more than a template that must be filled with the link's operating point parameters s, t in order to provide the correct measure of effective usage. Furthermore, experimentation has revealed that the values of s, t are, to a large extent, insensitive to variations of the traffic mix (percentage of different traffic types) (Courcoubetis, Siris and Stamoulis 1997). Since during different times of the day the traffic mix at a given link is anticipated to remain relatively constant, we can assign particular pairs (s, t) to different periods of the day. These values can be computed off-line using (1) and (3), where the expectation in (1) is replaced by the empirical mean which is computed from traffic traces.

2.2   Charges based on effective bandwidths

We have argued above that effective bandwidths can provide a way to assess resource usage, and hence can be used for constructing the usage-based component of the charge. There are two extreme methods by which this can be done.

Consider sources of type j, where "type" is distinguished by parameters of the traffic contract and possibly some other static information. The network could form the empirical estimate $\alpha'_j(s, t)$ of the expectation appearing in formula (1), as determined by past connections of type j. A new connection of type j would be charged at an amount per unit time equal to $\alpha'_j(s, t)$. This is the charging method adopted in an all-you-can-eat restaurant. At such a restaurant each customer is charged not for his own food consumption, but rather for the average amount that similar customers have eaten in the past. Under such a charging scheme, each customer may as well use the maximum amount of network resources that his contract allows, which will result in $\alpha'_j(s, t)$ eventually becoming the largest effective bandwidth that is possible subject to the agreed policing parameters. Customers who have connections of type j, but whose traffic does not have the maximal effective bandwidth possible for this type, will not wish to pay as if they did, hence will seek network service providers using a different (more competitive) charging method.

At another extreme, one might charge a customer wholly on the basis of measurements that are made for his connection, i.e., charge the value of the effective bandwidth of the traffic actually sent. This has a conceptual flaw which can be illustrated as follows. Suppose a customer requests a connection policed by a high peak rate, but happens to transmit very little traffic over the connection. Then an *a posteriori* estimate of quantity (1), hence his charge, will be near zero, even though the *a priori* expectation may be much larger, as assessed by either the customer or the network. Since tariffing and connection acceptance control may be primarily concerned with expectations of *future* quality of service, the distinction matters. This is the case because such a charging scheme does not account for the resources reserved at call setup, which is unfair for the network operator.

Our approach lies part way between the two described above. We construct a charge that is based on the effective bandwidth, but which is a function of both *static* parameters (such as the peak rate and leaky bucket parameters) and *dynamic* parameters (these correspond to the actual traffic of the connection, the simplest ones being the duration and volume of the connection); we *police* the static parameters and *measure* the dynamic parameters; we bound the effective bandwidth by a linear function of the measured parameters, with coefficients that depend on the static parameters; and we use such linear functions as the basis for simple charging mechanisms. This leads to a charge with the right incentives for customers, which also compensates the network operator for the amount of resources reserved.

2.3   Charges linear in time and volume

Suppose that a connection lasts for epochs $1, \ldots, T$ and produces load $X_1, \ldots, X_T$ in these epochs. Imagine that we want to impose a *per unit time* charge for a connection of type j that can be expressed as a linear function of the form

$$f(X) = a_0 + a_1 g(X), \tag{4}$$

where $g(X)$ is the measurement taken from the observation $X = (X_1, \ldots, X_T)$ corresponding to $(1/T)\sum_{i=1}^{T} X_i$. In other words, the total charge is simply a function of the total number of cells carried, and, through a_0, the duration of the connection. This is practically the simplest measurement we could take and leads to charging schemes based on just time and volume.

We argued in Section 2.2 that the usage-based charge of a connection should be proportional to the effective bandwidth $\alpha(s,t)$ of the connection, for appropriate s, t. Next we describe how linear functions of the form (4) can be constructed so that the expected charge bounds the effective bandwidth.

Let $\bar{\alpha}(m, \mathbf{h})$ be an upper bound for the greatest effective bandwidth possible subject to constraints imposed by the traffic contract $\mathbf{h}$, while the mean rate is m. Consideration of $\bar{\alpha}(m, \mathbf{h})$ is partly motivated by the remark that this is what we would charge to a customer with mean rate m who makes maximal use of his traffic contract.

We define our tariffs in terms of the charging function f parameterized with $m, \mathbf{h}$. Mathematically, this corresponds to the tangent of $\bar{\alpha}(m, \mathbf{h})$ at m:

$$f(m, \mathbf{h}; X) := \bar{\alpha}(m, \mathbf{h}) + \lambda_m(g(X) - m), \tag{5}$$

which is of the form $a_0 + a_1 g(X)$, where $a_0[m, \mathbf{h}] = \bar{\alpha}(m, \mathbf{h}) - \lambda_m m$, $a_1[m, \mathbf{h}] = \lambda_m = \frac{\partial}{\partial m}\bar{\alpha}(m, \mathbf{h})$. These coefficients depend on the customer's choice of m. Because $\bar{\alpha}(m, \mathbf{h})$ is concave in m (Courcoubetis, Kelly and Weber 1997), one can show that the expected value of the charging rate for this connection is $Ef(m, \mathbf{h}; X) \geq \bar{\alpha}(Eg(X), \mathbf{h})$, with equality if $m = Eg(X)$ (the actual mean rate of the connection). Hence, the customer minimizes his expected charge if he chooses the tariff $f(Eg(X), \mathbf{h})$.

As we intended, the coefficients $a_0[m, \mathbf{h}], a_1[m, \mathbf{h}]$ depend upon both static information, as well as the customer's expectation regarding his mean rate (which is measured by the network). The dependence of the charge on m provides the customers with the right incentives for avoiding the "all-you-can-eat restaurant" effect mentioned before.

Approximations for $\bar{\alpha}(m, \mathbf{h})$

Let m be the mean rate of a source, and $\bar{X}[0, t]$ be the maximum amount of traffic produced in a time interval of length t. Since the source is policed by parameters (ρ_k, β_k), $k \in K$, we have

$$\bar{X}[0, t] \le H(t) := \min_{k \in K} \{\rho_k t + \beta_k\} . \qquad (6)$$

The last constraint together with the convexity of the exponential function implies that

$$\bar{\alpha}(m, \mathbf{h}) \le \frac{1}{st} \log \left[1 + \frac{tm}{H(t)} \left(e^{sH(t)} - 1 \right) \right] = \tilde{\alpha}_{\mathrm{sb}}(m, \mathbf{h}) . \qquad (7)$$

We call the right hand side of the above equation the "simple" approximation. This equation is illuminating for the effects of leaky buckets on the amount of resource usage. Each leaky bucket (ρ_k, β_k) constraints the burstiness of the traffic in a particular time scale. The time scale of burstiness that contributes to buffer overflow is determined by the index k which achieves the minimum in (6).

If t=1, then the bound (7) reduces to

$$\tilde{\alpha}_{\mathrm{pm}}(m, \mathbf{h}) = \frac{1}{s} \log \left[1 + \frac{m}{h} (e^{sh} - 1) \right] , \qquad (8)$$

which is appropriate when the buffers are small and the argument minimizing expression (6) corresponds to the peak rate h. We refer to this as the "peak/mean" bound. Charges based on this bound have been considered in (Kelly 1994).

In many cases (Courcoubetis, Kelly and Weber 1997), the worst case traffic (for given values of s, t) consists of blocks of an inverted T pattern repeating periodically or with random gaps. In this paper we consider the periodic pattern shown in Figure 1, which gives the following effective bandwidth approximation (referred to as the "inverted T" approximation):

$$\tilde{\alpha}_{\perp}(m, \mathbf{h}) = \frac{1}{st} \log E \left[e^{sX_{\perp}[0, t]} \right] , \qquad (9)$$

where $X_{\perp}[0, t]$ denotes the amount of load produced by the inverted T pattern in a time interval of length t. The expected value in the right-hand side of (9) can be computed analytically.

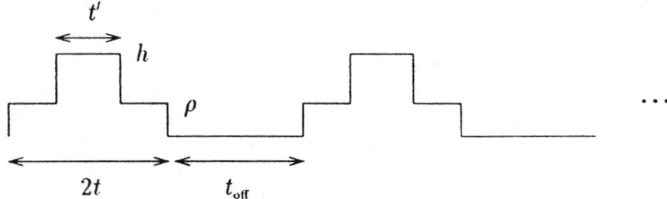

Figure 1 Periodic pattern for the inverted T approximation. $t' = \frac{\beta}{h - \rho}$, $t_{\mathrm{off}} = \frac{(2t - t')\rho + t'h}{m} - 2t$

3   EVALUATING THE CHARGING SCHEME

In Section 2.3 we introduced the class of tariffs $f(m, \mathbf{h}; X) = \bar{\alpha}(m, \mathbf{h}) + \lambda_m(g(X) - m)$, where $\mathbf{h}$ are the policing constraints in the traffic contract, $g(x)$ is the measured mean rate of the connection, and m is the anticipated value of this mean rate by the customer. For simplicity we assume that the customer knows his mean rate, hence his charge will be equal to $\bar{\alpha}(m, \mathbf{h})$, which can be approximated by (7), (8), or (9). In this section, we evaluate the performance of these approximations.

One important criterion for a pricing scheme, which is based on some approximation $\tilde{\alpha}$ of the bound $\bar{\alpha}$, is *fairness*. Ideally we would like the relative charges using $\tilde{\alpha}$ to be as close as possible to those using the actual effective bandwidth α. Hence, if (with a slight abuse of notation) we denote by $\tilde{\alpha}(x)$ and $\alpha(x)$ the corresponding charges for a connection x, then we would like to have $\tilde{\alpha}(y)/\tilde{\alpha}(x) \approx \alpha(y)/\alpha(x)$, for any two connections x, y. A reasonable measure of the *unfairness* of an approximation for a set of connections is the standard deviation of $\tilde{\alpha}(x)/(\mu\alpha(x))$, where μ is the average of $\tilde{\alpha}(x)/\alpha(x)$ as x ranges over the connection set. We will refer to this as the *unfairness index* $\mathcal{U}$. For example, an approximation that consistently overestimates the true effective bandwidth by some constant will have $\mathcal{U} = 0$, hence would be preferable than some other approximation which, on the average, is closer to the true effective bandwidth, but whose ratio $\tilde{\alpha}(x)/\alpha(x)$ varies (hence $\mathcal{U} > 0$).

We have done extensive experimentation involving the three approximations introduced in Section 2.3, with different types of traffic (e.g., MPEG video). In this paper we consider the case of Internet Wide Area Network (WAN) traffic using the Bellcore Ethernet trace BC-Oct89Ext[§] (Leland and Wilson 1991), which has a duration of 122797 seconds. We assume that a customer is policed by two leaky buckets $\mathbf{h} = \{(h, 0), (\rho, \beta)\}$, and initially assume that traffic is shaped in a 200 ms buffer. This reduces the peak rate to $h = 0.88$ Mbps. The pairs (ρ, β) for which no traffic is discarded by the policer corresponds to the indifference curve G (Figure 2). Finally, we assume that all users are "rational", i.e., they select the pair (ρ, β) that minimizes their charge.

From the initial Bellcore trace we created a set of 15 non-overlapping trace segments, each with duration 8186 seconds (approximately 2.5 hours). For this set, we wish to compare the three different charging schemes based on approximations (7), (8), and (9) according to the unfairness index $\mathcal{U}$ defined above.

As discussed in Section 2.1, the parameters s, t characterize the link's operating point. We consider a link with capacity $C = 34$ Mbps and a target overflow probability equal to 10^{-6}, and use equations (1) and (3) to compute "typical" values of s, t, where the expectation in (1) is replaced by the empirical mean which is computed from the trace.

Figure 3 shows that the unfairness for the simple bound and inverted T approximations is close, and much smaller than that for the peak/mean bound. Furthermore, while the unfairness for the former two approximations decreases when the buffer size increases, this is not the case for the peak/mean bound. This is expected because the peak/mean bound becomes accurate for small values of t, which are realized for small buffer sizes.

Figure 4 shows the unfairness for the three approximations in a neighborhood of values for s, t when $B = 0.25 \times 10^6$ bytes. Observe that both the simple bound and the inverted T approximations are fairer and more robust (the surface is "flatter") compared to the

[§]Obtained from The Internet Traffic Archive. <http://www.acm.org/sigcomm/ITA/>.

peak/mean bound. Furthermore, increasing the link capacity and buffer size increases the fairness and robustness of the schemes.

4 INCENTIVE COMPATIBILITY

As we have already mentioned in the introduction, the operating point of the link and the posted tariffs are interrelated in a circular fashion. The network operator posts tariffs that have been computed for the current operating point of the link, expressed through the parameters s, t. These tariffs provide *incentives* to the customers to change their contracts in order to minimize their anticipated costs. Under these new contracts, the operating point of the system will move, since the network operator must guarantee the performance requirements of these new contracts. Hence, the network operator will calculate new tariffs for the new operating point. This interaction between the network and the customers will continue until an equilibrium is reached. We validate below, for a simple example, that if the network operator uses our charging approach, then an equilibrium does exist and that it is a point maximizing social welfare, as measured in this example by the number of customers admitted to the system.

For simplicity, we assume that all customers have identical profiles, are policed with a single leaky bucket (ρ, β), and have identical indifference curves $G = \beta(\rho)$. We assume that G is convex, tends to infinity when ρ goes to the mean rate m, and is zero for $\rho = h$. The network consists of a shared link with capacity C and buffer B, and uses deterministic multiplexing for loading the link.

In the case of deterministic multiplexing (zero cell loss), our effective bandwidth theory suggests that the value of the parameter s should be ∞ (this follows from (2) when $\gamma = \infty$), and that the effective bandwidth of a connection policed with (ρ, β) is $\alpha_j(\infty, t) = \bar{X}[0, t]/t = \rho_j + \frac{\beta_j}{t}$ for $t > 0$ and $\alpha_j(\infty, 0) = \beta_j$. Simple algebra shows that the acceptance region A (one-dimensional in our case) is defined by the constraints

$$\sum_j \rho_j \leq C \text{ and } \sum_j \beta_j \leq B, \tag{10}$$

on which the effective bandwidth is defined for $t = \infty$ and $t = 0$, respectively.

We assume that the system proceeds in lock-step and customers have identical requirements. Hence, at any point in time their choices will coincide. Due to this, the above constraints become $n\rho \leq C$ and $n\beta \leq B$, where n is the total number of customers.

Consider the point $Q \in G$ where the two constraints coincide and the number of customers n is maximized (welfare optimum). This point is also defined by the intersection of the line with slope B/C (that passes from the origin) with G. One can easily see that for any point M in G (i.e., initial choice of (ρ, β) by customers) which is below Q, the system will fill so that the active constraint will have a corresponding value $t = \infty$ for the calculation of the effective bandwidth, whereas if M lies above Q, then $t = 0$.

Assume now that our charging approach is used by the network. If the customers choose a point M below Q, then the first constraint will be active ($t = \infty$) and the charge will be proportional to ρ; this will guide customers to reduce ρ and move towards Q. If the customers choose a point M above Q, then the second constraint will be active and the charge will be proportional to β; this will guide customers to reduce β and move towards

B (bytes)	Deterministic mult.			Statistical mult.		
	ρ (Mbps)	β (bytes)	n_{max}	ρ (Mbps)	β (bytes)	n_{max}
0.5×10^6	0.615	10600	33	0.475	29100	1530
1×10^6	0.553	18300	54	0.399	52800	1650
5×10^6	0.373	62500	80	0.202	175500	2070
10×10^6	0.285	95500	105	0.162	341100	2170

Table 1 Equilibrium under deterministic and statistical multiplexing.

Q. Assuming that, in order to avoid oscillations, customers are allowed to make small changes to their traffic contracts, the point Q will be eventually reached. At Q, since both constraints are active, the charge will be proportional to a linear combination $\lambda_1\rho + \lambda_2\beta$ of the effective bandwidths corresponding to the active constraints at Q (i.e., both ρ and β), where λ_1, λ_2 are the shadow prices of the optimization problem which maximizes the number of users under constraints (10). One can check that the above charges correspond to the tangent of G at Q, hence Q is an equilibrium since the user minimizes his charge by remaining there.

In the case of statistical multiplexing, the above arguments can be extended to show a similar user-network behavior. We have calculated such equilibria for a range of buffer sizes and for a target overflow probability 10^{-6} (Table 1). As expected, the utilization in the case of statistical multiplexing is much higher than in the case of deterministic multiplexing.

Effects of traffic shaping

We are now in a position to make some interesting observations about the effect of customers delaying their traffic into the network. As we will argue, for the anticipated buffer sizes, shaping has a surprisingly small effect on the overall multiplexing capability of the network.

First, observe in Figure 2 that for large values of β, the indifference curve $G(d)$ is not greatly affected when the shaping delay is smaller than 500 msec . Second, observe in Table 1 that in the case of statistical multiplexing and for buffer sizes greater than 1×10^6 bytes, at the equilibrium we have $\beta > 50000$ bytes. Combining these two observations we see that for buffer sizes greater than 1×10^6 bytes, the equilibrium point will not be affected by traffic shaping, when the shaping delay is less than 500 msec .

Of course a customer can use shaping to make a contract with a lower peak rate. However, contrary to the intuition, this will not affect his effective bandwidth as seen by the network, since the time parameter t at the equilibrium is always large enough so that $ht > \rho t + \beta$. In this case, the effective bandwidth is determined largely in terms of the values (ρ, β) (e.g., if the customer sends traffic close to the maximum amount allowed by the simple bound (7)) which, as argued previously, remain practically unaffected by shaping.

The above discussion demonstrates how the theory described in Section 2 clarifies the effects of various time scales and the importance of the various traffic and network parameters on the amount of resources used by connections.

5 CONCLUSIONS AND OPEN QUESTIONS

This paper has dealt with one important part of the charging activity: the part which aims to access a connection's network resource usage. In this direction, we have provided a framework for constructing incentive compatible charges that reflect effective resource usage. Our charging schemes are based on bounds on the effective bandwidth and involve only measurements of the duration and volume of connections. The schemes are simple in the sense that they are easily understood by the customers. Furthermore, they can be cast in the same formats that are used today, namely, charges depend on static contract parameters (e.g., access line speed, leaky bucket policing parameters, anticipated average rate), and on dynamic parameters of a connection (e.g, actual average rate). We have displayed the incentive compatibility of the proposed schemes through an example involving deterministic multiplexing, and have presented numerical results, with real broadband traffic, that display the fairness of the schemes. It is important to note that our approach is quite general and can be used to charge for effective usage at many levels of network access, ranging from individual users to large organizations. It can be applied to any packet switching technology and can be used under both deterministic and statistical multiplexing.

The extension of our approach to networks consisting of more than one link raises several further issues which we hope to treat in the future. Important choices concern whether a user sees a single charge from its immediate service provider, or whether a user might see several charges arising from various intermediate networks. We simply note here that charges linear in time and volume remain so under aggregation.

REFERENCES

Cocchi, R., Shenker, S., Estrin, D. and Zhang, L. (1993) Pricing in computer networks: Motivation, formulation, and examples. *IEEE/ACM Trans. on Networking*, 1, 614–627.

Courcoubetis, C., Kelly, F. P. and Weber, R. (1997) Measurement-based charging in communications networks. Technical Report 1997-19, Statistical Laboratory, University of Cambridge.

Courcoubetis, C., Siris, V. A. and Stamoulis, G. D. (1997) Many sources asymptotic and effective bandwidths: Investigation with MPEG traffic. Presented at the *2nd IFIP workshop on traffic management and synthesis of ATM networks*, Montreal, Canada, September 1997. Extended version submitted for publication.

Courcoubetis, C. and Weber, R. (1996) Buffer overflow asymptotics for a switch handling many traffic sources. *Journal of Applied Probability*, 33, 886-903.

Gupta, A., Stahl, D. O. and Whinston, A. B. (1994) Managing the Internet as an economical system. Technical report, University of Texas, Austin.

Kelly, F. P. (1994) On tariffs, policing and admission control for multiservice networks. *Operations Research Letters*, 15, 1–9.

Kelly, F. P. (1996) Notes on effective bandwidths. In F. P. Kelly, S. Zachary and I. Zeidins, eds., *Stochastic Networks: Theory and Applications*, pp. 141–168. Oxford University Press.

Leland, W. E. and Wilson, D. V. (1991) High time-resolution measurement and analysis of LAN traffic: Implications for LAN interconnection. In *Proc. of IEEE INFOCOM'91*, pp. 1360–1366.

Low, S. H. and Varaiya, P. P. (1993) A new approach to service provisioning in ATM networks. *IEEE/ACM Trans. on Networking*, **1**, 547–553.

Mackie-Mason, J. K. and Varian, H. R. (1995) Pricing congestible network resources. *IEEE J. Select. Areas in Commun.*, **13**, 1141–1149.

Mackie-Mason, J. K. and Varian, H. R. (1995) Pricing the Internet. In B. Kahin and J. Keller, eds., *Public Access to the Internet*. Prentice Hall, Englewood Cliffs, NJ.

Parris, C., Keshav, S. and Ferrari, D. (1992) A framework for the study of pricing in integrated networks. Technical Report TR-92-016, International Computer Science Institute, Berkeley, CA.

Sairamesh, J., Ferguson, D. F. and Yemini, Y. (1995) An approach to pricing, optimal allocation and quality of service provisioning in high-speed packet networks. In *Proc. of IEEE INFOCOM'95*.

Shenker, S., Clark, D., Estrin, D. and Herzog, S. (1996) Pricing in computer networks: Reshaping the research agenda. *ACM Computer Communication Review*, **26(2)**, 19–43.

Wang, Q., Peha, J. M. and Sirbu, M. A. (1996) The design of an optimal pricing scheme for ATM integrated-services networks. In J. P. Bailey and L. Mcknight, eds., *Internet Economics*, Massachusetts, 1996. MIT Press.

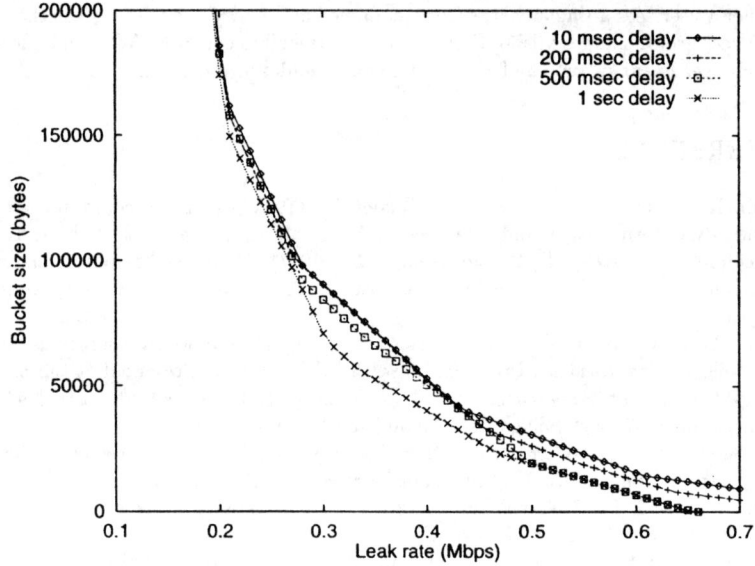

Figure 2 Indifference curve $G(d)$ for the Bellcore trace.

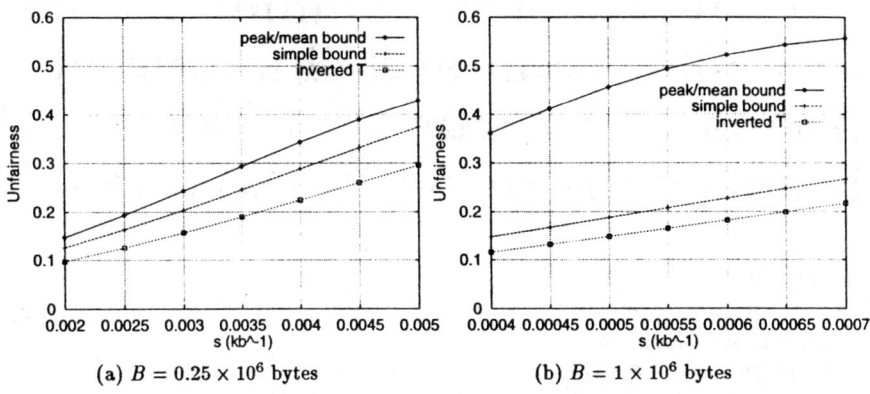

(a) $B = 0.25 \times 10^6$ bytes (b) $B = 1 \times 10^6$ bytes

Figure 3 Unfairness for $C = 34$ Mbps and two buffer sizes.

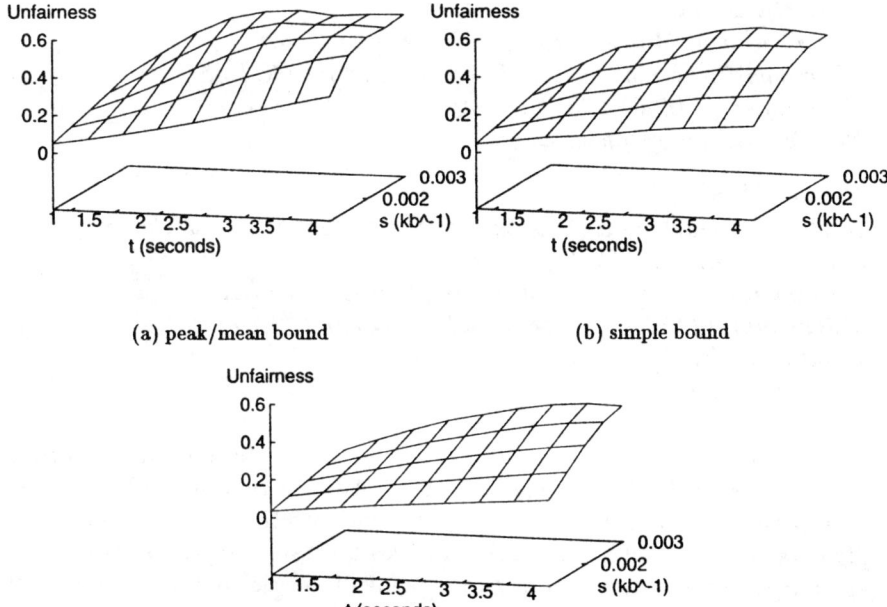

(a) peak/mean bound (b) simple bound

(c) inverted T

Figure 4 Unfairness for $C = 34$ Mbps, $B = 0.25 \times 10^6$ bytes.

18

A game theoretic framework for rate allocation and charging of Available Bit Rate (ABR) connections in ATM networks

H. Yaiche
Département de Génie Electrique et Génie Informatique, Ecole Polytechnique de Montréal
C.P. 6079, succ. A, Montreal, Quebec, Canada H3C 3A7
Telephone: +(514) 340-4123, Fax: +(514) 340-4562,
Email: yaiche@comm.polymtl.ca

R.R. Mazumdar
Department of Mathematics, University of Essex
Wivenhoe Park, Colchester, United Kingdom CO4 3SQ
Telephone: +(1206) 873-032, Fax: +(1206) 873-043,
Email: mazum@essex.ac.uk

C. Rosenberg
Nortel plc, Harlow, and Department of Electrical and Electronic Engineering, Imperial College, U.K.
London Road, Harlow, United Kingdom CM17 9NA
Telephone: +(1279) 403 889, Fax: +(1279) 402 400,
Email: caro@nortel.co.uk

Abstract

In this paper we present an abstract, rigorous game-theoretic framework for rate-based control of Available Bit Rate (ABR) services in ATM networks. The framework is based on the idea of the Nash arbitration scheme in cooperative game theory which not only provides the rate settings of users which are Pareto optimal from the network point of view of but are also consistent with the fairness axioms of game theory. We first consider the centralized problem and then show that this procedure can be de-centralized such that users can perform the optimization locally such that the overall rates are network optimal. We then consider the problem of charging of ABR connections considering users' valuations for the best-effort service. We propose a flat-rate charging policy such that the total network revenue is maximum. we show that the above arbitration set-up can be used to characterize a rate alloca-

Broadband Communications   P. Kühn & R. Ulrich (Eds.)
© 1998 IFIP. Published by Chapman & Hall

tion policy which takes into consideration users' bandwidth requirements and users' valuations in a fair way.

Keywords

ABR capability, explicit rate control, game theory, fairness, willingness-to-pay

1 INTRODUCTION

The ITU-T [10] has introduced a new ATM Transfer category (ATC) called the Available Bit Rate (ABR) ATC to support data applications which cannot be efficiently supported by existing bandwidth guaranteed ATC's. The ABR ATC has been targetted to support highly bursty applications which have no way of predicting their traffic requirements in advance but which have well-defined cell loss requirements and can tolerate time-varying and unpredictable transfer delays. Another characteristic of these applications is that they are able to modify their data transfer rates according to network conditions. This encapsulates the notion of *elastic traffic* services by which the source rates are adjusted according to the network conditions in order that the network can carry or pack as many revenue generating connections subject to some minimal guarantees.

These applications are expected to ride "on top of" bandwidth guaranteed connections and utilize any excess bandwidth. Since the available bandwidth will change depending on the amount of "background" bandwidth guaranteed services being carried the incoming elastic sources will have to continually change their rates based on some notification by the network on the available bandwidth. Thus the notion of *rate control* of sources arises.

The study of rate control in the context of ABR services has been receiving much attention recently since the success of ATM networks will be crucially dependent on the ability of such networks to carry internet data type applications. Since there will be many sources which will be competing for the use of the available bandwidth there are several issues which arise and must be dealt with to have an efficient means of allocating available network resources. These are: 1) efficient bandwidth allocation to the different sources taking into account their different performance requirements 2) the crucial notion of fairness 3) the notion of decentralized or local rate control in order to facilitate the control to be implemented on a per-connection or local procedure 4) and finally the charging and its relationship with bandwidth allocation.

The problem of rate control in the context of ABR services has been studied in terms of incremental schemes whereby the rate of sources is adjusted by an additive increase factor in case of notification by the network that more bandwidth is available and is decremented by a multiplicative factor on notification by the network that it is becoming congested. A recent paper by Hernandez-Valencia *et al* [9] presents a survey of the current state of the art.

The issue of fairness is addressed in that the rate settings are max-min fair. Other approaches to the problem can be found in the recent special issue [7].

In a recent important paper, Kelly [11] addresses the bandwidth allocation or the rate settings of the sources by considering explicit utility functions for the performance. In particular the paper addresses two important issues : those of optimal rate settings from a network point of view which corresponds to a constrained optimization of the sum of the utility functions as well as the problem of local user optimization whose solution corresponds to the network optimum. By considering logarithmic utility functions Kelly then goes on to show that such a scheme results in rate settings which are proportionally fair (i.e. if a source needs twice a given amount of bandwidth compared to another source it will receive twice the amount of bandwidth which is allocated to achieve the netwprk optimum). In a follow-up paper, Kelly *et al* [12] show that a real-time algorithm can be developed which converges to the required rate settings which can be used by individual sources and provide a detailed probabalistic sample-path analysis. The basic idea is that the equilibrium state of the differential system is precisely the required rates and they show that the basic property of the algorithm is that it acts as a Lyapunov function for the system.

This paper is motivated by the issues raised in the paper of Kelly [11]. In particular, by drawing upon ideas from cooperative game theory [8] it is shown that the idea proposed by Kelly is in fact a Nash arbitration scheme [14] which has the property that it is a Pareto optimal and hence qualifies as a network optimal scheme while satisfying certain axioms of fairness. Once this observation is made the abstract framework then allows us to address the ABR problem with non-zero minimum cell rate (MCR) requirements while also accounting for peak cell rate (PCR) constraints of sources. This allows us to address the rate allocation problem from the point of view of performance characteristics rather than abstract utility functions. In particular we show that a local procedure can be devised such that the solution of the local optimization corresponds to the arbitrated solution in the Nash sense. We then consider the problem of charging ABR connections considering users' valuations. In this context, we propose a network rate allocation policy which takes into account users' bandwidth requirements and users' valuations in a fair way.

The idea of the Nash arbitration scheme in the context of telecommunication networks is not new. This was first presented in the context of packet-switched (data) networks by Mazumdar *et al* [13]. The properties of Pareto optimality as well as the development of local optimization procedures which lead to Pareto optimal solutions (the local procedures being greedy schemes) was studied in a series of papers by Douligeris and Mazumdar [4], and [5] in the context of data networks. This paper is thus an extension of those ideas as well as a new approach in the context of ABR (or elastic) services in ATM networks.

The outline of this paper is as follows: in Section 1 we present the salient facts about the Nash arbitration scheme which provides the framework of the solution. Section 2 considers the optimal and fair rate allocation problem for ABR connections in which we discuss both the centralized (network optimality) as well as the connection based (local algorithm) contexts. In section 3 we introduce the charging framework. Section 4 concludes the paper. Throughout we omit details of proofs since they readily follow from standard results in optimization theory.

2 NASH ARBITRATION SCHEMES

In this section we present the salient concepts and results from cooperative game theory and the Nash arbitration schemes which are used in the sequel. For details we refer the reader to the book by Fudenberg and Tirole [8] and the paper of Nash [14].

The basic setting of the problem is as follows: there are N users (connections) which compete for the use of a fixed resource (bandwidth). Each user i ($i \in \{1..N\}$) has a performance function f_i and an initial performance u_i^0. Each performance function is defined on a subset of $\mathcal{R}^N$ termed X which is the set of game strategies of the N users. In a context of network resource allocation X could represent the space of allocated rate vectors. The initial performance of each user represents a minimum performance that a user wants to achieve and will not enter the game if it is not possible to realize it. Therefore, we will assume throughout our theoretical framework that each user involved in the game can achieve its initial performance. In other words, there exists at least a vector in X such that the value at that point of the agregate performance function, f, $((f_1, ..., f_N))$ is superior or equal to the initial performance vector, u^0.

In our framework we adopt the following mathematical assumptions regarding a user's performance function and the space of strategies. Indeed, X is assumed to be nonempty, convex, and closed set. The functions f_i are assumed to be real-valued, upper-bounded, and concave functions. Finally a word about notation: given two vectors $u, v \in \mathcal{R}^N$ we say $u \geq v$ if $u_i \geq v_i \; \forall i \in \{1, 2, ..., N\}$.

Let $U \subset \mathcal{R}^N$ be a nonempty convex closed and upperbounded set. Let $u^0 \in \mathcal{R}^N$ such that $U_0 = \{u \in U / u \geq u^0\} \neq \emptyset$. Let $G_k = \{(U, u^0)/U \subset \mathcal{R}^N$ is a nonempty convex closed and upperbounded set and $u^0 \in \mathcal{R}^N$ such that $U_0 \neq \emptyset \}$. G_k is a set of pairs. Each pair is characterized by a set and an initial point.

We first define the notion of Pareto optimality in the context of multiple-criteria objectives which occurs in the typical game setting with multiple players.

Definition 2.1 *The point $u \in U$ is said to be Pareto-optimal if for each $u\prime \in U$, $u\prime \geq u$ it implies that $u\prime = u$.*

The interpretation of a Pareto optimum is that it is impossible to find another point which leads to strictly superior performance for all the players simultaneously. In general in a game with N players (or equivalently for a set of N-objectives) the Pareto optimal points form a N dimensional hypersurface which implies that there are an infinite number of points which are Pareto optimal. From the definition of Pareto optimality it is clear that a network optimal operating point must be a Pareto optimal point. The question that arises is which of (the infinitely many) Pareto optimal points must the network be operated?

One way in which we can define suitable Pareto optimal points for operation is by introducing further criteria. From the perspective of resource sharing one of the natural criteria is the notion of fairness. This in general is a loose term and there are many notions of fairness. One of the commonly used notions is that of max-min fairness which corresponds to a saddle-point for the game and is most commonly used in the context of ABR control [9]. However a much more satisfactory approach is to use the fairness axioms from game theory as the fairness criteria [14].

We now define the Nash arbitration scheme which encapsulates the above requirements.

Definition 2.2 *A mapping $S : G_k \to \mathcal{R}^N$ is said to be a* Nash arbitration scheme *if:*

1. $S(U, u^0) \in U_0$.
2. $S(U, u^0)$ *is Pareto-optimal.*
3. *S satisfies linearity axiom; If $\phi : \mathcal{R}^N \to \mathcal{R}^N$, $\phi(u) = u\prime$ with $u\prime_j = a_j u_j + b_j$, $a_j > 0$, $j = 1, ..., N$ then $S(\phi(U), \phi(u^0)) = \phi(S(U, u^0))$.*
4. *S satisfies irrelevant alternatives axiom; If $U\prime \subset U$, $(U\prime, u^0) \in G_k$, and $S(U, u^0) \in U\prime$ then $S(U, u^0) = S(U\prime, u^0)$.*
5. *S satisfies symmetry axiom; If U is symmetrical with rspect to a subset $J \subseteq \{1, ..., N\}$ of indices (i.e. $u \in U$ and $i, j \in J$, $i < j$ imply $(u_1, ..., u_{i-1}, u_j, u_{i+1}, ..., u_{j-1}, u_i, u_{j+1}, ..., u_N) \in U$), and if $u_i^0 = u_j^0$ $i, j \in J$ then $S(U, u^0)_i = S(U, u^0)_j$ $i, j \in J$.*

Having defined the Nash arbitration scheme we define the optimal point as follows:

Definition 2.3 *Let $U = \{u \in \mathcal{R}^N / \exists x \in X$ such that $f(x) \geq u\}$. Let u^* be given by $S(U, u^0)$. Then u^* is the (Nash) arbitration point and $f^{-1}(u^*)$ is called the set of the (Nash) arbitrated solution.*

Remark 2.1 *The items 3, 4 and 5 above are the so-called axioms of fairness. The linearity property of the solution implies that the arbitration scheme is scale invariant i.e. the arbitrated solution is unchanged if the performance objectives are scaled. The irrelevant alternatives axiom states that the arbitration point is not affected by enlarging the domain while the symmetry property states that the arbitration point does not depend on the specific labels i.e. users with the same initial points and objectives will realize the same performance.*

The following result due to Stefanescu [15] provides for a characterization of the Nash arbitration point and will form the basis for the results in the sequel.

Theorem 2.1 *Let* $U = \{u \in \mathcal{R}^N / \exists x \in X$ *such that* $f(x) \geq u\}$. *Denote by* $X(u) = \{x \in X / f(x) \geq u\}$. $X_0 = X(u^0)$, *the subset of strategies that enable the users to achieve at least their initial performances.*

Then there exists a unique arbitration scheme and a unique arbitration point u^*. *Moreover the set of the arbitrated solution* $(f^{-1}(u^*))$ *is determined as follows:*

Let J *be the set of users able to achieve strictly better than their initial performance i.e.,* J *is defined as* $\{j \in \{1..N\}/\exists x \in X_0, f_j(x) > u_j^0\}$. *Each strategy vector* x *in the arbitrated solution set verifies* $f_J(x) > u_J^0$ *and solves the following maximization problem* (P_J):

$$(P_J) \qquad Max \quad \prod_{j \in J}(f_j(x) - u_j^0) \qquad x \in X_0$$

Hence, u^* *satisfies that* $u_j^* > u_j^0$ *for* $j \in J$ *and* $u_j^* = u_j^0$, *otherwise.*

Remark 2.2 *Note that for each* $j \in \bar{J}$, $\forall x \in X_0$ $f_j(x) = u_j^0$. *Also, it can be readily shown that if each function* f_j $(j \in J)$ *is injective on* X_0 *then the arbitrated solution set is a singleton and therefore there exists a unique Nash arbitrated solution strategy vector.*

We now state an equivalent optimization problem which will also result in a Nash arbitration scheme and which we will consider in context of rate allocation for ABR connections.

Theorem 2.2 *In addition to the assumptions in 2.1, let* $\{f_j\}$; $j \in J$ *be injective on* X_0.

Consider the two maximization problems (P_J) *and* $(P\prime_J)$:

$$(P_J) \qquad Max \quad \prod_{j \in J}(f_j(x) - u_j^0) \qquad x \in X_0$$
$$(P\prime_J) \qquad Max \quad \sum_{j \in J} \ln(f_j(x) - u_j^0) \qquad x \in X_0$$

Then:

(i) (P_J) *has a unique solution; The arbitrated solution set is a singleton.*

(ii) $(P\prime_J)$ *is a convex program and has a unique solution.*

(iii) (P_J) *and* $(P\prime_J)$ *are equivalent. Hence, the unique solution of* $(P\prime_J)$ *is the arbitrated solution.*

Remark 2.3 *In [11] Kelly considers the centralized optimization criterion as the weighted sum of the logarithmic utility functions and hence the corresponding optimal solution is a Nash arbitration solution in light of the above result.*

3   OPTIMAL AND FAIR RATE ALLOCATION FOR ABR CONNECTIONS

We propose a scheme that results in an optimal and fair share of the available bandwidth between ABR connections competing for network resources. The scheme is based on the game theory framework presented above. Each user (connection) has a performance function to be maximized. Each function is defined on the space of the allocated rate vectors. We characterize an optimal and fair operating point as a solution of the optimization of a global network's objective function.

It is natural to adopt a game theory approach to model and address the issue of network resource allocation. In the context of flow control in packet-switched networks many schemes were based on the use of game theory and gave a characterization for some candidate points. Some of them considered Nash equilibrium points [2] [5] and others considered Pareto-optimal points [6]. In [13], the Nash arbitration point was proposed as a suitable solution for the design of an optimal and fair flow control.

As in [13] we consider the Nash arbitration point as a desired point for the operation of the network. This is due to the Pareto optimality and fairness property associated with Nash arbitration schemes.

The definition of a Nash arbitration point is highly dependent on the consideration of an initial performance point (termed u^0 in the previous section). It represents a minimum performance that a user wants to achieve and the user will not enter the game if it is not possible. In the context of Available Bit Rate service (ABR), for each connection (user) the initial performance can be viewed as a performance achieved by the Minimum Cell Rate (MCR) guaranteed by the network.

We now introduce a utility function for each user (connection), U, that depends on the allocated rate vector. It represents user's satisfaction from a particular performance level achieved through an allocated rate vector. The determination of the Nash arbitrated solution give us a natural candidate for a utility function. Indeed, we assume that a user i with a performance function f_i and a minimum desired performance u_i^0 has a utility function U_i defined as follows:

$$U_i(x) = \ln(f_i(x) - u_i^0) \quad x \in X_0. \tag{1}$$

The network's global objective function to be maximized is the overall satisfaction (or the sum of the utilities) of the users able to achieve strictly better than their initial performance (u_i^0 for user i). The choice is motivated by the fact that the maximization of the considered global function leads to the desired fair and optimal operating point; the Nash arbitration point.

The results presented in this section are meant to be applied for the implementation of a rate-based control mechanism for Available Bit Rate (ABR) transfer capability. We propose a criterion for optimal and fair rate allocation based on the Nash arbitration scheme which takes into account both zero and non-zero-minimum-cell-rate ABR connections. This formulation copes with the limitations of the max-min fairness schemes, [1], which apply unambiguously only to ABR connections with minimum cell rate equal to zero.

Firstly, we present a centralized model in which network resources are the available link capacities and each ABR connection aims at maximizing its allocated rate beyond its minimum cell rate. The centralized model identifies a global optimization problem from which the Nash arbitrated solution (allocated rate vector) emerges as the unique solution. Secondly, we propose a decomposed or decentralized model in which each connection and the network provider are separate entities and have their own optimization criteria. We show that by appropriate choice of network parameters the Nash arbitrated solution of the centralized model is an optimal allocated rate vector for each connection and the network provider. The decomposed model is a necessary step to develop a distributed algorithm of a rate-based control mechanism implementing the Nash arbitrated solution criterion.

3.1 Centralized model

We consider a static model for the centralized (network) problem in which N connections are established. Each connection corresponds to an ABR connection with a Peak Cell Rate (PCR), a Minimum Cell Rate (MCR) guaranteed by the network, and an assigned path. Connections compete for available bandwidth resources within the network. These resources are network link available capacities and they are assumed to be fixed (non-time-varying). With respect to the abstract framework already presented the allocated rate vector space, X, is determined by network capacity constraints and connections' peak cell rates. It is defined as follows:

$$X = \{x \in \mathcal{R}^N / x \geq 0 \;\; x \leq PCR \, and \, Ax \leq C\}. \tag{2}$$

where C is the vector of link capacities , PCR is the vector of peak cell rates of the connections, and $A = (a_{lp})_{l,p}$ is a LxN incidence matrix i.e. a_{lp} is equal to 1 if the link l belongs to the path p and 0 otherwise.

In the context of Available Bit Rate it is natural to assume that each connection aims to maximize its throughput (and so its allocated rate) beyond its minimum cell rate. Therefore, with respect to the abstract framework the performance function, f_i, for a user i is simply defined as x_i. Moreover, MCR_i represents the initial and the minimum performance desired by user i.

We assume that the initial performance vector (the MCR_i's) is achievable which means that on each network link the spare capacity is superior to the sum of the MCR_i's of the connections crossing this link. The set of the achieving rate vectors, X_0, is characterized as follows:

$$X_0 = \{x \in \mathcal{R}^N / x_i \geq MCR_i \;\; x_i \leq PCR_i \; and \, Ax \leq C\}. \tag{3}$$

For simplicity and without loss of generality, we assume that on each link the spare capacity is strictly superior to the sum of the MCR_i's of the connections crossing this link. If this assumption is not valid then our model and results are still valid for the subset of connections to which we can allocate more than the corresponding minimum cell rate. One can show that this assumption ensures that X_0 has a nonempty interior.

With respect to the abstract framework, the Nash arbitrated solution of the centralized model is an optimal and fair rate allocation of network available capacities to the N considered ABR connections. It is the unique solution of the following global optimization convex problem (S):

$$\begin{cases} Max_{\{x\}} \prod_{i=1}^{N}(x_i - MCR_i) \\ x \in X_0 \end{cases}$$

3.2   Decomposed or decentralized model

In the previous section we formulated and solved the centralized network optimal rate allocation problem. In general this will involve centralized co-ordination amongst the ABR connections. Thus a challenging issue is that can such a problem be decentralized by which connections perform only a local optimization such that the locally optimized rates allocated are optimal in a global or network sense?

We propose a decomposed model in which each user (connection) and the network provider is a seperate entity. In this model, each connection can vary its connection rate, freely. The rate for the connection is bounded from below by the MCR and from above by the PCR. Each connection has a utility function measuring the satisfaction achieved through a particular rate. Moreover each user's global satisfaction experiences a decrease as a result of accessing and using network resources. Indeed, we introduce N positive network parameters, α_i's, which represent the decrease per unit rate of the global satisfaction of the N users given that they share the resources. α_i is also the benefit per rate unit realized by the network provider as a result of user i accessing its resources. Therefore, the objective of each user is to maximize its net satisfaction which is, for a particular rate, the difference between the utility and the cost of accessing the network. Hence, each user i solves the following strictly convex problem (U_i):

$$\begin{cases} Max_{\{x_i\}} U_i(x) - \alpha_i\, x_i \\ x_i > MCR_i \\ x_i \le PCR_i \end{cases}$$

The network provider aims to determine the optimal rate allocation to users that maximizes its total benefit. Hence, it has to solve the following convex problem (N):

$$\begin{cases} Max_{\{x\}} \sum_{i=1}^{N} \alpha_i\, x_i \\ x \in X_0 \end{cases}$$

The following proposition shows that by appropriate choice of network parameters, the α_i's, the Nash arbitrated solution of the centralized model maximizes each user's net satisfaction and the network total benefit.

Proposition 3.1 *Let x be the unique Nash arbitrated solution of the centralized problem (S). Then, there exist positive real numbers α_i ($i \in \{1..N\}$) such that x is the unique solution of the problems (U_i) and x is a solution of the problem (N).*

It is interesting to notice that if we view a user's utility function in terms of monetary value, then we obtain an economic and a pricing framework of rate allocation. In this framework, a utility function will have the interpretation of a willingness-to-pay for a particular rate. The α_i's will stand for the prices of a rate unit charged to users. Each user will aim to maximize its net benefit and the network (provider) will aim to maximize its total revenue.

In the following section we show how the game theoretic framework can be used to derive a charging mechanism for ABR and an allocation policy taking into account users' valuations in a fair way.

4 A CHARGING SCHEME FOR ABR USERS

We propose to address the issue of rate allocation together with the charging issue in the context of ABR considering users' bandwidth requirements and users' valuations for the best-effort service. Each user informs the network of its valuation either at subscription time or at connection set-up. Indeed, a user chooses a price it is willing to pay for any excess allocated bandwidth beyond the guaranteed minimum cell rate. The price may be chosen from a given set of values or from a set of pre-determined tariffs published by the network provider. Different users with different peak and minimum cell rates can choose from different sets of price values. The choice of a user reflects the value it attaches to any excess bandwidth allocated by the network. There is an agreement between users and the network by which the network assures that its allocation policy will allocate bandwidth optimally and fairly taking into account their bandwidth constraints as well as their willingness-to-pay.

We propose a network allocation policy and a two-components charging policy. First, a user is charged for its guaranteed minimum cell rate according to a tariff function which may depend on both minimum and peak cell rate parameters. The second component is a flat-rate charge that a user pays for any excess allocated bandwidth. The network chooses to charge each user its willingness-to-pay as in this case the total revenue is the maximum possible that the network can get from any bandwidth allocation.

The network adopts an allocation policy which takes into account users' willingness-to-pay (β_i for user i) for excess bandwidth in a fair way. This particular policy satisfies a desirable property in that the more a user is willing to pay for the excess bandwidth the more allocated bandwidth it gets. The rate settings is characterized as the Nash arbitrated solution of a cooperative game with the user performance objectives defined as the power function of the rate allocated above the MCR for each user with exponent β_i (assumed to be between 0 and 1). The reason behind the consideration of such allocation policy is that it incorporates nice fairness properties with regard to users' valuations. Considering a model of N users (connections) similar to the one described in section 3.1, the rate settings is determined as follows:

- If $\beta_i = 0$ then $x_i = MCR_i$ otherwise
- If $\sum_{l=1}^{L} \mu_l\, a_{li} > 0$ then $x_i = MCR_i + MIN\left(PCR_i - MCR_i\,,\ \dfrac{\beta_i}{\sum_{l=1}^{L} \mu_l\, a_{li}}\right)$.
- If $\sum_{l=1}^{L} \mu_l\, a_{li} = 0$ then $x_i = PCR_i$.

Where the μ_l's ($l \in \{1..L\}$) stand for the link shadow prices associated with the game-underlying optimization problem. If we view these prices as the allocation costs of a bandwidth unit to users then it can be readily seen that the allocation policy has the following nice characteristics:

- If a user i pays nothing for the share of bandwidth beyond the minimum cell rate then its allocated rate is the minimum cell rate.
- If a user i pays for the share of bandwidth beyond the minimum cell rate then its allocated rate is greater than the minimum cell rate.
- If the nework's resources along a user's path are free then its allocated rate is the peak cell rate.
- If the nework's resources along a user's path are not free and but the user's valuation exceeds the network path cost per unit of bandwidth by more than a factor of $(PCR - MCR)$ then the user is allocated its peak rate.
- If the nework's resources along a user's path are not free and the user's price is less than the path cost per bandwidth unit then the user is allocated a cell rate between the minimum and peak rate proportional to the valuation of the user. As a result, if two users share the same resources and one of them is willing to pay the double of the other, then the user receives double the share of bandwidth beyond minimum cell rate.
- If two users share the same resources, have the same maximum excess bandwidth (difference between peak and minimum cell rates), and are willing to pay the same price then they get the same share of excess bandwidth.

5   CONCLUDING REMARKS

In this paper we have presented a game theoretic framework for the allocation of optimal rates to ABR connections which share common bandwidth. This framework allows us to go further in showing how we can come up with a charging scheme and an allocation policy which presents nice fair properties. A point to note is that the framework is a static one. The main aim has been to point out how the problem of rate allocation can be addressed within the framework of the Nash arbitration scheme. An important issue to be addressed in the future is the development of a distributed algorithm which implements the Nash arbitrated allocation policy using the ABR Explicit Rate (ER) mechanism.

Acknowledgement

The authors would like to acknowledge useful insights provided by Prof. F. P. Kelly. This research has been supported in part by contracts from the Centre National d'Etudes des Telecommunications (CNET), France through the Consultations Thématiques (CTI) programme and by the CANCAN project of ACTS by the European Economic Community.

REFERENCES

[1] D. Bertsekas and R. Gallager: *Data networks.* Prentice-Hall, 1987.

[2] K. Bharathkumar and J.M. Jaffe: *A new approach to performance oriented flow control.* IEEE Trans. Comm., vol. COM-29, April 1981, pp. 427-435.

[3] F. Bonomi and K. W. Fendick: *The rate-based flow control framework for available bit rate ATM services,* IEEE Network, March/April, 1995.

[4] C. Douligeris and R. Mazumdar: *On Pareto-optimal flow control in an integrated environment.* Proc. 25th Allerton Conf., Univ. Illinois, Urbana, October 1986.

[5] C. Douligeris and R. Mazumdar: *User optimal flow Control in an integrated environment.* Proc. Indo-U.S. Workshop on Syst. Signals, Bangalore, India, January 1988.

[6] P. Dubey: *Inefficiency of Nash equilibria.* Math. Oper. Res., vol. 11, no. 1, pp. 1-8, 1986.

[7] European Transactions on Telecommunications, *Focus on Elastic Services over ATM networks,* R. Mazumdar and B. Doshi eds., ETT Vol. 8, N0. 1, 1997.

[8] D. Fudenberg and J. Tirole: *Game theory,* MIT Press, Cambridge, Ma., 1995.

[9] E. J. Hernandez-Valencia, L. Benmohammed, S. Chong and R. Nagarajan: *Rate-control algorithms for ATM ABR service,* in [10], pp. 7-20.

[10] ITU-T Recommendation I.371: *Traffic control and congestion control in B-ISDN,* Geneva, June 1996.

[11] F. Kelly: *Charging and Rate Control for Elastic Traffic.* in [10], pp. 33-37

[12] F. Kelly, A. Maulloo and D. Tan: *Rate control in communication networks: shadow prices, proportional fairness and stability,* pre-print, Statistical Lab., Cambridge, 1997

[13] R. Mazumdar, L. Mason, and C. Douligeris: *Fairness in network optimal flow control: optimality of product forms.* IEEE Transactions on communications, vol. 39, no. 5, 1991, pp. 775-782.

[14] J. Nash: *The bargaining problem,* Econometrica, Vol. 18, 1950, pp. 155-162.

[15] A. Stefanescu and M.W. Stefanescu: *The arbitrated solution for multiobjective convex programming.* Rev. Roum. Math. Pure. Appli., vol. 29, pp. 593-598, 1984.

19

Multipath FEC Scheme for the ATM Adaptation Layer AAL5

G. Carle
GMD FOKUS
Kaiserin-Augusta-Allee 31
Berlin, Germany
carle@fokus.gmd.de

S. Dresler, J. Schiller
Institute of Telematics
University of Karlsruhe
Karlsruhe, Germany
j.schiller@ieee.org
stefan.dresler@acm.org

Abstract

One approach to facilitate statistical multiplexing of bursty sources in ATM networks is dispersion of the traffic over independent paths. It has been shown within the literature that by dispersion of traffic over disjoint paths it is possible to increase the Quality of Service (QoS) (see e.g., (Maxemchuck, 1993), (Lee, 1993)). In multipath schemes without additional error control, the resulting QoS depends on the worst-case path. By using a multipath scheme with FEC (Forward Error Correction), it is possible to compensate cell losses. Using error correcting codes like Reed-Solomon-Erasure Codes (RSE Codes, (McAuley, 1990)), these schemes add redundancy and disperse the traffic using several transmission paths (Ding, 1995).

Until now, protocols and implementation concepts suitable for a widespread use of multipath schemes are still missing, in particular for the widely used adaptation layer AAL5. We present a novel protocol MP-FEC-SSCS for AAL5 for employing a redundant multipath communication scheme especially suited to the UBR (Unspecified Bit-Rate) service. This presentation is complemented by detailed results for software implementations based on C++ and first results using Java, targeted for an active network implementation.

Keywords

ATM, AAL, SSCS, FEC, Reed-Solomon-Erasure Code, Striping, Multipath Communication Scheme, Dispersion Routing

Broadband Communications   P. Kühn & R. Ulrich (Eds.)
© 1998 IFIP. Published by Chapman & Hall

1   INTRODUCTION

ATM networks are designed to allow an efficient use of network resources by statistical multiplexing of different types of traffic streams, such as data, audio, and video communication. A key challenge arises from the need to handle both highly bursty and unpredictable sources, along with relatively smooth sources. In the presence of highly bursty sources it proved to be difficult to achieve a statistical multiplexing gain while ensuring a high quality of service (QoS) for all connections.

One approach to facilitate statistical multiplexing of bursty sources it to use a multipath communication scheme and to disperse the traffic over independent paths. It has been shown within the literature that by dispersion of traffic over disjoint paths it is possible to increase the Quality of Service (QoS) (see e.g., (Maxemchuck, 1993), (Lee, 1993)). This improvement of QoS is possible because splitting of traffic leads to a reduced burstiness of the individual substreams. By using a Forward Error Correction (FEC) scheme that transmits redundant data over one or more additional paths, a receiver is able to reconstruct the original data even in case of losses. This is particularly attractive as ATM services with lower reliability can be provided at lower costs. Walker et. al. (Walker, 1997) e.g. propose a set of ATM tariffs, where the price per volume for VBR (Variable Bit Rate) traffic is two orders of magnitude higher than the price per volume for UBR (Unspecified Bit Rate) traffic. This price difference gives a high incentive to select the UBR service class whenever possible. While UBR services frequently will show high cell loss rates, the combination of UBR services with an FEC scheme allows to achieve relatively high reliability with very low costs.

1.1   AAL-level realization of a Multipath FEC Scheme

Dispersion of traffic over multiple paths and adding of redundancy can be realized within an ATM network at different layers. The FEC scheme of (Ohta, 1991) for example has been designed for being applied at VP (Virtual Path) level. An integration of a multipath FEC scheme at the ATM layer would require hardware modifications of switches and end systems. An alternative approach of integrating a multipath FEC scheme above the Common Part Convergence Sublayer (CPCS) of AAL5 can be realized in software, without hardware modifications.

Reducing the effective bit error rate and cell loss ratio for the layer above the adaptation layer (e.g., IP) will significantly improve the overall quality of service. This is especially true for applications that require the provision of a highly reliable service. A number of previous researchers (Shacham, 1990), (Ohta, 1991), (Biersack, 1993), (Ayanoglu, 1993) have shown how the use of FEC can improve end-to-end ATM performance in terms of effective throughput and latency.

1.2   Potential gain by striping

The potential gain obtained by using a striping mechanism is explained by the following example. Figure 1 shows an application sending a message with a size of six cells. From these six cells, three redundancy cells are derived using an error correcting code. This is achieved by lining up cells in a matrix consisting of m (m=2 here) rows of length n (n=3 here). Choosing the number of redundancy rows to be h=k-m=3-2=1, a single row of redundancy cells is computed. In our example, a simple code like an XOR code is sufficient which computes the bitwise XOR of each column in the matrix. User as well as redundancy cells are then transmitted on, say, three paths, e.g., by assigning each path a row of the matrix. Different assignments are possible, each with its characteristic strengths suited to specific loss assumptions. The receivers reconstructs lost user cells and reassembles the original message. Details on this process and the cell and frame format used are explained below.

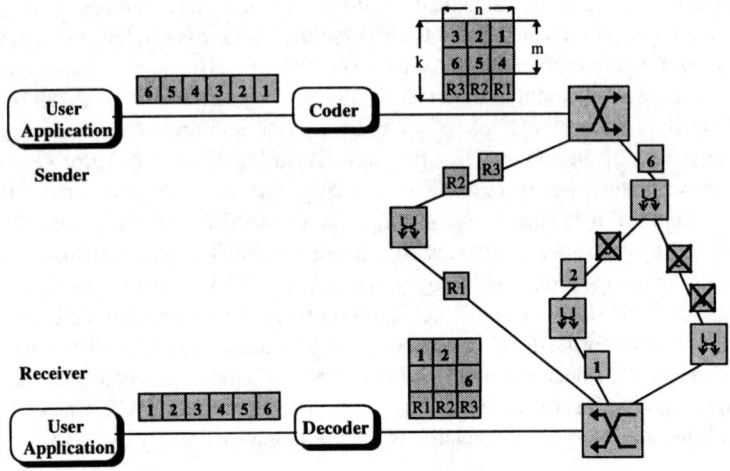

Figure 1 Multipath communication in ATM Networks.

By appropriately choosing the parameters it is possible to construct a coding scheme that allows for the reconstruction of user messages even if one of the paths suffers from the loss of all cells belonging the message, as long as no other cells are lost.

2   PROTOCOL ISSUES

This chapter presents details on the protocol on which our redundant multipath communication scheme is based.

2.1 Multipath-FEC-SSCS

Data transmissions over ATM currently usually use AAL5 due to its small bandwidth overhead for protocol fields, and due to its low processing requirements. It was chosen to develop the multipath FEC scheme as a Service Specific Convergence Sublayer (SSCS) of the ATM Adaptation Layer AAL5 and to name it MP-FEC-SSCS.

For reconstruction of lost cells belonging to an AAL frame it is necessary to identify cells by cell sequence numbers. Assuming that no successful reconstruction of a frame will be required in cases where more than 15 cells get lost consecutively on a single path, a cell sequence numbers of 4 bits (counting modulo 16) can be used. If all parameters (number of columns, user data rows and redundancy rows, etc.) of an SSCS session are fixed, no additional per-frame information fields are necessary. Adding an

FEC-SSCS frame header containing the parameters of a session provides for additional flexibility, however, at the cost of a small bandwidth and protocol processing overhead.

Figure 2 gives an overview of the PDU structure used in the Multipath-FEC-SSCS protocol. User data forms the upper part of a matrix, and redundancy is located in the lower part of the matrix. Every matrix element has a length of 46 bytes, leaving 2 bytes of the payload of an ATM cell for per-cell protocol information of MP-FEC-SSCS. This protocol information, called the MP-FEC-SSCS cell header, contains the 4 bit cell sequence number, one bit which is alternated between successive frames to allow for a simple receiver state machine, one bit indicating whether a cell contains user or redundancy data, and a 10 bit checksum (CRC-10).

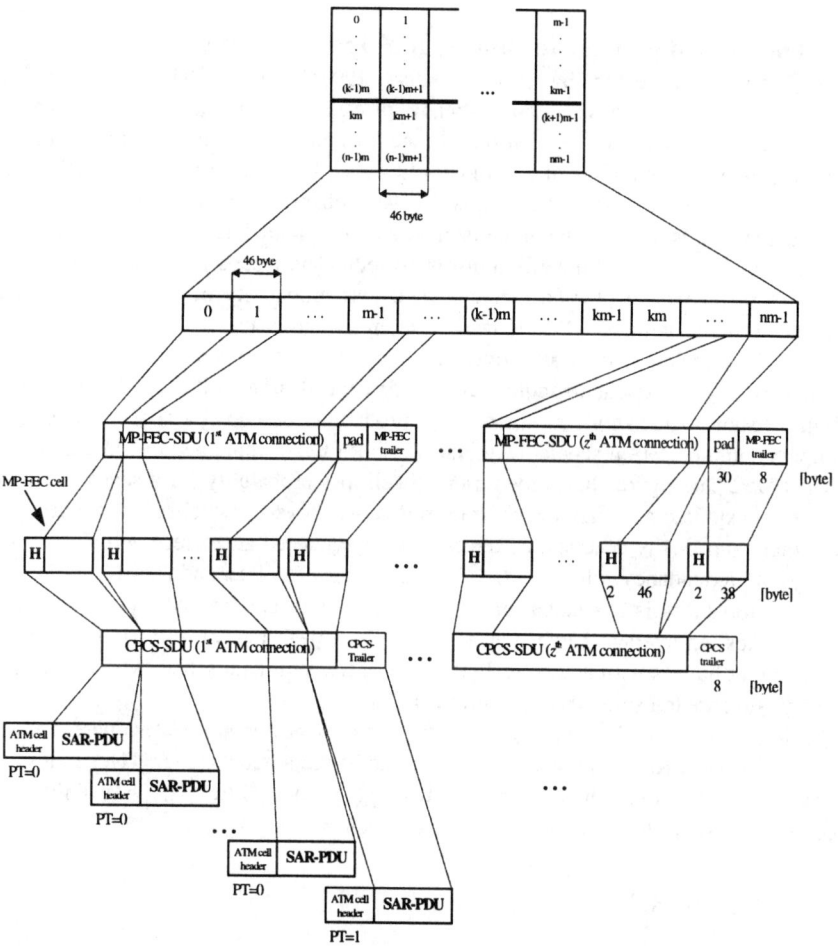

Figure 2 Mapping of MP-FEC-PDUs onto ATM Cells.

The last cell of an MP-FEC frame consist of the 2 byte MP-FEC-SSCS cell header, a 30 byte SSCS padding field, an 8 byte MP-FEC-SSCS-Trailer, and the 8 byte AAL5-CPCS-Trailer. The MP-FEC-Trailer contains fields for the number of MP-FEC cells in a matrix, the number of redundancy cells per matrix, a frame sequence number, and the length of the AAL-SDU in bytes. Also included is a bit indicating the type of data carried (user/redundancy data or peer-to-peer signaling information).

3   PERFORMANCE EVALUATION

This chapter presents a performance evaluation of MP-FEC-SSCS by a theoretical analysis, based on simulation, and based on implementations in C++ and Java.

3.1   Theoretical gain achievable by multipath schemes

Existing results presented by (Maxemchuck, 1993), (Lee, 1993), (Ding, 1995), (Brendan, 1995), (Adiseshu, 1996) already proved that multipath communication schemes with and without FEC can reduce latency and cell loss probability for bursty sources. For a multipath scheme without FEC, the dispersion of bursty traffic over multiple paths allows to reduce cell loss, as the probability for queue overflow at an individual switch is reduced in comparison with a non-parallel scheme. For a multipath scheme with FEC, the additionally transmitted redundancy cells increase the load and are therefore responsible for an increased cell loss probability in comparison with a multipath scheme without FEC. However, in many cases the reduction of cell loss because of dispersion of the traffic over multiple paths will be significantly larger than the increase of cell loss due to additional redundancy cells of a multipath FEC scheme.

Figure 3 shows an example for the cell loss probability of a non-parallel scheme and a multipath communication scheme with and without FEC, as obtained by simulations in (Ding, 1995). The figure shows the observed cell loss probability for a scenario where the traffic is splitted at a first switch, transmitted over several switches, aggregated at a last switch, and finally decoded, corrected, and resequenced at the receiver. Curves are given for no redundancy cells (R=1.0), 10% redundancy (R=1.1), and 20% (R=1.2). The loss probability is evaluated for on-off sources with exponentially distributed on- and off-states, an average batch size of 200 cells, a peak rate of 150 Mbit/s, and for switch output buffers with 800 cells. In the case of the multipath schemes, the data cells of every burst are transmitted over 5 different paths.

The following formulas show how adding redundancy significantly increases the probability that an AAL5 frame is successfully delivered at the receiver. The cell loss probability of the individual links is called q_i. Without FEC, the probability for successful delivery of a frame with a length of k cells is given by

$$1-Q = \prod_{i=1}^{N}(1-q_i)^k , \tag{1}$$

with Q denoting the frame loss probability. For identical cell loss probabilities $q = q_i$ on all links, (1) can be simplified to

$$1-Q = (1-q)^k . \tag{2}$$

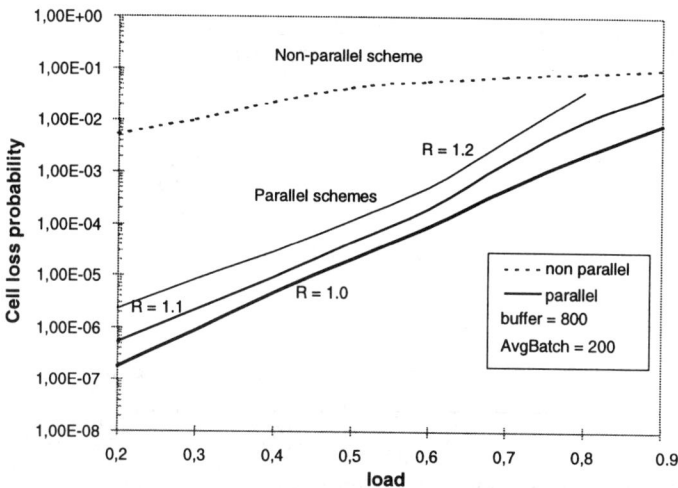

Figure 3 Cell loss probability as function of the load.

This calculation assumes statistically independent cell losses. Independent losses lead to a higher frame loss probability Q than statistically dependent cell losses. The assumption of cell losses occuring in a single burst per frame or at least in different columns of the coding matrix — based on the observation that cell losses occur in bursts (Ohta, 1991) — leads to the following estimation of the frame loss probability. In case of a multipath communication scheme with FEC, k information cells are protected by h redundancy cells. The frame loss probability decreases from the frame loss probability $Q(k)$ of a scheme without FEC to $Q(n,k)$ for the hybrid scheme, with the length of a frame with h redundancy cells being $n = k + h$ cells. The frame loss probability $Q(n,k)$ can be evaluated to

$$Q(n,k) = 1 - \sum_{i=0}^{h} \binom{n}{i} q^i (1-q)^{n-i} = \sum_{i=h+1}^{n} \binom{n}{i} q^i (1-q)^{n-i}. \tag{3}$$

Figure 4 shows the impact of FEC on the resulting frame error rate for AAL5 frames with a payload of 7.5 Kbyte, equivalent to 160 data cells ($k=160$), and for a redundancy of 0 ($R=1.0$), 10% ($R=1.1$), 20% ($R=1.2$) and 30% ($R=1.3$). For the example shown in Figure 3, the additional redundancy cells of a multipath FEC scheme with 20% redundancy increases the cell loss rate by approximately an order of magnitude. At the same time, the FEC scheme reduces the resulting frame loss rate already for 10% redundancy and up to a cell loss rate of 10^{-3} by more than two orders of magnitude, even more for 20% redundancy.

Deployment of FEC in multipath scenarios has thus shown to be useful in order to reduce the frame loss rate at the receiver(s). The slightly larger amount of traffic for FEC is dispersed over several connections.

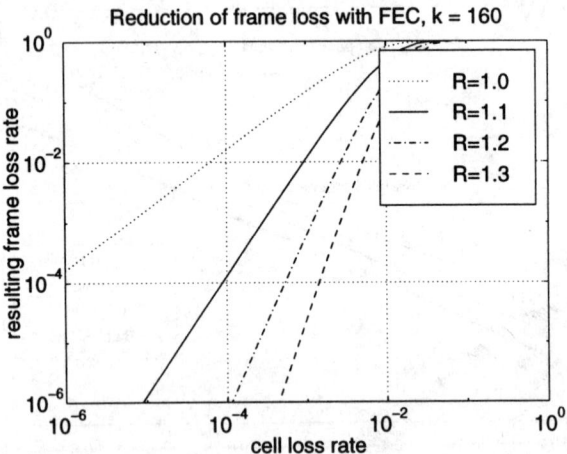

Figure 4 Impact of FEC onto resulting frame error rate.

3.2 Implementation Issues

MP-FEC-SSCS has been implemented using object-oriented programming techniques on the basis of C++ and the runtime library Channels 1.0 developed by Siemens (Böcking, 1995), using a SUN SPARC 10 with SunOS 4.1.3. The ATM connections were emulated by socket communication on top of IP over ATM. A SUN SPARC 20 workstation running Solaris 2.5.1 linked back-to-back with the SunOS station was used for emulation of an ATM switch dropping cells with predefined loss characteristics.

In addition to the implementation using the programming language C++, we implemented the performance critical FEC coding and decoding functions using Java. While today the performance of Java is not satisfying, the promising properties like platform independence and built-in security can be useful already. One example is the use of MP-FEC-SSCS within the Active Network approach, where nodes inside the network perform computation besides the pure routing and forwarding, such as coding and billing (Active Networks, 1997). Using Java, our code can also be pushed onto such nodes making the dissemination of the protocol much simpler. The performance problems of today's Java implementations can be overcome using specialized Java processors or the ever increasing power of standard processors together with just-in-time compilers (IEEE, 1997).

3.3 MP-FEC-SSCS Processing Delay

The processing delay introduced by the respective functions are listed in Table 1. The numbers are the maximum and minimum values of 100 measurements. The long time to receive an MP-FEC cell and to store it in the buffer is due to non-optimal memory allocation at the receiver. The table shows that the sender side needs 94 to 110 µs to create an MP-FEC message, and the receiver side 163 to 468 µs to process an incoming MP-FEC message.

Table 1 Execution times for sending and receiving

	execution step	*execution time (μs)*	
		min	*max*
sender	generate MP-FEC cell header	23	25
	generate MP-FEC cell (incl. cell header)	40	50
	generate MP-FEC trailer	31	35
receiver	detection of MP-FEC-PDU boundaries (per cell)	20	25
	reassemble MP-FEC cell and store in receive buffer	71	355
	decode and reassemble cell header incl. detection and localization of cell loss	20	35
	delete redundancy from cell matrix and store AAL-SDU in buffer	52	53

For two redundancy rows per matrix, amounting to 10% redundancy, and 6 connections, the coding and decoding delays are shown in Figure 5.

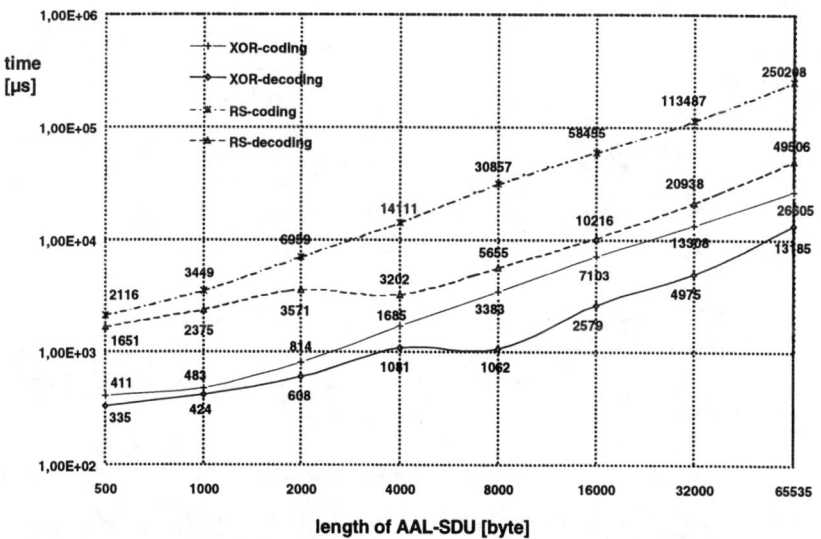

Figure 5 Time for en-/decoding using XOR and Reed-Solomon coding (C++).

The time to encode or decode a message increases (in large parts linearly) with its size in bytes, for both the simpler (and faster) XOR coding scheme, and the more powerful (and slower) RSE code. Enoding takes more time on average since it is always necessary to compute the redundancy cells at the sender, whereas the decoding process only has to be invoked if cells of a message are missing, and then only on those columns of the matrix that are affected. If a sender and a receiver are located on the same host and an SDU size of 65535 byte is chosen, usage of the RSE coding slows down transmission to at most 1 / ((250208 µs + 49508 µs)/65535 byte) = 218.7 kbytes/ sec.

The XOR coding achieves a rate of up to 1 / ((26605 µs + 13185 µs)/65535 byte) = 1.647 Mbytes/sec under these conditions (SUN Sparc10, SunOS 4.1.3).

Implementing the RSE coding in Java on a PentiumPro with 200 MHz, transmission rates of up to 260 kbit/s can be achieved with the just-in-time compiler. Using the interpreted version of Java, only the much simpler XOR-scheme reaches 280 kbit/s loading the CPU completely. Main reasons today are a very inefficient memory management and limited capabilities for fast copying of data.

Figure 6 shows the time needed for encoding and decoding using Reed-Solomon and XOR for 8000 byte AAL-SDUs depending on the redundancy vector *h*. In comparison with XOR it can be shown that Reed-Solomon coding is not the better solution for software implementations in praxis. With a growing *h* the average encoding and decoding time also increases drastically, while the times for XOR increase only slightly.

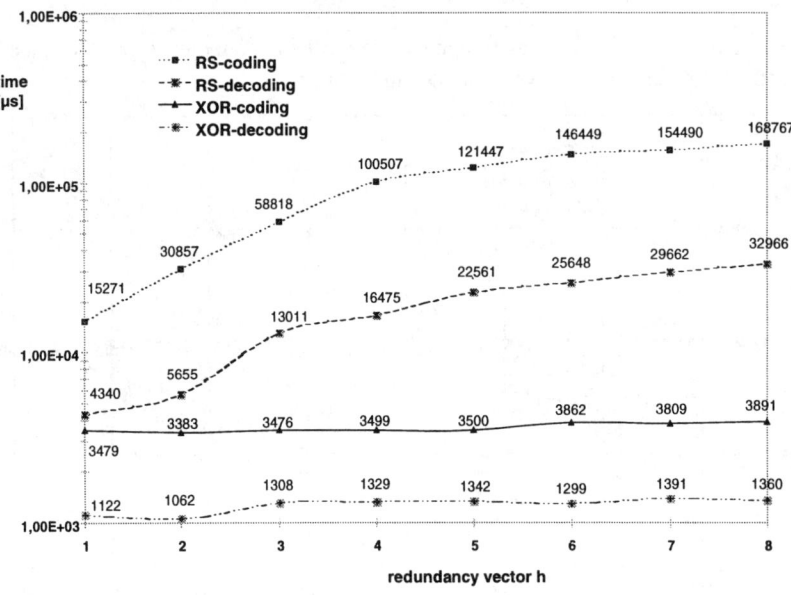

Figure 6 Time for coding and decoding depending on the redundancy vector.

Using a two-state Markov approach to model the loss characteristics of an ATM connection, a simulation was performed to obtain the probability of a successful reconstruction of a coding matrix at the receiver side. The probabilities P_1 = "*a successfully received cell is followed by another successfully received cell*" and P_2 = "*a lost cell is followed by another lost cell"* were varied from connection 1 (with (P_1, P_2) = (0.97, 0.7), corresponding to a cell loss rate of 9.1e-2 = 9.1%) to connection 7 (with (P_1, P_2) = (0.9997, 0.3), corresponding to a cell loss rate of 4.23e-4).

Again, it can be seen that RSE codes are more powerful than XOR schemes. Furthermore, it can be derived from the figure that in scenarios of connections with differing loss the multipath FEC scheme is still able to recover a substantial amount of messages that otherwise would have been erroneous because at least one cell of it got lost.

A protocol that introduces too much overhead compensates the gain it achieves by increasing the network load beyond a useful point. As described above, MP-FEC-SSCS adds an MP-FEC trailer and a 40 byte padding to each MP-FEC-SDU, i.e., for each connection. Furthermore, it adds cell headers of 2 bytes each to every 46 bytes of a message. Finally, AAL5 adds its CPCS-Trailer to each CPCS-SDU. Figure 7 shows the protocol overhead depending on the number of parallel connections and on the length of the AAL-SDU for a redundancy amount $r = 0.1$, i.e., the value of the quotient of redundancy cells and total cells. For long SDUs, the overhead approaches r. An AAL-SDU of size 16000 byte across 4 parallel connections has an overhead of approx. 15.6% with redundancy chosen as $r=0.1$.

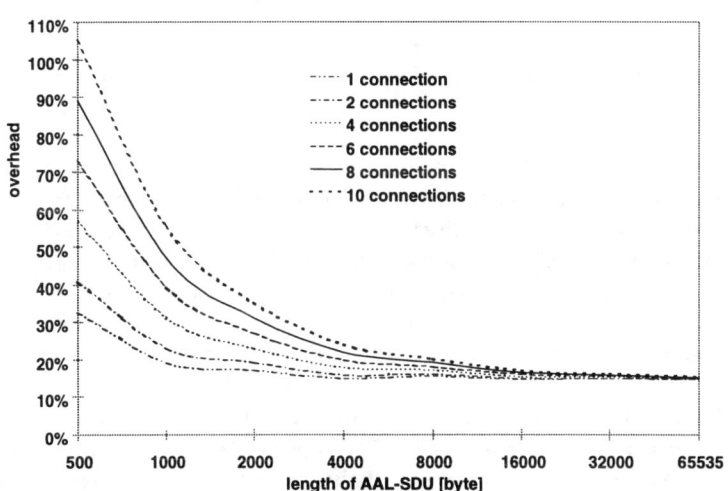

Figure 7 Protocol overhead depending on the number of parallel connections.

3.4 Resulting tradeoffs

Using redundancy, whether on a single connection or on several, involves increased bandwidth requirements compared to unsecured transmission. Figure 3 showed that a multipath transmission protocol experiences an increase in cell loss by about a magnitude if 20% of redundancy is added. This increase is more than compensated for by the cell reconstruction abilities of MP-FEC-SSCS, as shown in Figure 4. An AAL-SDU of 8000 byte length, corresponding to 167 cells in AAL5, e.g., has an overhead of less then 20% (compare Figure 8) in MP-FEC-SSCS while at the same time the protocol is able to reconstruct most of the cells lost (Figure 6).

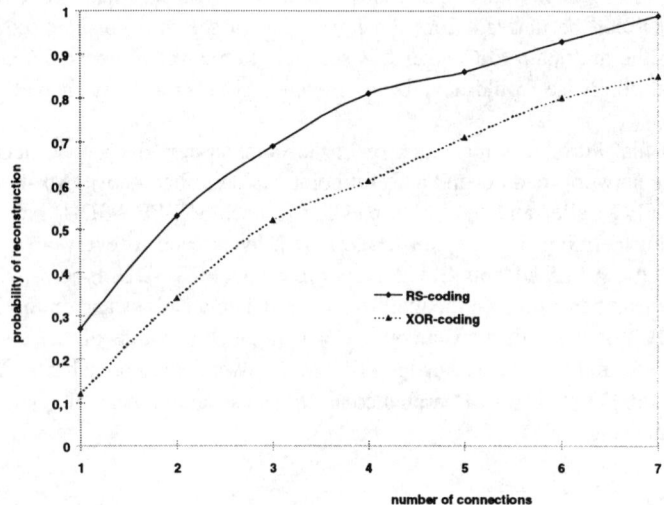

Figure 8 Probability of reconstruction depending on the number of connections.

4   CONCLUSIONS

Applications with bursty traffic characteristics that require certain QoS like a maximum delay and an upper bound for the error rate may significantly benefit from using a multipath FEC scheme. In contrast to error control with retransmissions, FEC allows for error control with low delay. By dispersion of traffic over several connections, both the burstiness of a traffic source can be reduced, and the negative impact of additional traffic of an FEC scheme can be compensated, and the benefits of FEC for error recovery can be fully exploited. The MP-FEC-SSCS described has shown to increase the QoS while keeping the network overhead small. Together with the platform independence of the Java implementation, dissemination of the approach within Active Networks is doable in the near future. With the implementation platforms used, only the C++ implementation allows to achieve the desired performance and proofs to be readily applicable.

5   REFERENCES

Active Networks, http://www.sds.lcs.mit.edu/darpa-activenet/

Adiseshu, H., Parulkar, G., Varghese, G. (1996) A Reliable and Scalable Striping Protocol. Proceedings of ACM SIGCOMM Conference, Vol. 26, No. 4

Ayanoglu, E., Gitlin, R.D., Oguz, N.C. (1993) Performance Improvement in Broadband Networks Using Forward Error Correction, *Journal of High Speed Networks*, **2**, 287-304.

Biersack, E. (1993) Performance Evaluation of Forward Error Correction in an ATM Environment. *IEEE Journal on Selected Areas in Communication*, **11**(4), 631-640

Böcking, S., Seidel, V., Vindeby, P. (1995) CHANNELS - A Run-Time System for Multimedia Protocols. 4th Intl. Conference on Computer Communications and Networks (ICCCN), Las Vegas.

Brendan, C., Traw, S., Smith, J. (1995) Striping Within the Network Subsystem. *IEEE Network*, **4**

Carle, G., Dresler, S. (1996) High Performance Group Communication Services in ATM Networks, in *High-Speed Networks for Multimedia Applications* (eds. W. Effelsberg, O. Spaniol, A. Danthine, D. Ferrari), Kluwer Academic Publishers, Boston/ Dordrecht/London

Carle, G., Schiller, J. (1995) Enabling High Bandwidth Applications by High-Performance Multicast Transfer Protocol Processing. 6th IFIP Conference on Performance of Computer Networks, PCN95, Istanbul

Carle, G., Esaki, H., Guha, A., Dwight, T. (1995) Necessity of Cell-Level FEC Scheme for ATM Networks. ATM Forum Technical Committee ATMF95-0325

Carle, G., Esaki, H., Guha, A., Dwight, T., Tsunoda, K., Kanai, K. (1995) Draft Proposal for Specification of FEC-SSCS for AAL Type 5. ATM Forum Technical Committee ATMF95-0326

Ding, Q., Liew, S. (1995) A Performance Analysis of a Parallel Communications scheme for ATM Networks. IEEE Globecom

Gustavson E., Karlsson G. (1997) A Literature Survey on Traffic Dispersion. *IEEE Network*, **2**, 28-36

IEEE Micro (1997) Making Java work. *IEEE Micro features Java*, **3**

Lee, T., Liew, S. (1993) Parallel Communication for ATM Network Control and Management. IEEE Globecom

Lin, S., Costello, D. (1983) Error Control Coding: Fundamentals and Applications. Prentice Hall, New Jersey

Maffeis, S. (1997) iBUS – The JAVA Intranet Software Bus. Technical Report, Olsen& Associates Zurich

Maxemchuk, N. (1993) Dispersity Routing on ATM Networks. IEEE Infocom, San Francisco, pp. 347-357

McAuley, A.J. (1990) Reliable Broadband Communication Using a Burst Erasure Correcting Code. ACM SIGCOMM, pp. 297-306

Ohta, H., Kitami, T. (1991) A Cell Loss Recovery Method Using FEC in ATM Networks. IEEE JSAC, **9**, 1471-1483

Shacham, N., McKenny, P. (1990) Packet Recovery in high-speed networks using FEC coding. IEEE INFOCOM, San Francisco

Walker, D., Kelly, F., Solomon, J. (1997) Tariffing in the New IP/ATM Environment. Telecommunications Policy, **21**, 283-295

Network Planning and Management

20

Redundancy Domains - a novel Approach for Survivable Communication Networks

Andreas Iselt
Lehrstuhl für Kommunikationsnetze, Technische Universität München
80290 München, Germany, tel. +49 89 289-23504, fax. -23523
iselt@lkn.e-technik.tu-muenchen.de

Abstract

To maintain a high availability of communication networks, traditionally two separate mechanisms are used, redundancy within network elements and network-wide redundancy. Usually these mechanisms are applied independently. In this paper a new technique is presented, which allows the redundant provision of simple non-redundant network elements instead of internally redundant elements. For that purpose new devices are proposed, which implement the protection functionality outside the usual network elements. Two different types of these devices are described, that allow non-stop, hitless operation in the case of a failure, using 1+1-redundancy or diversity techniques.

Keywords

Reliability, Survivability, Redundancy, Diversity

Broadband Communications   P. Kühn & R. Ulrich (Eds.)
© 1998 IFIP. Published by Chapman & Hall

1   INTRODUCTION

The technological progress in the fields of semiconductors, communications and networking allows modern broadband networks to transmit growing amounts of data. The high bandwidths of links and crossconnect systems lead to a reduced number of both of them with constant or enhanced network capacity. Unfortunately, faults in these systems affect an increased number of end-to-end network connections and lead to more loss of data. Consequently, it is important to maintain a very high availability for every part of the communication network.

Conventional approaches achieve this goal using multiple mechanisms. One mechanism is to keep the availability of the network elements at a high level. This is accomplished using redundancy mechanisms inside the network elements (e.g. redundant switch matrices). Another mechanism to increase the network survivability, is to use network wide procedures (e.g. automatic protection switching). Usually these different methods work uncoordinated.

In this paper a new approach is shown, that uses simple network elements and protects them externally to achieve a high reliability. The partitioning of the network in so-called redundancy domains with individual fault tolerance competence is the basic concept and has already been claimed for patent [St95].

Only few of the currently available techniques have the property of hitless operation in the case of failures. However, these methods require 100% redundancy of the protected entities. To achieve a reduction of the required bandwidth, diversity techniques are proposed in this paper, that are mainly known from higher protocol layers.

2   CONVENTIONAL APPROACHES

To minimize the effect of failures in communication networks and to be able to guarantee a continuos operational state, different fault tolerance mechanisms have been proposed and partly implemented. Breakdowns of network elements (e.g. crossconnects, switches, multiplexers) can be reduced by providing internal redundancy in the network element. In the case of link failures or network element failures, networkwide procedures can recover from the error.

2.1 Network Element Reliability

Network elements are composed of a number of different modules. Figure 1 shows the structure of a typical network node, composed of line interface modules, switch matrices, controller and power supply. Each module may fail independently of the others, although the node may also fail totally, e.g. as a consequence of fire or sabotage. To be able to tolerate the failure of single entities in the node, spare modules are provided.

In nodes with high reliability demands, the switch matrix is usually duplicated, and running in hot standby mode with both matrices working concurrently, thus allowing a hitless recovery from failures (e.g. [FFGL91]). Other approaches use the effect, that the switch matrix is often spread over several switching modules and form switching networks within the network element, that may be protected using m:N-redundancy for the switching modules ([It91], [YaSi91]). These fault tolerant switch matrices cannot operate hitless, since a reconfiguration is necessary after the detection of faults. Power supplies are often provided redundantly even in simple low cost switching systems, since they do not require very much effort and offer a simple means to improve a nodes availability. For highly reliable systems, redundant controllers may be provided as sketched in Figure 1. If the hardware of the controller is foreseen redundantly, a synchronization of micro instructions is necessary to achieve non stop operation in the case of a breakdown of one controller. Software errors cannot be overcome even with multiple redundant hardware modules. To tolerate these, separately developed software would be necessary. Micro instruction synchronization isn't possible with this kind of redundancy, thus leading to interruptions when one processor takes over the operation of the other. Very important for the availability of network elements are the interface modules. These modules cannot be protected by automatic mechanisms within the network element and are protected together with the network links, as described later.

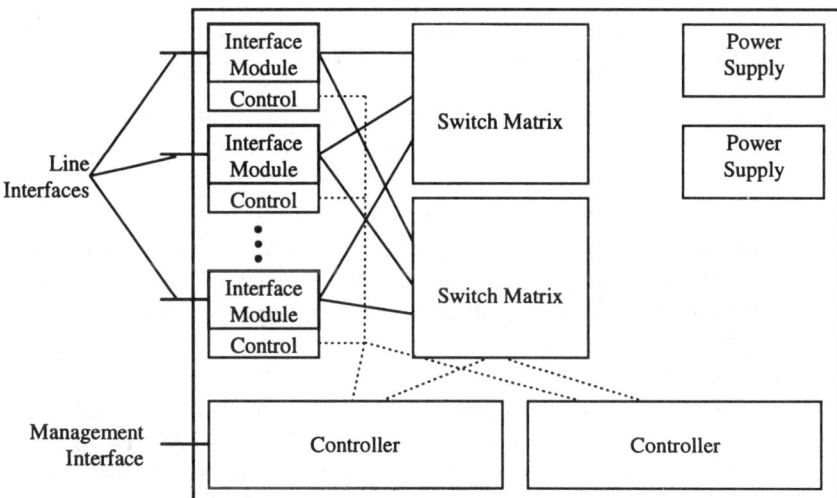

Figure 1 Internally redundant Crossconnect System.

Hence, it is clear, that the design of reliable network elements is not trivial and requires an fairly increased effort compared to simple, non-redundant network elements, like they are mainly used for private network applications.

2.2 Network Survivability

In addition to the methods which increase the network elements' availability internally, a large variety of network wide approaches to maintain a high level of survivability exists. They are used to recover from link failures or from node failures, if the internal mechanisms are not able to fix the problem. Figure 2 shows a common classification of recovery techniques, based on [I.311] and [Ed95].

The advantage of the "Centralized Rerouting" approaches is, that the central network management system has a global view over the network and may achieve a new optimal routing in the case of failures. Due to the high signaling effort and the difficult calculation of the new routing these approaches are rather slow and may lead to enormous loss of data or even drop of calls.

The "Selfhealing Networks" implement distributed algorithms to reroute the connections in case of failures. This distributed determination of new paths speeds up the recovery, but still a considerable loss of data and call dropping occur.

	Protection *Redundant paths predefined*	*Restoration* *Protection paths determined during recovery*
Distributed	Protection Switching, Diversity	Selfhealing Networks (distributed rerouting)
Centralized	-/-	Centralized Rerouting

Figure 2   Classification of Fault Recovery Techniques

The fastest recovery is possible with the "Protection Switching" methods. Using the terms "1:1", "1:N" and "m:N", methods are classified, that have 1 or N protection paths for 1 or m working paths. If a working path fails, it may be switched to a protection path. Therefore not more than N working paths may be recovered at the same time. In the case of no failure having occurred, the protection paths may be used for low priority traffic. Since these methods require the source and the sink of the protected path to be switched to the protection path, requiring an amount of time (usually some milliseconds), this procedure cannot operate hitless and a break in the transmitted signal still occurs.

With 1+1-redundancy, data is always transmitted on two redundant paths. At the receiving side one path is chosen depending on the failure state of the two paths. This approach has the advantage, that only the receiver has to switch to the protection path and no switching operation is necessary at the sender. But therefore 100% redundancy is necessary and no low priority traffic may be transmitted on the protection path.

With diversity techniques it is possible to reduce the required bandwidth by distributing the traffic on multiple links with lower bandwidth and protecting only against single link failures with support for hitless recovery. A more detailed investigation is given later in this paper. An overview on diversity techniques can be found in [GuKa97].

2.3 Evaluation of known Survivability Techniques

In the redundancy approaches presented up to now, the following disadvantages may be identified. They do not offer an integrated approach for network redundancy and network elements' redundancy. The application of the fault recovery techniques on an freely chosen partition of the network is not supported. Most approaches do not provide non-stop operation in case of failures. The integration of redundancy in the network elements may be uneconomical. Further, no existing approach integrates fault recovery techniques with fault correction techniques, although both may be based on a common redundant path.

3 NOVEL REDUNDANCY TECHNIQUE

3.1 Redundancy Domains

To circumvent the prerequisite, that the protection switching operation has to be implemented at topological and physical predefined places (in general in the network elements) the concept of partitioning the network in redundancy domains is proposed. This makes it possible, to provide the whole network element redundantly. A redundancy domain is a subnetwork consisting of two or more redundant network elements, links or subnetworks. These redundant parts coexist independently of each other. They are connected only at so called redundancy domain gateways (RDGs) at their common border. In these new devices the fault detection and protection switching functions are implemented.

Figure 3 shows an example for the redundancy domain principle. Domain D2 is composed of two network elements (e.g. crossconnect systems). At the border of the domain, the redundancy domain gateways connect the two redundant elements. Domain D1 shows, how the same approach can also be applied to network links.

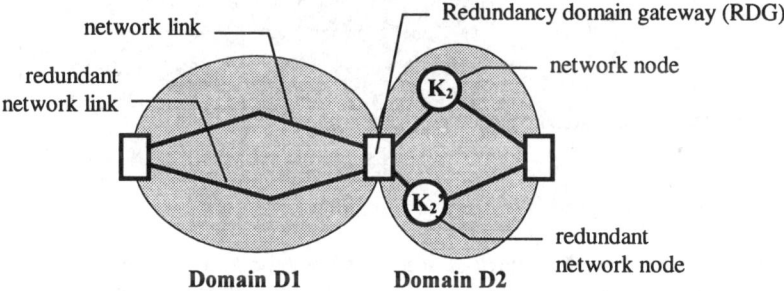

Figure 3 Redundancy domain principle

The use of redundancy domains as protection mechanism is equal to subnetwork protection in the sense of [G.805]. The main difference is, that up to now the protection switching function has always been collocated with the normal switching function. In general the switch matrix of existing network elements is also used for the protection switching operation. With the new approach the protected domains may cover any part of a network and the protection function has not to be physically located at existing network elements (e.g. crossconnect systems). Thus, the network element itself can be protected without the need for internal redundancy. Low-cost, non-redundant devices (e.g. equipment for private-network application) may be util-

ized. Furthermore, network links can also be protected at other points than crossconnect or switching systems, allowing even to split long links in several domains.

In principle the redundancy domain view can also be applied to existing redundancy
approaches. In Figure 4a an example is shown, where a network connection is protected using redundant paths with separate network elements (K_1' protects K_1, K_2'
protects K_2). In this case, the protection function has, up to now, always been collocated with a network node (K_3). The protection of this network node poses a problem,
since redundancy may only be integrated internally and no protection against total
network element failure is provided.

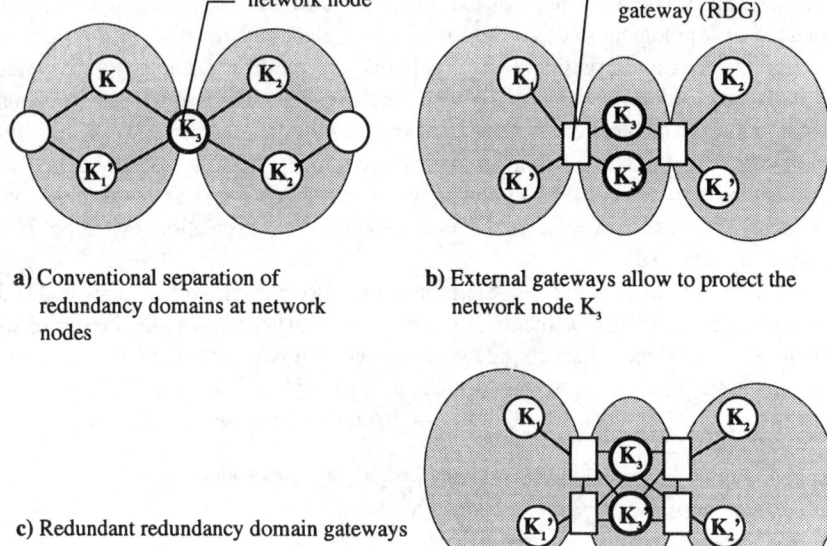

a) Conventional separation of
   redundancy domains at network
   nodes

b) External gateways allow to protect the
   network node K_3

c) Redundant redundancy domain gateways
   may protect the gateways

Figure 4   Node protection with redundancy domains

With the use of the new RDG devices, it is possible to build an own domain around
the network element, which can now be provided redundantly (Figure 4b). Obviously
the problem is now transferred to the gateway elements which now represent so called
single points of failure. With the use of redundant RDGs this problem can be solved
(Figure 4c).

Within the domains, in principle any type of protection may be applied. In this paper
only protection mechanisms with hitless operation are investigated, namely 1+1-
redundancy and diversity. For both types of protection the corresponding new RDG
devices are described in the following sections.

3.2   Redundancy Domains with 1+1-Redundancy

With 1+1-redundancy, the data streams are duplicated when entering a domain and
transmitted on two different paths. Before leaving the domain one data stream has to

be selected. The redundancy domain gateway for 1+1-redundancy is called DS-gateway (Duplicate/Select). Block diagrams for the duplicator and the selector part of the gateway are shown in Figure 5. It is important to notice, that besides the duplication, synchronization and selection of the user data, the protocol overhead (e.g. OAM messages) also has to be processed.

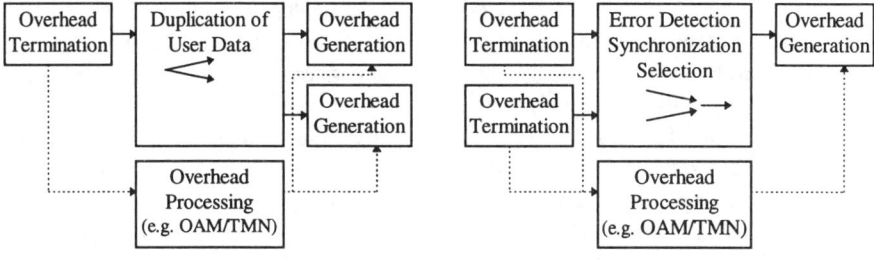

a) Block diagram of duplicator device   **b)** Block diagram of selector device

Figure 5 Block diagrams for Duplicator and Selector

The duplication function is rather simple. Protocol overhead from incoming data streams is stripped off and the user data is forwarded to two outputs where new protocol overhead is added. Depending on the protocol layer in which the duplication is realized, more or less effort for the overhead processing has to be spent. Duplication in the physical layer for example may be implemented using passive optical splitters, whereas duplication in the ATM layer requires the processing of all the layers beneath. On the other hand, using higher layers makes it possible to selectively duplicate only logical connections with high reliability constraints.

For the selection operation also the overhead processing functions have to be provided. In the same sense as stated above, lower layers mean less effort for overhead processing. In some cases it might be strictly necessary to use higher layers, if switching nodes of the respective layer reside within the protected domain. Otherwise the sequencing of data units on redundant paths might not be identical and synchronization becomes impossible. For example, if ATM-VP-crossconnects reside within the domain, the multiplexing of cells from different ATM-VP-connections on the respective outputs of two redundant switches may create different cell sequences, due to different internal states and buffers. Solely within the VPs the cell sequence is guaranteed. Therefore, in this example it would be necessary to implement the selection at the RDG for the ATM-VP-layer and process each VP separately.

Besides the overhead processing, the selection functionality consists of two main tasks, namely synchronization of the data streams and selection of one of them. To provide hitless recovery from failures a permanent synchronization of the redundant data streams is necessary. Different approaches may be identified to achieve this synchronization. Traditionally extra sequence information is added to the data stream, for example sequence numbers or synchronization cells. Besides the need for additional bandwidth, this also requires a structure of the data units that accepts this sequence information (for example the small header in ATM cells does not offer the possibility to add sequence information). Furthermore, special devices to insert this extra information are necessary at the sending side, too. A more sophisticated way to achieve synchronization is to correlate the user data of the redundant data streams. So no extra bandwidth and no extra functionality at the sender have to be provided. The whole functionality is concentrated at the receiver. In [OhUe93] an algorithm for this purpose has been proposed. The algorithm has the drawbacks, that it takes rather long to recognize the synchronization and it does not tolerate the loss of single data units (cells). An improved algorithm has been developed and is currently being patented.

For the selection of one of the synchronized data streams, two different approaches exist. Selection of the leading data stream results in a minimal propagation delay, whereas selection of the lagging data stream allows the correction of loss of single data units, since data is not forwarded until it has been received from both data streams. Depending on the requirements one approach might be chosen.

Conventional recovery techniques use distributed error detection and signaling mechanisms (mostly OAM mechanisms, e.g. AIS, RDI) to trigger the protection switching operation. This leads to interruptions in the data stream due to time constants in the detection process and propagation delay of the signaling messages. In this paper it is proposed that the error detection is placed directly at the switching function. Corrupted data units are detected using CRC check and removed if they cannot be corrected. Loss of single data units might be tolerated or corrected. Lost data units are detected by the synchronization mechanism. Full breakdown of one link is assumed, if more than one data unit in direct sequence is missing or in error. An extension to tolerate or correct even more loss is conceivable. With this approach, it becomes possible to switch to the protection channel with zero loss.

As already mentioned before, the 1+1-redundancy technique for non-stop operation may be applied to different network layers. In Figure 6 an overview of the properties of the application on different layers is given. Obviously, the number of terminated OAM channels increases with the higher layers. Also increasing is the selectivity and the multiplexing factor, especially in the ATM layer. This leads to a high implementation effort in the higher layers, since every virtual connection has to be processed individually. Looking at today's available semiconductor technology, it should be possible to integrate the selection functionality for the layers up to the ATM-VP-layer in a single chip.

For the application of the new technique on existing networks and with existing equipment it is important to regard the interactions with common protection switching mechanisms. Obviously, no problem arises, if the new protection technique resides in a layer below the existing mechanism. For example a duplication-selection-redundancy domain might be applied to a SDH subnetwork without affecting an DRA (distributed restoration algorithm) in the VP layer. Conversely, if lower layers are protected using a conventional mechanism and higher layers (e.g. ATM-VP) use the redundancy domain approach, it might come to states where the two mechanisms influence each other disadvantageously. For example a redundancy domain in the VP layer might recover instantly from a crossconnect failure without loss, while in the SDH layer rerouting is started. Therefore, it would be necessary to coordinate the recovery techniques.

	AAL-Link	VC-Link	VP-Link	TP-Link	MS-Link	RS-Link	PS-Link
Terminated OAM Channels	F5	F4	F3	F2	F1	-	-
Selectivity	ATM-VC		ATM-VP	VC-4	none		
Multiplexing	1	max. 2^{16}	max. 2^{12}	max. 64	1		
Implementation Effort	very high	high	medium			low	very low
	No single chip solution possible	VLSI Single chip solution possible					
Interaction with existing recovery mechanisms	Interactions with SDH protection switching must be considered				Transparent for SDH protection switching		
	Interactions with ATM-VP- protection switching must be considered			Transparent for ATM-VP protection switching			

Figure 6   Properties of DS-RDGs in different network layers

3.3 Redundancy Domains for Diversity

Another approach to guarantee non-stop-operation in the case of a failure are diversity techniques. As already introduced, they reduce the required bandwidth by distributing the data stream on multiple paths and transmitting redundant information on supplementary paths (Figure 7).

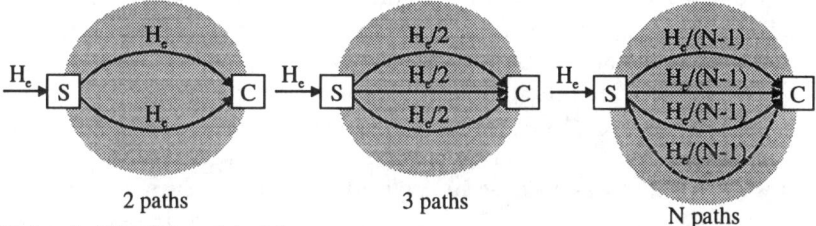

2 paths   3 paths

N paths

H_e=bandwidth of the original data stream

Figure 7   Protection with Diversity

Although the availability may be slightly reduced compared to 1+1-redundancy, a hitless recovery can be guaranteed for all recoverable errors. The required bandwidth B_{total} and the resulting availability A_{total} for this (N-1)-of-N system in dependence of the number N of redundant paths are

$$B_{total} = B_e \frac{N}{N-1} \quad , \qquad A_{total} = NA_S^{N-1} + (1-N)A_S^{n},$$

with B_e = bandwidth of the complete input path, A_s = availability of one single path.

Figure 8 shows the interdependence of unavailability and relative bandwidth for different numbers of redundant links. Obviously, already with a few paths, a lot of bandwidth can be saved compared to a full redundant transmission on two paths, while the reduction of the unavailability is less than one order of magnitude.

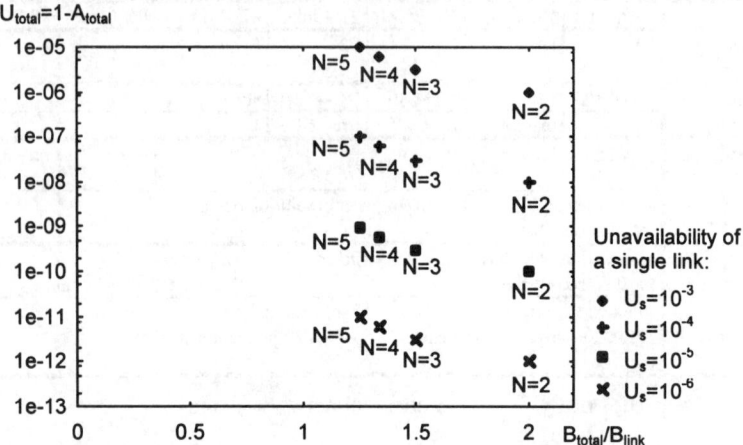

Figure 8   Interdependence of unavailability and relative bandwidth for different numbers of redundant links

For the application of the diversity technique to redundancy domains, the data streams have to be split and sent on several paths when entering the domain. Before leaving the domain the data streams from these paths have to be combined and the original data has to be reconstructed. The principle structure of the RDGs for diversity protection is very similar to that for DS-gateways. The protocol for the underlying layers has to be processed before and after the splitting or combination function.

Two main challenges have to be solved for redundancy approaches: coding/decoding and synchronization of multiple data streams. The coding operation should work very fast and it should be applied to short data units to minimize additional delay. Further, it must not be too complex, to simplify implementation and maintain a high availability of the coder and decoder. Distributing data units on the paths in a round robin fashion and transmitting parity information on an additional path is a simple approach. The second challenge is the synchronization of multiple data streams. With additional synchronization information, for example sequence numbers, a simple operation is possible. The effort for synchronization based on the user data, as described for the synchronization of two data streams, raises exponential with the number of parallel paths and is therefore hardly appropriate. With a combination of both approaches the advantages can be combined. Simplified synchronization detection using sequence numbers in the search phase and exact synchronization using user data correlation in the working phase. Such algorithms are currently investigated and developed.

3.4 Redundancy for RDGs

In general, the two parts of a gateway (e.g. duplication and selection) are combined in a single device. Figure 9a shows a DS-gateway. The selection element in this non-redundant simple DS-gateway selects one incoming data stream from the redundant paths and transfers it to the duplication block, where it is again duplicated and sent towards two different outputs. Similarly in the SC-gateway (Figure 9b) first the incoming data streams are combined and then split to be sent on several separate paths.

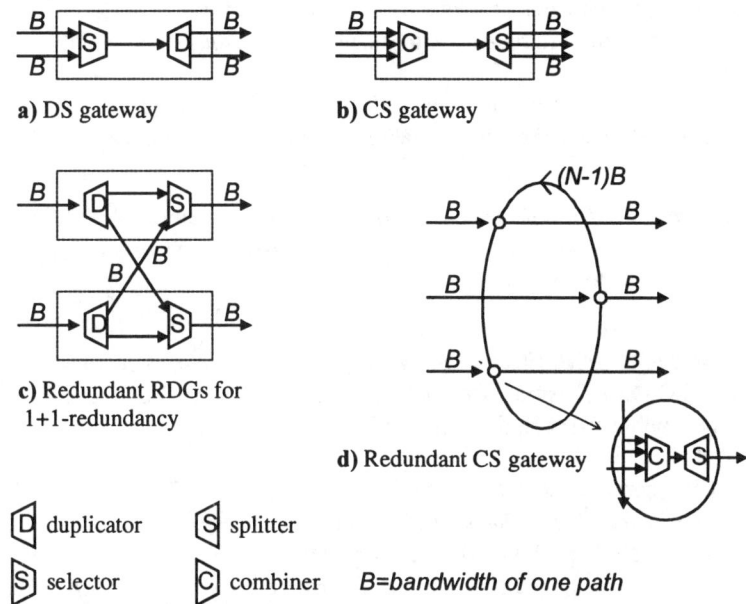

a) DS gateway   **b)** CS gateway

c) Redundant RDGs for
1+1-redundancy

d) Redundant CS gateway

duplicator   splitter

selector   combiner   *B=bandwidth of one path*

Figure 9   Redundancy for redundancy domain gateways

As already mentioned before, the fault tolerance of the RDGs itself plays a major role. Redundant provision of RDGs has already been sketched in Figure 4. A more detailed schematic is shown in Figure 9c. Two devices, each consisting of a duplicate and a select function, are connected via cross links. This allows to survive the failure of a single device or parts of it.

For the case of diversity, the redundancy for the gateways becomes more complex. One possible approach is shown in Figure 9d, where for each parallel path one RDG is foreseen. The gateways are logically fully meshed using a physical ring structure. Each gateway forwards one path. If the input path fails, the data can be reconstructed using the information of the other paths.

4   CONCLUSION

Redundancy domains are a new architectural concept for recovery in fault-tolerant networks, that allows to implement the recovery functionality outside the usual crossconnects or switches and thus to protect them externally instead of the provision of internal redundancy. Redundancy domains may cover any part of a network and may therefore not only be used for the external protection of single nodes but also for the protection of links or whole subnetworks independently of the physical location of crossconnect or switching systems.

To achieve non-stop-operation for connections in the case of node or link failures, two alternative approaches have been investigated concerning their application on these redundancy domains. While 1+1-redundancy is easy to implement and only requires two redundant paths, it uses a lot of bandwidth, whereas with diversity techniques the bandwidth requirements can be reduced, leading to a slightly increased unavailability..

Currently, our work is ongoing in the refinement of the coding and synchronization for diverse transmission. Further, availability measures for redundancy domains are investigated.

The work presented in this paper has been funded by Siemens AG, Munich. The author would like to thank Dr. K.-U. Stein and Dr. M-N. Huber for their helpful comments and valuable discussions.

References

[Ed95] Edmaier B.: Pfad-Ersatzschalteverfahren mit verteilter Steuerung für ATM-Netze, PhD thesis, Institute for communication networks, Technische Universität München, published at Herbert Utz Verlag Wissenschaft, München 1996, ISBN 3-89675-112-3

[FFGL91] Fischer W., Fundneider O., Göldner E.-H., Lutz K.A.: *A Scalable ATM Switching System Architecture*, IEEE Journal on Selected Areas in Communications, Vol. 9, Nr. 8, p.1299-1307, October 1991

[GuKa97] Gustafsson E., Karlsson G.: *A Literature Survey on Traffic Dispersion*, IEEE Network, p.28-36, March/April 1997

[G.805] ITU-T Specification G.805: Generic Functional Architecture of Transport Networks

[I.311] ITU-T Specification I.311: B-ISDN General Network Aspects

[I.610] ITU-T Specification I.610: B-ISDN Operation and Maintenance Principles and Functions

[It91] Arata Itoh: *A Fault-Tolerant Switching Network for B-ISDN*, IEEE Journal on Selected Areas in Communications, Vol. 9, Nr. 8, p. 1218-1266, October 1991

[Kr95] Krishnan P.: *An Efficient Architecture for Fault-Tolerant ATM-Switches*, IEEE/ACM Transactions on Networking, Vol. 3, No. 5, p. 527-537, October 1995

[KrDP95] Krishnan K.R., Doverspike R.D., Pack C.D.: *Improved Survivability with Multi-Layer Dynamic Routing*, IEEE Communications Magazine, July 1995

[OhUe93] Ohta H., Ueda H.: *Hitless Line Protection Switching Method for ATM Networks*, International Conference on Communications, Proceedings ICC '93, p.272-276, 1993

[St95] Stein K.: *Redundant cell or packet oriented Network*, application for a patent, Siemens AG, 1996

[YaSi91] Yang S.C., Silvester J.A.: *Reconfigurable Fault Tolerant Networks for Fast Packet Switching*, IEEE Transactions on reliability, Vol. 40, Nr. 4, p.474-478, 1991

21

Time analysis of fault propagation in SDH-based ATM networks

K. Van Doorselaere (INTEC), K. Struyve (INTEC), C. Brianza (ITALTEL), P. Demeester (INTEC)
Department of Information Technology (INTEC/IMEC),
University of Ghent
St-Pietersnieuwstraat 41, B-9000 Ghent, Belgium,
Tel : + 32 9 264 33 24, Fax : + 32 9 264 35 93,
Email : kvdoor@intec.rug.ac.be

Abstract
This paper provides a detailed analysis of the fault propagation in an SDH-based ATM network from a network view. Both transient and steady state conditions are considered and their influence on the triggering of recovery mechanisms.

Keywords
Faultpropagation, recovery triggering, SDH, ATM

1   INTRODUCTION

Telecommunications networks are subject to random failures caused by natural disasters, wear out, overload, human error, etc. and intentional failures caused by sabotage. A failure affects transmission and/or switching facilities which in turn disrupts residential and business user traffic. Preventive actions such as fire safety plans, armoured cables, etc. try to reduce the failure occurrence frequency. Reactive actions on the other hand heal the network after failure occurrence. Reactive actions can be categorised into three phases, i.e. fault detection, notification and localisation actions, recovery actions and repair actions. Recovery

Broadband Communications   P. Kühn & R. Ulrich (Eds.)
© 1998 IFIP. Published by Chapman & Hall

involves, preferable autonomous, rerouting of disrupted traffic bypassing the failed network facilities, whereas repair involves substitution and mending of failed network facilities.

Consider an SDH-based ATM network transporting ATM traffic via SDH connections. A root or *primary* failure such as a cable cut event causes an avalanche of consequent or *secondary* failures such as disrupted ATM Virtual Paths (VP). More precisely ITU-T recommendation M.20 defines a *failure* as 'the termination of the ability of an item to perform a required function' and a *fault* as 'the inability of an item to perform a required function'.

Operation and maintenance functions implemented in the distributed network elements *detect* these faulty items resulting from the corresponding failures, and *notify* downstream and upstream network elements as well as TMN. Based on this possibly overwhelming number of fault notifications TMN tries to *localise* the primary fault cause.

Standard bodies have defined maintenance signals and fault detection and notification mechanisms among others for SDH (ETS 300 417, 1996), (ITU-T G.783, 1993), (ITU-T G.783-draft, 1996) and ATM (ITU-T I.610, 1995), (ITU-T Q19/13, 1997). Similar concepts and mechanisms for WDM are currently under study. These recommendations define the individual fault detection and notification functional components from a network element view but do not provide a network level view. This paper addresses therefore the timing of fault detection and notification from a network perspective rather than from a network element view and pays special attention to fault propagation across network layers, i.e. Physical Section (PS) layer, Regenerator Section (RS) layer, Multiplex Section (MS) layer, Higher Order Path (HOP) layer, ATM Virtual Path (VP or Avp) layer, for an SDH-based ATM network. Clarification of these mechanisms is not a goal in itself but rather a pre-requisite to study among others survivability strategies for multi-layer networks (Nederlof, Struyve., O'Shea., Misser., Du., Tamayo, 1995). A possible survivability strategy for an SDH-based ATM network, known as 'interworking at the lowest layer' (ACTS-PANEL project, 1997) (Demeester et al., 1997), involves recovery at the SDH layer in case of a physical or SDH layer primary failure and recovery at the ATM layer in case of an ATM layer primary failure. Note that an ATM layer failure such as an ATM switch breakdown cannot be recovered at the lower SDH layer, and hence demands recovery at the ATM layer. Although ATM layer recovery may also resolve physical and SDH layer failures, this strategy uses SDH layer recovery since less entities with a coarser granularity have to be rerouted and thus a higher recovery speed is expected. Correct and fast fault propagation plays clearly an important role in this strategy. Indeed, to minimise service impact recovery must be triggered and completed, either at the SDH layer or at the ATM layer dependent on the primary failure, in the range of milliseconds to seconds.

The rest of the paper is organised as follows: in the first section the elementary fault detection and propagation mechanisms are explained, used to notify the down- and upstream network elements of the failure; in the second section these mechanisms are applied on a SDH based ATM network scenario: two different fault cases are presented and analysed in detail and some issues are pointed out; in the third section some generalised conclusions are derived and finally the influence of the fault detection and propagation mechanisms on the triggering of recovery mechanisms is explained. Used abbreviations are explained in an appendix.

2   CONSIDERATIONS

As mentioned above, a root failure causes an avalanche of secondary failures, notified in the downstream network elements. The detection of these secondary failures is performed in atomic functions within the network elements. An *anomaly* is the first indication that there is something wrong in the network. Some examples of anomalies are : the detection of the Out Of Frame (OOF) condition within the adaptation sink function between the Physical Section and Regenerator Section layer (PS/RS_A_Sk function), the detection of the Out of Cell Delineation (OCD) condition within the adaptation sink function between the SDH VC4 layer and the ATM VP layer (S4/Avp_A_Sk function), the reception of a single all-ONEs signal or a single invalid AU-pointer within the adaptation sink function between the Multiplex Section layer and the VC4 layer (MS/S4_A_Sk function). If there is sufficient density of anomalies for a short period of time (defined as detection time) a *defect* is declared (defect filter F1, see figure 4). The detection time depends on the defect type and may even vary for a specific defect type. For instance the Loss Of Signal defect (dLOS) detection time is dependent among others on the power level of the input signal before the failure (e.g. the fibre cut) and on the receiver technology being used. APD/SAW filter receivers could detect LOS in about 50 µs, but some experiments have proved that pin-fet/PLL receivers may need several milliseconds before detecting dLOS when the input signal before the break is at maximum power level. Thus sometimes it may also happen that the Loss Of Frame defect (dLOF) is detected before dLOS within an equipment. The dependency of the fault propagation mechanism on the detected defects and their required detection times makes that this paper distinguishes two realistic and important fault scenarios, supposing a fibre cut, which resolves in a quick or slow dLOS-detection.

The detection of an anomaly doesn't imply any *consequent action* within an atomic function (consequent action filter F2). Thus while an atomic function is integrating an anomaly into a defect, it passes through the input data under all conditions. As such, an STM-frame filled with corrupted bytes, will present this invalid information to all downstream atomic functions more or less at the same time (see note about transfer delays later in this section). When an atomic function detects a defect, the incoming data signal is no longer forwarded and an Alarm Indication Signal (AIS) (this is an all-ONEs data stream in SDH) is propagated towards the next atomic function. Each downstream atomic function passes through these all-ONEs until this atomic function declares a defect and inserts new AIS-signals (either new all-ONEs, either VP-AIS cells). Remark that, complementary to the all-ONEs signal, the defect condition is also forwarded to the next atomic function by a Characteristic Information Server Signal Fail (CI_SSF) in case an adaptation sink function detects the defect and by an Adapted Information Trail Signal Fail (AI_TSF) in case a trail termination sink function detects the defect.[1] In this next atomic function, possibly not performing AIS-detection, the CI_SSF and AI_TSF represent the defect condition and can thereby possibly, among other consequent actions, control AIS-insertion. As such, the insertion of all-ONES in the Administrative Unit (AU)-pointer within the adaptation source function between the Multiplex Section layer and the SDH VC-4 layer (MS/S4_A_So function) is not controlled by the presence of all-ONES in the VC4 path layer data signal, but is controlled by the CI_SSF emitted by the MS/S4_A_Sk function as a consequence of the Administrative Unit Loss Of Pointer defect (AUdLOP) or Administrative Unit Alarm Indication Signal defect (AUdAIS) detection (ETS 300 417-3-1, 1996). Within two frames (read : "at most the next second frame") from the reception of the CI_SSF signal, the MS/S4_A_So function will replace the outgoing AU pointer with an all-ONES pointer (AU-AIS).

Each atomic function correlates its detected defects (fault cause filter F3) and the result, a *fault cause*, is reported to the Equipment Management Function (EMF) of the network element (ETS 300 417, 1996). When a fault cause persists for X seconds (default value for X = 2.5 ± 0.5 s) in the failure filter F4 (see figure 4), it is reported to TMN as a *failure* (ETS 300 417, 1996). In TMN these failures are correlated to identify the root failure. A correct correlation in the atomic functions requires thus that the fault cause filter F3 in each atomic function rejects secondary defects and holds back the primary defect. As such, the LOF fault cause (cLOF) will not being reported to the EMF on an incoming AI_TSF, indicating that dLOS was detected. Some of the secondary defects are autonomously cleared on receipt of AIS, SSF or TSF. For example, defects as MS-DEG, MS-EXC, AUdLOP are being cleared on incoming all-ONEs signalling, indicating that the failure was already noticed more upstream. Some atomic functions, not detecting the all-ONEs, clear some of their defects on the reception of a CI_SSF or AI_TSF signal. For example, the S4_TT_Sk function, not performing all-ONEs detection, will clear its defects (dUNEQ, dTIM, etc...) on an incoming CI_SSF. In a similar way, a correct correlation in the S4/Avp_A_Sk requires that the LCD fault cause (cLCD) is cleared on an incoming AI_TSF, indicating an ATM physical layer trail signal fail condition. Remark that this is not yet standardised.

For a detailed time analysis of the fault propagation also transfer delays should be taken into account. Within an equipment transfer delays may be introduced by

---

[1] CI_SSF and AI_TSF are functional signals and realised often as either hardware or software signals and are thus implementation dependent.

buffers (e.g. pointer buffer, mapping and demapping buffers, jitter reduction buffers), switch fabric, frame alignment circuits, encoders and decoders, series/parallel and parallel/series convertors and other equipment specific processes. For a VC-4 the transfer delay within an equipment is about 10 μs, which is quite low in comparison with the frame repetition time (125μs) and therefore neglected in this paper. Another transfer delay is added in the fibres, depending on the length of the fibre : e.g. a signal needs 250 μs to propagate through a fibre of 50 km.

The next section discusses in detail the fault detection and propagation mechanism for an SDH based ATM network scenario. Two fault scenarios are considered : dLOS is detected quick or slow after a fibre cut.

3   QUICK AND SLOW LOSS OF SIGNAL DEFECT DETECTION

3.1 Network scenario

Figure 1 presents a multilayer (ATM-VP layer over SDH HOP layer) network scenario, consisting of some basic SDH and ATM equipment, which is connected by bidirectional lines (i.e. a fibre in each direction). In the scenario a fibre cut has been supposed (root failure). The high level sequence of events in steady state conditions is itemised below.
1.   LOS is detected in the regenerator.
2.   MS-AIS is generated and sent to DXC-AU_1.
3.   MS-RDI is generated and sent in the upstream direction.
4.   MS-AIS is escalated to AU-AIS, which is sent to DXC-AU_2.
5.   AU-AIS is forwarded to ATM-VPXC_1.
6.   HP-RDI is generated and sent in upstream direction.
7.   VP-AIS is sent downstream for each VP connection supported by the affected SDH path.

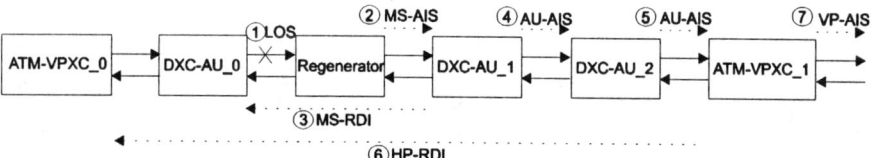

Figure 1   SDH based ATM network scenario.

3.2   Quick LOS defect detection

In this fault scenario it is assumed that the fibre cut leads to a quick LOS detection in the regenerator (and thus before the detection of other defects in the network). The fault propagation mechanism is illustrated in figure 2a, while the timing aspects are represented in figure 2b.

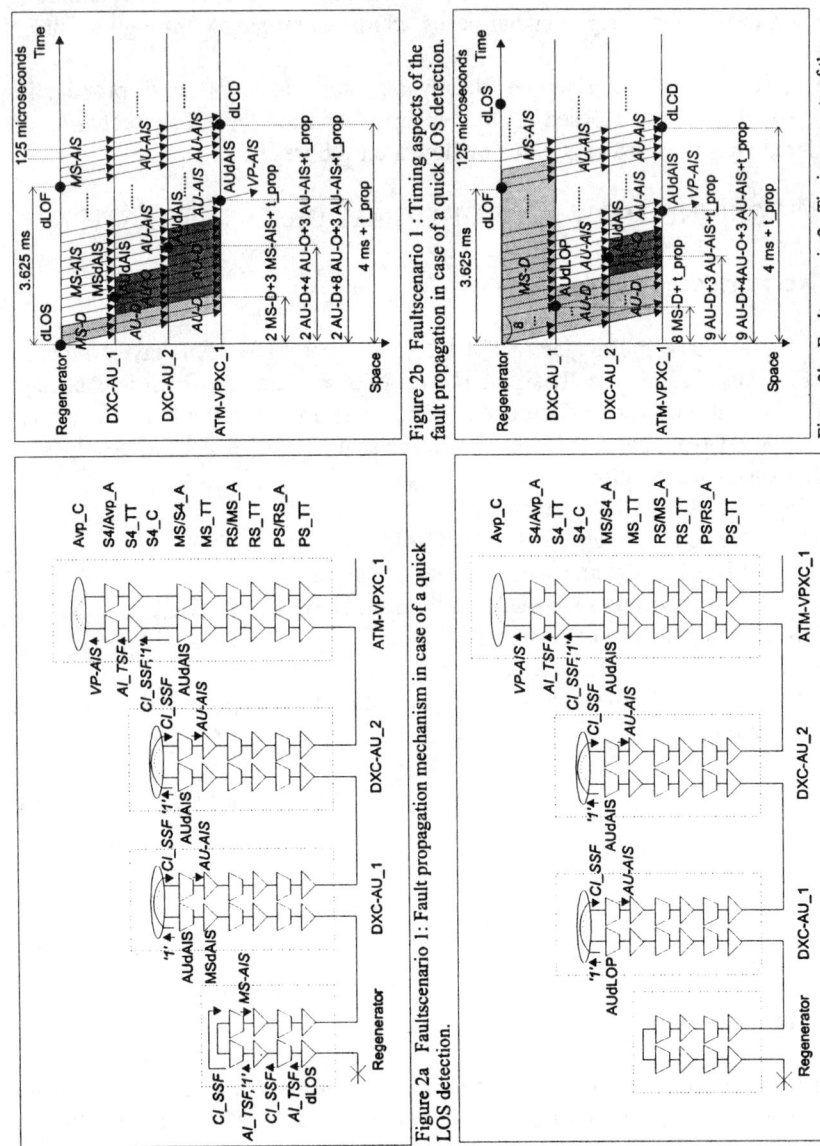

Figure 2b Faultscenario 1 : Timing aspects of the fault propagation in case of a quick LOS detection.

Figure 3b Faultscenario 2 : Timing aspects of the fault propagation in case of a slow LOS detection.

Figure 2a Faultscenario 1 : Fault propagation mechanism in case of a quick LOS detection.

Figure 3a Faultscenario 2: Fault propagation mechanism in case of a slow LOS detection.

The Trail Termination sink function in the Physical Section layer (PS_TT_Sk function) of the regenerator notices that the physical input signal is no longer valid and declares the dLOS-defect. In the meantime this atomic function generates 0/1 bits in a random way (depending on the hardware technology) and inserts them in the downstream direction. Each downstream atomic function passes through this digitised noise until a defect is detected and AIS signals (all-ONEs/VP-AIS cells) are inserted. Upon the declaration of dLOS an AI_TSF is activated in the PS_TT_Sk function towards the PS/RS_A_Sk function. This signal is refreshed in each of the next atomic functions, either as CI_SSF either as AI_TSF and resolves finally in the RS/MS_A_So function within 2 frames into the insertion of MS-AIS towards DXC-AU_1, containing a valid Regenerator Section Overhead (RSOH) and an all-ONEs regenerator payload. Each downstream atomic function passes through these all-ONEs until an AIS defect or other local defect is detected and new AIS signals (all-ONEs/VP-AIS cells) are inserted. In summary, after the fibre cut the regenerator outputs 2 frames downstream, containing a valid RSOH but digitised noise in the regenerator payload (later on referred to as MS-D frames), before generating MS-AIS signals.

The digitised noise in the MS-D frames initiates in the DXC-AU_1 the detection of anomalies, such as the MS BIP-24N violations in the MS_TT_Sk function and invalid pointers in the MS/S4_A_So function. The MS-AIS signalling (containing all-ONEs in the regenerator payload) however suppresses these anomalies and results instead in the declaration of two AIS-defects : the MS_TT_Sk function declares the MSdAIS defect after having received for X (X is in the range 3 to 5) consecutive frames '111' in bits 6,7,8 of byte K2, while the MS/S4_A_Sk declares the AUdAIS defect after having received for 3 consecutive frames '11111111' in the H1 and H2 bytes (AU-pointer bytes). Within 2 frames after the MSdAIS-detection the upstream direction is notified by the insertion of an MS-RDI signal. Within 2 frames after the declaration of the AIS defect (MSdAIS, AUdAIS), the MS_TT_Sk function respectively the MS/S4_A_Sk function renew the all-ONEs stream. In the meantime the incoming all-ONEs-stream is passed through transparently. The detection of AUdAIS activates however also a S4_CI_SSF towards the MS/S4_A_So function, which adds within 2 frames an all-ONEs AU-pointer to the Higher Order Virtual Containers (HOVC). Meanwhile normal AU-pointers are written to the HOVCs in the generated STM-frames. In summary, after the root failure the DXC-AU_1 outputs 2 AU-D frames (no standardised definition) (containing the same noise in the HOVC as the MS-D frames, but a correct AU-pointer and new Section Overhead (SOH), 4 AU-O frames (no standardised definition) (containing the same all-ONEs in the HOVC as the incoming MS-AIS, but a new valid AU-pointer and new SOH) and only then AU-AIS (with a new all-ONEs HOVC, an all-ONEs AU-pointer and correct SOH).

In DXC-AU_2 no anomalies, except for AUdAIS, are observed because of the renewed overhead of the STM-frames in DXC-AU_1. When the AU-AIS frames enter DXC-AU_2, 3 consecutive frames with the all-ONEs AU-pointers are required to declare AUdAIS and 2 more frames to output AU-AIS. The AU-AIS signal is thus again delayed for 4 frames (compared to the HOVC payload in the STM-frame) and the DXC-AU_1 will output meanwhile 4 AU-O frames.

In ATM-VPXC_1 no Physical -, Regenerator - or Multiplex Section anomalies are noticed. AUdAIS is again detected after 3 consecutive AU-AIS frames. Making the summation, 2 AU-D frames + 4 AU-O frames + 4 AU-O frames + 3 AU-AIS frames entered ATM-VPXC_1 since the fibre cut until AUdAIS declaration. The AUdAIS detection time depends thus on the number of intermediate AU-pointer processing functions.

The detected AUdAIS activates a CI_SSF and this signal, refreshed in the S4_TT_Sk function as AI_TSF, causes thereby in the S4/Avp_A_Sk function the insertion of VP-AIS, the first as soon as possible and then every second one. In the S4_TT_Sk function the CI_SSF suppresses the detection of defects and generates within 2 frames a HP-RDI to notify the upstream network elements. In the S4/Avp_A_Sk function the AI_TSF suppresses the fault cause cLCD but not the defect dLCD. After receiving thus for 4 ms invalid VP-cell headers, containing random bits or all-ONEs, dLCD is declared, activating also VP-AIS cells.

As already highlighted, the detection time of AUdAIS in ATM-VPXC_1 depends on the number n of intermediate AU-pointer processing functions before this network element. Consider therefore the network scenario presented in figure 2c, consisting of a regenerator and n DXC-AUs before ATM-VPXC_1.

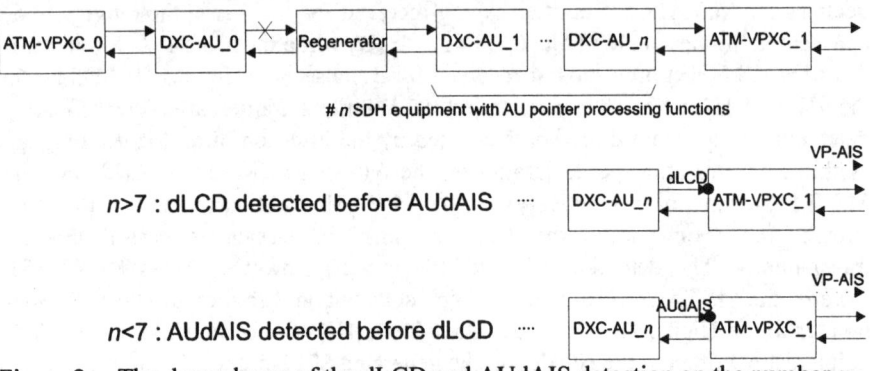

Figure 2c   The dependency of the dLCD and AUdAIS detection on the number n of SDH equipment with AU-pointer processing functions in case of a quick LOS.

Reusing the reasoning above, 2 AU-D frames + $n*4$ AU-O frames + 3 AU-AIS frames entered ATM-VPXC_1 since the fibre cut until the detection of the AUdAIS defect. **If $n>7$, the defect dLCD is detected before the defect AUdAIS in ATM-VPXC_1** (remark that the defect dLCD is detected wihin 4 ms, equalling 32 frames). Supposing $n>7$, it is the defect dLCD that activates the insertion of the VP-AIS cells in ATM-VPXC_1. When finally AUdAIS is detected, this sends a CI_SSF/AI_TSF towards the S4/Avp_A_Sk function, which clears the fault cause cLCD. Because the fault causes are integrated on a X seconds basis in the EMF of the network element, the correct failure will be passed to TMN.

3.3 Slow LOS defect detection

In this fault scenario the receiver technology notices very slowly the fibre cut, and thus secondary defects such as dLOF will be detected before the primary defect dLOS.

As a consequence of the slow reacting receiver technology, the PS_TT_Sk function assumes that the input signal is still valid and generates thus incorrect decoded bits in the downstream direction. As such, the PS/RS_A_Sk function declares, while hunting for the A1A2-bit pattern to align the frames, after 5 frames with incorrect A1A2-bytes an OOF-anomaly, which results 3 ms later in the defect dLOF. Within 2 frames after dLOF-declaration, MS-AIS-signals are inserted to inform DXC-AU_2. When finally dLOS is detected, the generated AI_TSF clears in the PS/RS_A_Sk function the fault cause cLOF. In summary, after the fibre cut the regenerator outputs 30 MS-D frames (containing a correct RSOH but incorrect decoded bits in the regenerator payload) before generating MS-AIS frames.

The incorrect regenerator payload of the MS-D frames causes also in the other atomic functions in the chain the detection of anomalies. In the MS/S4_A_Sk of DXC-AU_1 this results in the declaration of the defect AUdLOP after the detection of 8 consecutive invalid AU-pointer-anomalies. Within 2 frames after AUdLOP the MS/S4_A_Sk inserts all-ONEs in the downstream direction. At the same moment results the CI_SSF, emitted by the MS/S4_A_Sk, in the MS/S4_A_So in the writing of all-ONEs AU-pointers to the incoming HOVCs (containing the all-ONEs) and generates thus AU-AIS signals. In conclusion, after the fibre cut the DXC-AU_1 has output 9 AU-D frames (containing the same incorrect HOVC payload as the MS-D frames but with a new correct AU-pointer and SOH) before generating AU-AIS signals. **Thus it is the detection of a defect in a downstream equipment (i.e. AUdLOP) that initiates the insertion of AIS-signals.** When finally MS-AIS signals reach the DXC-AU_1, both AIS-defects are detected again : the detected MSdAIS suppresses thereby the MS BIP-24N violations and generates also within 2 frames in the upstream direction an MS-RDI, while the detected AUdAIS clears the AUdLOP defect.

For the fault detection and propagation mechanism downstream of DXC-AU_1 can be referred to the fault scenario with a quick LOS. As thus, after the fibre cut 9 AU-D frames + 4 AU-O frames + 3 AU-AIS frames must enter ATM-VPXC_1 before declaring the AUdAIS defect. The AUdAIS is thus detected first in ATM-VPXC_1 and activates the insertion of the VP-AIS cells.

Considering now again the network scenario of figure 2c, ATM-VPXC_1 will detect AUdAIS after it received 9 AU-D frames + $(n-1)*4$ AU-O frames + 3 AU-AIS frames since the fibre cut. **Thus if $n>6$, the defect dLCD-defect is detected before the defect AUdAIS in the ATM-VPXC.**

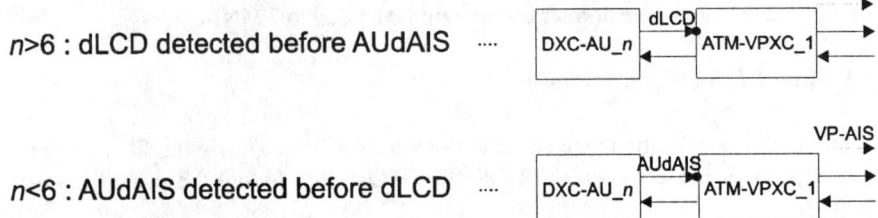

Figure 3c   The dependency of the dLCD and AUdAIS detection on the number n of SDH equipment with AU-pointer processing functions in case of a slow LOS.

4   GENERALISED CONCLUSIONS

The fault scenarios analysed in the previous subsections show that during transient conditions after failure occurrence the network sometimes behaves in a strange way. Such strange behaviour manifests itself principally in the following events:

- detection of defects within the downstream equipment quite before the detection of the primary defect (e.g. AUdLOP detection in DXC-AU_1 before dLOF detection in the upstream regenerator - see 2.3).
- anticipated manifestation of an upper layer defect which is detected before the correlated lower layer defect within the same equipment (e.g. dLCD defect detection before the AUdAIS defect detection in an ATM VPXC - see 2.2 and 2.3).

Table 1 summarises for the explained fault scenarios the first detected defect in the network and explains the dependency of the AUdAIS and dLCD defect detection in ATM-VPXC_1 on the number n of AU-pointer processing functions in front of ATM-VPXC_1.

Table 1: Side effects during the transient period of the defect detection and propagation process in an SDH based ATM network.

Fault scenario	First defect detected in the network	x frames entered ATM-VPXC_1 since root failure till AUdAIS-detection, with x:	dLCD detected before AUdAIS in ATM-VPXC_1 if n>
quick LOS	dLOS	$2 + n*4 + 3$	7
slow LOS	AUdLOP	$9 + (n-1)*4 + 3$	6

TMN however will never notice these side during the transient period of the defect detection and propagation, due to the fault cause integration for X seconds (X : 2.5 ± 0.5) in the EMF before reporting them as a failure to TMN.

After the root failure is repaired, valid data will be transmitted again through the network. Only when the declared defects in each atomic function are cleared, it will stop the insertion of the all-ONEs signal and pass through the valid data. In consequence, the defect clearing is performed in a serial way and takes thus much longer than the defect propagation. As a result, supposing a temporary failure, the defect clearing will never catch up the defect propagation. Due to the fault cause integration of X seconds, no failure notification reports will be generated on a temporary fault.

5 GUIDELINES

In case the recovery strategy is defined as 'interworking at the lowest layer' (ACTS-PANEL project, 1997) (Demeester et al., 1997), a distributed recovery mechanism in e.g. the ATM VP layer (distributed restoration/protection) should only be triggered on ATM VP layer primary failures or on non recoverable server layer failures and provide thus non redundant protection. For the network scenario of figure 2c, this implies that a fibre cut in the span of DXC-AUs should be recovered within this span and not trigger ATM VP layer recovery mechanisms in the ATM-VPXCs. As the presented fault scenarios made clear, the ATM-VPXCs will always detect a defect in the range of a few milliseconds, whereas the SDH recovery will require at least 50 ms (protection). The ATM-VPXC may thus not trigger an ATM-VP layer recovery mechanism on the simple detection of a defect and requires thus more defined conditions. Steady state fault causes can provide thereby useful information : e.g. the steady state fault cause cLCD in an ATM-VPXC indicates that the S4/Avp_A_So function of the upstream ATM network element has failed, which can not be recovered in the lower SDH layer and demands thus recovery in the ATM-VP layer. To assure however that the transient conditions are elapsed, a reasonable persistency time of the fault causes is required, to be defined on the basis of the values suggested by the presented fault scenarios. Remark that this recovery strategy requires also other mechanisms, besides the use of the fault causes, to control a correct triggering of the recovery protocols. These mechanisms are currently under study in the PANEL-project. Figure 4 illustrates the most relevant fault management filters used in this paper and proposes some parts of a triggering architecture for the 'interworking at the lowest layer'-strategy.

ACKNOWLEDGEMENT

This work has been partly funded by the ACTS project AC205 - PANEL : "Protection Across Network Layers". During the study of this topic, M.P.J. Vissers (Lucent Technologies Network Systems Nederland) has provided valuable information about the fault management mechanisms.

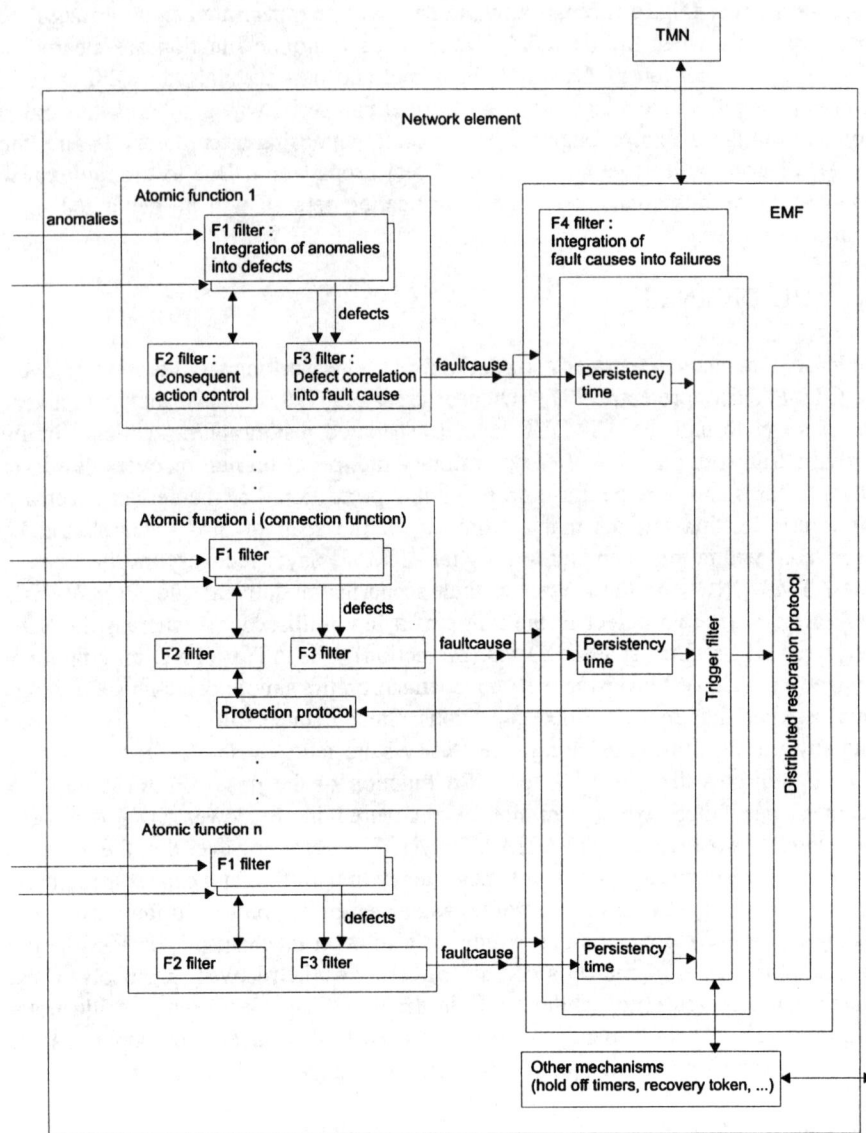

Figure 4   Recovery triggering architecture within a NE.

APPENDIX

This appendix explains used abbreviations and acronyms.

- *Layers :* PS (Physical Section); RS (Regenerator Section); MS (Multiplex Section); S4 (SDH VC-4 path layer); VP or Avp (ATM Virtual Path layer).

- *STM-frame structure* : AU (Administrative Unit); HOVC (Higher Order Virtual Container); MSOH (Multiplex Section Overhead); RSOH (Regenerator Section Overhead); SOH (Section Overhead).
- *Atomic functions* : X/Y_A_Sk : Adaptation Sink function between client layer Y and server layer X; X/Y_A_So : Adaptation Source function between client layer Y and server layer X; X_C : Connection function in layer X; X_TT_Sk : Trail Termination Sink function in layer X; X_TT_So : Trail Termination Source function in layer X.
- *Anomalies* : OCD (Out Of Cell Delineation); OOF (Out Of Frame).
- *Defects* : dAIS (Alarm Indication Signal defect); dDEG (degraded defect); dLCD (Loss of Cell Delineation defect); dLOF (Loss Of Frame defect); dLOP (Loss Of Pointer defect); dLOS (Loss Of Signal defect); dPLM (Payload Mismatch defect); dRDI (Remote Defect Indicator defect); dTIM (Trace Identifier Mismatch defect); dUNEQ (unequipped defect); Exc (Excessive error defect).
- *Alarm signals* AIS (Alarm Indication Signal); AI_TSF (Adapted Information Trail Signal Fail); CI_SSF (Characteristic Information Server Signal Fail); RDI (Remote Defect Indicator).

REFERENCES

ACTS-PANEL project (1997). *Deliverable D2a : Overall Network Protection-Version 1*.

Demeester et al., (1997) PANEL - Protection across network layers. *NOC '97, European Conference on Networks and Optical Communications*, Antwerp, Belgium.

ETSI Recommendation (1996). *ETS 300 417 : Transmission and Multiplexing (TM); Generic functional requirements for Synchronous Digital Hierarchy (SDH) equipment*.

ITU-T Recommendation (1993). *G.783 : Characteristics of Synchronous Digital Hierarchy (SDH) Equipment Functional Blocks*.

ITU-T Recommendation (1995). *I.610 : B-ISDN Operation and Maintenance Principles and Functions*.

ITU-T Recommendation (1996). *G.783-draft : Characteristics of Synchronous Digital Hierarchy (SDH) Equipment Functional Blocks*.

ITU-T Q.19/13 Rapporteur's Meeting (16-20 June 1997). *AIS in the ATM Network Functional Model*. Turin, Italy.

Nederlof L., Struyve K., O'Shea C., Misser H., Du Y., Tamayo B. (September 1995) End-to-end survivable broadband Networks. *IEEE Communications Magazine*.

22

Resource management for service accommodation in optical networks

Admela Jukan
Vienna University of Technology
Institute of Communication Networks
Gusshausstrasse 25/388, A-1040 Vienna, Austria
Tel. +43-1-58801-3993, Fax +43-1-587 05 83
E-mail: admela.jukan@tuwien.ac.at

Abstract

In this paper, a novel approach for service accommodation in WDM networks is proposed. It is based on the classification of optical network services according to Quality-of-Service (QoS) to be guaranteed for a client network (typically, SDH or ATM). This principle is applied to node architectures and node main functions as well as to wavelength assignment. Various resource management heuristics for WDM optical networks are presented, where shifting of physical and logical resources in dynamically routed optical networks is minimised. Simulation results for an example network are presented for a demonstration of applicability of the proposed methods.

Keywords

Optical network service, QoS-routing, node architectures, resource management, blocking probability

Broadband Communications   P. Kühn & R. Ulrich (Eds.)
© 1998 IFIP. Published by Chapman & Hall

1   INTRODUCTION

In multi-client optical networks, it is advantageous to classify each client according to the Quality-of-Service (QoS) to be guaranteed for that client. This is because end-to-end parameters like transmission performance, throughput, or survivability of different client networks are not required to be equal. In spite of the fact that this principle is generally applicable to multi-client networks, only wavelength division multiplexing (WDM) networks are considered in this paper.

Client networks, i.e. typically synchronous digital hierarchy (SDH) and asynchronous transfer mode (ATM) networks, are connected to optical nodes for access, where one or more WDM channels are assigned to them. The wavelength channels are circuit-switched and with a certain grade of transparency handled within an optical network. The QoS-guarantee refers to a dedicated wavelength channel, i.e. to the comprehensive client network, but not to a single end-user of that network in terms of bit patterns, audio or video quality.

In this paper, service accommodation is proposed based on classifying optical network services according to QoS-guarantee for a client network for which a service is provided. This principle will be applied to node architectures and node main functions as well as to routing and wavelength assignment. The resource management heuristics for wavelength assignments in optical networks are presented, where the number of service quality degree shifting, resource shifting as well as the number of wavelength shifting required is minimised. These methods are applied to dynamically routed optical networks. Finally, simulation results for an example network with meshed topology and various service quality degrees and heuristics are presented.

2   QUALITY-OF-SERVICE IN OPTICAL NETWORKS

Instead of considering QoS-attributes like data patterns, video- or audio- quality, in optical networks we deal with QoS-attributes that refer to many electronically multiplexed end-users simultaneously. Therefore, "Quality-of-*optical-network-*Service" (QoonS) was defined in (Jukan, 1997). An optical network service (ONS) handles optical signals originating from optical network clients. The most important optical network clients today are networks based on SDH and ATM.

The resource management in an optical network decides about necessary and available resources for optical network service accommodation according to the quality attributes of a certain service. The basic factors which define these quality attributes are: complexity of access interfaces, transmission performance, cross-connecting and routing functions, network supervision and management, optical network service restoration, and economic efficiency (Gruber, 1986). The performance objectives of each singular client differ. High throughput may be required for a client handling data, while video communication may require higher signal-to-noise ratio. The selection of the network route for a particular client must take into account which route is the best suited for providing the necessary quality.

There are two possibilities how quality attributes are associated to routing (Jukan, 1997). Either routing is separated from the handling of quality attributes or the quality attributes directly take part in routing by influencing the cost of links. In both of these methods, optical network services are classified according to the quality degree required. The blocking probability is here defined as probability that for a certain service a sufficient quality cannot be provided.

3   NODE ARCHITECURE

Optical networks considered here are capable of providing various services for client networks, like access, transmission, routing, or alternative path searching. They consist of optical fibres, which connect optical nodes. The access for optical network clients is provided by means of client network access (CNA). Optical nodes contain QoS-selective switches realised with controllable wavelength and/or fibre switches. The functions of CNA and QoS-selective switches will be discussed in more detail. Nevertheless, other node functions like de/multiplexing, regeneration, or control, which are not discussed here, must also be realised on the basis of QoS-guarantee.

The architecture of a QoS-selective switch for an optical node is shown in Figure 1. In this figure, only the functionality of each part, but not the exclusive implementation methods is shown. A QoS-selective switch is implemented for routing of multi-channel WDM signals and for routing of single channels being demultiplexed from a WDM signal or being newly assigned from a CNA. Space switching is provided on a fibre-to-fibre basis. In Figure 1, two functional blocks for switching called Multi-Channel Shifting and Single Channel Shifting are shown. Multi-Channel Switching is capable of shifting a complete incoming WDM signal or a part of that signal containing more than one channel. This is achieved by means of Bandwidth Shifting (BS). This function can be realised with converter arrays designed for wide bandwidth conversion (rb) (Iannone, 1996). Otherwise, a multi-channel signal can be bypassed (rb_0).

Single channels can be handled as follows. They can be routed to a CNA if the affected node is the destination node (r_3). Within the same service quality degree, single channels can be converted into another wavelength within a Wavelength Shifting (WS) array (r_1). They can also be routed to a Quality Degree Shifting (QDS) stage for re-transmission on a new wavelength (r_2). As previously shown, they can be bypassed, too (r_0).

Both functions of bypassing (r_0 and rb_0) might be necessary for WDM signals carrying one or more single channels of such a quality degree that need not be or can not be processed by the affected node. This is an even bigger grade-of-transparency in optical networks.

The common architecture for CNA and QDS can be seen in the bottom of Figure 1. A client network input/output is connected to a configurable point-to-point client switch (service switch). According to the service quality degree to be configured for an attached client the appropriate input/output of that switch is chosen dynamically. This input/output is furthermore connected to laser/receiver arrays. Laser/receiver arrays are supposed to be different for different quality degrees, which is in accordance with the design of commercial devices.

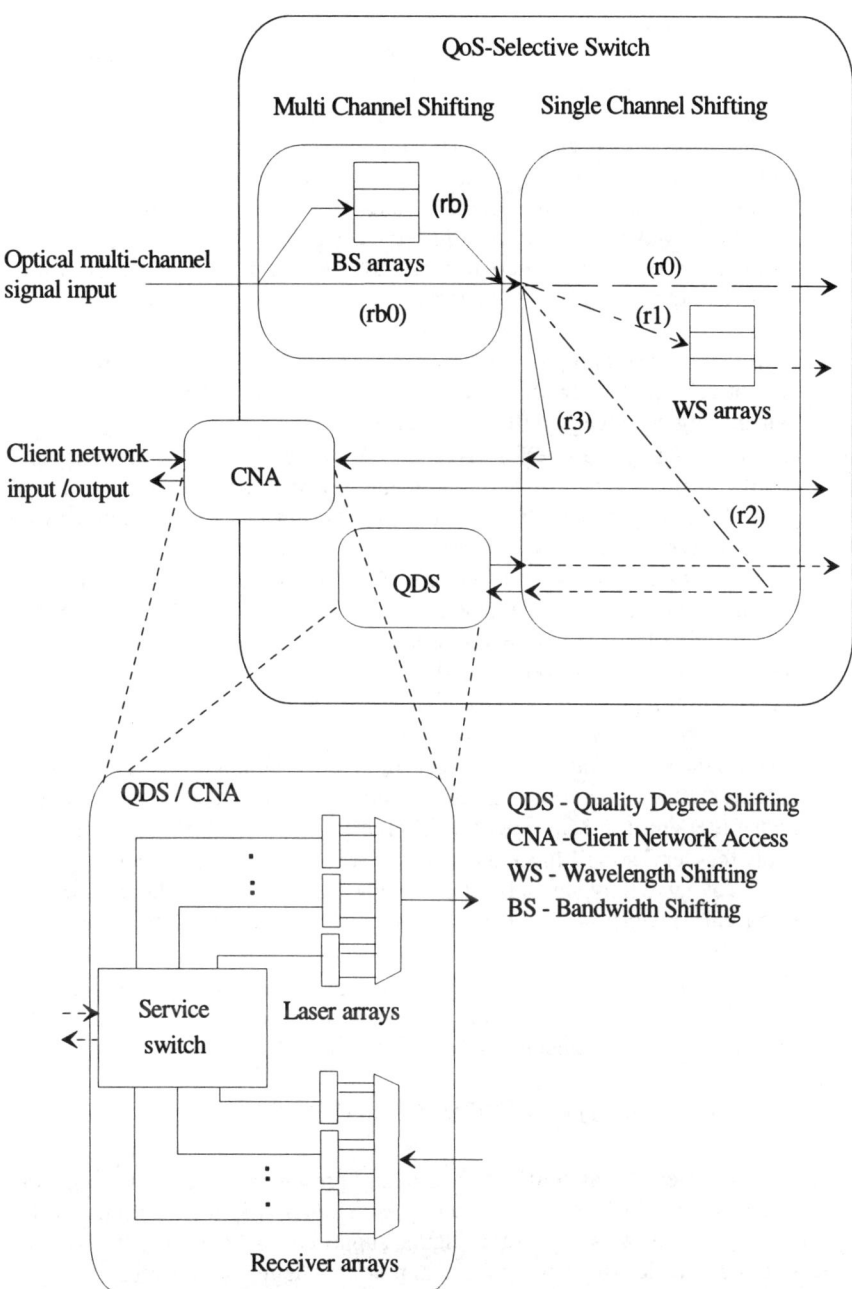

Figure 1. QoS- Selective switch with CNA and QDS architectures. Routing of optical multi-channel signals: with BS within the same service quality degree (rb) or without BS (rb$_0$). Routing of single optical channels: (r$_0$) bypassing, (r$_1$) all-optical conversion (wavelength shifting) within the same service quality degree, (r$_2$) QDS for re-transmission on a new wavelength, and (r$_3$) receiver arrays in CNA.

With the CNA architecture as shown in Figure 1, it is not necessary to provide wavelength conversion for accessing signals, because the best transmitting /receiving wavelength can be arbitrarily chosen. This feature has large importance for wavelength continuous optical networks, where the choice of transmitting wavelength has a large impact on performance.

The QDS-architecture is almost the same as the CNA-one, only the client input/output part is missing. Here, a single channel is transparently re-transmitted on a wavelength reserved for services of higher quality degree.

Summarised, service accommodation provided by the architecture as shown in Figure 1 yields the following advantages:

- Wavelength adaptation for routing is performed at CNA, which is important for optical networks without wavelength conversion where the choice of transmitting wavelength influences performance.
- Network management effort is reduced because less management and control functions are to be implemented when lower quality is to be guaranteed. Regarding economic constraints, cheaper components can be used for service classes where less quality is required.
- Different functions like wavelength shifting or network access can be partially (per service) implemented according to the particular requirements on a certain class of service which is of great importance for modular and cost effective design of optical nodes.
- Integrated components like arrays of lasers, converters, filters or receivers can be used. They are generally designed to provide a certain set of performance in defined ranges (for example, technologies for components used for handling various kinds of optical signals, like 10 Gb/s optical transmitter and 155 Mb/s one, are different).
- With the service shifting designed with the re-transmission by means of electrical switching, an advantage might be an easy implementation of electrical regenerators. This might be unavoidable for long-haul connections.
- This architecture is suitable for operating many corporate optical networks over the same physical topology where users or group of users with different QoS requirements can be accommodated separately.

4   WAVELENGTH ASSIGNMENT HEURISTICS

For the wavelength assignment heuristics presented here, the following definitions are made. The *primary wavelength set* (PWS) for a certain service quality degree contains the wavelengths supposed to be primarily assigned before other wavelengths belonging to higher service quality degrees are taken. *Resource shifting (RS)* occurs when a called service is not using resources from its own PWS. *Wavelength shifting (WS)* refers to wavelength shifting on a link-to-link basis only between wavelengths belonging to the same PWS. *Quality degree shifting (QDS)* occurs on a link-to-link basis, too, and changes the current service quality degree by means of wavelength shifting into another PWS.

The basic ideas for the wavelength assignment heuristics discussed here are to minimise wavelength shifting, quality degree shifting, and/or resource shifting.

Wavelength shifting minimisation might be necessary in order to improve transmission performance, if the cascadeability of wavelength converting is critical. When minimising quality degree shifting, less additional hardware and fewer service resources of higher quality degree are needed. Resource shifting minimisation improves network performance since each service uses its own resources.

Assume a network with N nodes and L links for which a wavelength set Λ and a service set S are defined as $\Lambda=\{\lambda_1, \lambda_2,..., \lambda_W\}$ and $S=\{S_1, S_2, ... S_P\}$, respectively. The basic assumption is that each service can be provided at least by one and maximally by all wavelengths from Λ (Jukan, 1997). Service quality degrees are defined in terms of wavelength usage. The number of wavelengths limits the maximal number of services (generally $P \leq W$). Assume that a particular service $S_i \in S$, $i \leq P$ may assign any wavelength from the wavelength set $W[S_i]=\{\lambda_{w1}, \lambda_{w2}, ..., \lambda_{Wi}\}$, generally $W_i \leq W$. That means for every service quality degree D_i corresponding to the service S_i a set of pre-reserved wavelengths with $W[S_i]$ is defined as described above. The primary wavelength set is defined as $W_{prim}[S_i]=\{\lambda_{w1}, \lambda_{w2}, ... \lambda_{Np}\}$, generally $N_p \leq W_i$.

The problem of minimising resource-, quality degree- or wavelength shifting is here mapped to the one of finding the shortest path with least costs by using the following cost definitions. The main definitions used in equation (1)-(4) are shown in Table 1.

Cost of resource shifting (C_{rs}) is defined for a service requested (called) by a client S_{call} (where *call* is the index for quality degree D_{call}). If S_{call} is using the wavelength λ^k at the kth link l_k from the primary wavelength set of the higher service quality S_j (quality degree D_j), then:

$$C_{rs}\left[S_{i=call}(\lambda^k), S_j(\lambda^k)\right] = \begin{cases} (j-i)*D_d & \text{for } i \leq j \\ \infty & \text{for } i > j \end{cases} \qquad (1)$$

A constant $D_d \geq 0$ is defined *a priori*. The expression (1) refers to the usage of network resources with respect to the called service.

Cost of quality degree shifting (C_{qds}) is defined for the case that a service S_i, instead of using the wavelength λ^{k-1} (used on the link l_{k-1}) on the succeeding link l_k, is re-transmitted on a λ^k, $\lambda^{k-1} \neq \lambda^k$, from the PWS of higher service quality degrees.

$$C_{qds}\left[S_j(\lambda^{k-1}), S_j(\lambda^k)\right] = \\ \left| C_{rs}\left[S_{i=call}(\lambda^k), S_j(\lambda^k)\right] - C_{rs}\left[S_{i=call}(\lambda^{k-1}), S_j(\lambda^{k-1})\right]\right|, \text{ for } \lambda^{k-1} \neq \lambda^k \qquad (2)$$

Cost of wavelength shifting (C_{ws}) from λ^{k-1} to λ^k (from the links l_{k-1} and l_k, respectively) within the same primary wavelength set $W_{prim}[S_i]$, $S_i \in S$, of any service quality degree D_i, is defined for every constant $C_{conv} \geq 0$ as:

$$C_c \left[S_i(\lambda^{k-1}), S_i(\lambda^k) \right] = \begin{cases} C_{conv} & \lambda^{k-1} \neq \lambda^k, \text{ for } \lambda^{k-1}, \lambda^k \in W_{prim}[S_i] \\ 0 & \lambda^{k-1} = \lambda^k, \text{ for } \lambda^{k-1}, \lambda^k \in W_{prim}[S_i] \\ \infty & \lambda^{k-1} \neq \lambda^k, \text{ for } \lambda^{k-1}, \lambda^k \notin W_{prim}[S_i] \end{cases} \qquad (3)$$

Table 1 Basic definitions used in equations (1)-(4)

$S_{call}(\lambda^k)$	service called for connection set-up using λ^k wavelength at the link l_k
$S_h(\lambda^k)$	highest possible service quality degree allowed to use λ^k at the link l_k
$C^k \left[S_{call}(\lambda^k) \right]$	minimum cost of establishing called service S_{call} over the first k network links, if wavelength used at the kth link is λ^k
$C_0(l_k)$	basis cost of using link l_k for any service
$C_{ws} \left[S_{call}(\lambda^{k-1}), S_{call}(\lambda^k) \right]$	cost of wavelength shifting between the links l_{k-1} and l_k
$C_{qds} \left[S_{call}(\lambda^{k-1}), S_{call}(\lambda^k) \right]$	cost of service quality degree shifting between the links l_{k-1} and l_k
$C_{rs} \left[S_{call}(\lambda^k), S_h(\lambda^k) \right]$	cost of resource shifting at the link l_k

Here, wavelength assignment heuristic returns the minimum link cost for every connection request of the service S_{call} by assigning the optimal wavelength Λ_{opt} on every link l_k as following (Kovacevic, 1996):

$$C^k \left[S_{call}(\Lambda^k_{opt}) \right] = C^{k-1} \left[S_{call}(\Lambda^{k-1}_{opt}) \right] + C_{ws} \left[S_{call}(\Lambda^{k-1}_{opt}), S_{call}(\Lambda^k_{opt}) \right] +$$
$$+ C_{qds} \left[S_{call}(\Lambda^{k-1}_{opt}), S_{call}(\Lambda^k_{opt}) \right] + C_{rs} \left[S_{call}(\Lambda^k_{opt}), S_h(\Lambda^k_{opt}) \right] \qquad (4)$$
$$+ C_0(l_k)$$

By applying the wavelength assignment for every link of a certain connection according to (4), the number of wavelength-, resource-, and service quality degree- shifting can be minimised. This is performed in the following steps:

1. transform the network graph with the new cost scheme by considering relevant cost functions;
2. find the shortest path between **Source** and **Destination** for the requested connection of the service requested by user;
3. minimise the number of wavelength-, resource and/or service quality degree- shifting for the path (generally, n-link one) found by Step 2.

It is not possible to optimise all of the members of the expression (4) simultaneously. This is shown in more detail in the next section.

5   PERFORMANCE STUDY AND RESULTS

In the performance study shown here, the following is assumed:

- three different quality degree service classes (high – H, medium – M, and low – L) for a meshed 40-node example network, where 8 wavelengths per link (131 links) are used for unidirectional connections;
- network load is defined as the ratio between arrival- and duration rate;
- wavelength primary sets per service are shown in Figure 2;
- wavelengths for lower quality services are generally allowed to use higher quality resources;
- uniform traffic distribution, Poisson service request arrivals normalised to the mean call duration, exponentially distributed call duration times, no queuing of connection requests;
- wavelength shifting is possible only within the same service quality degree, otherwise it is referred to as quality degree shifting.

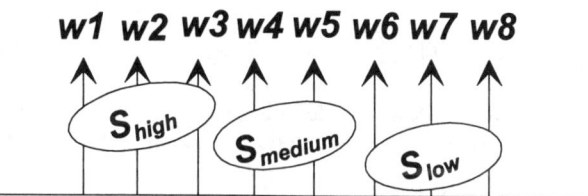

Figure 2. Primary wavelength sets for the services in the example network

For the simulation results shown, the following algorithms are applied:
- minimisation of WS only (mWS);
- minimisation of QDS (mQDS), with an additional mWS applied afterwards;
- minimisation of RS (mRS), with an additional mWS applied afterwards.

The blocking probability per service is shown in Figure 3. In this figure, it can be seen that the blocking probability of all three kinds of services increases with increasing network load. We modified the methods mRS and mQDS, so that after the lowest possible class of service is chosen, we still try to remain on the same wavelength even within a primary wavelength set (additional mWS). For that reason, the blocking probability is of the same order for the methods mRS and mWS. The blocking probability of H-services is larger for mWS/mRS with respect to mQDS. The contrary is true for L-services, because the usage of higher quality resources in the methods mWS and mQDS increases. Therefore, a trade-off between mWS/mRS and mQDS must be found for maintaining the performance of high quality services.

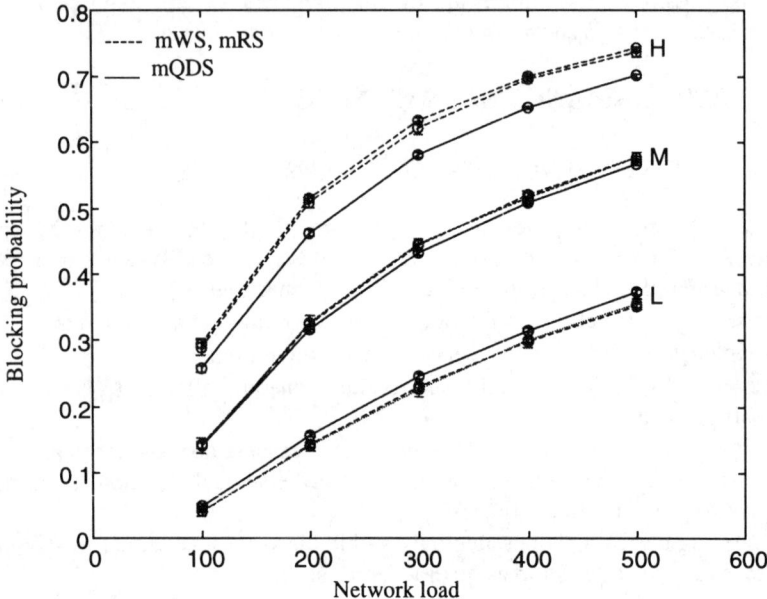

Figure 3. Blocking probability per service for each single service (H – high, M - medium, L - low quality service) with the methods of minimal wavelength shifting (mWS), minimal quality degree shifting (mQDS), as well as for minimal resource shifting (mRS), calculated for an example network.

The WS-, QDS- or RS-utilisation per path length, $u_{WS,QDS,RS}$, is defined as:

$$u_{WS,QDS,RS}\,[\%] = \frac{\text{number of necessary WS, QDS or RS}}{\text{path length in number of links}} \qquad (5)$$

The parameter u has a practical significance, because u_{WS} corresponds to the percentage of the wavelength converting devices used per link, while the parameter u_{QDS} refers to the number of wavelength re-transmitting line-cards. The parameter u_{RS} refers to the percentage of resources of higher quality then necessary used for a connection. The methods of mWS, mRS and mQDS are compared for parameter u (Figure 4). The wavelength conversion usage is of the same order for the methods mRS and mWS, particularly for higher loads (Figure 4a). The usage of wavelength conversion significantly decreases by applying the method mQDS. In Figure 4b, it can be seen that quality degree shifting is mostly performed when minimising resource- and wavelength- shifting (mRS and mWS), since in both cases the same service quality degree is preferably kept. In Figure 4c, it is shown that the method mRS indeed yields the minimal resource shifting, which can be used when the blocking probability of high-quality services must be improved.

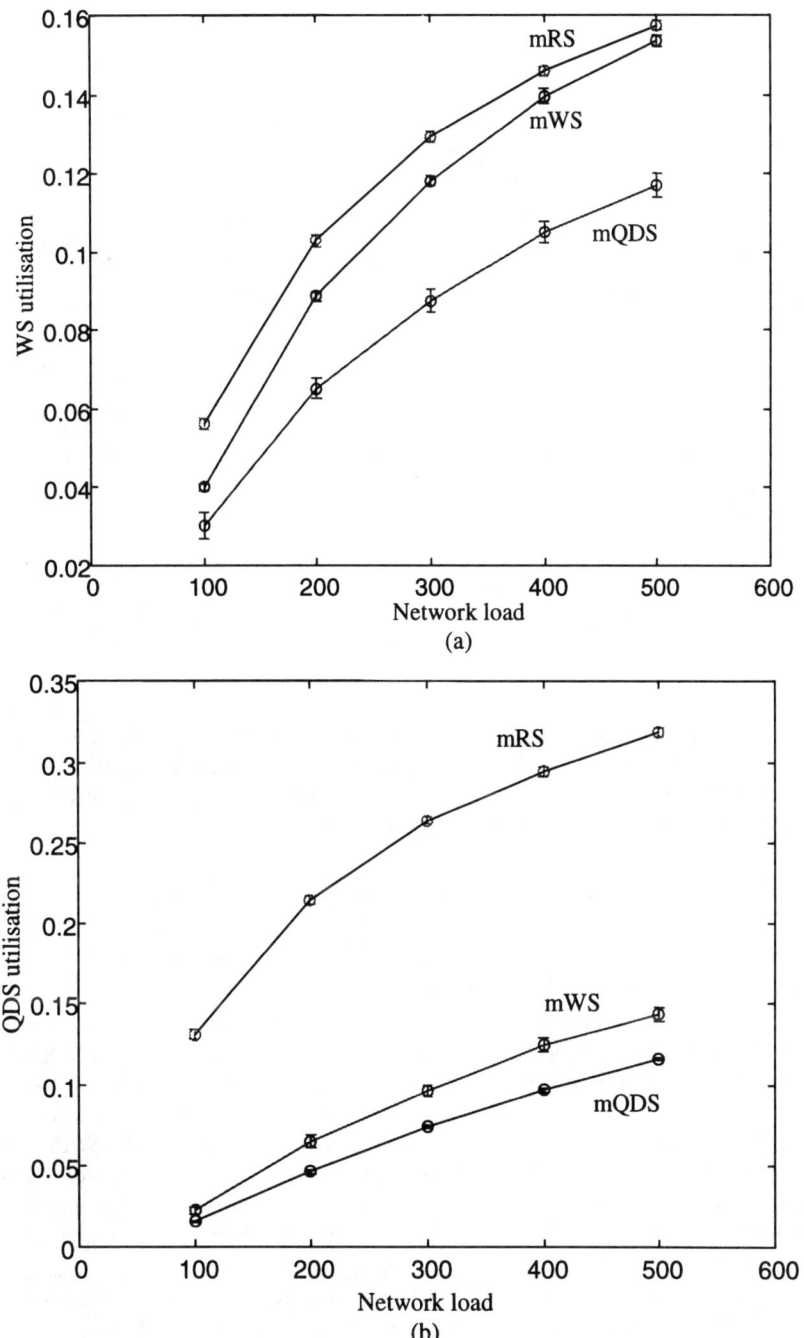

(a)

(b)

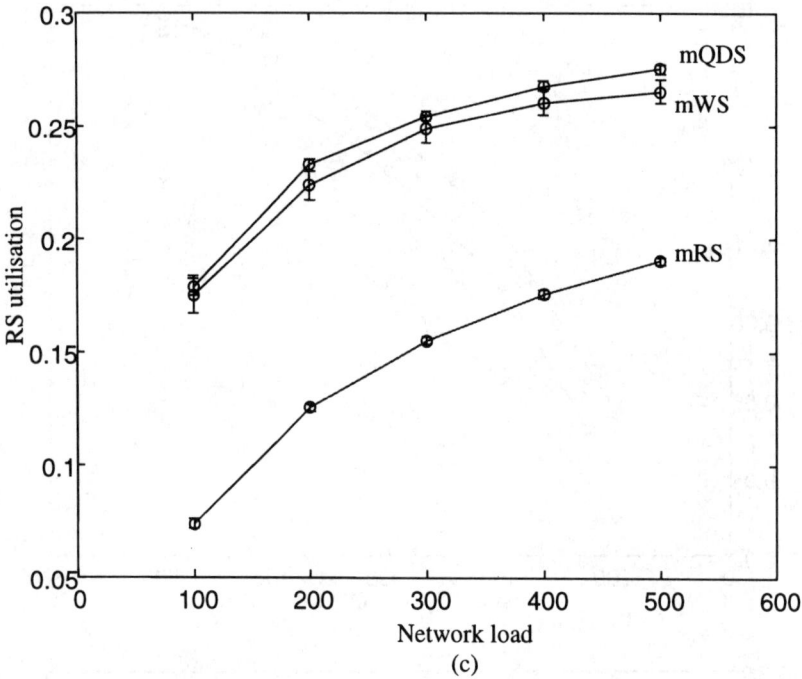

Figure 4. Utilisation of wavelength (a), quality degree (b) and resource (c) shifting for each single service (H – high, M - medium, L - low quality service) for minimal wavelength shifting (mWS), quality degree shifting (mQDS), as well as for minimal resource shifting (mRS).

6   CONCLUSION

Based on the idea that multi-client optical networks must provide many particular services for satisfying performance requirements of each single client, a novel architecture of optical nodes, using QoS-selective switches, is proposed. The main advantages of this architecture are predictable in service-specific usage of wavelength shifting, better wavelength adaptation, reduced network management, and usage of integrated components. For this architecture, three different heuristics for wavelength assignment are proposed. With these heuristics, a route with minimal number of wavelength-, resource-, and/or quality degree- shifting is found for every service request. By means of numerical examples, it is shown that with the method of minimal resource shifting blocking probability of high-quality services can be improved. If the transparency in optical networks must be maintained or a limited number of optical converting devices or wavelength re-transmitting devices can be implemented, the minimisation methods for wavelength shifting and quality degree can be successfully implemented.

Acknowledgement: Arnold Monitzer and Harmen R. van As are acknowledged for their support and helpful discussions.

7 REFERENCES

J.G. Gruber, et al (1986), Quality-of-Service in Evolving Telecommunication Networks, *IEEE Journal of Selected Areas in Communication, Vol. SAC-4, No. 7, 1986, pp. 1084-1089.*

E. Iannone and R. Sabella (1996), Optical Path Technologies: A Comparison Among Different Cross-Connect Architectures, *IEEE Journal of Lightwave Technology, Vol. 14, No. 10, Oct. 1996, pp. 2184-2196.*

A. Jukan and H.R. van As (1997), Networking based on Quality-of-optical-network-Service, *Proceedings NOC'97, Antwerp, Belgium, pp. 18-24.*

M. Kovacevic and A. Acampora (1996), Electronic wavelength translation in optical networks, *IEEE Journal of Lightwave Technology, Vol. 14, No. 6, June 1996, pp. 1161-1170.*

8 BIOGRAPHY

Admela Jukan received the B.S. and M.S. degrees from the Faculty of Electrical Engineering and Computer Science in Zagreb and Polytechnic of Milan (CEFRIEL), respectively. She is currently working towards PhD degree in electrical engineering with the Institute of Communication Networks at Vienna University of Technology in Austria. Her research area focuses on QoS-based methods of resource allocation in optical networks.

ATM Interworking

23

Design and Implementation of an ATM Cell Controller for FR/ATM Interworking System

Do-Yeon Kim and Jung-Sik Kim
Electronics and Telecommunications Research Institute
161 Kajong-Dong, Yusong-Gu,Taejon, 305-350, KOREA
Tel : +82 42 860 5349   Fax : +82 42 860 6224
E-mail : dykim@nice.etri.re.kr

Abstract

In this paper, we described on the design and implementation of an ATM cell controller for a high speed interface between FRIM (Frame Relay Interworking Module) and ALS (ATM Local switching Subsystem). In upstream direction, the ATM cell controller receives a cell stream from 16 Frame Relay subscriber access boards, and performs UPC (Usage Parameter Control), cell head translation, EHEC generation, AAL type 5 segmentation, and ATM switch interface function. In reverse direction, the ATM cell controller classifies the received cell stream through an ATM switch into user cells and IPC (Inter Processor Communication) cells, and performs 64/53 octets conversion, AAL type 5 cell reassembly, HEC generation, and handling of user cells, IPC cells, and OAM cells. And these functions are implemented by using FPGA, and we have verified the whole ATM cell controller functions by connecting protocol tester, Frame Relay subscriber board and an ATM switch.

Broadband Communications   P. Kühn & R. Ulrich (Eds.)
© 1998 IFIP. Published by Chapman & Hall

Keywords

ATM, Frame Relay, Interworking, ATM Cell Controller

1   INTRODUCTION

Early ATM systems serve as a backbone mainly for data communications. Therefore existing or upcoming data services, such as X.25, Frame Relay, and SMDS (Switched Multi-megabit Data Service) have to be supported. Frame Relay is an enhanced packet-type service. Higher throughput and less delay are achieved by reducing error control and forgoing end-to-end flow control. Frame Relay is a connection-oriented service offering bit rates from some Kbit/s up to 2 Mbit/s or possibly higher.

The merit of an ATM backbone network for a network operator is that a common, unique network infrastructure can be deployed flexibly to support all the existing and future services. New switching systems are thus needed to handle not only new ATM-based services but also existing services, such as FRS (Frame Relay Service), POTS and N-ISDN (Narrowband Integrated Services Digital Network) services. Recently with these trends, the concept of FROA (Frame Relay Over ATM) has been actively discussed.

In Korea, Frame Relay commercial services were started last year. And we are constructing a national high speed network with ATM. But actual services on the ATM network are uncertain except existing Frame Relay. This forces ATM switching system to include Frame Relay interworking function. For interworking implementation, we refer to implementation agreements of Frame Relay Forum and ITU-T recommendation.

The content of this paper is organized as follows. In section 2, we describe the overall system configurations for the Frame Relay service over ATM networks. In section 3, we show the results of estimating the number of connections in FRIM. In section 4, we describe an ATM cell controller structure. In section 5, we describe an ATM cell controller functions. And we finish our paper with conclusions in section 6.

2   SYSTEM CONFIGURATION

2.1 Overall system architecture

Figure 1 shows the configuration of ATM switching system with FRIM implemented for frame relay service access. The switching system has 3-stages configuration as ALS-ACS-ALS. Here ACS (ATM Central switching Subsystem) plays a central role of whole system such as loading and management and supports to expand system capacity, and ALS (ATM Local switching Subsystem) plays a role of local switch with complete signaling processors and usually accommodates UNI/NNI. The operation terminal (WS) and the OMP (Operation and Maintenance

Processor) are connected to each other by an Ethernet bus. The operation terminal provides operator with various graphical input/output functions. FRIM is a kind of IWF (Inter Working Function) and it is connected to ALS via IMI (Inter Module Interface) as a part of switching system. FRIM supports all Frame Relay interworking functions except call control signaling. The IMI is common bus between subsystem and functional module and it is used to transfers IPC cells as well as user cells with 155Mbps rate. The internal cell has 3 bytes routing tag in front of usual ATM cell to indicate the port of ALS/ACS switching system. Our internal cell format has 56 bytes/cell logically but 64 bytes/cell physically. The routing tags are assigned by CCCP (Call Connection Control Processor) of ALS at call setup. The call control signaling including Q.933-to-Q.2933[7] mapping are supported by the CCCP.

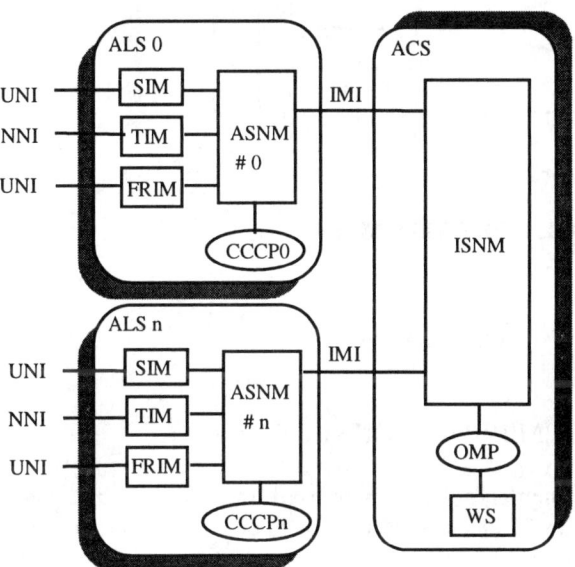

Figure 1   ATM switching system for FR/ATM interworking

2.2 Frame relay interworking module

FRIM consists of three kinds of functional boards such as ATM cell control board, system clock distribution board and Frame Relay subscriber access board as shown in Figure 2. ATM cell control board performs cell multiplexing and de-multiplexing function, AAL type 5 function for IPC between upper processors, and other ATM common functions such as UPC, VPI/VCI translation and OAM cell detection/generation. Frame Relay subscriber access board performs channelized DS1/E1 interfaces, Q.922 core function, frame to ATM cell conversion function, network layer protocol processing, management translation between OAM and

PVC status, and Frame Relay/ATM interworking core function. The board is used at UNI and NNI. System clock distribution board performs module clock generation from system clock which is derived from 155.52 MHz. Within FRIM, cell bus is implemented by 16 bits width FIFO communication to transfer user cells. IPC bus is implemented by 16 bits width DPRAM to transfer module maintenance information including board status, DS1/E1 alarm status, hardware configuration data, connection setup data and status change of each connection.

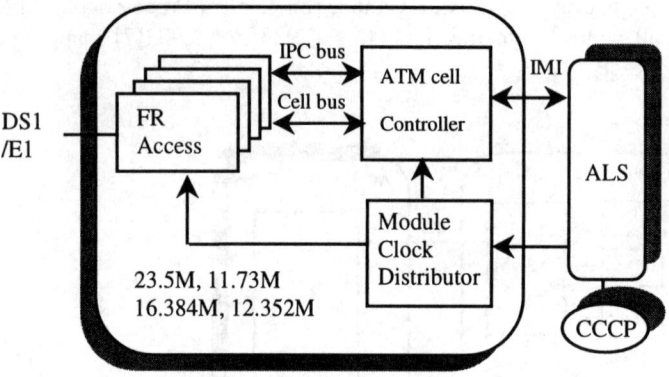

Figure 2 Frame relay interworking module

3   ESTIMATING NUMBER OF CONNECTIONS

FRIM has 64 DS1/E1 interfaces. The multiplexed traffic is 128Mbps that is less than STM1. Each E1 has 31 channels of 64kbps in channelized Frame Relay. We assign 16kbps to minimum CIR (Committed Information Rate) for PVC. This means FRIM accommodates maximum 7936 connections. But the number may be reduced to some reasonable amount like 4096. This is dependent on how wide the user traffic are. The user traffic trends are getting wide in accordance with diverse multimedia services. On the other hand, 1024 connections are regarded reasonable for STM1 ATM UNI. Supporting more connections in our FRIM is the capacity of ATM cell controller. We need to calculate the number of Frame Relay PVC connections in our system. The number of connections are calculated by the following equation.

$$\text{Number of Connections} = \frac{\text{Traffic}}{\sum_{i} P(i) \times CIR(i)}$$

where

Traffic=64 x 31 x 64 kbps
P(i)=probability of CIR(i)
CIR(i)=CIR value for i-th item

We obtained PVC distribution among CIR from Frame Relay Forum of Japan. We used it for an initial value. To estimate the distribution of near future, we used the iteration scheme as following equation.

P(n, m) = (1-w) x P(n,m-1) + w x P(n-1,m-1)

where

P(n,m) = Probability of CIR(n) at m-th iteration
P(n,m-1) = Probability of CIR(n) at (m-1)-th
iteration
P(n-1,m-1) = Probability of CIR(n-1) at (m-1)-th
iteration
w = weighting factor

This equation assumes that some connections enlarge their bandwidth to the next CIR at each iteration. Figure 3 shows the estimated number of Frame Relay connections with w=0.1 in our FRIM. The CIR ranges from 0 to 1.5 Mbps. With the present distribution, over 4800 connections are required. At iteration 6 we gained 4096 connections, and its dominant CIRs range from 32 kbps to 256 kbps. The number of connection may become less, if we give constraint to the allowed CIR such as 16 kbps, 32 kbps, 64 kbps, 128 kbps, 192 kbps, 256 kbps, 384 kbps, 512 kbps, 768 kbps, 1 Mbps and 1.5 Mbps.

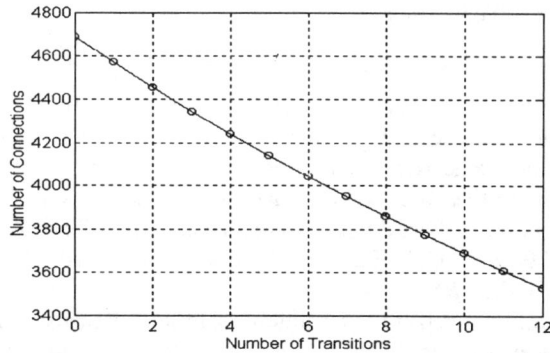

Figure 3 Frame relay connections for 64 El(w=0.1)

4 THE STRUCTURE OF ATM CELL CONTROLLER

The structure of an ATM cell controller is shown in Figure 4. It is composed of a CBRI (Cell Bus Receive Interface), two CAMs (Content Addressable Memory), a

UPC processor, a DPRAM, HCTI (Head translation and Cell Transmission IC), an AAL type 5 Segmentation/Reassembly Device, 256 Kbyte Packet/Control Memory, 512 Kbyte program memory, 4 Mbyte data memory, a MPU, a BAAI (Bus Arbitration and Address decoding IC), a LTRI (Link Transmitter/Receiver IC), a Clock Receiver, an ALI (ATM Link Interface IC), an IPC cell receive FIFO, and others.

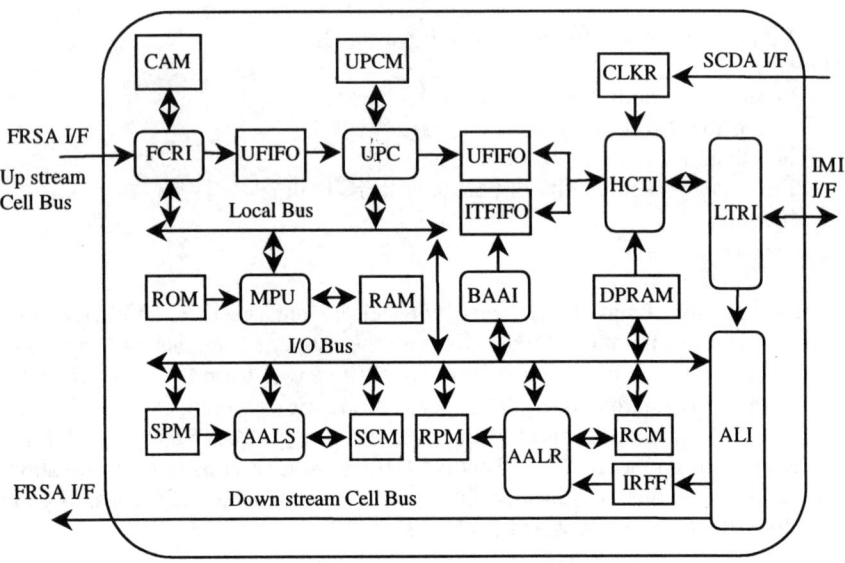

Figure 4 The structure of ATM cell controller

The ATM cell controller has 4 interfaces. They are a cell bus interface, an IPC bus interface, an IMI (Inter Module Interface), and a clock interface. The CBRI reads data from synchronous FIFO in Frame Relay subscriber access board when the CBRI detects cell available flag. In this case, cell boundary is classified by start of cell signal. Incoming cell stream has 20 variable bits in cell head, which are BD3-BD0, LI1-LI0, SB4-SB0, DL9-DL0, PTI, CLP, HEC, and cell payload. The incoming cell format is shown in Figure 5.

We need to translate incoming cell to connection identifier. Because of incoming cell stream has too many variable bit, the number of connection is dependent on how wide the user traffic are. In section 3, we estimated the number of connections. By the result, an ATM cell controller supports maximum 4,096 connections. For this we used two 2K x 64 bits CAMs (Content-addressable Memory). The CAM, also known as Associative Memory, operate in the converse way to Random Access Memories. In a RAM, the input to the device is an address, and the output is the data stored at that address. In a CAM, the input is a data sample and the output is a flag to indicate a match and the address of the matching data.

	7	6	5	4	3	2	1	0
1	X	X	X	X	X	X	BD3	BD2
2	BD1	BD0	LI1	LI0	X	SB4	SB3	SB2
3	SB1	SB0	DL9	DL8	DL7	DL6	DL5	DL4
4	DL3	DL2	DL1	DL0	PTI			CLP
5	HEC							
	Payload 0							
	Payload n							
53	Payload 47							

BD3-BD0 : Board Identifier(0 to 15)
LI1-LI0 : Link Identifier(0 to 3)
SB4-SB0 : Subscriber Identifier(0 to 30)
DL9-DL0 : DLCI(0 to 1023)
X : Not used

Figure 5 Incoming cell format

We use CAM mode because the CAM mode allows large address space to be searched rapidly and efficiently. Also, the CBRI provides CAM control signal and it writes ATM cell into UPC FIFO. In this case, the VPI/VCI of cell is connection identifier generated by the CAM. Figure 6 shows UPC input and output cell format.

	7	6	5	4	3	2	1	0
1	Not used							
2	Not used							
3	CI11	CI10	CI9	CI8	CI7	CI6	CI5	CI4
4	CI3	CI2	CI1	CI0	PTI			CLP
5	HEC							
	Payload 0							
	Payload n							
53	Payload 47							

CI11-0 : Connection identifier (0 to 4095)

Figure 6 UPC in/out cell format

UPC processor checks incoming connections for violations of negotiated traffic parameters, and selectively discards cells in violation or tags cells in violation with Cell Loss Priority (CLP) =1 on a per connection basis. The UPC processor transfer cells through standard UTOPIA FIFO interface and maintains the following counts per connection : total cells, CLP = 0 cells, and cells in violation. The UPC processor supplies two GCRA (Generic Cell Rate Algorithms) engines. The GCRA is a continuous leaky bucket algorithm in which cells leak from the bucket at a continuous rate. The bucket state reflects the amount of time required for the bucket to empty. The bucket size is the maximum amount of time allowed for the bucket to empty. If cells arrive faster than they leak out, the level of the bucket rises. Once the bucket becomes full, arriving cells are determined to be in violation. If cells arrive slower than they leak out, the level of the bucket sinks toward empty. The transferred user data is saved into UFIFO (User FIFO). When 53 octets user data is saved, almost full flag of UFIFO is activated to low. IPC FIFO saves OAM cell and IPC cell. Both OAM cell and IPC cell are segmented by AAL type 5 segmentation device. The AAL5 segmentation device performs the functions necessary to segment frames into ATM cells.

HCTI saves a cell to temporary register when an almost FIFO full flag of either user FIFO or IPC FIFO is detected. If the almost UFIFO full flag is received, the HCTI reads a cell with 8 bit data format from UFIFO and makes address using VPI/VCI in cell. The address is pointer to read head translation table in DPRAM. The head translation table has 3 bytes routing tag and 4 bytes outgoing VPI/VCI as shown Figure 7. After head translation, the HCTI sends reformatted cell to the LTRI (Link Transmitter and Receiver IC). If the almost FIFO full flag is not received, the HCTI sends IDLE pattern to the LTRI. The HCTI also provides EHEC (Extended HEC) code, odd parity bit to the LTRI. The HCTI performs EHEC calculation over 7 octets (3 bytes routing tag + 4 bytes cell header) of internal cell header for the verification of the header information. The EHEC is inserted into the HEC field of the cell header. The polynomial for EHEC is X^8+X^2+X+1.

Connection ID x 4	15	8	7	0
0	Not used		Routing Information 0	
1	Routing Information 1		Routing Information 2	
2	Not used	Outgoing VPI		
3	Outgoing VCI			
n	Not used		Routing Information 0	

Figure 7 The translation table of cell header

LTRI performs IMI (Inter Module Interface) functions. The IMI has functions such as digital phase alignment, elastic buffer, cell synchronization, encode/decode, parallel to serial conversion and serial to parallel conversion, link redundancy control, and maintenance and administration to obtain highly reliable interface. It also checks both EHEC (Extended HEC) bit and parity bit, and then reports the results of the check to the HCTI. Clock receiver receives 46.9494 MHz and 11.7374 MHz clock which is generated by the system clock distribution board and provides it to LTRI. Using divider, the clock receiver also provides 23.4747 MHz clock to the CBRI, HCTI etc. At this time the 23.4747 MHz clock is given by 155.52 MHz x 64/53octets. MPU registers the incoming VPI/VCI into CAM to search connection identifier and it loads 3 bytes routing tag and 4 bytes outgoing VPI/VCI in DPRAM. BAAI performs data bus arbitration between processor and AAL type 5 device and address decoding function.

ALI receives a cell from LTRI. The received cell format is shown in Figure 8. When a cell is received, ALI checks MSB of the first octet. If the MSB is 0, ALI classifies either a user cell or a IPC cell checking third octet. If incoming cell is user cell, ALI checks PTI field of cell header. If the PTI is b"100" (segment OAM) or b"101" (end to end OAM), incoming cell is written into IPC receive FIFO. If the PTI is neither b"100" nor b"101", incoming cell is transferred to cell bus. The cell bus is connected to Frame Relay subscriber access board. Also, IPC cell is written into IPC receive FIFO. And ALI performs HEC calculation over 4 octets of incoming cell header for the verification of the header information. The AAL5 reassembly device performs the functions necessary to make ATM cells into segment frames.

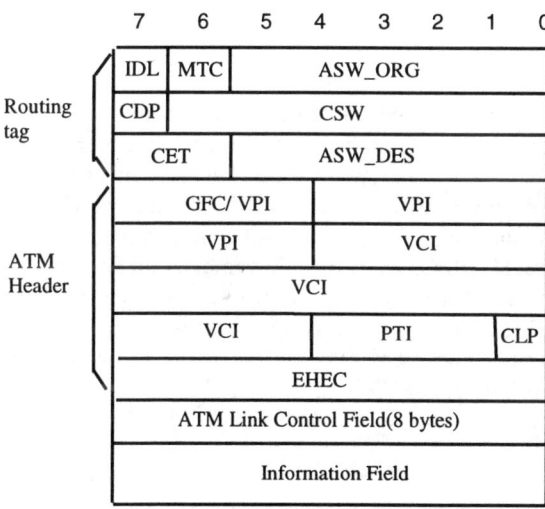

IDL : Idle cell ID(0:Busy, 1:Idle)
MTC : Multicast cell(1:p-t-p, 0:p-t-mp)
CDP : Cell Delay Priority ID
CET : Cell type ID(00:User information cell,
        01, 10: Controller cell, 11: IPC cell)
ASW_ORG : Access Switch Origin address
CSW : Central Switch physical address
ASW_DES : Access Switch Destination address
MCN : Multicasting Channel Number

Figure 8 Internal cell format

The CBRI, HCTI, BAAI, and ALI was implemented by using a FPGA. It was described using HDL code and their logic gates were generated by a synthesis tool. They were simulated by a logic simulation tool.

5   The fuctions of ATM cell controller

An ATM cell controller contains the following functions.

- Up/Down Cell bus interface
- Make connection identifier
- Usage Parameter Control
- ATM Cell Header translation
- AAL type 5 cell segmentation and reassembly
- IMI interface
- Clock receive
- 64octets/53octets cell conversion
- Cell classification
- HEC generation

6   CONCLUSION

We have implemented an ATM cell controller to provide a high speed interface between FRIM and ALS. The ATM cell controller performs cell multiplexing and de-multiplexing functions, AAL type 5 function for IPC between upper processors, and other ATM common functions such as UPC, VPI/VCI head translation and OAM cell detection/generation. The ATM cell controller is composed of FPGA, CAM, UPC processor, DPRAM, FIFO, AAL type 5 segmentation and reassembly device, program/data memory, MPU, LTRI, ALI, BAAI, and others. We have verified the whole ATM cell controller functions by connecting a protocol tester, Frame Relay subscriber board and an ATM switch. Also, this ATM cell controller may be applicable to the development of VTOA.

7.   REFERENCES

Rainer Handel, Manfred N. Huber, and Stefan Schroder (1994) ATM Networks: Concept, Protocols, Applications, Addison-Wesley Publishing Company, Wokingham.
ITU-T Rec. I.555 (1995) Frame Relaying Bearer Service Interworking.
David Ginsburg (1996) ATM, Solutions for enterprise internetworking, in Addison Wesley, pp.269-275.
Frame Relay Forum (1994) Frame Relay/ATM PVC Network Interworking Implementation Agreement, Frame Relay Document FRF.5.

Frame Relay Forum (1994) Frame Relay/ATM PVC Service Interworking Implementation Agreement, Frame Relay Document FRF.8.

Hong-Shik Park, Yool Kwon, Young-Sup Kim, and Seok-Youl Kang (1997) ATM Interface Technologies for an ATM Switching System, ETRI Journal, Vol. 18, No. 4.

ITU-T Rec. Q.2933 Draft (1991) BISDN-DSS2-Signalling Specification for Frame Relay Service.

24

High Speed TCP/IP Experiment over International ATM Test Bed

Toru Hasegawa*, Mitsuru Yamada*, Kanji Hokamura*,
Kohei Yoshiizumi*, Teruyuki Hasegawa*, Toshihiko Kato*,
Linda Galasso**, and Hiroyuki Fujii***
* KDD, ** AT&T, *** NTT
* KDD (Kokusai Denshin Denwa Co., Ltd.) R&D Labs.,
2-1-15, Ohara, Kamifukuoka-shi, Saitama, Japan.
phone: +81-492-78-7368 fax: +81-492-78-7510
e-mail: hasegawa@hsc.lab.kdd.co.jp

Abstract

AT&T, KDD and NTT conducted a joint ATM trial called Multimedia Application Project (MAP) for the purpose of testing and validating network-based broadband multimedia applications and services between US and Japan. Although the Internet based network is promising for broadband multimedia communications, there is a performance problem that TCP is the throughput bottleneck in an international ATM network with large propagation delay. In MAP, the high speed TCP/IP experiment was jointly performed by the three companies with the leadership of KDD, for the purpose of validating the high speed TCP/IP infrastructure using TCP gateway proposed by KDD. In this infrastructure, a pair of TCP gateways are introduced in international ATM networks, and they improve the TCP throughput by introducing a link-by-link flow control and a sufficient window size between gateways, without any modifications to end terminals.

In the high speed TCP/IP experiment, the throughput of TCP communications with and without TCP gateways were measured under various conditions. This paper describes the outline and the results of this experiment. The experiment was successful and the TCP gateways greatly improved the throughput of TCP based applications. For example, the response time of WWW access retrieving a 600 Kbyte image file was improved from 13.5 seconds to 2.6 seconds, and the ftp throughput was improved from 1.5 Mbps to 22 Mbps.

Keywords
ATM, TCP/IP and International ATM Networks

Broadband Communications   P. Kühn & R. Ulrich (Eds.)
© 1998 IFIP. Published by Chapman & Hall

1 INTRODUCTION

Recent technological advances in the fields such as ATM (asynchronous transfer mode) and SDH (synchronous digital hierarchy) have realized a high speed digital transport network. Many communication providers have started ATM network services for domestic and international environments, for the purpose of allowing broadband multimedia applications and services to be deployed in wide area networks. However, further investigations are required to understand the behaviors of multimedia applications over ATM networks and to identify advanced ATM network functionality supporting multimedia services.

From these backgrounds, AT&T, KDD and NTT conducted a joint ATM trial called *Multimedia Application Project* (MAP), between July 1996 and February 1997. This trial created an international ATM test bed for validating network-based broadband multimedia applications and services between US and Japan, and performed four kinds of broadband multimedia experiments[1]. The high speed TCP/IP experiment is one of the experiments and was jointly performed by the three companies with the leadership of KDD.

The Internet based network is one of the most promising network configurations, because it allows the use of the huge legacy of existing application software. However, there is a performance problem that TCP is the throughput bottleneck in a network with large *bandwidth-delay products*, such as an international ATM network (Stevens, 1994). That is, the throughput is limited by the window based flow control because the maximum value of TCP window size is limited by 64 Kbyte and most applications use smaller values. Against this problem, RFC 1323 proposed the TCP extensions including the TCP window scale option which uses the window sizes larger than 64 K bytes (Jacobson, 1992). However, there are still some problems such that RFC 1323 is supported by only limited computers, and that the performance is also degraded due to the TCP congestion control (Kato, 1997). Therefore, it will be an effective approach for an wide area ATM network to provide the functionality to attain a high speed TCP/IP communication.

KDD has been studying the realization of a *high speed TCP/IP infrastructure* over an international ATM network, and has proposed a *TCP gateway approach*[2]. Here, a pair of gateways are introduced in an international ATM network, and each of them works as an intermediate system along a TCP connection and supports both the IP function and the transport protocol handling (Kato, 1997 and Hasegawa, 1995). As for the transport protocol, the gateway uses TCP to communicate with an end terminal, and a KDD proprietary high speed transport protocol with a sufficient window size to communicate with the other TCP gateway. Using these two transport protocols, the acknowledgment and retransmission of data segments are performed on a link-by-link basis, i.e. between

---

[1] *MAP focused on broadband PC-VMS (Virtual Meeting Service) (Beaken, 1995) led by AT&T, high speed TCP/IP (Kato, 1997 and Hasegawa, 1995) led by KDD, Virtual LAN (Hariu, 1997) led by NTT and broadband imaging (Nakabayashi, 1995) led by NTT.*

[2] *TCP gateway is KDD proprietary.*

the gateway and an end terminal or between the gateways. As a result, it is possible to increase the end-to-end throughput without giving any modification to the end terminals and, furthermore, without making the end terminals aware of the existence of the TCP gateways.

In the high speed TCP/IP experiment, the three companies collaboratively evaluated a high speed TCP/IP infrastructure using TCP gateway over an international ATM test bed. In this experiment, the throughput of TCP/IP communications with and without TCP gateway was measured using various applications under various conditions. This paper describes the results of the high speed TCP/IP experiment conducted in MAP. In section 2, we describe the problem of TCP and the structure of high speed TCP/IP infrastructure. In section 3, we describe the network configuration of international ATM test bed used in this experiment. In section 4 , we describe the experiment results.

2 HIGH SPEED TCP/IP INFRASTRUCTURE USING TCP GATEWAY
2.1 Problems of TCP over Wide Area ATM Networks
TCP performs the flow control based on the sliding window mechanism between end terminals on an end-to-end basis. In order to obtain a high throughput corresponding to the network bandwidth, it is required that the TCP window size is large enough to transmit data segments continuously during one RTT (round trip time). As shown in Figure 1, if the window size is insufficient, the throughput will be degraded because the sender waits for ACKs (acknowledgments) after it finishes sending all data segments within the window. In this situation, the throughput Th [bps] is given by Equation (1). In this equation, W [byte], Rtt [sec], M [byte] and C [bps] indicate the TCP window size, RTT, the maximum transmission unit size and the bandwidth respectively.

$$Th = W * 8 / (Rtt + M * 8 / C) \qquad (1)$$

Many applications over TCP use window size such as 8 K bytes. In the case of W = 8 K bytes, M = 9188 bytes and C = 6 Mbps, Equation (1) gives the result that Th is roughly 310 Kbps and 110 Kbps when RTT is 200 ms and 600 ms, respectively. This throughput does not increase no matter how large bandwidth the wide area ATM networks provide.

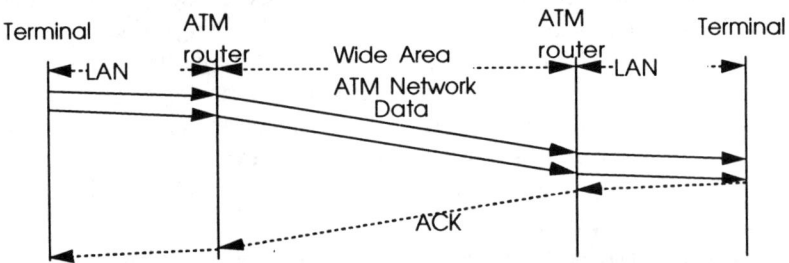

Figure 1 Sequence of End-to-End Flow Control with Insufficient Window Size.

2.2 Structure of High Speed TCP/IP Infrastructure Using TCP Gateway

Figure 2 shows a configuration of the high speed TCP/IP infrastructure over an international ATM network. As described in this figure, a TCP gateway works as an adjunct equipment of an international ATM network and a pair of gateways collaborate with each other to provide a high speed TCP/IP for customers. The functions of TCP gateway are summarized as follows.

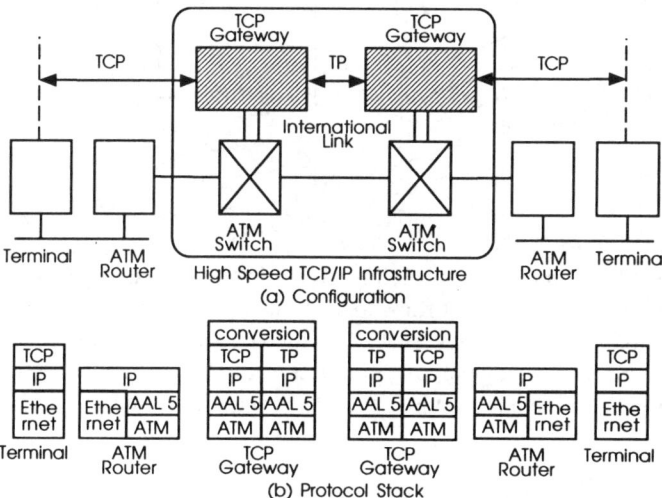

Figure 2 High Speed TCP/IP Infrastructure.

(1) TCP gateway provides both IP functions such as IP routing and the gateway function at the transport layer. It applies IP and gateway functions to TCP segments.

(2) TCP gateway introduces a KDD proprietary high speed transport protocol, called *TP*. It is designed so as to be suitable for wide area ATM networks.

(3) TCP gateway uses one VC (virtual channel) with a customer, and another VC with the other TCP gateway. When more than one customers are connected to a TCP gateway, the data from the VCs between the gateway and the customers are aggregated into the VC between the gateways. TCP gateway communicates with end terminals using TCP, and communicates with the other gateway using TP.

(4) TCP gateway performs the protocol conversion between TCP and TP in the following way.

- The establishment and release of TCP connections and TP connections are performed together on an end-to-end basis.
- The data transfer function such as the flow control and the error recovery is performed on a link-by-link basis. For the link between two TCP gateways, TCP gateway allocates sufficient window sizes for the long propagation delay.

- In order to relay data segments, TCP gateway transfers them according to the flow control individually for the customer side and the TCP gateway side, buffering them if required by the flow control, and sends ACKs individually for the customer side and the TCP gateway side.

A communication sequence during the data transfer phase is illustrated in Figure 3.

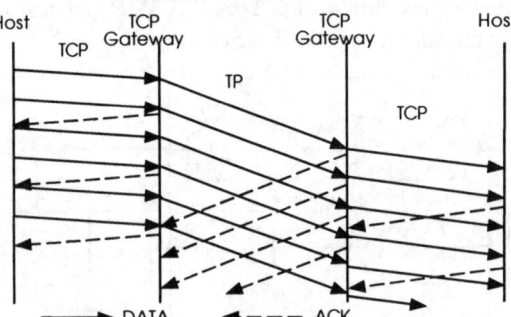

Figure 3 Communication Sequence During Data Transfer Phase.

(5) TP introduces two levels of flow control, the flow control for an individual TP connection, and that for the aggregate TP connections over VC between the TCP gateways. The purpose of the latter flow control is used for the aggregated TP traffic not to excess the bandwidth of the VC between the gateways.

3   INTERNATIONAL ATM TESTBED

In MAP, an international ATM test bed was established among AT&T (Holmdel in US), KDD (Shinjuku in Japan) and NTT (Musashino and Yokosuka in Japan) through the Pacific Ocean. Figure 4 illustrates the network configuration for the high speed TCP/IP experiment, which is a part of the whole test bed. The features are as follows:

(1) Network Facility

- A 45 Mbps transmission facility (DS-3) was used for an international link. The traffic was routed on DS-3 facility in TPC-4, an optical fiber submarine cable between US and Japan.
- A 45 Mbps transmission facility (DS-3) was used for a US domestic network.
- A 155 Mbps transmission facility (STM-1) was used for a Japanese domestic network.
- RTT between AT&T and KDD, and RTT between KDD and NTT Yokousuka, measured by ping commands, were 175 ms and 2 ms, respectively.

(2) VP (Virtual Path) Characteristics

A virtual path network was established among the four ATM VP cross-connects at AT&T Holmdel, KDD Shinjuku, NTT Musashino and NTT Yokousuka. VP connections were established between every pair of the cross-connects.

AT&T and KDD networks (International network and US domestic network) use VBR (variable bit rate), and NTT networks use CBR (constant bit rate).

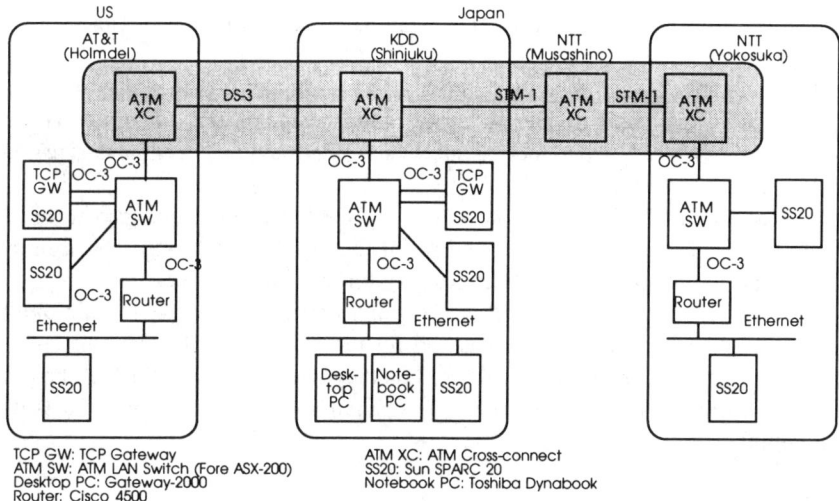

Figure 4 International ATM Test Bed.

(3) TCP Gateways
The TCP gateways were installed at AT&T Holmdel and KDD Shinjuku. A VC connection was established between two TCP gateways.
(4) LANs and Terminals
At AT&T, KDD and NTT Yokosuka, both ATM LAN and Ethernet LAN were installed as customer premises networks. Ethernet LANs were accommodated to ATM network using ATM routers.

 TCP terminals were installed at AT&T Holmdel, KDD Shinjuku and NTT Yokosuka. TCP terminals at AT&T Holmdel and KDD Shinjuku communicate with the TCP gateway at the same location. TCP terminals at NTT Yokosuka communicate with the TCP gateway at KDD Shinjuku.

4   EXPERIMENT RESULTS
4.1 Methodology
In order to validate the performance of the high speed TCP/IP infrastructure using TCP gateway, the following approaches are taken in the experiment.
(1) Both the application level throughput of widely used applications, such as WWW (world wide web) and ftp, and the TCP level throughput are measured. The application level throughput includes the processing overhead of application protocols and the overhead of accessing disks and displaying. The TCP level throughput is measured by memory to memory copy through TCP protocol, and indicates the pure protocol processing overhead. A free software called ttcp is used for measuring the TCP level throughput.
(2) As TCP terminals, both workstations and personal computers are used in the experiment. By using different kinds of terminals, it is possible to test different

cases for processing capability of terminals, and to test different TCP implementations.

(3) As a customer premises network, both ATM LAN and Ethernet are used in the experiment. This can evaluate different transmission speed of terminals, and different values of TCP parameters such as MSS (maximum segment size) and window size.

(4) The number of TCP connections between TCP terminals is changed from one to eight. This can evaluate the performance of TCP gateway for multiple connections, and the fairness of TCP gateway for individual connections.

(5) As described in section 3, TCP gateway is installed at only KDD Shinjuku in Japan side. The performance is measured for the case that TCP terminals are co-located in KDD Shinjuku and the case that they are located in NTT Yokosuka. This can evaluate how the domestic propagation delay in Japan affects the throughput.

4.2 Test Conditions

(1) SS-20 SPARC station with 60 MHz SuperSPARC and SunOS 4.1.3 is used for TCP gateway. It has 64 M byte main memory, and two Fore SBA-200E ATM NICs (Network Interface Cards).

(2) The following workstations and personal computers are used as TCP terminals.

- Workstation: SS-20 SPARC station with 150MHz HyperSPARC, 32 M byte main memory, a Fore SBA-200E ATM NIC, and Solaris 2.5.1.
- Notebook Personal Computer: Toshiba Dynabook with 90MHz Pentium, 40 M byte main memory, and Windows 95.
- Desktop Personal Computer: Gateway-2000 with 200MHz Pentium Pro, 64 M byte main memory, and Windows 95.

(3) The window sizes of high speed transport protocol between the TCP gateways are 1M bytes and 256 K bytes, when ATM LAN and Ethernet are used as customer premises networks, respectively. These values of window size are large enough for 45 Mbps and 10 Mbps, the values corresponding to the bandwidth of bottleneck networks, respectively.

Table 1 Combination of Computers and Customer Premises Network.

	WWW	ftp	ttcp
SS-20 -> SS-20 (ATM LAN)	done	done	done
SS-20 -> SS-20 (Ethernet)	done	done	done
SS-20 -> Notebook PC (Ethernet)		done	
SS-20 -> Desktop PC (Ethernet)		done	
Notebook PC -> SS-20 (Ethernet)	done	done	
Desktop PC -> SS020 (Ethernet)	done	done	

(4) All the computers connected to ATM networks perform traffic shaping with 30.0 Mbps as a peak cell rate. This corresponds to 27.1 Mbps as an ATM payload transmission rate.

(5) In the experiment, throughput is measured for WWW, ftp and ttcp communication. Table 1 indicates the combinations of computers and customer premises networks used for individual throughput measurements. It should be noted that the arrow between computers indicates the direction from client to server.

4.3 Results of WWW and FTP Throughput Measurements
4.3.1 WWW Throughput Measurements
A WWW server is installed in the SPARC station at AT&T Holmdel, and it stores twelve image files formatted in GIF (graphics interchange format) whose sizes are from 321,051 bytes to 572,101 bytes. A WWW client using Netscape Navigator 3.0 is installed in KDD Shinjuku, and it retrieves image files and displays the received data. As an application level throughput of WWW communication, the response time is measured, which is the time between when the retrieval request is invoked and the time when the retrieved data is fully displayed. The response time is measured for the cases with and without TCP gateway using the combinations of computers and customer premises networks described in Table 1. For all the measurements, TCP gateway improves the response time.

Figure 5 shows the measurement results of response time using a SPARC station as a WWW client and ATM LANs as customer premises networks in both client and server sides. For example, while the response time for 572,101 byte file without TCP gateway is about 13.5 seconds, that with the gateway is just 2.6 seconds. For the cases that a personal computer is used as a WWW client and that Ethernet is used as a customer premises network, the measurements give similar results for response time. As the results of this measurement indicate, it can be said that the high speed TCP/IP infrastructure using TCP gateway increases an application level WWW performance more than five times.

4.3.2 FTP Throughput Measurements
The throughput is also measured between an ftp server installed at AT&T Holmdel and an ftp client at KDD Shinjuku. In the measurement, an ftp client gets a binary

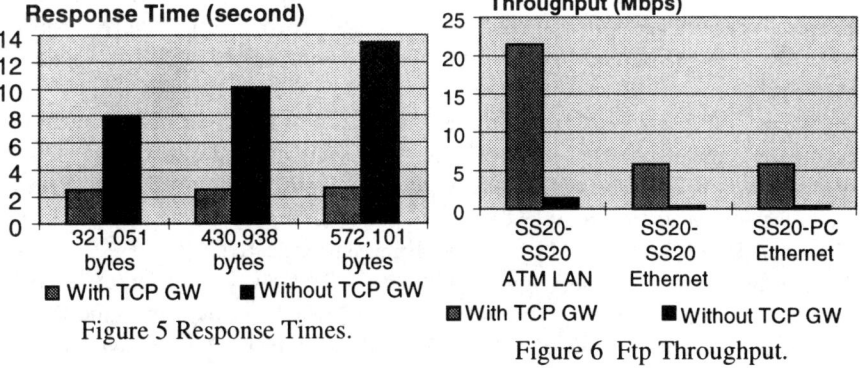

Figure 5 Response Times.

Figure 6 Ftp Throughput.

file stored in a disk whose size is 100 Mega bytes. The throughput is measured for the cases with and without TCP gateway using the combinations of computers and customer premises networks described in Table 1. As a result, TCP gateway improves the throughput for all the measurements.

Figure 6 shows the results for some cases of measurements. For the case that ATM LAN is used as customer premises network, the high speed TCP/IP infrastructure using TCP gateway improves the throughput more than ten times, and for the case of Ethernet, it improves the throughput more than eighteen times.

4.4 Results of Ttcp Throughput Measurements

In order to evaluate the TCP level performance of the high speed TCP/IP infrastructure, the throughput measurement using ttcp software is performed using SPARC stations. The improvement of the TCP level throughput by TCP gateway is evaluated for the cases that one TCP connection is established and that more than one TCP connections are established simultaneously. The measurement also focuses on the case that there is some propagation delay between the TCP terminal and the TCP gateway in Japan.

The TCP parameters used in this measurement are summarized as follows.

- Nagle algorithm is used.
- Ttcp writes 8192 byte data at a time.
- The values of MSS are 9148 bytes and 1460 bytes when the customer premises network is ATM LAN and Ethernet, respectively. It should be noted that, in this measurement, SPARC stations running Solaris 2.5.1 sent data segments whose length is 8192 bytes for the ATM LAN customer premises network.

The send and receive socket buffer sizes are default values used in Solaris 2.5.1. They are 36,592 bytes and 8760 bytes for the ATM LAN and the Ethernet customer premises networks, respectively.

4.4.1 Throughput Measurement for One TCP Connection

Figures 7 and 8 show the TCP level throughput with and without TCP gateway, when the number of TCP connections are changed for the ATM LAN customer premises network and Ethernet customer premises network, respectively.

For the case of one TCP connection, the following results and discussions are obtained.

(1) When TCP gateway is not used, the throughput of one TCP communication is 1.42 Mbps and 0.32 Mbps for ATM LAN and Ethernet, respectively. On the contrary, when TCP gateway is used, the throughput is improved more than ten times faster than the original TCP throughput. The improved throughput is 21.2 Mbps and 5.69 Mbps for ATM LAN and Ethernet, respectively.

(2) The reason why the throughput without TCP gateway is different for ATM LAN and Ethernet is that the socket buffer sizes, which correspond to the window sizes, are different. In the case of ATM LAN customer premises network, four 8192 byte segments are sent continuously corresponding to one window size. The

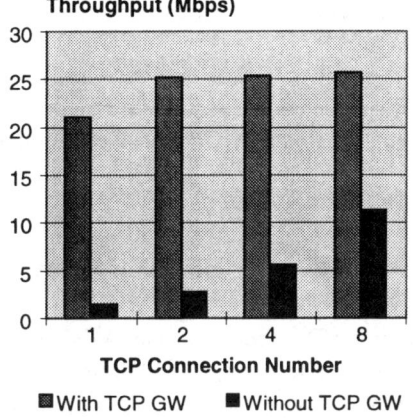

Throughput (Mbps)

TCP Connection Number

■With TCP GW ■Without TCP GW

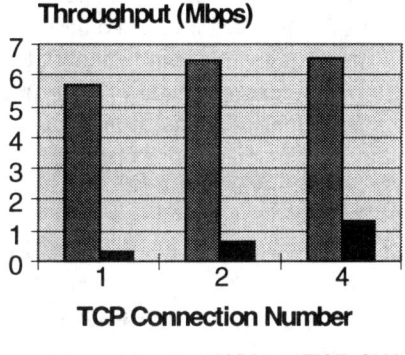

Throughput (Mbps)

TCP Connection Number

■ With TCP GW ■ Without TCP GW

Figure 7 TCP Level Throughput for ATM LAN Customer Premises Network.

Figure 8 TCP Level Throughput for Ethernet LAN Customer Premises Network.

throughput without TCP gateway, which can be calculated by equation (1), is 1.47 Mbps. In the case of Ethernet customer premises network, five 1460 byte segments are sent for one window size, and the throughput without TCP gateway is 0.33 Mbps. The obtained results match these calculated values.

4.4.2 Throughput Measurement for Multiple TCP Connections

Figures 7 and 8 also give the TCP level throughput with and without TCP gateways, when TCP connections are used between multiple pairs of TCP terminals. For multiple TCP connections, the following results and discussions are obtained.

Table 2 TCP Throughput of Individual TCP Connection.

Connection	1	2	3	4	5	6	7	8
Num. of Connection								
2	12.64	12.64	-	-	-	-	-	-
4	6.36	6.34	6.34	6.34	-	-	-	-
8	3.25	3.22	3.21	3.20	3.20	3.20	3.19	3.19

Throughput : Mbps

(1) When TCP gateway is not used, the total throughput of multiple TCP connections increases in proportion to the number of connections. However, the total throughput is still much lower than the throughput when TCP gateway is used.

(2) When TCP gateway is used, the total throughput of multiple TCP connections is more than 25 Mbps for ATM LAN customer premises network, and is more than 6.5 Mbps for Ethernet customer premises network. Since the ATM payload

transmission rate is 27.1 Mbps as described in section 4.2, the total throughput for ATM LAN is considered to be limited by this rate.

(3) The throughput of individual TCP connections is also measured for ATM LAN customer premises network. The results are listed in Table 2, and the throughput is the same for individual connections. This means that TCP gateway handles multiple TCP connections fairly.

4.4.3 Throughput Measurements for Domestic Propagation Delay in Japan

Figure 9 shows the TCP level throughput with TCP gateway and ATM LAN customer premises network for the case that TCP gateway and TCP terminals are co-located in KDD Shinjuku (without domestic delay), and the case that TCP gateway in KDD Shinjuku and TCP terminals in NTT Yokosuka (with domestic delay). Figure 10 shows the corresponding throughput when Ethernet is used as a customer premises network. As for one TCP connection, the TCP throughput with domestic delay is a little worse than that without the delay. The throughput degradation by the local access line delay is more significant for ATM LAN than for Ethernet. On the contrary, when multiple TCP connections are used, the total throughput is almost the same for both cases, as shown in Figs. 9 and 10.

5   CONCLUSION

In this paper, we described the result of the high speed TCP/IP experiment jointly performed in Multimedia Application Project conducted by AT&T, KDD and NTT. The experiment focused on validating the high speed TCP/IP infrastructure using TCP gateway over an international ATM network. In the experiment, the

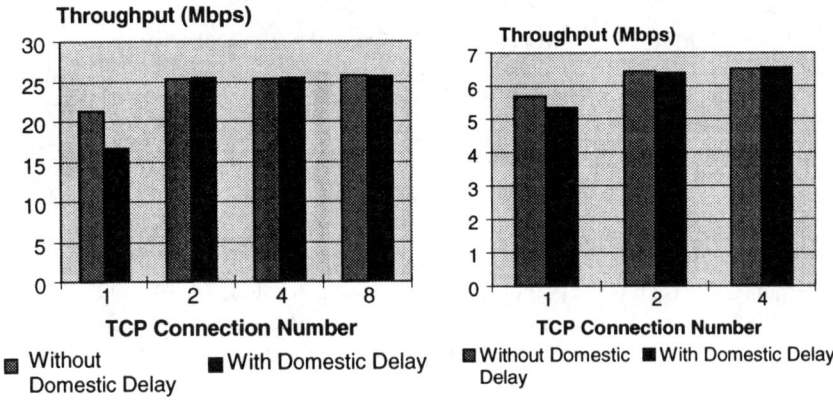

Figure 9 TCP Level Throughput with Domestic Delay for ATM LAN Customer Premises Network.

Figure 10 TCP Level Throughput with Domestic Delay for Ethernet Customer Premises Network.

throughput of TCP communications with and without TCP gateway were measured under various conditions. The experiment results are summarized as follows.

- TCP gateway greatly improved the throughput of TCP based applications. For example, the response time of WWW access retrieving a 600 Kbyte image file was improved from 13.5 seconds to 2.6 seconds, and the ftp throughput was improved from 1.5 Mbps to 22 Mbps.
- The throughput improvement by TCP gateway was confirmed for different TCP terminals and customer premises networks, and for different number of TCP connections.
- When multiple TCP connections were established simultaneously, TCP gateway improved the throughput of each connection fairly.
- If TCP gateway and terminal were located in different locations and if one TCP connection was used, the TCP throughput was a little worse than if they were co-located. On the contrary, when multiple connections were used, the delay did not degrade the total TCP throughput at all.

As described above, the high speed TCP/IP experiment was successful and has proved that TCP gateway can realize a high throughput infrastructure across an international ATM network for Internet based broadband applications and services. multimedia

6   REFERENCES

Stevens, W. R. (1994), TCP/IP Illustrated, Volume 1 [The Protocols]. Addison-Wesley.

Jacobson, V., Braden, R. and Borman, D. (1992), TCP Extensions for High Performance. RFC 1323.

M. Beaken, M. (1995), The Provision of Intelligent Agent-Based Enhanced Multimedia Network Services. AT&T TECHNICAL JOURNAL, pp.68-77.

Kato, T., Hokamura, K., Yamada, M., Hasegawa, T., Hasegawa, T. and Sawada, K. (1997) TCP Gateway Improving Throughput of TCP/IP over Wide Area ATM Networks. ISS'97 (International Switching Symposium), pp. 35-42.

Hasegawa, T., Hasegawa, T., Kato T. and Suzuki, K. (1995) Implementation and Performance Evaluation of TCP Gateway for LAN Interconnection through Wide Area ATM Network (in Japanese). IEICE Trans. Comm., vol.J79-B-I, no.5, pp.262-270.

Hariu, T. and Tanimoto S. (1997), 'Logical Office' Service: A Mobile VLAN Service for a Mobile Computing Environment. APSITT'97 (Asia-Pacific Symposium on Information and Telecommunication Technologies).

Nakabayashi, K. (1995) An Intelligent Tutoring System on World Wide Web: Towards an Integrated Learning Environment on a Distributed Hypermedia. Proc. of ED-MEDIA95, June 1995.

Acknowledgments

We wish to thank all the members of KDD, AT&T and NTT who participated in Multimedia Application Project. We would like to express special thanks to Dr. Y. Hatori, KDD, Dr. D. Rajala and Mr. R. Ramamurthy, AT&T, and Dr. S. Hatano and Mr. Y. Harada, NTT for their valuable advice.

25

Performance measurements in local and wide area ATM networks

R. Eberhardt, C. Rueß, R. Sigle
Daimler-Benz AG, Research and Technology
P.O. Box 2360, 89013 Ulm, Germany
Tel: ++49-731-505-2103, Fax: ++49-731-505-4110
{eberhardt, ruess, sigle}@dbag.ulm.DaimlerBenz.COM

Abstract

Daimler-Benz Research and the International Computer Science Institute (ICSI) participated in one of the first transatlantic Asynchronous Transfer Mode (ATM) field trials. We describe the intention of the project Multimedia Applications on Intercontinental Highways (MAY), its infrastructure and results of various performance evaluations based on Classical IP over ATM (CLIP) and LAN Emulation (LANE). In addition the commonly used videoconferencing tool "vic" was evaluated in various measurements. We discuss the state of the art of wide area ATM networks and their limitations. Finally, solutions to some of the problems are outlined.

Keywords

ATM, ATM LANs/WANs, LAN Emulation, IP over ATM, Field Trial

1   INTRODUCTION

Today we can observe the evolution from export oriented to global operating companies. In order to open up new markets their special needs have to be taken into account during the development of products. This requires besides global production worldwide research and development. Integrated solutions including the areas Computer Supported Cooperative Work, Concurrent Engineering and High Performance Communications are needed to support the collaboration of engineers

Broadband Communications   P. Kühn & R. Ulrich (Eds.)
© 1998 IFIP. Published by Chapman & Hall

spread all over the world. Therefore more flexible network infrastructures providing higher bandwidth than nowadays corporate networks are demanded.

ATM is the first networking technology which allows high speed communication for both local and wide area networks using the same network infrastructure. It is ideally suited for all kinds of applications, contrasting todays situation where different services use different network technologies. ATM has the potential to embrace both corporate and public networks, voice and data communication.

The project MAY examined if ATM can satisfy these requirements. The MAY project was initiated by T-Berkom, a subsidiary of Deutsche Telekom. The global ATM network for the MAY testbed was provided by Deutsche Telekom, Sprint, GlobalOne and Teleglobe.

This infrastructure was used by several companies and research organizations to test and develop new applications and communication protocols. Several applications in the areas cooperative engineering, distance learning, and telemedicine were envisaged. In section 2 we describe the MAY network infrastructure. We then present in sections 3 network and in section 4 application performance measurements. Finally we draw several conclusions.

2   MAY NETWORK INFRASTRUCTURE

The connection between ICSI Berkeley (California, USA) and Daimler-Benz Corporate Research Ulm (Germany) was composed of several parts made available by the service providers mentioned in Figure 1.

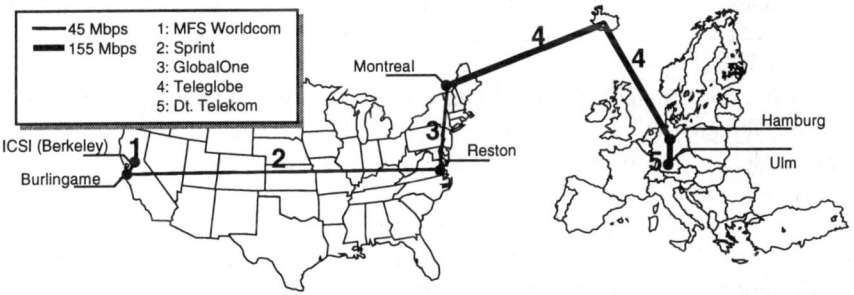

Figure 1: ATM Connection between ICSI (Berkeley, USA) and Daimler-Benz Research (Ulm, Germany)

Due to the lack of standards and implementations of B-ICI (Broadband Intercarrier Interface), PNNI-1 (Private Network Node Interface) and Q.2931 no signalling is available in current wide area ATM networks. The sole possibility to interconnect local ATM networks through wide area ATM networks is to use Virtual Paths (VP). The end systems (PCs and workstations) in both locations were connected via local ATM switches to the wide area ATM network.

The network providers Sprint, GlobalOne and Teleglobe used spare capacity on their ATM links. The VP from Berkeley to Hamburg was permanently configured. In order to setup an end to end ATM connection a VP from Ulm to Hamburg had to be reserved and configured by Deutsche Telekom's ATM network management center in Cologne. The local ATM switches and end systems had to be configured on both sites.

To be able to reuse existing software without any changes we used Classical IP over ATM [RFC1483, RFC1577] and LAN-Emulation Version 1.0. Because the network providers only offered a constant bit rate (CBR) service, we had to use traffic shaping in the end system ATM adapters.

3   NETWORK PERFORMANCE MEASUREMENTS

In order to examine which kind of applications can be run on top of the transatlantic testbed we made several performance measurements. We tested the throughput and round-trip delay of TCP and UDP using CLIP and LANE in the local and transatlantic ATM network. For the transatlantic connection an extended TCP implementation [RFC1323] was used which solved the problem of "Long Fat Networks (LFN)".

3.1   Benchmark programs and utilities

The public domain tool "netperf" [np] from Hewlett-Packard allows several kinds of performance measurements. Of special interest are measurements of the bulk data transfer performance and request/response delay of network connections using TCP and UDP.

To measure the throughput netperf enables the variation of socket and message sizes on the sender and receiver side. In the case of datagram oriented connectionless transport protocols like UDP the loss rate is of special significance. All measurements were performed over a period of 60 seconds.

To measure the CPU load of the end systems, the tool "sar" was used which is included in the Solaris operating system. The CPU load was measured every five seconds during the sending/receiving period.

To get information about the AAL5 PDUs sent/received by the Fore ATM adapter, the tool "atmstat" was used. Atmstat is included in the Fore driver software.

3.2   Network configuration

As end systems several SUN workstations with varying performance were used: From Sparc 5 to Sparc Ultra 2 with dual CPU. All the workstations have a FORE SBA-200 ATM adapter. Tests were performed running Solaris 2.5 and 2.6 beta operating systems. The ATM Switch in Ulm is a Fore ASX 200 and in Berkeley a Bay Networks LattisCell switch. Figure 2 shows the test configuration.

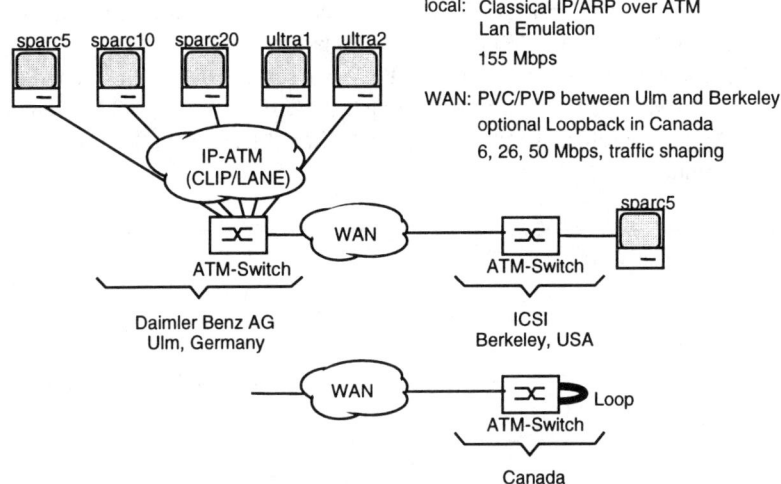

sparc5 sparc10 sparc20 ultra1 ultra2

local: Classical IP/ARP over ATM
        Lan Emulation
        155 Mbps

WAN: PVC/PVP between Ulm and Berkeley
     optional Loopback in Canada
     6, 26, 50 Mbps, traffic shaping

IP-ATM
(CLIP/LANE)

WAN   sparc5

ATM-Switch   ATM-Switch

Daimler Benz AG   ICSI
Ulm, Germany   Berkeley, USA

WAN

Loop
ATM-Switch

Canada

Figure 2: ATM test environment

The workstations use CLIP or LANE. In the wide area network permanent virtual connections (PVCs) or Permanent Virtual Paths (PVPs) were established. Optionally a loop was configured on a switch in Canada connecting hosts in Germany with about the same delay. The PVCs were configured with a bandwidth of 6, 26 and 50 Mbps. Using the loop over Canada also simplified the task of configuring the end systems, as they were at the same location. The transmitting end systems used traffic shaping to avoid cell loss in the interim switches of the wide area network providers.

3.3 Performance in local ATM networks

This chapter summarizes the results of the performance measurements in the LAN. First the TCP performance of Classical IP over ATM was evaluated. In a second step TCP using LAN Emulation was considered. The chapter closes with some UDP measurements using CLIP.

TCP performance using Classical IP over ATM

The TCP throughput using Classical IP was measured between the Sun workstations sparc10 and sparc20 with varying message sizes. The buffer sizes of both sockets were set to 64 kByte.

The message size was increased in steps of 1024 Bytes. During the measurement the workstations and the ATM network weren't used otherwise.

Due to the lack of the computing power the workstations couldn't make use of the full bandwidth provided by the ATM adapters (Figure 3a). Without the overhead of several protocol layers (SONET-OC3, ATM, AAL5), AAL5 can provide a maximum bandwidth of 135.6 Mbps [atmarp], [krivda]. CLIP and TCP overhead are reducing this value because of SNAP header, AAL5 trailer and padding to 135.1 Mbps using the default MTU size of 9180 bytes.

The ATM adapters have an onboard RISC processor and special hardware to process the ATM cells. They are providing an interface to the ATM adaptation layer 5 (AAL5). Therefore the processing of ATM cells doesn't add further load to the host computers.

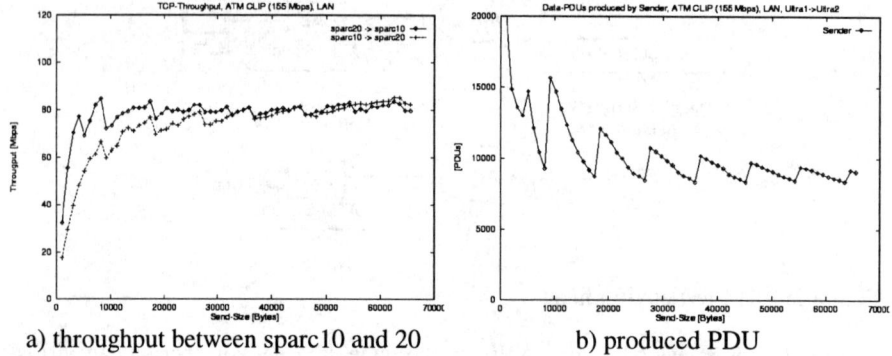

a) throughput between sparc10 and 20   b) produced PDU

Figure 3: TCP measurements in a LAN environment

The transfer rate decreases regularly about every 8 kByte. This is due to the MTU size of 9180 bytes in CLIP. When a TCP packet larger than the MTU size is transmitted it has to be fragmented at the sender side and reassembled at the receiver side. This operation is done within the processor of the end system and decreases the overall performance. With bad fragmentation (e.g. 9216 or 18432 bytes) the protocol overhead increases additionally [mol].

Figure 3b shows, how the number of PDU increases when the packet size skips above a multiple of the MTU size. The size of one PDU is limited by the MTU-size. Therefore an additional PDU is necessary each time the packet size steps over a multiple of the MTU-size. Figure 3b relies on the results of the measurements with Ultra Sparc workstations.

These workstations seem to have sufficient CPU performance to make use of the available bandwidth. A throughput of up to 130 Mbps was achieved.

The CPU load during this test was about 20% for the Ultra 2 with dual CPU and 40% for the Ultra. For small packet sizes the CPU load is significantly higher, e.g. for a packet size of 1024 Byte the CPU of the Ultra 1 was at its limits (100%).

To get a direct comparison of the CPU load of the sender and the receiver, a TCP transmission to "localhost" was measured. The tool "sar" only shows the total CPU load of a host. To get a process related value for the CPU load, we used the publicly available tool "top". Figure 4 shows the result of the measurement for three different Send-Sizes.

For very small packet sizes, the sending process needs more CPU power than the receiver. For large packet sizes, the load gets more and more equalized.

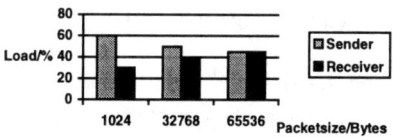

Figure 4: CPU load of sender and receiver (TCP)

TCP performance using LAN Emulation

For LAN Emulation similar measurements were made. In comparison to Classical IP, LAN Emulation adds an additional protocol layer and thus additional overhead to the communication. The default MTU size for (Ethernet-) LAN Emulation is 1500 bytes.

The throughput of LANE for a TCP stream between Sun Sparc Ultra computers reaches about 110 Mbps and is close to the theoretical upper limit. For small packet sizes the throughput decreases to 60 Mbps (1024 bytes packet size).

The CPU load for these measurements is in the range of 80-100% (Ultra 1) depending on the packet sizes. Compared to the CLIP results, the load is about twice as high.

The number of PDUs generated with LANE is about five times (50000) as high as with CLIP. This is due to the reduced MTU size for LANE

UDP measurements using CLIP

The UDP measurements show that the maximum transfer rate is limited by the processing power of the end systems. Due to the lack of flow control in the UDP protocol a slow receiver can be overloaded by a fast sender.

In our test case we send 62 Mbps from a SUN SPARCstation 10 to a SUN SPARCstation 20. In the other direction the SPARCstation 20 could send 102 Mbps but the slower receiving workstation wasn't able to process this overload leading to a packet loss rate of almost 100 percent. Similar results are expected for UDP on top of LANE (not measured).

3.4 Performance in a transatlantic ATM network

For the WAN measurements, Classical IP was used. Most of the examined effects correspond to the TCP protocol. LAN Emulation would therefore make no difference. The following configurations were setup:

First we tested a default implementation of TCP between two hosts in Ulm and Berkeley. The available bandwidth for this connection was set to 6 and 26 Mbps; the roundtriptime was about 190 ms.

In a second test, we made similar measurements with an extended implementation of TCP supporting RFC1323. These tests were performed between hosts in Ulm using the loopback in Canada. The roundtriptime of this connection was also about 190 ms. In order to measure the performance in the wide area network a PVC was setup between two workstations (Figure 2) and cell rates of 6, 26 and 50 Mbps were used.

To avoid cell loss in the ATM switches in the wide area network the traffic was shaped on the sender side to the peak cell rate [atmarp].

TCP performance with standard TCP implementation

For the TCP tests the message sizes were varied and socket sizes were set to the maximum value of 64 kByte. With the 6 Mbps as well as with the 26 Mbps connection only a fraction of the provided bandwidth could be used (approximately 2.2 Mbps). The increase in bandwidth from 6 to 26 Mbps had almost no effect to the results.

Sun's standard TCP implementation in Solaris 2.5 cannot exploit the available bandwidth in an environment with a high-speed network connection and large round-trip delay. The problem of a high path capacity[1] is already wellknown. A sender of TCP packets has to wait until the transmitted packets are acknowledged by the receiver (sliding window protocol [comer, brzi96]). Assuming the sender waits for an acknowledgment after it has sent the maximum window size of 64 kByte means that a maximum of 64 kByte could be on the wire. With a round-trip delay of 190 msec the maximum throughput is 64 kByte / 190 msec ≈ 2.7 Mbps. Several TCP extensions have been suggested to solve this problem [RFC 1323], some are available in beta implementations (see below).

TCP with extensions for Long Fat Networks (LFN)

RFC1323 suggests an extension to the TCP protocol to support window sizes greater than 64 kByte. An implementation of this extended TCP version should be capable of using the bandwidth of ATM connections with high delay (Long Fat Networks). A beta release version of the Solaris operating system (2.6) which implements the extensions of RFC1323 was used during these tests.

To enable the operating system to use the "Big Window" option, a couple of "ndd" commands have to be issued. Ndd is a Solaris command to manipulate driver parameters. In particular the buffer space for the socket can be adapted according to the bandwidth*delay product of the connection

Table 1 shows the results of the WAN measurements. For the 6 Mbps connection the roundtriptime was approximately 190 ms, which leads to a bandwidth*delay product of about 140 kByte. After adapting the TCP parameters with the "ndd" program, we achieved a throughput of 5.05 Mbps. An increment of the maximum window size to 1 MByte did not make any difference. For this data rate the Send-Size did not affect the throughput.

For 50 MBps the bandwidth*delay product is about 1.2 MByte. With buffers of 1.5 MByte we could only use 60% of the available bandwidth. To investigate this

---

[1] bandwidth and roundtrip delay product

problem, measurements in the LAN environment were carried out, especially concerning the traffic shaping capability of the Fore adapter.

Table 1: WAN measurements with TCP supporting RFC1323

Avail. BW	buffer	send size	Throughput	CPU send	CPU recv
6 Mbps	140000	1k-64k	5.04 Mbps	4%-1%	1%
50 Mbps	1572864	1024	31.45 Mbps	28%	5%
		32k, 64k	~29.83 Mbps	~13%	3%
	524288	1024	21.35 Mbps	18%	4%
		32k	19.17 Mbps	6%	2%
25 Mbps	1048576	1024	18.29 Mbps	17%	5%
		32k, 64k	~17.64 Mbps	~8%	~2%

In the LAN environment the traffic shaping of the Fore adapter was set to different values using the "atmarp" command. The possible throughput was measured for a send size of 32768 bytes. The results are shown in Figure 5. The theoretical boundary is also given. The traffic shaping parameter of "atmarp" corresponds to the 48 byte payload of the ATM cells. As the figure shows, the characteristics of the traffic shaper is very inefficient for bandwidths greater than 40 Mbps.

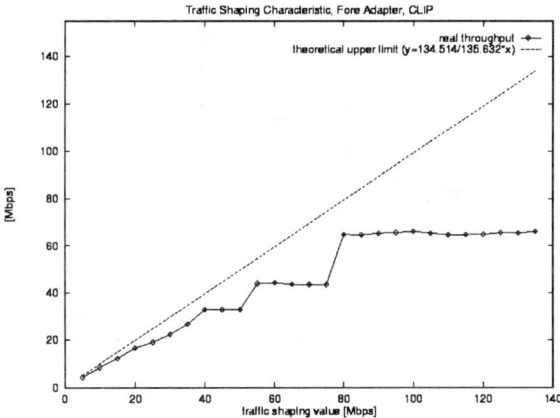

Figure 5: Traffic shaping characteristic of Fore adapter

You can also recognize that the WAN results (Table 1) are as good as the results from the LAN case. The problem of the bad throughput results from the traffic shaper in the workstation and not from the TCP implementation.

UDP results in WAN environment
For an available bandwidth of 6 Mbps there were no packet losses.
The 26 Mbps measurements were carried out from a SUN SPARCstation 20 to a SUN SPARCstation 5. The sending computer was able to send with the full

bandwidth. Only for small packet sizes losses could be observed at the receiver side. In this test case the ATM connection was monitored to guarantee that these errors weren't caused by cell losses of the ATM connection.

Rather it looks like the receiving workstation was already at its limits. In this case the receiver is much more loaded due to the processing of the UDP packets. Additionally every received packet releases an interrupt at the receiver side.

4   PERFORMANCE OF THE VIC APPLICATION

The MBone tool "vic" [vic] is widely used for video conferencing [MBoneApps, kumar]. In this section the mesurements concerning the tool vic are presented.

The MBone tools support a wide range of compression schemes and workstations. If the workstation doesn't have a special hardware support for the used compression scheme most of the operations have to be done in software. Especially for video, this causes a nonneglectible CPU consumption. Therefore we carried out several measurements in order to evaluate the bandwidth and computation requirements using the vic application.

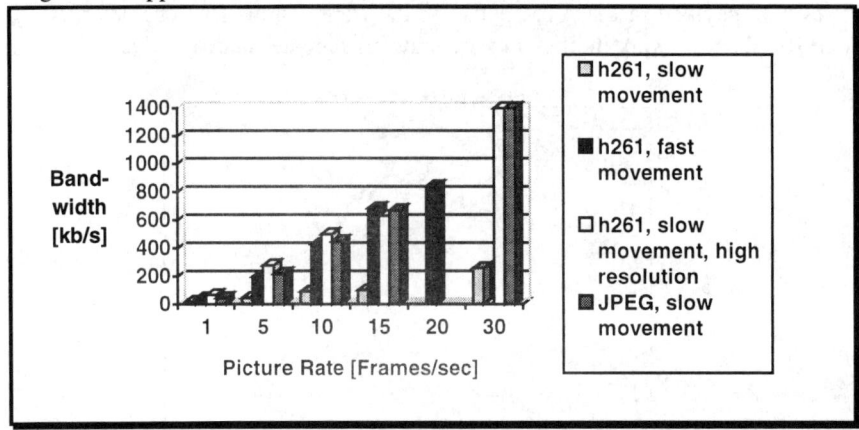

Figure 6: Bandwidth Requirements

For our tests we used two Ultra Sparc high performance workstations. To separate the effects caused by encoding and decoding of the video we first used unidirectional video transmission. We used h261 and JPEG video encoding (CIF format) and run tests with slow and fast moving people in front of the camera (slow and fast changing scene). We also run a test with high resolution video transmission. In these tests we varied the picture rate between one and thirty frames per second. We evaluated the bandwidth requirements using the measurement embedded in vic. Figure 6 shows the results of these tests.

The needed bandwidth increases linearly with the picture rate, as expected. As h261 uses an inter picture encoding mechanisms, the bandwidth requirements for slow changing scenes of h261 are much lower than this of JPEG. It can be seen, that the bandwidth requirements of h261 increases significantly when the movement of the people in front of the camera becomes very fast. Moreover, due to end system limitations we couldn't send video at 30 frames per second using h261 encoding in the fast moving case (maximum picture rate: 20 frames per second). As the results from CPU measurements show, this limitation was caused by I/O operations the application has to wait for. The bandwidth requirements of high resolution h261 encoded video for slow changing scenes is similar to the requirements of JPEG with medium quality.

Parallel to the bandwidth measurements we observed the CPU consumption on the sender and receiver side using the sar tool (Figure 7 and Figure 8, respectively).

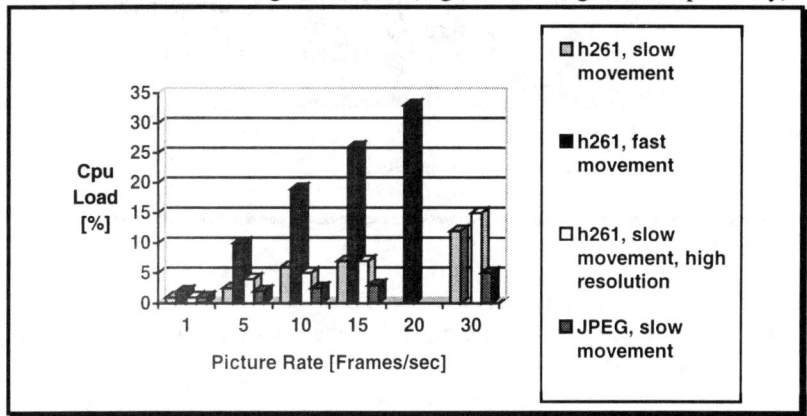

Figure 7: CPU load at the Sender

Like the bandwidth the CPU load increases almost linearly with the picture rate. The measurements show that the sender's CPU is much more loaded when using h261 than using JPEG. In the slow movement case the CPU load at the receiver is smaller for h261 than for JPEG, but this changes in the fast movement case. The figures also show that with h261 faster changes in the scene have stronger influences on the CPU load than changing the picture resolution.

Our bidirectional tests (results not presented here) show, that the CPU load of sender and receiver of the unidirectional case must simply be added to get the figures for the bi-directional case. Our measurements show, that software encoding and decoding cause significant CPU load which can't be used for other applications (e.g. application sharing) running in parallel. Videoconferences with more than two participants or between lower performing workstations increase the problem further. Nevertheless the MBone tools are applicable in cases where video is just used in the starting phase of a conference with little number of participants (i.e. for greeting purposes) when no other application is running. They also satisfy the needs when only small picture rates are requested.

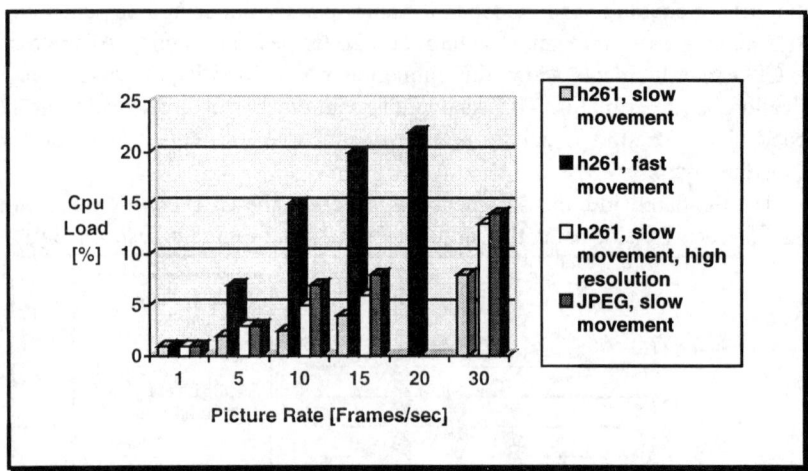

Figure 8:CPU load at the Receiver

To use the end system and network resources in an optimal way multimedia applications need mechanisms to throttle down the sender's transmission rate if the receiving end system is already saturated or if the network can't guarantee the requested quality of service anymore [AlSi96].

5   CONCLUSIONS AND FUTURE WORK

Our experiences with ATM in global networks show that up to now ATM's potentials are only used to a limited extend. To exploit them, enhancements in the provided services and protocols are needed. Especially signalling in local and wide area networks and means to monitor connections must be made available by the manufacturers and network providers. New lightweight protocols that take advantage of the ATM QoS guarantuees must be developed. Further reasearch for efficient multicasting over ATM is required. Also transport layer protocols need to be adapted to environments with high bandwidth delay products. In order to support worldwide collaboration of people new applications must be developed or adapted to a global high speed network environment [EbRuSi97]. Future work is dealing with the provision of QoS to the applications. This can be done using for example the WinSock 2 API or by using RSVP in conjunction with ATM [BrSt97].

6 REFERENCES

[AlSi96]   Alfano M., Sigle R., *Controlling Resources in a Collaborative Multimedia Environment,* Proceedings 5th IEEE International Symposium on High-Performance Distributed Computing (HPDC-5), Syracuse, USA, 1996

[atmarp]   Manual page atmarp, Fore Systems, 2/1996

[BrSt97]   Braun T., Stüttgen H.J., *Implementation of an Internet Video Conferencing Application over ATM,* IEEE ATM'97 Workshop, Lisbon, Portugal, 1997

[brzi96]   Braun T., Zitterbart M., *Hochleistungskommunikation, Band 2: Transportdienste und -protokolle,* Oldenbourg 1996, ISBN 3-486-23088-3

[comer]   Comer, Douglas E., *Internetworking with TCP/IP,* Volume I, Second Edition, Prentice Hall 1991, ISBN 0-13-468505-9

[EbRuSi97]   Eberhardt R., Rueß C., Sigle R., *Multimediale Anwendungen in globalen ATM-Netzen,* Proceedings of GI/ITG Kommunikation in Verteilten Systemen (KiVS'97), Braunschweig, Germany, 1997

[krivda]   Krivda, Cheryl D., *Analyzing ATM Adapter Performance, The Real World Meaning of Benchmarks,* Efficient Networks Inc., 1996, http://www.efficient.com/doc/EM.html

[kumar]   Kumar V., *MBone: Interactive Multimedia On The Internet,* Macmillan Publishing, Simon & Schuster 1995

[MBoneApps]   Kumar V., *Mbone Desktop Applications,* http://www.best.com/~prince/techinfo/mc-soft.html

[mol]   Moldeklev, Kjersti., *The effect of end system hardware and software on TCP /IP throughput performance over a local ATM network,* Telektronik, 1995

[np]   Netperf, www, http://www.cup.hp.com/netperf/NetperfPage.html

[RFC1323]   Jacobson V., Braden B., Borman D.: *TCP Extensions for high-performance,* Request for Comments 1323, May 1992

[RFC1483]   Heinanen J., *Multiprotocol Encapsulation over ATM Adaptation Layer 5,* Request for Comment 1483, Juli 1993

[RFC1577]   Laubach, M., *Classical IP and ARP over ATM,* Request for Comments 1577, Januar 1993

[vic]   LBNL, *The Video Conferencing Tool vic,* http://www-nrg.ee.lbl.gov/vic

Traffic Control, Modelling and Analysis

ATM Connection Admission Control

26

Can Equivalent Capacity CAC Deal with Worst Case Traffic in GCRA–Policed ATM Networks?

Kai Schmidt, Maike Wichers, Ulrich Killat
University of Technology Hamburg–Harburg
Digital Communication Systems
21071 Hamburg, Germany,
Phone +49-40-7718-3049, email: killat@tu-harburg.de

Abstract

In order to manage modern ATM networks various methods for connection admission control (CAC) have been proposed. In addition, usage parameter control (UPC) algorithms supervise the transmitted traffic to prevent the user from exceeding the negotiated traffic parameters. This paper addresses the question, whether the broadly accepted Equivalent Capacity CAC algorithm can guarantee the quality of service asserted to the subscribers provided that the algorithm has only knowledge of the specified GCRA parameters used for UPC. We describe two types of worst case traffic conforming to the Generic Cell Rate Algorithm and show the performance of Equivalent Capacity CAC in various traffic scenarios.

Keywords

ATM Network, CAC, UPC, GCRA, Worst Case Traffic, Simulation

1 INTRODUCTION

Since the Asynchronous Transfer Mode (ATM) has been proposed for the up-coming Broadband ISDN by the ITU–T, a considerable number of methods for connection admission control (CAC) have been scrutinised. Reviews like [4, 5, 6, 7] provide a comprehensive overview. Each CAC algorithm requires a CAC specific set of traffic parameters describing the statistical properties of the cell streams to be dealt with by the CAC. Based on these traffic parameters a traffic contract is made if resources are available to accommodate the requested connection. In order to protect the network from overload and to guarantee the quality of already established connections it is necessary to control whether the admitted cell streams meet their respective traffic contracts. The usage parameter control (UPC) identifies cells of a certain cell stream as being conforming or non–conforming to the traffic contract. An in-

Broadband Communications   P. Kühn & R. Ulrich (Eds.)
© 1998 IFIP. Published by Chapman & Hall

tricate problem is that the UPC in general will not operate on the same traffic parameters as the CAC does. Therefore, a sometimes ambiguous translation of the CAC traffic parameter set to a set of UPC traffic parameters becomes necessary.

The key idea pursued in this paper is the following:

The network must be prepared for the situation in which the user tries to transmit a maximum of data conforming to the specified UPC parameters. Now, this can be done by submitting a constant rate cell stream or as another extreme by submitting bursty traffic streams. In this context worst case behaviour turns out to be periodic bursty traffic with a maximum admissible burst size.

In this paper we will identify two types of worst case traffic which are described by a set of GCRA [1] parameters. These traffic streams that are fully conforming to their respective GCRAs have two remarkable properties: They to show worst case behaviour with respect to buffer demands and their parameter set can unambiguously be translated to an Equivalent Capacity [3] parameter set.

In [2] B. T. Doshi shows in a qualitative analysis that for certain traffic streams the Equivalent Capacity CAC will be too optimistic. We will address the question whether the combination of connection admission control based on the Equivalent Capacity Algorithm and usage parameter control based on the GCRA will be able to protect the network from overload and evaluate the performance of the Equivalent Capacity Algorithm in a realistic network environment.

2   WORST CASE TRAFFIC

Having specified an algorithm for usage parameter control – the Generic Cell Rate Algorithm – the question arises, how a worst case cell stream could look like. From the network operators point of view worst case traffic will be most demanding according to the buffer occupancy inside a multiplexor. We will assume a single ATM multiplexor having a number of N input lines. The two types of worst case traffic which we will describe in the following will be determined by a given set of GCRA parameters so that all cells are conforming to this parameter set. Thereby, the cell streams are constructed in such a way that the superposition of these traffic streams will require a maximum number of buffers within the multiplexor.

Following the ATM–Forum Traffic Management Specification [1] we consider the four parameter $GCRA(T,\tau,T_s,\tau_s)$ equivalent to a two stage leaky bucket. $GCRA(T_s,\tau_s)$ representing the second leaky bucket controls the mean cell rate where T_s and τ_s correspond to the buckets leak and size, respectively. Mutatis mutandis the $GCRA(T,\tau)$ is applied for peak cell rate policing. Since we do not want the peak cell rate to be exceeded $\tau = 0$ (Peak Rate Enforcement).

2.1 Worst Case Traffic A – Greedy ON/OFF

Intuitively the worst case traffic should be a traffic where leaky bucket equivalent to the GCRA is filled with a burst of cells emitted at peak cell rate followed by a silence period until the leaky bucket is again empty. Then the next burst will follow. We will call a source emitting this type of worst case traffic as *greedy ON/OFF source* emitting *worst case traffic type A (WCT-A)*.

As an example in Figure 1 the worst case traffic is described by the GCRA parameter set $(T, \tau, T_s, \tau_s) = (2, 0, 8, 23)$. The peak cell rate of each source is 50%, the mean cell rate 12.5%, and the maximum burst size (=mean burst size) is 4 cells according to equation (7). The cell stream is periodic with a period of $\ell_A = \lceil MBS \cdot T_s \rceil$ because the GCRA requires T_s time slots for each cell in the peak cell rate burst to recover.

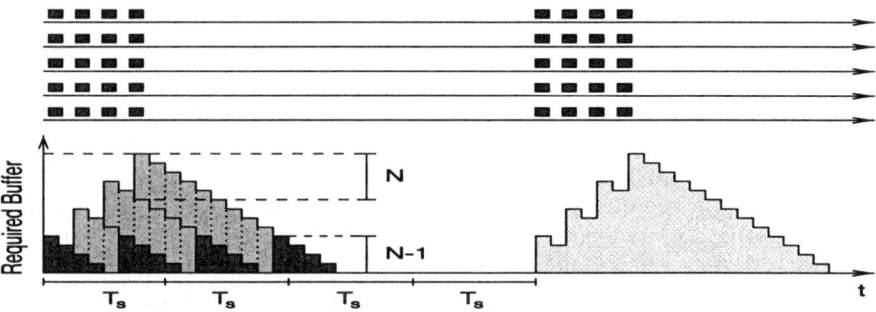

Figure 1 *Synchronous Greedy ON/OFF Streams*

In Figure 1 the buffer occupancy in the switch is shown, when $N = 5$ lines of synchronised worst case traffic streams are superposed in a multiplexer. The necessary buffer size $K_A = 13$ can be calculated directly from the number of cell streams N and the worst case traffic parameter set:

$$K_A = MBS \cdot N - \lceil (MBS - 1) \cdot T \rceil - 1 \qquad (1)$$

Even if the buffer size is sufficient for the superposition of the cell streams, it is important to know the remaining number of buffers for additional traffic provided the minimum buffer size K_A has been implemented. Figure 1 shows a graphical approach to calculate the number of occupied buffers for the greedy ON/OFF traffic. Parting the graphic in two areas (light grey and dark grey) makes it easy to derive an analytical expression for the mean number of free buffers. The total amount of available buffer during one period is determined by the period length ℓ_A and buffer size K_A. For a given value of K_A the mean number of free buffers κ_A is

$$\kappa_A = K_A - \frac{1}{\ell_A} \left\{ \frac{MBS \cdot N}{2} [MBS(N - T) + T - 1] \right\} \text{cells/time slot} \quad (2)$$

The smaller the mean number of free buffers is, the more likely cells will be lost if additional traffic is passing the buffer. We will later on use κ to determine how much "worst case" a traffic stream is.

2.2   Worst Case Traffic Type B

The greedy ON/OFF traffic described above fully exploits the amount of traffic the UPC allows. However, the greedy ON/OFF traffic is not the only type of traffic that transmits the maximum number of cells during a period of time. We will present the worst case traffic type B as such a type of traffic, and we will show that this type of traffic can be more demanding with respect to both the necessary buffer size and the mean number of occupied buffers per time slot.

Similar to the greedy ON/OFF traffic the empty leaky bucket is filled with a burst at peak cell rate but now the peak rate burst is followed by the emission of one cell at the earliest possible time slot.

To understand why this traffic behaviour is worse than the greedy ON/OFF traffic, we first look at the end of the peak cell rate burst. The peak cell rate phase stops as soon as the next cell will cause a leaky bucket overflow. Once the leaky bucket has been completely filled with the last cell we will have to wait T_s time slots before the next cell can arrive. However, we also have to face the situation where the leaky bucket has been filled with the last cell in such a kind that there is remaining space for *nearly* one cell. But now, already in the next few time slots a new cell can fit into in the leaky bucket. This situation is shown in Figure 2 and it can be seen that the necessary buffer is larger than for greedy ON/OFF traffic. (Figure 2 shows an example for the superposition of $N = 5$ WCT-B cell streams again with the GCRA parameter set ($T = 2$, $\tau = 0$, $T_s = 8$, $\tau_s = 23$).) It is important to know that the larger buffer is only needed when the cell following the peak cell rate phase is emitted in a specific window of time after the last cell of the peak cell rate burst: The earliest time slot is the $T + 1$st time slot, the latest time slot is the $N - 1$st time slot after the last cell of the peak cell rate burst. Therefore, the maximum buffer that is required depends on whether the peak buffer occupation is reached at the end of the peak cell rate phase or at the arrival of the following cell:

$$K_B = \max(K_A, K_A + N - [T + MBS(T_s - T) - \tau_s]) \quad (3)$$

In total the WCT–B cell stream emits $MBS + 1$ cells in each period. There-

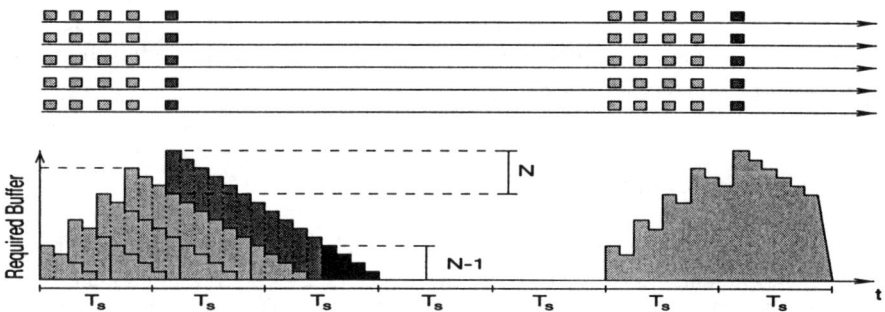

Figure 2 *Synchronous WorstCaseTraffic–B Streams*

from the period length is determined to be $\ell_B = \ell_A + T_s = \lceil MBS + 1 \rceil \cdot T_s$ time slots

The maximum required buffer K_B varies with the value of τ_s. Therefore, the calculation of the mean number of free buffers κ_B is not as simple as for the greedy ON/OFF traffic. In Figure 2 the light grey areas represent the buffers occupied by cells of the peak cell rate burst which already has been calculated in equation (2) while the darker areas show the additionally occupied buffer space. The graphical approach leads to:

$$
\begin{aligned}
\kappa_B \;=\; & \frac{1}{\ell_B} \cdot \{ K_B \cdot \ell_C \\
& - \frac{1}{2} MBS \cdot N[(N-1) + (N-T)(MBS-1)] \\
& + N \cdot [MBS(N - T_s) + \tau_s] \\
& + \frac{1}{2} N(N-1) \} \text{ cells/time slot}
\end{aligned}
\tag{4}
$$

3 TRAFFIC PARAMETERS FOR UPC AND CAC

Both the connection admission algorithm and the usage parameter control need a set of parameters describing the statistical properties of the cell streams announced by the user. Ideally, the CAC would operate on exactly the same set of parameters advertised by a source to the UPC, because this set of parameters is enforced at the network boundary. Unfortunately, some of the most popular CAC methods were developed independent of the (later starting) discussion on UPC and therefore use other traffic parameters than those specified for the UPC. That's the case for our chosen pair of UPC and CAC as well. In general there is no unique transformation of the traffic parameters used by the two algorithms.

However, concentrating on the presented "worst case traffic" patterns, which have to be anticipated by a CAC to be on the safe side, the parameter translation between UPC and CAC becomes unambiguous and therefore the question is well defined whether or not the CAC does its job properly under all traffic scenarios of "worst case" sources.

3.1   Usage Parameter Control

Following the interpretation in the Traffic Management Specification [1] there is a correspondence between the parameter set $(T, \tau(= 0), T_s, \tau_s)$ and the peak cell rate (PCR), the sustainable cell rate (SCR), and the maximum burst size (MBS):

$$PCR \ = \ 1/T \tag{5}$$

$$SRC \ = \ 1/T_s \tag{6}$$

$$MBS \ = \ \left\lfloor 1 + \frac{\tau_s}{T_s - T} \right\rfloor \tag{7}$$

where $\lfloor \bullet \rfloor$ means the integer part of $\bullet$.

3.2   Connection Admission Control

Among a multitude of CAC algorithms proposed [4, 5, 6, 7] the Equivalent Capacity Algorithm has both the virtue of broad acceptance and characterisation by three parameters, which brings it close to the three–parameter approach adopted for UPC in chapter 3.1:
Peak Bit Rate R_{peak}, Mean Bit Rate ρ, and Mean Burst Size b.

The latter two quantities are defined as mean values and it is not obvious in which way these means have to be determined: Can mean values be determined for connections lasting only a fraction of a second and can an application be aware of the duration of the connection (and the resulting averages) beforehand?

The way out of these conceptional problems is to consider each connection to be of "worst case" type – an assumption which might become reality with more sophisticated ATM communication boards, anyhow.

3.3   Parameter Transformation

The "worst case traffic" which we have presented is a periodic traffic. For this reason the calculation of the mean cell rate and the mean burst size is possible on the basis of a single period. Moreover, the mean cell rate coincides with the sustainable cell rate of the GCRA and, similarly, the mean burst size equals the maximum burst size and can be identified with this parameter of the GCRA setting.

4 GCRA POLICER AND EQUIVALENT CAPACITY CAC

Our traffic model is the following: Worst case cell streams are generated according to a set of (T, T_s, τ_s) so that no cells of these cell streams violate the traffic contract. From the parameter set (T, T_s, τ_s) the input parameters for the Equivalent Capacity CAC (ρ, R_{Peak}, b) are derived. A number of N traffic streams will be multiplexed into a FIFO buffer with capacity K and the cell loss is investigated where N is the maximum number of connections that is accepted by the CAC algorithm. This means that only situations are investigated where the cell loss should not exceed the cell loss limit imposed on the CAC.

First we will present the synchronous superposition of cell streams where the burst phases of arriving cell streams all begin at the same time slot. Such a scenario must be seen as a worst case scenario when superposing cell streams. However, in general the cell streams will arrive with arbitrary phases. The impact of non–synchronised superposition of worst case cell streams is investigated in the second part of this section. Finally we extend our investigations from considering only one multiplexor to a simulation of worst case traffic in a 13 node ATM network model in section 4.3.

4.1 Synchronised Superposition

Using the Equivalent Capacity CAC that number of connections was determined for which the cell loss rate is estimated to stay below $\epsilon = 10^{-5}$. The evaluation of equation (1) and (3) reveals to which degree this target is met. Figure 3 shows results for peak cell rates of $R_{Peak} = 0.1 (T = 10)$, $R_{Peak} = 0.05 (T = 20)$, and combinations of τ_s and T_s in a range from $20 \ldots 250$ and $20 \ldots 40$, respectively.

We distinguish three possible effects: Grey Areas indicate that the CAC Algorithm could protect the buffer from overload. Bright areas indicate combinations of τ_s and T_s, where the CAC still can deal worst case traffic type A properly but not worst case traffic type B. Areas plotted dark grey represent worst case traffic, where the CAC underestimates the impact on the multiplexer buffer for all types of worst case traffic. In these areas at least one cell is lost in every period of the cell stream.

The graphics show that the synchronous superposition of worst case cell streams causes huge problems to the Equivalent Capacity Algorithm. In contrast to what one might expect, the CAC performance is reduced furthermore, if the peak cell rate is reduced. This is because the Equivalent Capacity Al-

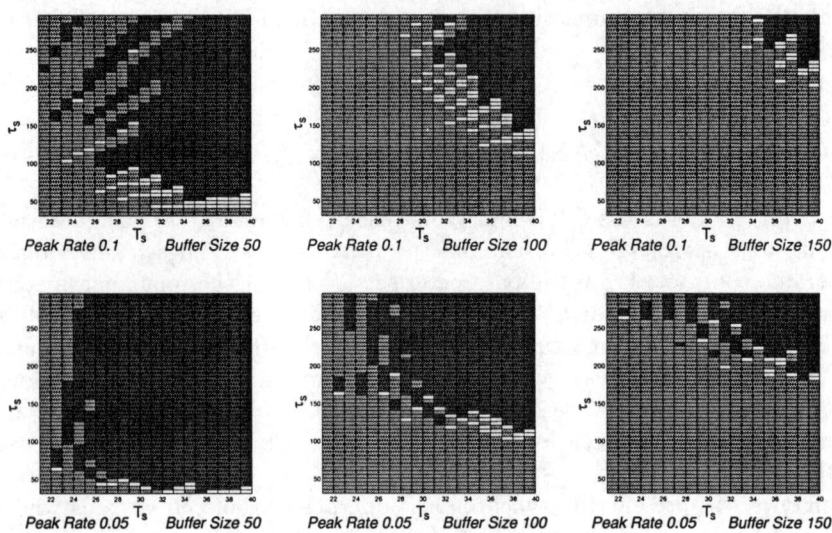

Figure 3 *Performance of EquCap CAC: "worst case traffic" load with 10% and 5% peak cell rate*

gorithm assumes a lower peak cell rate to be less demanding with respect to the necessary buffer size.

The second criterion to determine the "badness" of traffic streams is the amount of free buffer that is left for other traffic streams. We evaluate equations (2) and (4) for some example parameter sets which all lead to the burst size of 4 cells for WCT–A and 5 cells for WCT–B:

τ_s :	23	22	21	20	19	18	τ_s:	$[18 \ldots 24[$
K_B :	15	14	13	13	13	13	K_A:	13
κ_B	10.12	9.25	8.38	8.50	8.62	8.75	κ_A:	8.94

Table 1 K_A, κ_A, K_B, and κ_B for $T = 2$, $\tau = 0$, $T_s = 8$, and $\tau_s = 18 \ldots 23$

WCT–B is more demanding in terms of buffer required, although a higher buffer demand goes along with a greater average number of free buffers. In this context it is interesting to look to non worst case traffic streams. Based on $K = 13$ as a buffer size to achieve comparable results we find for the mean number of free buffers for the synchronous superposition of $N = 5$ constant cell rate streams ($T_s = 8$): $\kappa_{CBR} = \frac{8 \cdot 13 - 10}{8} = 11.75$ cells/time slot. Using the mean buffer occupation for N-Geo/D/1 models for $N = 5$ streams with geometrically distributed interarrival times $\kappa_{geo} = 13 - 1.024 = 11.98$ cells/time slot.

4.2 Non–Synchronised Superposition of Cell Streams

The synchronised superposition of cell streams represents the worst case traffic scenario, but it can be assumed that arbitrary phases of the cell streams will result in substantially lower cell loss rates. On the other hand, the number of phase constellations leading to unpredicted cell loss still may be very high. So far it is not easy to assess whether the overall cell loss rate will be lower or higher than the estimation of the CAC algorithm.

Due to the huge number of possible phase constellations an exhaustive analysis is not possible so far. E.g. a parameter set of Figure 3 ($T = 20, \tau_s = 150, T_s = 30$) leads to $7.1 \cdot 10^{38}$ different phase constellations.

Although the problem dimension is very big, a tendency can be observed, when the phases are selected randomly. For different combinations of (T, T_s, τ_s) we have simulated the superposition of the corresponding number of cell streams with random phases, using a target cell loss rate of $\epsilon = 10^{-4}$. Only the simulations with a 50 cell buffer have a target cell loss rate of $\epsilon = 10^{-5}$ in order to be comparable with the results shown in Figure 3.

In Table 2 some results are collected. We have investigated three types of dependencies: The cell loss effects for different buffer sizes, different values of τ_s, and different values of T_s. The respective parameter values are typed in bold letters.

In contrast to the synchronised superposition of the cell streams, the non-synchronised superposition in general does not lead to higher cell loss than the target cell loss rate used by the CAC. Only for very small buffers the observed cell loss rate is slightly higher than aimed at.

4.3 Extended ATM Network Model

The simulation results presented here are obtained with the ATM network simulation software SABINE, operating on a 125 processor massive parallel computer. This simulation engine has been developed at the University of Technology Hamburg–Harburg [8, 9]

Our realistic network model assumes an ATM backbone network for Germany consisting of 13 nodes interconnecting 13 larger cities. The number of links adapted to each ATM network node varies from node to node since the network topology follows geographic conditions. In our network the number of links per node varies between two and six. In addition to ATM links interconnecting the ATM network nodes, each node is supplied with an access link that is connected to an aggregate source model. The idea behind this type of

Buffer Size	T	T_s	τ_s	# Connections	Total Load	Loss Rate
5	20	40	69	22	0.55	$6.3 \cdot 10^{-4}$
10	20	40	69	25	0.625	$4.4 \cdot 10^{-7}$
15	20	40	69	27	0.675	$< 10^{-7}$
10	20	40	299	22	0.55	$< 10^{-7}$
10	20	40	149	22	0.55	$< 10^{-7}$
10	20	40	79	25	0.625	$5.2 \cdot 10^{-7}$
10	20	40	69	25	0.625	$4.4 \cdot 10^{-7}$
10	20	40	39	28	0.70	$5.2 \cdot 10^{-6}$
10	20	30	69	22	0.73	$1.0 \cdot 10^{-6}$
10	20	35	69	23	0.66	$2.9 \cdot 10^{-7}$
10	20	40	69	25	0.625	$4.4 \cdot 10^{-7}$
50	20	30	50	26	0.87	$< 10^{-7}$
50	20	30	250	22	0.73	$< 10^{-7}$
50	20	70	250	37	0.53	$1.3 \cdot 10^{-6}$
50	20	70	50	55	0.79	$< 10^{-7}$
50	20	50	150	33	0.66	$< 10^{-7}$

Table 2 *Cell loss simulation for random phase superposition of WCT–B. Target cell loss rate is* 10^{-5}.

source modelling is that subscribers from a local network not are connected directly to the ATM backbone network, but are sharing one access link to the ATM backbone network.

The simulation results presented here are based on a number of subscribers who all are policed by the same set of GCRA parameters. Hence, they all produce the same worst case traffic. We have assumed the subscribers to be independent of each other, therefore no synchronised superposition of worst case traffic streams occur in this simulation. Due to the inhomogeneous network topology, some traffic streams only will pass two network nodes, while other traffic streams are routed through up to six network nodes. Furthermore, each traffic stream is multiplexed with a varying number of other traffic streams, which themselves already have been multiplexed in the same manner. All this leads to an unpredictably complex load scenario, which can be seen as a good approximation to a realistic ATM network situation. Compared to a simple Peak Rate Allocation in our example the admitted network load was substantially higher showing the efficiency of the Equivalent Capacity CAC. Depending on the burst size a factor between 3 and 11 could be obtained.

The simulation of this scenario yields that no cell loss could be observed at all. We interpret the result as follows: Due to the deterministic source behaviour in this scenario, statistical effects are reduced to random phases of the traffic streams. The Equivalent Capacity CAC obviously did not make too optimistic decisions and the critical situation of synchronous superposition of worst case streams did not occur within the simulation time. We have simulated the entire ATM network with a number of 10^8 randomly selected phases of worst case traffic streams.

The simulation of a second scenario should consider statistical source behaviour. For this purpose we used statistical ON/OFF sources producing a basic traffic load. Onto this basic load a number of worst case traffic sources were loaded. It is expected that this number of worst case sources will lead to higher cell loss than the same number of statistical ON/OFF sources with the same load. The Equivalent Capacity CAC will not be able to distinguish between both types of traffic, so the interesting question is whether the same quality of service is observed or not. Simulation results reveal that the measured cell loss rate did not exceed the limit aimed at.

5 CONCLUSION

This paper was motivated by the question, whether the Equivalent Capacity Algorithm CAC and a GCRA based usage parameter control (UPC), in cooperation will protect the network from overload, even though, in general, they are based on different algorithms.

On the basis of the GCRA parameter set, we developed two types of worst case traffic that will pass the UPC but on the other hand will be most demanding for the buffer in a multiplexer.

In a first step, a fully synchronised superposition was assumed. Analytical expressions for the necessary buffer size at the input switch as a function of the GCRA parameters were derived. The analysis revealed, that the CAC algorithm was not able to protect the multiplexer against overload, and significant cell losses could be observed.

The synchronous superposition of cell streams is assumed to be very unlikely, although no one can guarantee that this situation will not happen. Hence, in a second step non synchronised cell streams were considered. An analytical approach for this scenario seems to be out of reach due to the huge number of possible phase constellations. Instead simulations were performed where the phases of the single cell streams were selected randomly. Compared with the synchronous superposition, the simulations of the non–synchronised cell streams yielded a much better performance of the CAC algorithm. The simulated cell loss rate for nearly all considered GCRA parameter sets was lower than the target cell loss rate used by the CAC. However, for small buffer sizes slightly higher cell loss rates were observed. This result was confirmed as the simulation was extended to a realistic network topology .

Our results show that if the source model is based on a two level leaky bucket with peak rate enforcement, Equivalent Capacity CAC and UPC can be reconciled on the basis of "worst case traffic" assumptions. Our experimental results show that in general the CAC will provide conservative but still sensible estimates of cell loss rates which allow loadings of the network in excess of 70%.

Part of the work presented in this paper was supported by Deutsche Telekom AG, Darmstadt.

REFERENCES

[1] *ATM Forum Traffic Management Specification 4.0*, December 1995, ATM Forum/95-0013R9, pp63–68.

[2] B. T. Doshi: *Deterministic Rule Based Traffic Descriptors for Broadband ISDN: Worst Case Behaviour and Connection Acceptance Control*. Proceedings of the 14th International Teletraffic Congress, Antibes Juan-les-Pins, France, 6–10 June, 1994, pp 591–600.

[3] Roch Guerin, Hamid Ahmadi, Mahmoud Naghshineh: *Equivalent capacity and its Application to Bandwidth Allocation in High Speed Networks*. IEEE Journal on Selected Areas in Communications, Vol 9, No 7, September 1991, pp 968–981.

[4] Harry G. Perros, Khaled M. Elsayed: *Call Admission Control Schemes: A Review*. IEEE Communications Magazine, November 1996, pp. 82–91.

[5] Nikolas M. Mitrou, Kimon P. Kontovasilis, Hans Kröner, Villy Bæk Iversen: *Statistical Multiplexing, Bandwidth Allocation Strategies and Connection Admission Control in ATM Networks*. ETT, Vol 5, No 2, Mar.-Apr. 1994, pp. 33/161–47/175

[6] Gagan L. Choudhury, David M. Lucantoni, Ward Whitt: *Squeezing the Most Out of ATM*. IEEE Transactions on Communications, Vol 44, No 2, February 1996, pp. 203–217

[7] Ruth Kleinewillinghöfer–Kopp, Birgit Kaltenmorgen: *Connection Acceptance Control in ATM Networks*. Studie des Forschungsinstituts der DBP Telekom, Forschungsbereich 5 'Vermittlung und Netze', Juni 1991, also presented at: TIMS/ORSA Joint National Meeting, Nashville, Tennessee, USA, May 1991.

[8] K. Schmidt, U. Killat, F. Mayer–Lindenberg, P. Kraft: *A Parallel ATM Network Simulator* Proc. XV. International Switching Symposium ISS '95, April 1995, Berlin, Germany.

[9] K. Schmidt: *Efficient ATM Network Simulation* Proc. EUROSIM 1995 Conference, September 1995, Vienna, Austria.

27

Traffic Management in an ATM Multi-Service Switch for Workgroups

Stefan Bodamer and Thomas Renger
University of Stuttgart
Institute of Communication Networks and Computer
Engineering, Pfaffenwaldring 47, D-70569 Stuttgart
e-mail: {bodamer, renger}@ind.uni-stuttgart.de

Georg Rößler
Bosch Telecom
UC-PN/EGH, Kleyerstraße 94, D-60326 Frankfurt am Main
e-mail: Georg.Roessler@pcm.bosch.de

Abstract

ATM switches have to provide traffic management functions to meet the QoS require-ments of different service categories. Among the traffic management functions we will focus on connection admission control (CAC) and priority control in this paper. Besides the mechanisms and algorithms behind these functions the issue of application to a con-crete architecture of an ATM multi-service workgroup switch is emphasized.

For CAC a new method is proposed that is based on a simple approximation of the effective bandwidth. The difficulties occurring if connections with different parameters are mixed together are solved by handling CBR traffic separately and dividing the parameter space for VBR connections into several regions. Within the VBR regions the linear method is applied while for connections associated to different regions a reduced service rate is considered.

The performance of both the priority control mechanisms and the CAC method is evaluated using analysis and simulation.

Keywords

ATM, Traffic Management, Connection Admission Control, Priority Control

Broadband Communications   P. Kühn & R. Ulrich (Eds.)
© 1998 IFIP. Published by Chapman & Hall

1   INTRODUCTION

Traffic management in ATM switches comprises several functions which act at all levels from cell level to connection level in order to make sure that the Quality of Service (QoS) requirements are met for all connections carried by the switch. Traffic management becomes especially demanding if traffic ranging from video conferencing over interactive applications like the World Wide Web (WWW) to bulk data transfer has to be handled by a switch with minimal interference between different traffic streams.

Video conference connections have stringent requirements in terms of delay and cell loss ratio which can only be fulfilled if the service rate closely matches the rate of the offered traffic. Data traffic, however, is very elastic in that it is not sensitive to delay and can be served at any rate the switch can provide. Of course, large buffers are needed to store bursts until they obtain service. Interactive applications like WWW generate traffic that is somewhat elastic but users naturally prefer short response times. Hence, the service rate for these connections should approximately follow the demand. The large spectrum of traffic characteristics and QoS requirements has led to the definition of five ATM service categories (CBR, rt-VBR, nrt-VBR, ABR, UBR) by the ATM Forum (1996) which largely correspond to the transfer capabilities specified by the ITU (1996), and the discussion about new categories is still ongoing.

The key areas of traffic management are connection admission control (CAC), usage parameter control (UPC), and priority control in the switching hardware, all of which have been extensively studied since many years (Kröner, 1995). Quite often, however, only one particular aspect is investigated and described in detail, whereas other parts of the system are neglected or approximated by some simplifying assumptions. In this paper, we concentrate on the ensemble of traffic management functions of an ATM Multi-Service Switch (MSS) which supports all traffic categories defined by the ATM Forum.

The next chapter presents the switching architecture of the ATM MSS and the priority control functions implemented in the system. Section 3 focuses on the CAC function of the MSS which takes into account that multiple service categories have to be supported. Results of both cell level simulations of the system and analytical as well as simulative investigations of the CAC are presented in Section 4.

2   SWITCHING ARCHITECTURE

2.1   Overview

The ATM MSS uses a folded bus as its switching fabric to which different types of units are connected (Figure 1). The bus is used for transmitting both user data and switch-internal control information. The main call processing functions are located on a Control Unit (CU), whereas terminal equipment is connected to Port Units (PU). Two types of PUs are currently defined: the PU25 contains twelve 25 Mbit/s ports, the PU155 has four 155 Mbit/s ports.

An ATM MSS consists of up to 16 units which share the bus with a total capacity of 1.6 Gbit/s (switch internal cells have a length of 56 octets).

All units are connected to the bus through the Bus Access Controller (BAC) which is realised as an ASIC. The BACs together with the bus implement the ATM layer in the

MSS. Physical layer chips and AAL chips exchange cells with the BAC via a standard Utopia interface (ATM Forum, 1995). The total capacity of this interface is 400 Mbit/s.

The input part of the BAC contains a small input buffer for 20 cells as well as control logic for a much larger External Input Buffer (EIB). An output buffer holds approximately 100 cells after the transfer via the bus until they can be forwarded to their destination ports. Rate matching buffers with small capacity for every port complete the picture of the queueing model. The resulting model of the ATM MSS, as it has been used for simulations, can be found in Figure 2.

The ATM MSS belongs to the class of ATM switches with buffers on both the input and output sides. The bus which implements the switching fabric is a shared resource, which means that blocking may occur when it is occupied by another unit.

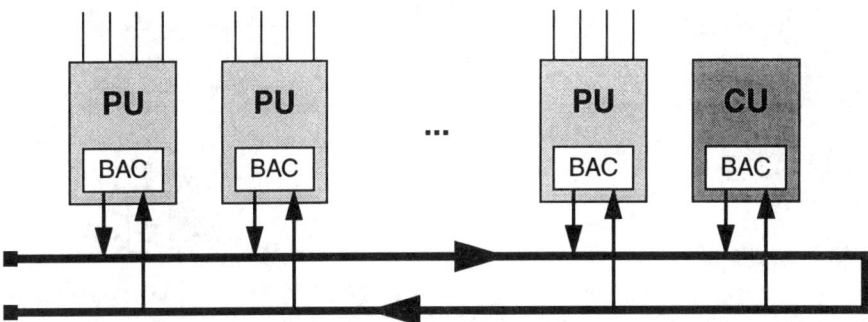

Figure 1 Internal switch architecture

2.2 Bus Access

Access to the bus is determined by a distributed bus access protocol which is implemented by the BAC. The bus consists of a data bus and a separate control bus used parallel to the data transmission in the current slot to determine the PU which will obtain the next slot. All BACs with a cell to send compete for the next slot by sending their unit identifier on the control bus. The PU with the dominant identifier receives back its own identifier and hence knows that it is allowed to use the next slot.

Fairness is ensured by a credit mechanism. Each PU uses one credit per transmitted cell. When the credits are exhausted, it may no longer compete for time slots. The unit continues to observe the control bus and regenerates a number of credits when it detects that no unit competes for the next slot which means that the other PUs either have no cells to send, or their credits are exhausted.

A compromise has to be found for the number of credits to regenerate. On one hand, the number of credits has to be small to limit the cell delay variation as well as to minimise the risk of cell loss. On the other hand, a larger number of credits reduces the overhead caused by the slot which is lost when all PUs regenerate credits. A good compromise is to generate in the order of 10 credits per PU for each cycle. Note that different credit values may be assigned to the PUs in order to prioritise some of them.

The BAC also implements a backpressure (BP) mechanism which avoids cell loss if the output buffer is occupied. In this case, the receiving BAC signals that it could not store the cell. The sending BAC then keeps the cell and schedules it for retransmission. Head-of-line blocking is avoided by a service discipline which is not strictly first-in, first out. Up to six cells at the head of the buffer can be considered for transmission on the bus, such that cells waiting behind blocked cells can be served in the meantime. Blocked cells are discarded after a number of failed transmission attempts.

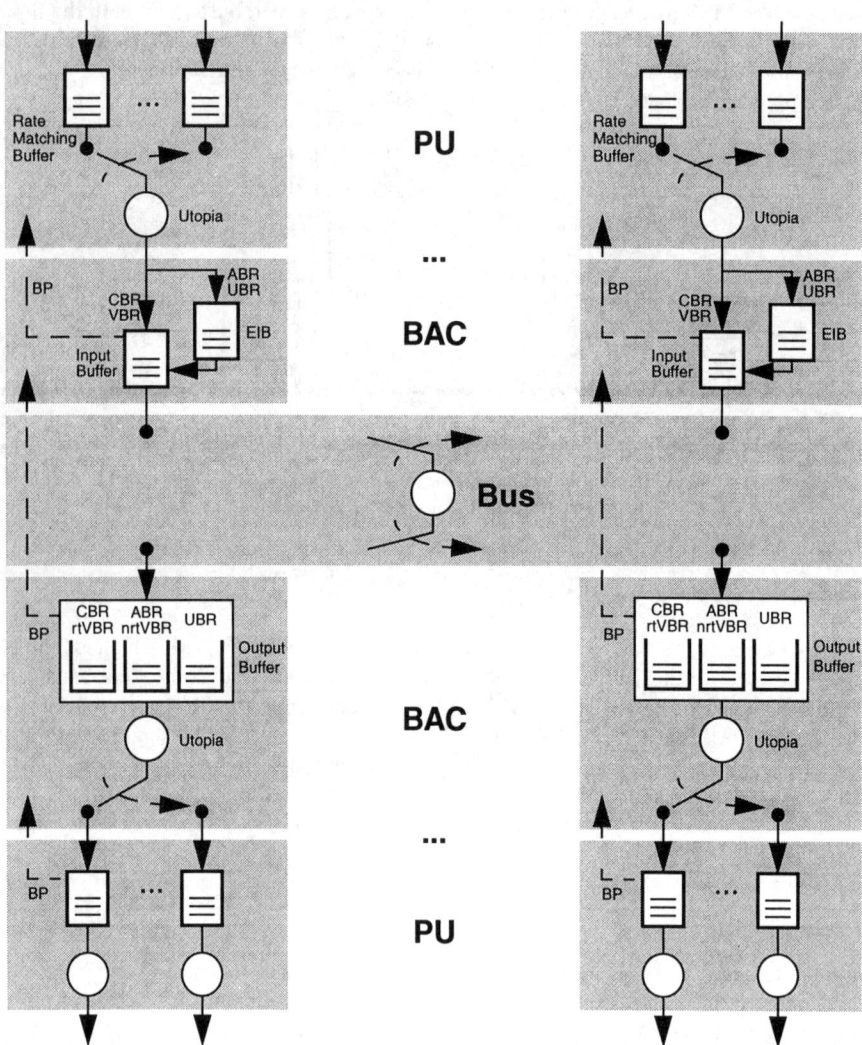

Figure 2 Queueing model of the switch

2.3 Priority Control

The purpose of priority control is to ensure that each service category obtains its required QoS. In the ATM MSS, priority control applies to both cell loss ratio and cell transfer delay.

On the input side of the BAC, UBR and ABR cells are immediately forwarded to the EIB which can store at least 2048 cells. The control logic for the EIB implements partial buffer sharing (Kröner et al., 1991) between ABR and UBR, such that UBR cells are discarded if the occupancy of the EIB becomes high while ABR cells can use the remaining buffer space. The control logic retrieves cells from the EIB only if the occupancy level of the input buffer drops below an adjustable threshold, hence VBR and CBR cells are prioritised.

The output buffer of the BAC implements three logical queues with statically assigned service priorities. CBR and rt-VBR cells are always served first in order to minimize their transit delay, then nrt-VBR and ABR cells, and finally UBR cells. Moreover, a pushout mechanism is implemented which discards a UBR cell if a cell having higher priority arrives from the bus. Thus the backpressure mechanism across the bus is activated only if the output buffer is completely occupied by VBR, CBR and ABR cells. The number of transmission attempts is configurable per service category, and the number is usually limited to one attempt for UBR cells because it doesn't make sense to keep UBR cells for retransmission if the load is already that high that the output buffer is completely occupied.

While the mechanisms described above are sufficient for prioritising VBR and CBR within a PU, they cannot effectively limit the amount of bandwidth on the bus that ABR and UBR traffic from other PUs can occupy. An enhancement of the bus access protocol ensures that the CAC can allocate a significant share of the bus capacity to VBR and CBR connections and that the capacity for this traffic can be guaranteed. Of course the administrator can limit the share of the bus capacity that the CAC allocates to VBR and CBR traffic.

The BAC implements two logical queues in the input buffer, one for VBR and CBR cells, the other for UBR and ABR. VBR and CBR cells always have higher priority, and the BAC competes for a slot if the credits are not exhausted. UBR and ABR cells are considered for transmission only if at least a certain number of credits are left, since part of the credits per cycle are reserved for the transmission of VBR or CBR cells. The number of credits per cycle that UBR or ABR cells may use can be limited to one by setting the limit to the same value as the number of credits to regenerate per cycle. In this case at least 80 percent of the bus capacity can be guaranteed for VBR and CBR traffic.

3 CONNECTION ADMISSION CONTROL

3.1 Requirements

Connection admission control (CAC) has to determine whether a new connection setup request can be accepted or should be rejected. This decision is based on the constraint to meet the negotiated QoS requirements of all existing connections as well as of the new connection. Another desirable goal is to allow for a statistical multiplexing gain, i.e. an efficient CAC method should accept as many connections as possible without violating any QoS guarantees.

CAC is performed as part of the connection setup process and generally runs as a software module on a switch control processor. Therefore, further evaluation criteria of CAC algorithms are processing time and memory requirements as well as the implementation and integration effort. As setup times and control processor load must be limited, time consuming and complex CAC algorithms are not appropriate for real systems.

The main focus of CAC is on guaranteed traffic, namely CBR and VBR connections and the minimum cell rate (MCR) share of ABR connections. Among that type of traffic statistical multiplexing is only possible for VBR connections. UBR connections are not necessarily controlled by CAC because they cannot influence the performance of other service categories due to priority-based buffer control mechanisms of the switch. The presented system, which is desgned for a workgroup environment, needs a CAC algorithm that is reasonably efficient but inexpensive with regard to processor resources.

Finally, a CAC algorithm has to be robust and should therefore not rely on a description given by the user but make a worst-case assumption according to the parameters negotiated during connection setup and monitored by the UPC function. Additionally the CAC may use measurements taken in the switch (e.g. for the actual mean rate of a connection) to improve effectiveness (Kröner et al., 1994; Antunes et al., 1997).

3.2   Classification and Characteristics of CAC Methods

Almost all CAC methods concentrate on the cell loss ratio (CLR) as the most crucial QoS parameter. Some algorithms also take into account delay requirements, e.g. the one presented by Antunes et al. (1997). However, this is not necessary here, as a limitation of transfer delay is guaranteed by the switch architecture with small buffers for real-time traffic and appropriate priority control mechanisms.

The CLR-based CAC methods make use of the results for the cell loss probability in an ATM multiplexer. Cell losses may occur due to congestion on cell level or on burst level. The cell level effects can be quite well described by an M/D/1-(S+1) model (Hui, 1988). The evaluation of this model shows that for a buffer size in the order of 50-100 cells the load should not exceed a value of about 85%. This result can be taken as a simple approximation to characterize the cell level effects.

For the characterization of the burst level effects two types of multiplexing strategies can be distinguished: rate envelope multiplexing (REM) considering the multiplexer to be bufferless and rate sharing where buffering is used to share the available rate among the connections (Roberts et al., 1996).

The switch described in Section 2 does not provide large buffers for VBR traffic which would be mandatory for queueing bursts. Therefore, in the following we will focus on the REM paradigm only. For this strategy the cell loss probability B is given by only considering the overload states where the aggregate rate $R(t) = R_1(t) + \ldots + R_N(t)$ of all N connections exceeds the service rate C:

$$B = \frac{E[(R(t) - C)^+]}{E[R(t)]}.$$ (1)

This leads to a convolution of the rate distributions of the single connections. We assume on-off traffic sources with parameters m_i (mean rate) and h_i (peak rate). In the homogeneous case where all sources have the same parameters m and h the convolution simplifies to a binomial distribution:

$$B = \frac{1}{m \cdot N} \cdot \sum_{i = \lfloor C/h \rfloor + 1}^{N} \binom{N}{i} \cdot \left(\frac{m}{h}\right)^i \cdot \left(1 - \frac{m}{h}\right)^{N-i} \cdot (i \cdot h - C). \qquad (2)$$

CAC methods which directly use the cell loss probability result have to evaluate the convolution or binomial distribution in real-time. Each time a new connection request occurs, the cell loss probability including the new connection must be computed and compared with the CLR objective. As this is a very time-consuming task for larger values of N, using effective bandwidths appears to be an appropriate method.

The idea is to calculate an effective bandwidth c_i for each connection, so that the CLR requirements are met as long as the following equation holds:

$$\sum_i c_i \leq C \qquad (3)$$

The difficulty here is, however, to find an appropriate expression for the effective bandwidth. In the case of homogeneous on-off sources a solution can be obtained by numerically inverting equation (2) to find the largest number of connections $\hat{N}$ maintaining a cell loss probability still below the CLR objective. The effective bandwidth is then given by $c = C/\hat{N}$.

Various approaches have been proposed that give an approximation of the effective bandwidth (Guérin, 1991; Kelly, 1991; Lindberger, 1991; Lindberger 1994, Sykas et al., 1992). Among them the solution presented by Lindberger (1994) which is also described in (Roberts et al., 1996) is promising as it yields a simple formula for computing the effective bandwidth based on the source parameters m and h, the service rate C and the CLR objective $\hat{B}$:

$$c = \begin{cases} a \cdot m \cdot (1 + 3 \cdot z \cdot (1 - m/h)), & z \leq \min(1, h/3m) \\ a \cdot m \cdot (1 + 3 \cdot z^2 \cdot (1 - m/h)), & 1 \leq z^2 \leq h/3m \\ a \cdot h, & \text{otherwise} \end{cases} \qquad (4)$$

with $a = 1 - \dfrac{\log \hat{B}}{50}$ and $z = \dfrac{-2 \cdot \log \hat{B}}{C/h}$.

In Figure 3 the results of this approximate formula are compared with those obtained by numerically inverting the binomial distribution. The approximation matches quite well at least in the region of large values of C/h. Figure 3 also reveals that the effective bandwidth approaches the mean rate with increasing service rate. The multiplexing gain which can be expressed by $G = h/c$ increases with the service rate C up to the burstiness h/m of the source.

While in the homogeneous case the maximum number of connections that can be accepted only depends on the source parameters and the service rate, it is also influenced by the actual traffic mix in the heterogeneous case. This can be described by a multi-dimensional acceptance boundary. The typical shape of such a curve is concave. That means the effective bandwidths are higher compared to the homogeneous case if connections of different types are mixed together.

The acceptance boundary may be linearly approximated by assuming that the effective bandwidths are independent of the traffic mix. For this approximation the acceptance region is specified by the condition

$$N_1 \cdot c_1(m_1, h_1, C, \hat{B}) + N_2 \cdot c_2(m_2, h_2, C, \hat{B}) \le C \tag{5}$$

where N_1 and N_2 are the number of type 1 and type 2 connections, respectively. The effective bandwidths c_1 and c_2 correspond to the values at the intersections with the axes. They can be computed either exactly or using an approximation, e.g. Lindberger's formula.

The linear approximation is quite good if the parameters of the source types do not differ too much. However, if for example CBR connections are mixed together with very bursty VBR connections the divergence of the acceptance boundaries is considerable (Figure 4). That means the linear approach underestimates the cell loss probability in a non-negligible way.

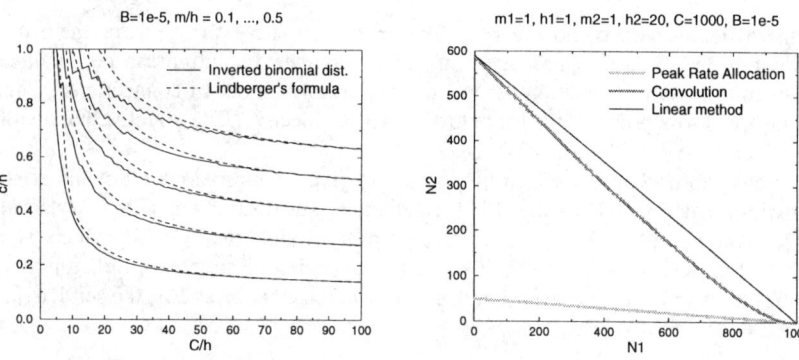

Figure 3 Effective bandwidth   **Figure 4** Boundary for two source types

3.3   Effective Bandwidths for Heterogeneous Traffic

As described in the previous section the linear CAC method is a simple and reasonably exact approximation as long as the multiplexed sources are not too different. Therefore the first idea is to separately handle CBR traffic (including also the MCR share of ABR connections), which causes most of the difficulties when put together with bursty VBR traffic, and to apply the linear method only to VBR connections.

For the computation of the effective bandwidth of a VBR connection with mean rate m_i and peak rate h_i a slightly modified version of Lindberger's formula is used where the effective bandwidth is limited to h_i. Instead of the full service rate C a reduced rate $C_{red} = C - C_{CBR}$ is given as parameter with C_{CBR} being the aggregate bandwidth of all CBR connections. Simply reducing the capacity available for VBR connections leads to an overestimation of the VBR loss probability as then the CBR traffic is assumed to be loss-free. This can be corrected by a modified CLR objective

$$\hat{B}^* = \hat{B} \cdot \frac{M_{VBR} + M_{CBR}}{M_{VBR}} \tag{6}$$

as parameter in the effective bandwidth formula. M_{CBR} and M_{VBR} are the aggregate mean rate of the CBR and VBR connections, respectively.

Partitioning the service rate into a VBR and a CBR share allows to handle the extremely concave behaviour in the acceptance boundary caused by CBR traffic. However, among the VBR connections there might still be differences which cause similar effects. Therefore, partitioning is extended to VBR connections by dividing the VBR source parameter space into K regions. An example with four VBR regions distinguished according to different values of burstiness and mean rate is depicted in Figure 5.

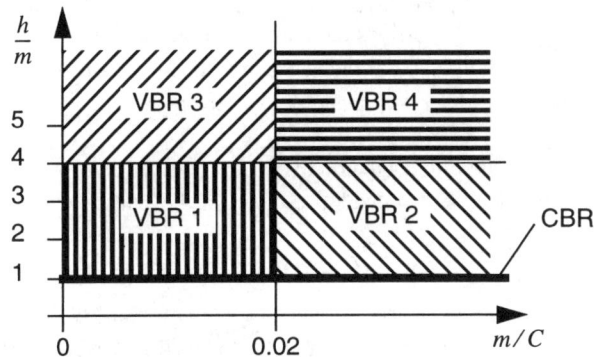

Figure 5 Example of VBR partitioning with four regions

The linear CAC method is now applied to connections of different types belonging to the same region. On the other hand, if two source types are located in different regions i and j, a reduced service rate $C_{red,i}$ is taken as a parameter for the effective bandwidth formula for connections in region i:

$$C_{red,i} = C - C_{CBR} - \sum_{\substack{k=1 \\ k \neq i}}^{K} C_{VBR,k}. \tag{7}$$

The service rate is reduced by the aggregate CBR bandwidth C_{CBR} and by the aggregate effective bandwidths of all VBR connections associated to regions different from i. A correction of the CLR objective corresponding to equation (6) is also performed:

$$\hat{B}_i^* = \hat{B} \cdot \frac{\sum_k M_{VBR,k} + M_{CBR}}{M_{VBR,i}}. \tag{8}$$

An alternative approach defined by Villen-Altamirano and Sanchez-Canabate (1997) is to subtract only the aggregate mean rates of the VBR connections in the other regions:

$$C_{red,i} = C - C_{CBR} - \sum_{\substack{k=1 \\ k \neq i}}^{K} M_{VBR,k}. \tag{9}$$

The partitioning of VBR sources into different regions depends among other parameters on the link rate and the application types using the system. Therefore it is a difficult task to find an optimal partitioning strategy. In general, it is important to keep the number of regions as small as possible to maintain efficiency on the one hand and to assign very different VBR sources to different regions on the other hand.

3.4 Proposed CAC Algorithm

If equation (7) is applied, variables C_{CBR}, M_{CBR}, $C_{VBR, k}$ and $M_{VBR, k}$ ($k = 1, ..., K$) have to be provided by the CAC. When a VBR connection request with parameters m_{new} and h_{new} arrives the following procedure is performed:

1. Determine the region k_{new} for the new connection by evaluating m_{new} and h_{new}.
2. Compute the reduced service rate $C_{red, k_{new}}$ for region k_{new} according to equation (7).
3. Calculate $\hat{B}^*_{k_{new}}$ applying equation (8).
4. Compute the effective bandwidth $c_{new} = c(m_{new}, h_{new}, C_{red, k_{new}}, \hat{B}^*_{k_{new}})$ using Lindberger's formula.
5. If $c_{new} + C_{CBR} + \sum_k C_{VBR, k} \leq C$ accept the connection, refuse it otherwise.
6. If the connection could be accepted, update $C_{VBR, k}$ and $M_{VBR, k}$ ($k = 1, ..., K$).

The parameters h_{new} and m_{new} of a VBR connection correspond to the negotiated traffic contract parameters PCR and SCR, respectively. This rather conservative approach has been chosen to guarantee robustness.

The update at the end of the acceptance procedure (6.) comprises the incrementation of $M_{VBR, k_{new}}$ and $C_{VBR, k_{new}}$ by m_{new} and c_{new}, respectively, but it is also necessary for the corresponding variables of the other regions. This is because $C_{red, k}$ and $\hat{B}^*_k$ change if a connection in any other region is added or removed. So the effective bandwidths of all existing connections have to be recalculated and summed up for each region to obtain $C_{VBR, k}$. This procedure has to be done iteratively until the variables $C_{VBR, k}$ converge to their exact values. In practice, however, it is sufficient to perform only a few iteration steps. If a CBR instead of a VBR connection is set up, an update in the same way is necessary.

If a connection with parameters m_{rem} and h_{rem} is removed, a procedure similar to the one described above has to be performed. The effective bandwidth for the removed connection is calculated and subtracted from the $C_{VBR, k_{rem}}$ value for the corresponding region. $M_{VBR, k_{rem}}$ is reduced by m_{rem}. Afterwards an update of all $C_{VBR, k}$ variables is performed in the same way as for a connection setup.

As the effective bandwidth can be computed very fast with Lindberger's formula, the update is not too time-consuming for a reasonable number of connections if an effective implementation is used. Nevertheless, there is a trade-off between necessary processing power and accuracy of the effective bandwidth calculation. It is justified to leave out updates if the traffic mix has not changed too much since the last update, i.e. if the $C_{VBR, k}$ show only moderate fluctuations. An appropriate solution is to carry out the update calculations not for every connection arrival or departure but in regular intervals, either timer-triggered (e.g. every 30 seconds) or on a counter basis (e.g. after every 10 setup or release requests).

3.5 Application to the Switch Architecture

The CAC algorithm presented in the previous section is applied to several locations in the switch. The local bus is shared by all connections. Here the largest multiplexing gain can be achieved due to the high bus service rate. Furthermore, CAC is necessary at the 25 Mbit/s and 155 Mbit/s output links. For each output link in the system a dedicated CAC entity using the presented algorithm has to be provided.

Additionally, an overload can occur at the Utopia interface on the input as well as on the output side of the switch if PUs with four 155 Mbit/s ports are used. However, here the maximum aggregate input rate exceeds the service rate of the Utopia interface only by a factor of about 1.6 so that statistical multiplexing is very limited. Therefore peak rate allocation is more appropriate than a more complex CAC algorithm at this location.

The CAC controlling the traffic over the bus must also take care of UBR traffic. As shown in Section 2 the influence of UBR traffic may be reduced by adapting the credit mechanism but it cannot be completely avoided. UBR cells consume a share of the bus bandwidth which heavily depends on the configuration and the traffic pattern. This effect is considered by a limit for the bus bandwidth that can be allocated by the CAC for CBR/VBR traffic. The limit depends on the number of units in the ATM MSS and the number of credits that are regenerated in each cycle. The limit can be further reduced if the operator wishes to reserve a larger proportion of the bus bandwidth for UBR/ABR traffic.

4 PERFORMANCE EVALUATION

4.1 Switch Architecture

The switch architecture presented in Section 2 has been extensively evaluated using a simulation program based on the queueing model depicted in Figure 2. Only some issues can be addressed in this paper. We will concentrate on effects related to the credit mechanism controlling bus access while investigations about buffer dimensioning and validating the buffer control strategies shall be omitted here.

As described in Section 2.3, the credit mechanism had to be changed to prioritise CBR/VBR cells at the bus and to reduce the influence of ABR/UBR traffic. If 10 is chosen to be the maximum number of credits allocated to each PU, on average 10 CBR/VBR cells per PU can be transmitted instead of one ABR/UBR cell. For a configuration with 14 PUs half of them sending VBR traffic contributing 50% to the total offered load while the other PUs only carry UBR traffic, the results depicted in Figure 6 have been obtained. The total load offered to the bus is increased while maintaining the proportion of VBR and UBR traffic unchanged. As soon as the offered load exceeds 1, the UBR share of the bus bandwidth decreases linearly until it reaches saturation at a value of $1/11$, which is directly determined by the number of credits each PU receives during the regeneration cycle.

Obviously, limiting the bandwidth that ABR/UBR traffic may occupy to an even lower value can be achieved by further increasing the maximum number of credits. However, large values of the credit per PU may potentially lead to unfairness. Assuming symmetric traffic from sources that are uniformly distributed over 14 PUs, we get the results shown in Figure 7. For large credit numbers one can see that PUs with lower

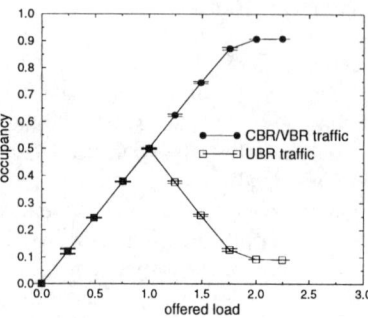

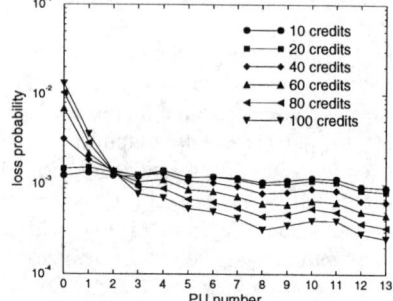

Figure 6 Influence of UBR traffic   **Figure 7** Influence of number of credits

identifiers suffer much more from losses. This effect is due to the fact that a credit regeneration in all PUs is performed if there is no BAC that has credits as well as cells in its input buffer. If the maximum credit value is greater than the size of the input buffer, PUs with lower identifiers cannot use up their credits because of the limited number of cells that they can accumulate until they receive the right to send. PUs with higher identifiers, however, can make better use of their credits because they can immediately send a newly arrived cell provided they still have credits. Thus, when credits are regenerated the PUs with higher numbers usually have exhausted their credits whereas the PUs with lower numbers still have credits but currently no cells to send. Hence the high priority PUs receive a larger share of the bus bandwidth which in fact reduces their cell losses. As a conclusion from that the number of credits is chosen to be in range of 10 to 20.

4.2   CAC Algorithm

The performance of the CAC algorithm presented in Section 3 heavily depends on the traffic pattern, the service rate, the CLR objective and the definition of the VBR regions. In the following we will concentrate on a traffic mix that consists of connections of two different source types:

- type 1 with parameters $m_1 = 1$ Mbit/s, $h_1 = 2$ Mbit/s and
- type 2 with parameters $m_2 = 1$ Mbit/s, $h_2 = 20$ Mbit/s.

The CLR objective is chosen to be $B = 10^{-5}$ for all case studies.

If the parameters of the source types belong to different VBR regions, partitioning is applied. The reduced capacity is calculated by either using effective bandwidths (equation (7)) or mean rates (equation (9)). In Figure 8 and Figure 9 the acceptance boundaries of both approaches are compared to those obtained by the convolution method and by peak rate allocation. For a service rate of $C = 1.37$ Gbit/s referring to the net transmission rate of the bus, the partitioning curves are rather closely tied to the ideal boundary. Here an enormous improvement compared to peak rate allocation can be achieved. For a smaller service rate of $C = 140$ Mbit/s that results as a net rate of a 155 Mbit/s output link the results are not that good. Partitioning yields acceptance boundaries that are still much better than peak rate allocation but quite far away from the ideal curve. However, this is not only the effect of partitioning but also of using Lindberger's effective

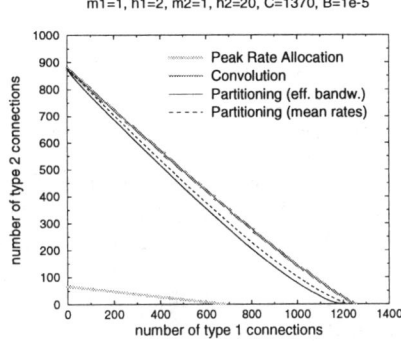

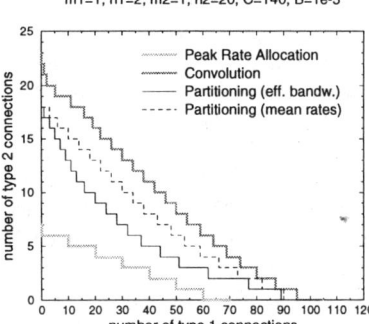

Figure 8 Boundary for C = 1.37 Gbit/s   **Figure 9** Boundary for C = 140 Mbit/s

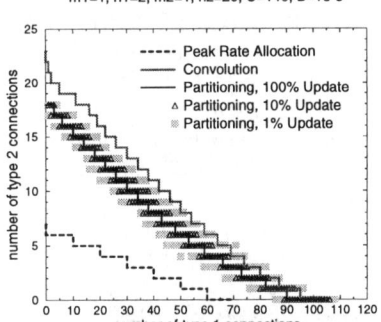

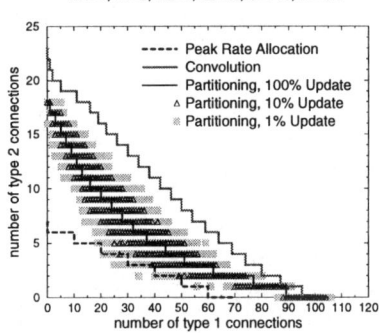

Figure 10 Update policies for partitioning   **Figure 11** Update policies for partitioning
using mean rates   using effective bandwidths

bandwidth approximation formula, as can be observed from the deviation at the axes intersections.

Comparing the two partitioning approaches one can see that the acceptance boundary for the method using aggregate mean rates for reducing the entire capacity is always closer to the ideal curve, especially for smaller service rates. However, this algorithm may be too optimistic if the ratio C/h is further reduced (Villen-Altamirano and Sanchez-Canabate, 1997). Furthermore, the effect of being quite conservative by using equation (7) may be compensated by the linear approach applied within the VBR regions.

As mentioned in Section 3.4, the update at the end of the acceptance procedure may be delayed in order to reduce the processor load caused by CAC. Two cases of performing an update only after every 10 and every 100 CAC procedure calls have been investigated. Thereby the processor load which is mainly due to the update is reduced to 10% and 1%, respectively. Simulation results in Figure 10 and Figure 11 show that the acceptance boundary curve changes to an area consisting of states in which a connection request either of a type 1 or a type 2 connection may be rejected. The rejection states have been collected from several simulations with different loads based on Markovian arrival and holding-time processes. The number of observed rejection states increases if the update rate is reduced. This is due to the fact that the effective bandwidth calculation for a new connection has to rely on possibly out-of-date parameters for the reduced bandwidth. Therefore the acceptance decision more and more depends on the random walk produced by the connection arrival and release processes.

The effect of "spreading" rejection states is much heavier for the approach using effective bandwidths for the reduced capacity calculation. An update rate of 1% would be not acceptable as it may be even less effective than peak rate allocation. For the method based on mean rates a very low update frequency may extend the acceptance region beyond the ideal curve leading to illegal states. An update frequency of 10% seems to be a good compromise for both CAC methods.

5   CONCLUSIONS

The design of traffic management functions for an ATM switch must precisely take into account the internal switch architecture as well as QoS requirements, utilization targets and implementation effort. Here an ATM Multi-Service Switch (MSS) for workgroups is considered which is built of several port units that are attached to an internal bus. In order to support all five ATM service categories, appropriate control functions are mandatory which guarantee QoS objectives and can be realized with acceptable implementation effort.

Priority control has to ensure that access to switch resources is given according to QoS requirements of different service categories. Especially the performance of the credit mechanism for the bus access has been investigated by simulation. The results confirm the ability of the protocol to limit the impact of ABR and UBR traffic on the guaranteed services as CBR and VBR to an acceptable level. Furthermore, it has been shown that fairness between different port units can be ensured by a proper setting of the number of credits regenerated per cycle.

While CAC for CBR services becomes trivial (peak rate allocation), for VBR type of traffic the effects of statistical multiplexing have to be taken into account. We found a simple formula for the calculation of an effective bandwidth for each VBR connection to be a good compromise between accuracy and implementation complexity. To overcome the typical problems when applying the concept of effective bandwidths in heterogeneous traffic environments, a classification into VBR traffic regions is proposed. This allows to consider the mutual influences between these regions by replacing the total link capacity with the capacity which is effectively available for each region. Obviously, this is a conservative approach but the performance study reveals that the deviation from the ideal boundary is acceptable. Finally, the MCR portion of an ABR connection is treated similarly to CBR, while UBR requests are never rejected.

The control framework comprising UPC, CAC and various priority control schemes has been tailored to the internal MSS architecture. Together with some other control functions which are not described in detail (e.g., backpressure mechanisms), the MSS is equipped with powerful traffic management capabilities allowing to effectively support a broad spectrum of applications and services.

REFERENCES

Antunes, N., Rocha, R., Pinto, P. (1997) "Analysis and Simulation of a Traffic Management Control Scheme for ATM Switches with Loose Commitments", *Conference on Communication Networks and Distributed Systems Modeling and Simulation CNDS '97*, Phoenix, Arizona.

ATM Forum - Technical Committee (1996) *Traffic Management Specification*, Version 4.0, af-tm-0056.000.

ATM Forum - Technical Committee (1995) *UTOPIA Level 2*, af-phy-0039.000.

Guérin, R., Ahmadi, H., Nagshineh, M. (1991) "Equivalent Capacity and Its Application to Bandwidth Allocation in High-Speed Networks", *IEEE Journal on Selected Areas in Communications*, Vol. 9, No. 7, pp. 968-981.

Hui, J.Y. (1988) "Resource allocation for broadband networks", *IEEE Journal on Selected Areas in Communications*, Vol. 6, No. 9, pp. 1598-1608.

ITU-T (1996) *Recommendation I.371: Traffic Control and Congestion Control in B-ISDN*, International Telecommunication Union - Telecommunication Standardization Sector, Geneva.

Kelly, F.P. (1991) "Effective bandwidths at multi-class queues", *Queueing Systems*, Vol. 9, pp. 5-15.

Kröner, H., Hébuterne, G., Boyer, P., Gravey, A. (1991) "Priority Management in ATM Switching Nodes", *IEEE Journal on Selected Areas in Communications*, Vol. 9, No. 3, pp. 418 - 427.

Kröner, H. (1995) *Verkehrssteuerung in ATM-Netzen - Verfahren und verkehrstheoretische Analysen zur Zellpriorisierung und Verbindungsannahme*, Ph. D. Thesis, University of Stuttgart.

Kröner, H., Renger, T., Knobling, R. (1994) "Performance Modelling of an adaptive CAC strategy for ATM networks", *Proceedings of the 14th ITC*, Antibes Juan-les-Pins, pp. 1077-1088.

Lindberger, K. (1991) "Analytical models for the traffical problems with statistical multiplexing in ATM networks", *Proceedings of the 13th ITC*, Copenhagen, pp. 807-813.

Lindberger, K. (1994) "Dimensioning and design methods for integrated ATM networks", *Proceedings of the 14th ITC*, Antibes Juan-les-Pins, pp. 897-906.

Roberts, J., Mocci, U., Virtamo J. (1996), *Broadband Network Teletraffic - Final Report of Action COST 242*, Springer, Berlin.

Sykas, E.D., Paschalidis, I.C., Mourtzinou, G.K., Vlakos, K.M. (1992) "Congestion Avoidance in ATM Networks", *Proceedings of IEEE INFOCOM '92*, Florence, pp. 904-914.

Villen-Altamirano, M., Sanchez-Canabate, M.F. (1997) "Effective Bandwidth dependent of the actual traffic mix: an approach for bufferless CAC", *Proceedings of the 15th ITC*, Washington D. C., pp. 47-57.

28

CAC investigation for video and data

E.Aarstad [a], S.Blaabjerg [b], F.Cerdan [c], S.Peeters [d] and K.Spaey [d]

[a] *Telenor Research & Development, P.O. Box 83, N-2007 Kjeller, Norway,egil.aarstad@fou.telenor.no*
[b] *Tele Denmark, Telegade 2, DK-2630 Taastrup, Denmark, sblb@dtk.dk*
[c] *Polytechnic University of Catalonia, Campus Nord. Modulo D6, E-08071 Barcelona, Spain, fernando@ac.upc.es*
[d] *University of Antwerp, Universiteitsplein 1, B-2610 Antwerp, Belgium, {speeters,spaey}@uia.ua.ac.be*

Abstract

A key objective of ATM-based networks is to provide at the same time guaranteed QoS to real time and non-real time services. This calls for thoroughly engineered traffic control methods for those service categories as well as for the overall integration strategy. The main objective of this paper is to investigate CAC in an ATM testbed with real switches and as realistic traffic as possible. The used traffic is video (MPEG model based on traces) and data modelled as traditional on/off sources. The video traffic is given priority over the data traffic. To complement and verify the experiments, two simulation tools using as input the artificial MPEG models and the real traces have been developed. The CAC boundary for the non-priority case has also been derived analytically. The experimental sources are modelled as discrete-time Markov sources. A matching method to avoid state space explosion for the superposition is applied on the source models.

Keywords

Connection Admission Control (CAC), delay priorities, multiplexing experiments

Broadband Communications   P. Kühn & R. Ulrich (Eds.)
© 1998 IFIP. Published by Chapman & Hall

1   INTRODUCTION

One of the most important objectives of ATM is the integration of different services into one multi-service network while providing differentiated Quality of Service (QoS). In order to meet this objective, an integrated traffic control framework which can give performance guarantees while ensuring efficient network utilization is required.

An important ATM traffic control function is Connection Admission Control (CAC). CAC determines whether a new connection can be accepted or not, depending on the availability of the necessary resources. Since the traffic types differ not only in their peak and mean rate but also in their burstiness, the CAC function has to be designed carefully to make correct bandwidth allocation decisions in order to avoid wasting bandwidth.

With the increasing interest in multimedia and the tremendous growth in the Internet traffic it is expected that a significant part of the future broadband traffic will consist of video as well as data traffic. Most literature on video and data so far considers statistical multiplexing (by simulation and analytically) of video and data separately, see e.g. (Maglaris, 1988), (Reininger, 1994), (Rose, 1997), (Roberts, 1996) and (Leland, 1994).

In this paper we investigate a CAC solution with delay priorities for integrating real time and non-real time Variable Bit Rate (rt-VBR and nrt-VBR) services. We present results from experiments performed within the framework of the ACTS (Advanced Communications Technologies and Services) EXPERT project at the EXPERT Testbed in Basle, Switzerland.

Section 2 presents an artificial MPEG traffic model and the on/off data model, a description of the configuration and the obtained CAC boundaries. Additional to the experiments a simulator has been developed. The results of the simulations and a comparison with the experimental results is also given in section 2. Section 3 presents a tool which enables a study of multiplexing the real traces and on/off data sources. In section 4, a CAC boundary for the non-priority case is analytically derived by describing the sources used in the experiments as Discrete-time Batch Markovian Arrival Processes (D-BMAPs) and approximating their superposition by a new D-BMAP.

2   EXPERIMENTAL CAC BOUNDARIES

2.1 Modelling video and data

In order to reduce the required bandwidth to acceptable levels, several coding schemes for the compression of video streams have been developed, with MPEG as the most promising scheme. It comes in two versions, MPEG-I (see (Le Gall, 1991) and (ISO/IEC, 1993a)) and MPEG-II (see (ISO/IEC, 1993b)) where the MPEG-I functionality is a subset of that of MPEG-II. In this paper it is MPEG-I which is dealt with. MPEG encoding of a video sequence is realised using three different

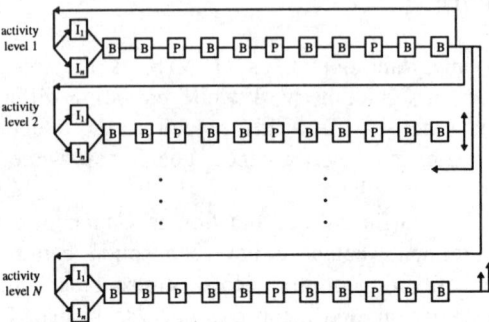

Figure 1: Structure of the artificial MPEG source

compression levels. Thereby three different frame types, I-, P- and B, are generated. A periodic generation of a pattern of these frame types is the result.

From a traffic modelling point of view, MPEG has traditionally either been modelled as Markovian (Rose, 1997), autoregressive (Ramamurthy, 1992) or more recently fractal (Beran, 1995). Since the traffic generators available at the EXPERT platform implement a Markovian model at the burst level, we have chosen to follow the Markovian approach of B. Helvik (Helvic, 1993). No intention to capture long range correlation is made, mainly due to our multiplexing scenarios where only short or moderate sized buffers are used. The periodic generation is compromised in the sense that the duration in individual states is assumed exponential instead of constant (a requirement if more than one MPEG source on each Synthesized Traffic Generator (STG) is to be generated). The sum for B- and P-frames is approximated by using four or five load levels and the I-frame size distribution is further refined by defining two sublevels within each load level (see Figure 1). The duration of each frame is 45 msec. Parameters in the model are the transition probabilities between the load levels as well as the I, B and P rate at each load level. A method developed by B. Helvik (and quite similar to the simple Markov method presented in the video modelling section in part I of (Roberts, 1996)) to generate these parameters from an arbitrary MPEG video trace has been used. For more details we refer to (Helvic, 1996). Using artificial MPEG sources enables us to investigate traffic scenarios with more sources than would have practically possible using only real MPEG sources and makes our results repeatable.

The two MPEG sources used are based on 24 minutes traces from the Bond movie "Goldfinger" and from an Asterix cartoon, which both have been made available by O. Rose at the Institute of Computer Science, University of Wuerzburg. The source for the Bond movie is implemented as a 65 state model and has a mean rate of 0.59 Mbps, while the Asterix cartoon has 52 states and a mean rate of 0.51 Mbps. Thereby we are able to multiplex both MPEG models in one STG without exceeding the upper limits for what the equipment can handle.

The on/off data source used in the experiments has a peak rate of 7.78 Mbps, a mean rate of 3.89 Mbps and a mean on- and off duration of 50 msec. This source represents a high bandwidth source which is not so bursty. The duration in both on- and off states is exponentially distributed and mutually independent (traditional 2-state Markov source).

2.2 Experimentally derived boundaries

2.2.1 Experimental configuration

Figure 2 shows the experimental configuration involving a Cisco LS1010 ATM switch and a test instrument called ATM-100 which gives the possibility to generate and analyse quite general random traffic. The ATM-100 is equipped with two STGs (Helvic, 1993) which generate the artificial MPEG and the on/off data traffic. Due to hardware constraints in the traffic generators a pacing rate function has been used to limit the output port capacity to 37.44 Mbps, thereby reducing the number of sources required to adequately load the system. The traffic is multiplexed in an output port of the Cisco switch and the resulting traffic stream is analysed in the ATM-100, permitting cell loss and delay measurements.

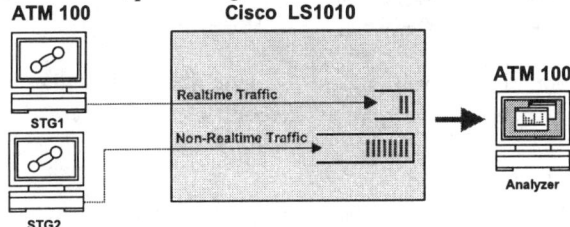

Figure 2: The experimental configuration

The buffer architecture of the Cisco switch implements delay priorities to protect the real-time (rt) traffic from influence of the non-real time (nrt) traffic. A short buffer has been used for the rt-traffic. For the nrt-traffic, which can tolerate longer delays, a longer buffer has been used. The service discipline is such that the rt-traffic is served as long as there are cells in the short buffer. Only when the short buffer is empty, the nrt-traffic will be served.

2.2.2 Measurement results

In the first two sets of experiments a number of MPEG sources (Bond and Asterix) and on/off sources has been multiplexed in a large FIFO buffer of 1260 cells and a small one of 256 cells. In the third set of experiments the MPEG traffic has been multiplexed in a small high priority buffer (256 cells) while the on/off traffic is multiplexed in a larger low priority buffer (1260 cells). The number of sources has been varied to obtain a cell loss ratio (CLR) below, but as close to 10^{-4} as possible. By changing the traffic mix it has been possible to obtain the CAC admission boundary for each of the three cases.

Figures 3 and 4 give the 2-dimensional CAC boundaries for Bond + on/off and Asterix + on/off. Experiments with a mix of the three types of sources have also been performed. From these measurements some points on the plane (the CAC boundary of MPEG-Bond + MPEG-Asterix + on/off) have been derived. Figure 5 gives the three-dimensional admission boundary for the non-priority case with a small buffer. The three-dimensional CAC boundary when delay priorities are used, can be found in Figure 6.

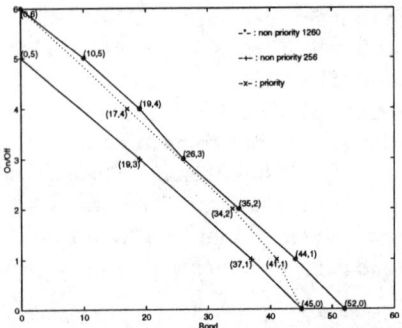

Figure 3: CAC boundaries for Bond + on/off.

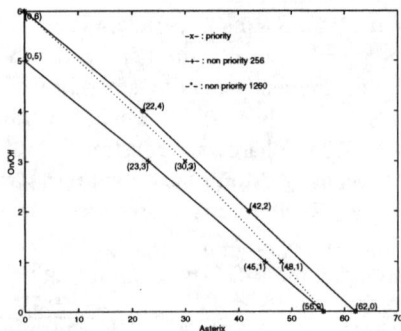

Figure 4: CAC boundaries for Asterix + on/off.

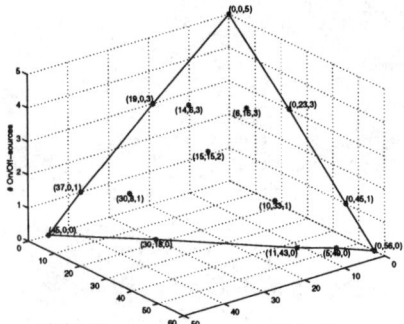

Figure 5: CAC boundary for the non-priority case with buffer = 256 cells.

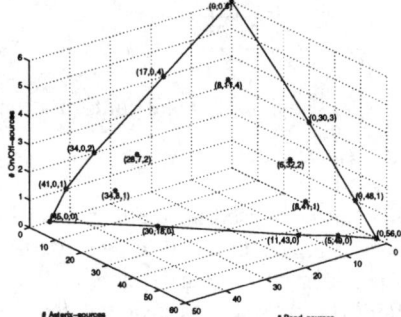

Figure 6: CAC boundary for the delay priority case.

In the non-priority case, the two dimensional boundaries are close to linear. The small deviations from linear are due to general uncertainties which exist in any measurement. This linearity supports the feasibility of the concept of effective bandwidth (Kelly, 1991). As expected, we can allow more MPEG sources in the delay priority case than when we use a small common buffer. Almost the same admission boundaries as in the non-priority case with large buffer come out, except that the number of MPEG sources that can be supported is less. The reason is that in the delay priority case the multiplexing of MPEG takes place in a buffer of size 256 while in the non-priority case the size of the buffer is almost 5 times larger and when no or only a small amount of on/off traffic is present this large pool can be used by the MPEG traffic. This of course at the expense of larger delay and larger variations in the delay.

2.2.3 Simulation results

Some aspects related with the STG have to be taken into account before comparing experimental and simulation results. The STGs have some hardware limitations which can make the implemented Markov models slightly different from the exact Markov models:

- only transition probability values in integer multiples of 1/256 are possible
- the peak rate must divide the link rate such that the interarrival time between cells in a given state is always the same integer number of slots.

The simulation tool used can provide any transition probability value as well as any peak bit rate over a period of observation. Depending on the number of sources generated, these slight differences may become important. They give rise to differences in the mean bit rate (in Mbps) for the MPEG sources (with a 95% confidence level measured on 30 independent sources): Bond: 0.59221 (STG) vs. 0.59764±0.003 (simulator) and Asterix: 0.51318 (STG) vs. 0.54943±0.002 (simulator). Thus, the exact model for Asterix sources generates 0.03625 Mbps more than the implemented model, i.e. for a certain experimental point the CLR obtained by simulation can be worse depending on the number of Asterix sources used, since in a congestion situation small increases in load can lead to big differences in CLR. Nevertheless, in those points where Bond sources have more influence, negligible differences between experimental and simulated CLR should be seen. For all cases presented in the Tables 1, 2 and 3 a 90 % confidence interval was used.

Table 1: Non-priority case with 256 cells buffer size

Bond	Asterix	On/off	Experimental CLR [x 10^4]	Simulated CLR [x 10^4]
14	6	3	0.5024	0.8295±0.0891
30	8	1	0.5637	0.8172±0.1503
45	0	0	0.5001	0.5747±1.1058
19	0	3	0.6423	0.7780±0.1035
0	23	3	0.3759	1.0975±0.1020
30	18	0	0.6477	0.7412±0.1608

Table 2: Non-priority case with 1260 cells buffer size

Bond	Asterix	On/off	Experimental CLR [x 10^4]	Simulated CLR [x 10^4]
0	0	6	0.3798	0.3532±0.1416
10	0	5	0.9742	0.9032±0.2498
52	0	0	0.3090	0.8954±0.0913
0	22	4	0.6150	2.2484±0.3441

Table 3: Priority case 256/1260 cells buffer size

Sources			Experimental CLR [x 10⁻⁴]		Simulated CLR [x 10⁻⁴]	
Bond	Asterix	On/off	Bond-Asterix	On/off	Bond-Asterix	On/off
0	30	3	0.0	0.6831	0.0	1.9±0.12
17	0	4	0.0	0.3230	0.0	0.72±0.19
8	11	4	0.0	0.8740	0.0	0.91±0.05
34	8	1	0.0134	0.8700	0.0225±0.0109	1.11±1.01

3   REAL MPEG TRACES AS INPUT TRAFFIC

To evaluate the accuracy of the artificial MPEG sources used in the experiments, a multiplexing scenario has been simulated where traces of the real MPEG sources are mixed with on/off traffic. In order to obtain a sufficient number of MPEG sources a video sequence was chopped into a number of equally sized subsequences implying that it is possible to multiplex a high number of MPEG sources on the expense of a shorter simulation time. To avoid dependence between consecutive pieces of the chopped video traces there is a limit on the number of pieces a video sequence can be chopped into. We have considered one minute as a desirable minimum and 20 seconds as an absolute minimum. Due to the periodic nature of the MPEG coding, multiplexing experiments are highly dependent on how the actual phasing between the MPEG sources is. To obtain random phasing between the MPEG sources, a relatively large number of simulation runs needs to be performed.

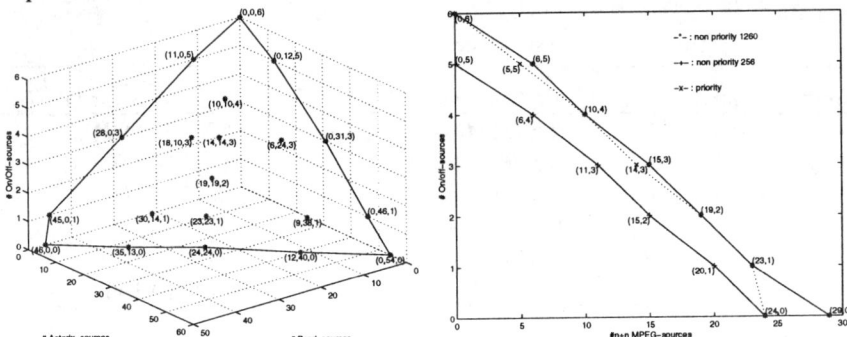

Figure 7: CAC boundary for the delay priority case

Figure 8: CAC boundaries for MPEG + on/off

Figure 7 shows the three dimensional admission boundary with a target CLR of 10^{-4} for the delay priority case. The results don't differ in a significant way from the experimental results which were based on the artificial MPEG sources. With a single exception, there is a good agreement between the CLR obtained in the trace driven simulations and the experimental results. Confidence intervals are quite big which is due to the tremendous influence the phasing of the individual MPEG sources has on the multiplexing buffer. Figure 8 shows a two dimensional admission boundary where we have constrained the number of Asterix and Bond sources to be equal. The boundaries have small discrepancies from linear which is due to the large confidence intervals in the simulations. However, the results do indicate that the admission set in the delay priority case , as a first approximation, can be determined as the intersection of the admission boundary in the non-priority case with large buffer and both types of traffic and the set determined by the high priority traffic multiplexed in the small buffer alone.

4   ANALYTICALLY DERIVED BOUNDARIES

A 10^{-4} CAC boundary is derived analytically for the non-priority case with small buffer. By a straightforward discretisation of time, the Bond and Asterix STG sources are modelled as D-BMAPs (discrete-time batch Markovian arrival processes). To handle the multiplexing without a state space explosion, the circulant matching method of Hwang and Li (Hwang, 1995) is adapted for D-BMAPs by which an accurate approximation of the superposition process without too many states is made possible. The determination of cell loss is then reduced to the solution of a D-BMAP/D/1/K+1 queue.

4.1   The discrete-time batch Markovian arrival process

A discrete-time batch Markovian arrival process (D-BMAP) $(\mathbf{D})_{n\geq 0}$ is defined as follows: the matrix $\mathbf{D} = \sum_{n=0}^{\infty} \mathbf{D}_n$ is the transition matrix of a discrete-time finite Markov chain. Suppose that at time k this chain is in some state i. At the next time instant $k+1$, there occurs a transition to another state j, with probability $(\mathbf{D})_{i,j}$ at which a batch arrival may or may not occur. The matrix $\mathbf{D}_0$ governs transitions that correspond to no arrivals while the matrices $\mathbf{D}_n$, $n\geq 1$, govern transitions that correspond to arrivals of batches of size n. For more details and properties, see (Blondia, 1993).

4.2   A method to approximate the superposition of D-BMAPs

Since the exact analytical description of the superposition of D-BMAPs, which is again a D-BMAP, has a state space size which is non tractable from a numerical point of view, it is necessary to replace the exact arrival process by another, but simpler process, which matches the exact one as close as possible for some important statistical functions. In (Hwang, 1995), it is proposed to replace the superposition of Markov modulated Poisson processes (MMPP) by a MMPP with a special structure, called circulant modulated Poisson process (CMPP). This process has a completely different Markovian structure as the superposition, but approximates two important statistical functions of the input rate process: the cumulative distribution representing the stationary statistics and the autocorrelation function in the time domain or equivalently the power spectral function in the frequency domain representing the second order statistics. Since our interest is in the superposition of D-BMAPs, the method is adapted for it and a circulant D-BMAP is used to replace the superposition. A circulant D-BMAP $(\mathbf{Q}_k)_{k\geq 0}$ of dimension N is defined by $(\mathbf{Q}_k)_{i,j}=a_{(j-i)\bmod N}\dfrac{(\gamma_i)^k e^{-\gamma_i}}{k!}$. The circulant D-BMAP is thus completely determined by $\mathbf{a}$, the first row of $\mathbf{Q}$ and a rate vector γ.

The different steps in the procedure to obtain a circulant D-BMAP $(\mathbf{Q},\gamma)$ as an approximation for the superposition of M independent D-BMAPs $(\mathbf{D}_k^{(i)})_{k\geq 0}$, and the extension of this method for periodic transition matrices are described in (Spaey, 1997). The steps are based on the analogous steps in the matching procedure described in (Hwang, 1995).

4.3 Modelling of the MPEG and on/off sources as D-BMAPs

The method developed by Helvik (Helvic, 1996) to generate the MPEG traffic source model used in the experiments gives us the following data as output:

- a periodic transition matrix which describes the transition probabilities between the different states of the model. The state sojourn time is 45 msec for all states (a frame duration)
- a table with a load (expressed in bits per frame) for each state which is generated while being in that state

A D-BMAP $(\mathbf{D})_{n\geq 0}$ can be gathered from this data by using the given transition matrix as the transition matrix $\mathbf{D}$ for the D-BMAP, and by transforming the loads in the load table into number of cells per frame. If this gives x cells per frame for state i, define then $\forall j: (\mathbf{D}_x)_{i,j}:=(\mathbf{D})_{i,j}, (\mathbf{D}_y)_{i,j}:=0 \ \forall y\neq x$. Application of the method for the superposition of D-BMAPs on the Bond, Asterix or a combination of those sources, gives a new D-BMAP $(\mathbf{Q},\gamma)$ which is an approximation for the superposition. The underlying time unit is still a frame length of 45 msec. Since this D-BMAP has to be used as input for a single server queue with a constant service time which equals the time needed to put one cell onto the outgoing link (=1 slot), $(\mathbf{Q},\gamma)$ is transformed into a D-BMAP $(\hat{\mathbf{Q}},\hat{\gamma})$ with one slot as the underlying time unit, supposing that the number of slots for being in a state is geometrically distributed with mean x, where x is the number of slots in a frame duration.

The on/off sources used in the experiments have the following characteristics:

- in the on state, data is generated at 7.78 Mbps,
- the mean sojourntime in on and off state is 50 msec

A D-MAP for this kind of sources is generated by supposing again that the number of slots for being in a state is geometrically distributed, with mean x, where x is now the number of slots in 50 msec and by asuming that while the source is in the on state, the probability of generating a cell in a slot is p. To describe the superposition of M independent on/off sources, the transition matrix **S** of the number of sources which are in the on state is composed:

$$(\mathbf{S})_{i,j} = \sum_{k=\max\{0,i+j-M\}}^{\min\{i,j\}} (1-\alpha)^{M-i-j+2k} \alpha^{i+j-2k} \binom{i}{k}\binom{M-i}{j-k} \quad \text{with} \quad \alpha = \frac{1}{x}$$

Since a source which is in the on state generates a cell in a slot with probability p, k sources in the on state generate t ($t \in \{0,...,k\}$) cells in a slot with probability $\binom{k}{t} p^t (1-p)^{k-t}$. From this, the matrices $\mathbf{S_0},...,\mathbf{S_M}$ can be composed.

For the superposition of a number of MPEG sources based on the Helvik model and some on/off sources, the new D-BMAP $(\mathbf{T_n})_{n \geq 0}$ is composed as the superposition of $(\hat{\mathbf{Q}}_n)_{n \geq 0}$ and $(\mathbf{S_n})_{n \in \{0,...,M\}}$.

4.4 The queueing model and numerical results

The multiplexer whose input consists of the superposition of Bond, and/or Asterix and/or on/off sources is modelled as a discrete-time D-BMAP/D/1/K+1 queue. The input D-BMAPs are obtained by the approximation method, K+1 is the systemlength and the servicetime equals one timeslot. In (Blondia,1992), the D-BMAP/D/1/K+1 queue was solved analytically and these results are applied to obtain the CLR.

All the results are obtained for a queue length of 100 cells and an outgoing link of 37.44 Mbps. This implies that one time slot equals 11.325 µsec. By transforming bits to number of cells, it is assumed that a cell of 53 bytes can contain 48 bytes of data. Figure 9 shows the analytical derived 10^{-4} CAC boundary for the non-priority case. As in the experiments, for a mix of two types of sources, the admission boundaries are close to linear. If the obtained points are compared with experimental points (see Table 4), which is not possible for all points since only a few experiments are performed with a buffer size of 100 cells, it is seen that all the analytical results are more conservative than the experimental results.

Table 4: Comparison between analytically and experimentally obtained results

Experimental points	(44,0)	(30,17)	(20,29)	(10,42)	(0,55)
Analytical points	(43,0)	(30,14)	(20,26)	(10,38)	(0,51)

Although the analytical obtained results are compared here with experimental CAC results, the matching method is basically a method used for avoiding the state space explosion when multiplexing D-BMAPs. Since the CAC experiments are based on multiplexing, these experiments were used to test the behaviour of the analytical approximation method for multiplexing D-BMAPs, with the cell loss ratio as used measure of performance.

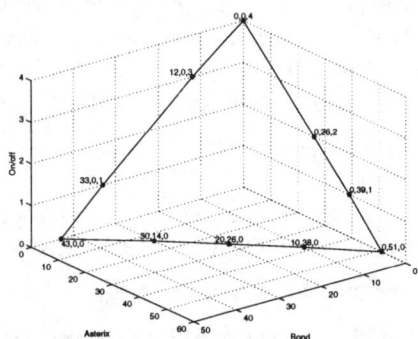

Figure 9: Analytically derived CAC boundary for a non-priority buffer of 100 cells

5   CONCLUSIONS

Extensive multiplexing experiments using model-based traffic generators have been carried out at the EXPERT ATM testbed in Basel to obtain CAC boundaries for traffic scenarios involving MPEG video and data traffic. Both FIFO as well as strict delay priority scheduling schemes have been studied, and the experimental results have been compared with simulation results and results based on an analytical method.

The flexibility of software simulations has been exploited to investigate any possible measurement error introduced due to differences between the derived MPEG models and MPEG models as implemented in the traffic generators. The simulation results support the validity of the traffic source implementation.

Trace driven simulations have been performed to compare the multiplexing behaviour of the MPEG sources with the video traces on which the artificial MPEG sources are based. Although this procedure is associated with some difficulties due to the dependency on phasing between individual traces, some evidence is given to the validity of the chosen MPEG modelling approach.

Furthermore, the non-priority case with a small buffer has been analysed using analytically derived boundaries based on D-BMAP models of the traffic sources. The matching method based on circulants has been applied to approximate the superposition of D-BMAP models by another D-BMAP model with much smaller state space thus making a queueing analysis feasible.

By combining results from hardware experiments, software simulations and analytical methods we have demonstrated a coherent approach for investigating multiplexed video and data traffic. Application of these methods can be used for developing efficient network resource management schemes.

Acknowledgements: The achievements being made within the ACTS project EXPERT are only possible with the conscientious co-operation of all partners and the support of the Commission of the European Union. The contribution of all these players is therefore hereby gratefully acknowledged. We also acknowledge O. Rose of the University of Wuerzburg for providing us the video traces, as well as H. Christiansen of the Technical University of Denmark for his trace driven program with which the results of section 3 have been derived.

6. REFERENCES

Beran, J., Sherman, R., Taqqua, M. and Willinger, W. (1995) Variable bit rate video traffic and long range dependence, *IEEE Trans. on Comm.,* **43**, 1566-79.

Blondia, C. and Casals, O. (1992) Statistical multiplexing of VBR sources: A matrix analytical approach, *Performance Evaluation*, **16**, 5-20.

Blondia, C. (1993) A discrete-time batch Markovian arrival process as B-ISDN traffic model, *Belgian Journal of Operations Research, Statistics and Computer Science*, **32**, 3-23.

Draft International Standard ISO/IEC DIS 13818-2 (1993a) Generic coding of moving pictures and associated audio Part 2 Video.

Helvic, B., Melteig, O. and Morland, L. (1993) The synthesized traffic generator: objectives, design and capabilities, *Proceedings of the IBCN&S.*

Helvik, B. (1996) MPEG source type models for the STG (Synthesized Traffic Generator), *SINTEF report STF40 A96016.*

Hwang, C.L. and Li, S.Q. (1995) On the convergence of traffic measurement and queueing analysis: A Statistical MAtch Queueing (SMAQ) tool, *Proc. of IEEE Infocom 95,* 602-12.

International Standard ISO/IEC DIS 13818-2 (1993b) Coding of moving pictures and associated audio for digital storage media up to 1.5 Mbit/s Part 2 Video.

Kelly, F.P. (1991) Effective bandwidth at multi-class queues, *Queueing Systems,* **9**, 5-16.

Le Gall, D. (1991) MPEG: A video compression scheme standard for multimedia applications, *Comm. ACM*, **34(4)**, 46-58.

Leland, W.E., Taqqu, M.S., Willinger, W. and Wilson, D.V. (1994) On the self-similar nature of Ethernet traffic, *IEEE/ACM Trans. on Networking*, **2**, 1-15.

Maglaris, B., Anastassiou D., Sen, P., Karlsson, G. and Robbins, J.D. (1988) Performance models of statistical multiplexing in packet video multiplexers. *IEEE Trans. on Comm,* **36**, 834-44.

Ramamurthy, G. and Sengupta, B. (1992) Modelling and analysis of a variable bit rate video multiplexer, *IEEE Infocom 92.*

Reininger, D., Melamed, B. and Raychaudhuri, D. (1994) Variable bit rate MPEG video: characteristics, modelling and multiplexing. *ITC-14*, 295-306.

Roberts, J., Mocci, U. and Virtamo, J. (ed.) (1996) Broadband network traffic, Final Report of Action Cost 242, Springer Verlag.

Rose, O. (1997) Discrete time analysis of finite buffer with VBR MPEG video traffic input, *ITC-15*, 413-22.

Spaey, K. and Blondia, C. (1997) Circulant matching method for multiplexing ATM traffic applied to video sources, submitted for publication.

Congestion Control

29

The MAPS control paradigm: using chaotic maps to control telecoms networks

L.G. Samuel, J.M. Pitts, R.J. Mondragón, D.K. Arrowsmith
Queen Mary and Westfield College
Mile End Road, London E1 4NS, United Kingdom, e-mail:
{L.Samuel, J.M.Pitts, R.J.Mondragon, D.K.Arrowsmith}
@qmw.ac.uk
Tel: +44-171-415-3756   Fax: +44-181-981-0259

Abstract

This paper proposes a new control method for telecommunications networks based on chaotic dynamics. Self-similar behaviour has been observed in a variety of services and networks. ATM is a unifying vehicle for transporting a wide range of traffic streams, and hence is a focus for assessing the impact of self similarity on networks. Chaotic maps have been used to model self-similar traffic. Using this approach, we present results that show the sensitivity of the Hurst parameter to the dynamical parameters. Conventional control techniques for ATM each address only a single time scale, and hence are less effective as a result of self-similarity in the traffic behaviour. We set out the MAPS control paradigm, using local control techniques in coupled map lattices, which aims to effect control over the range of relevant timescales.

Keywords
Self-similar traffic, ATM networks, chaotic map, chaotic control

Broadband Communications   P. Kühn & R. Ulrich (Eds.)
© 1998 IFIP. Published by Chapman & Hall

1   INTRODUCTION

ATM is becoming the transport vehicle for a wide variety of traffic streams, whether it is legacy traffic, LAN-LAN, Internet IP, multimedia, etc. ATM was created as a unifying transport mechanism. The mechanism provides the means to statistically multiplex variable and constant bit rate streams. One of the main features of ATM is its statistical multiplexing gain (Saito, 1994). Statistical multiplexing gain arrives out of multiplexing traffic streams where the sum of the individual peak bandwidths is greater than the capacity of a given link (Chen, 1995). This is possible because the peaks in the individual traffic streams seldom occur together. Therefore the statistical multiplexing effect relies on the condition that enough sources are multiplexed and that they are not correlated (de Prycker, 1991). The analysis of such a gain has been attempted under the assumptions of Poisson arrival processes and exponential distributed holding times (Saito, 1994). The implication of this type of analysis is that such traffic streams when aggregated tend to white Gaussian noise, i.e. the variation of the traffic would eventually smooth out (see Figure.1 after Figure 4 in Leland 1994). However, traffic measurements carried out in the late 1980's and early 1990's revealed that whereas the correlations of the traffic were thought to decay exponentially fast (Markovian in structure) the traffic measured in real networks possessed correlation structures which decayed much slower than exponentially (Fowler, 1991). This type of traffic has become known as Long Range Dependent (LRD).

Why should this be a problem? The answer lies in the correlation structure of the LRD traffic. Heuristically, one can view the individual traffic streams' correlation as overhanging each other when aggregated, causing an increase in the probability of the large aggregated bursts occurring. More importantly the aggregated traffic streams do not tend to white Gaussian noise (for a representation of these effects see Figure 1). In actual fact the aggregated traffic process tends towards a second-order statistically self-similar process which remains bursty over many time scales (Fowler, 1991). This feature of the traffic poses problems for the traffic control schemes designed for ATM. In ATM preventative congestion control is preferred over reactive congestion control schemes. This is because the reactive control becomes inadequate in terms of response times for the high bit rates used in ATM (Chen, 1995). The preventative measures are concentrated in the connection admission control schemes (CAC) used to make decisions on the acceptance of calls into the system. It now appears that CAC cannot minimise the congestion within the network and increasing buffer sizes also appear to have no effect (Leland, 1994). For this reason the time has arrived to consider control schemes which are based on totally different paradigms. In this paper we present such a paradigm based on chaotic dynamics.

The organisation of the paper is as follows: in section 2 we give a brief introduction on the nature and effects of self-similar traffic; in section 3 we outline how dynamical systems can be used to model teletraffic; and in section 4 we describe the principles of chaotic control.

Real Traffic   Markov Models

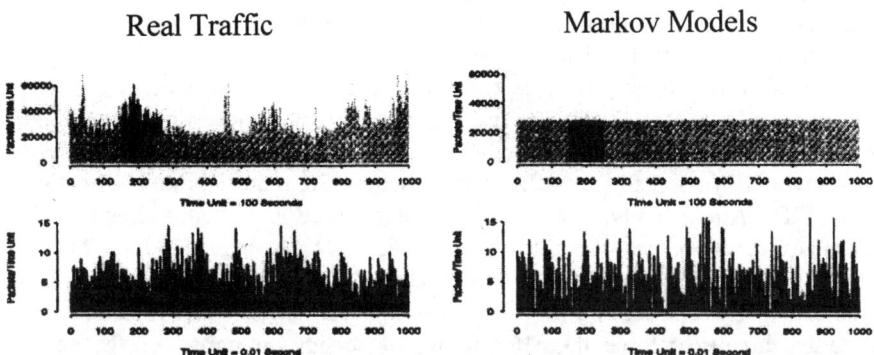

Figure 1 Real traffic trace against Markov model based trace for the same load. *(The picture is taken from figure 4 in* Leland (1994), *Reproduced with permission).*

2. THE IMPACT OF SELF-SIMILAR TRAFFIC

Fowler (1991) reported on studies conducted at the at the end of the 1980's and early 1990's that packet traffic exhibited burstiness over a large number of time scales. These bursts exist at every time scale, from milliseconds to days and they look similar independently of the time scale, i.e. the traffic is self similar. One characteristic of this self-similar traffic is that it is correlated at all time scales of engineering interest, i.e. the traffic has long range dependence (LRD). The self-similarity and the LRD are quantified by the Hurst parameter H $(\frac{1}{2} \leq H < 1)$. Large values of H correspond to larger fluctuations on the burst size and stronger correlations in the traffic. These large fluctuations manifest themselves as heavy tailed distribution in the LRD traffic (Leland, 1994) and has been linked to the probability of higher buffer occupancy (Norros, 1993). Practically this increased probability has a drastic effect on the buffer occupancy since providing more buffer space is not a solution to buffer saturation (Eramilli, 1996). Eventually the buffer will fill up. The implication that an increasing value of H leads to higher buffer state occupancy (Norros, 1993, Eramilli, 1996) and hence increased probability of network congestion is much higher than if low valued H traffic was present which is interesting from the point of developing a chaotic network control since a small adjustments in H can drastically alter the buffer occupancy.

A natural question to ask would be: if CAC is conservative (in the sense of admitting traffic to suit the bottle neck link) then why is it that cell loss still occurs? The answer to this question lies in results presented by Willinger (1997). They have shown that it is the aggregation of *ON-OFF* sources with *ON* and/or *OFF* periods which are long range dependent that causes the self-similarity present in the networks. It is the self-similarity and the LRD of the aggregated traffic which is perceived as the future cause of network congestion (Leland 1994, Fowler, 1991, Erramilli 1996). A first step in finding solutions to the problems caused by self-similar traffic was research on models which adequately capture the variability seen in real traffic. Conventional stochastic traffic models, based in Markovian traffic theory, describe real traffic only over a single timescale. They do not have LRD (see Figure 1). There exist alternative stochastic traffic models known as Fractional Brownian Motion (FBM) models (Mandelbrot, 1968, Norros, 1993) which describe all the traffic characteristics of real traffic, the self-similarity and LRD. Alternatively, there exist models based on chaotic maps which reproduce the properties of real traffic (Eramilli 1994a, Samuel, 1997b). The need for the new models was so that the impact of self-similar behaviour could be assessed against current and proposed provisioning practices for ATM networks. Broadly, this research has taken the form of assessing the impact on the buffers due to the self-similar traffic. The motivation for this research could be said to be an attempt to reconcile the control of this type of traffic into some queuing theoretic approach, which can then be used in "traditional ways" in order to solve network provisioning and CAC (in ATM terms) problems that arise. In the following paragraphs these approaches will be briefly outlined.

Certain approaches have been traffic class specific. For example traffic classes such as VBR_{RT} and CBR_{RT} are very delay sensitive and have used control paradigms based on buffer partitioning (Schormans, 1997) and priority scheduling (Schormans, 1993) to solve the congestion. That is to say, the play-out rate of the occupied portion of the buffer will be sufficiently fast enough as to make the Markovian model estimation of the resource requirement conservative enough to be useful.

One approach which one would have naturally thought that would have at least reduced the impact of self-similar traffic on the network would have been traffic shaping. One would have thought that spreading the burstiness of the individual traffic sources would have altered the characteristics of the traffic sufficiently to the point where individual traffic streams do not become a problem. Unfortunately this is not the case (Leland, 1994). Work undertaken recently by Molonár (1997) shows that shaping will not alter greatly the self-similarity present in the traffic. A robust indication of this could be implied for the work of Eramilli (1996) where experiments on reshuffled LRD data were undertaken. Essentially the entire order of a data stream had to be shuffled randomly before the LRD nature in the stream was lost. If all the LRD streams are shaped then all that is achieved is an extension over the period over which the self similar traffic is present. This is because shaping still preserves the order of the data.

Another approach had been to accept Markovian models as adequate and use large deviation techniques to asses the impact of the rare event "large bursts" on a queueing system. Here, large deviation theory is used to calculate the probability of buffer overflow. This information is then used to provision the network/accept calls accordingly. This too has led to the formulation of CAC algorithms based on this principle (Duffield, 1995).

Naturally there have been approaches which combine FBM modelling and large deviations theory in order to arrive at some qualitative assessment of the effects self-similar traffic has on network buffering systems. These approaches have led to the notion of cross-over effects when self-similar streams are multiplexed together (Krishnan, 1996 and Fan, 1997). This effect describes an increase in the multiplexing gain in the buffering system when streams of self-similar traffic are multiplexed. However, once beyond the cross-over point the self-similar traffic streams once again are detrimental to the network. The point in this approach is however, that Markovian models (those with self-similarity parameter $H=0.5$) provide good (conservative) estimates for the buffer sizes required by the system in order to cope with the self-similar traffic streams.

Having summarised very briefly the traditional (current) control methods we now look comparatively at the way in which control affects common areas where chaotic control could be applied. The control of network traffic could be viewed as:

1. **Call level control** - permits the traffic onto the network provided that there are enough resources on the requested path which permit the incoming traffic to propagate across the network with out causing congestion at the cell level.

2. **Cell level control** - allocates resources in the network (such as buffer space) in order to accommodate the call in terms of cells or allow some traffic loss according to some pre-agreed cell loss rate.

Both (1) and (2) have traditionally been handled on entry onto the network via CAC algorithms. Additionally (2) can be approached via the design/dimensioning of the switches and cell level control methods such as UPC. In MAPS control, call level control is termed "order". This is the selection of call or burst based on a weighted decision derived form the dynamics of the system (network). Cell level control is termed "procession" and is the effect of the control of the dynamical system which permits the transfer of the data between source and sink.

3. DYNAMICAL SYSTEMS APPROACH TO TELETRAFFIC

An *ON/OFF* traffic source can be modelled using the family of one dimensional chaotic maps (Erramilli, 1994a, Erramilli, 1994b, Pruthi 1995a)

$$x_{n+1} = F(x_n) = \begin{cases} F_1(x_n) = \varepsilon_1 + x_n + x_n^{m_1}(1 - \varepsilon_1 - d)/d^{m_1}, & 0 < x_n \le d \\ F_2(x_n) = \varepsilon_2 + x_n + (1 - x_n)^{m_2}(\varepsilon_2 - d)/(1 - d)^{m_2}, & d < x_n < 1 \end{cases} \quad (1)$$

with parameters $m_1, m_2 \in (3/2, 2)$, $\varepsilon_1, \varepsilon_2 \ll 1$ and $0 < d < 1$. The *ON-OFF* traffic source is simulated by the associated indicator random variable (see Figure 2)

$$y_n(x_n) = \begin{cases} 0, & 0 < x_n \leq d, & \text{passive state} \\ 1, & d < x_n \leq 1, & \text{a packet / cell is emitted (active state)} \end{cases} \tag{2}$$

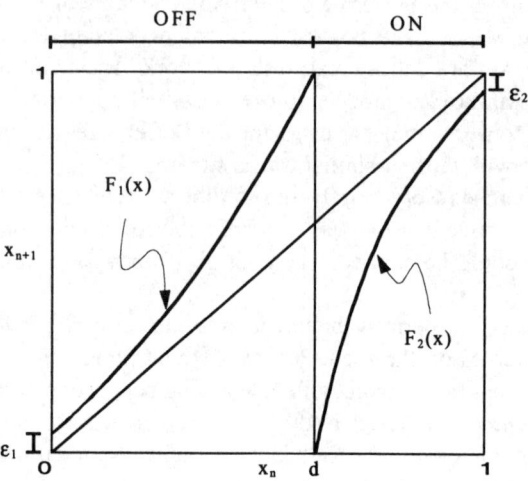

Figure 2 Chaotic Intermittency map showing parameters d, ε_1 and ε_2 and curve functions $F_1(x)$ and $F_2(x)$ (m_1 and m_2 describe the degree of polynomial for $F_1(x)$ and $F_2(x)$ respectively).

In common with measured traffic, the traffic simulated by these maps has a non-integer fractal dimension *($\frac{1}{2} \leq H < 1$)*, long range dependence (LRD) on the correlation and, if the outputs of several maps are aggregated, the simulated traffic tends to FBM. For a summary of the different map interpretations as a source model the interested reader is referred to the following references (Eramilli,1994a and 1994b, Pruthi, 1995b, Samuel, 1997a and1997b).

Traffic with different characteristics is labelled by the parameters of the maps. However, from a control point of view it is better to study the family of maps rather than any individual member of the family. To this end the Bernoulli shift map (which models Poisson-like behaviour, $m_1 = m_2 = 1$), the single intermittency map (which models traffic with LRD in either *ON* or *OFF* state, $m_1 = 1$, $m_2 > 3/2$ or $m_1 > 3/2$, $m_2 = 1$) and the double intermittency map (which models traffic with long range dependence in both *ON* and *OFF* states $m_1 > 3/2$, $m_2 > 3/2$) are all considered members of the same family. These different traffic models are consistent with the reported characteristics of *ON-OFF* sources by Willinger (1997).

As a first stage in developing a chaotic control for telecommunication networks we have studied how H changes with the map parameters. A decision to alter H based on the adjustment of these parameters can be interpreted as the controlling action. For example an alteration in the value of ε imposes an upper cut-off on the correlation, alteration in d changes the mean traffic load and changes in m_1 and m_2 changes the sojourn time of the *ON* and *OFF* states, i.e. the LRD of the traffic and its H value. Figure 3 shows two examples of the dependence of H with the map parameters.

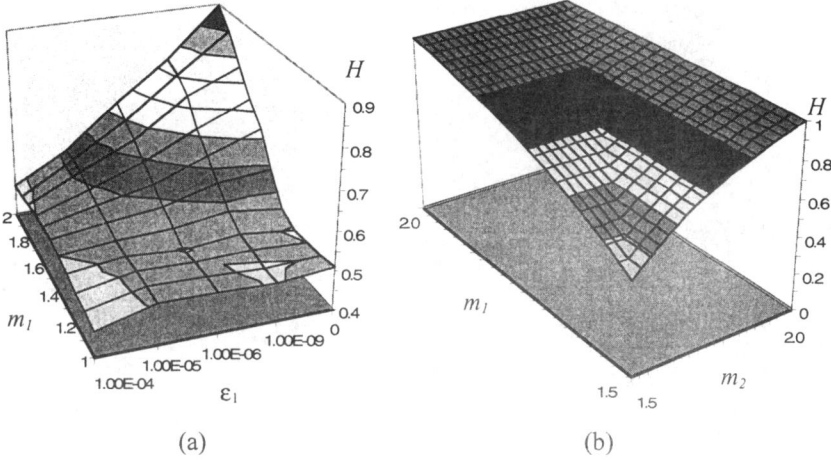

(a)   (b)

Figure 3 The dependence of the Hurst parameter on (a) m_1 and ε_1, and (b) m_1 and m_2. All other parameters are fixed.

Figure 3(a) shows that, for a single intermittency map, as ε_1 increases, the value of H tends to 0.5 and the long range dependence disappears. Figure 3(b) shows that, for a double intermittency map, the value of H depends on both m_1 and m_2. These results show Eramilli's conjecture (Eramilli, 1995, conjecture 3) that H is a function only of m_1 to be false.

As all sources can be modelled by a map, the next stage is to model the aggregate traffic produced by the individual source maps with a single "equivalent" map which preserves the traffic load and H. The parameterisation of this "equivalent" map is reported in Samuel (1997a and 1997b).

4. THE MAPS CONTROL PARADIGM

In the section 3 we have mention how chaotic maps can be used to provide aggregate models for self-similar traffic. We considered the case of aggregation at a single node (Samuel, 1997a). A natural extension of nodal models is to couple them together to form networks and then attempt to model and control the characteristics of these networks. In such a model the nodes would be the switching sites and the couplings between the nodes would represent the links between the switching sites. The simplest mathematical models which resemble such a construction are Coupled Map Lattices (CML) where a dynamical system at each node will produce the local traffic input. Coupling will be provided by external input from neighbouring nodes due to the queueing (aggregation) and switching. This is shown schematically in Figure 4.

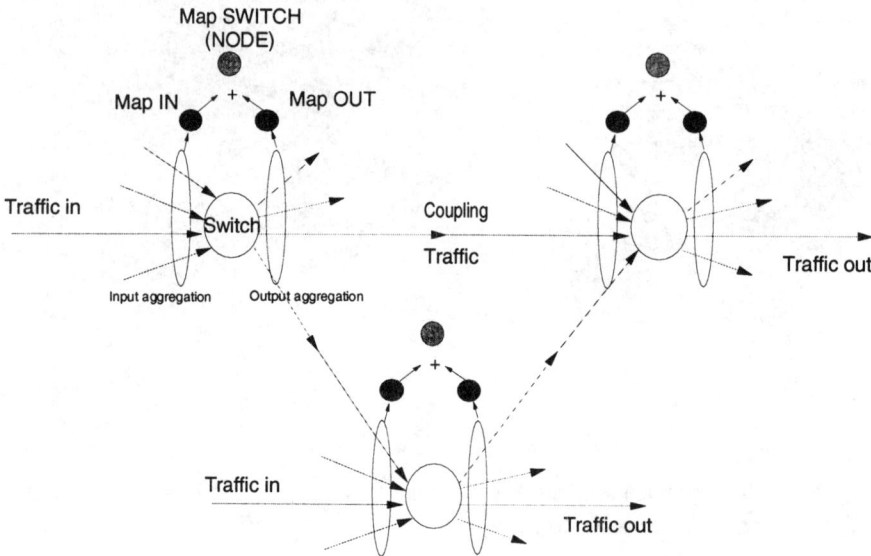

Figure 4 Lattice interpretation of a telecommunications network.

Investigations have been made into regular lattice structures where coupling exists between nodal sites.   An important property has been the discovery that global control across all nodes can be obtained via local control at each node (Mondragon, 1997a and 1997b, Arrowsmith, 1996, Ott, 1990).

In these investigations the dynamical behaviour of each nodal site is modelled by a chaotic map and the coupling to the neighbouring nodes imparts perturbations into the orbits of the dynamical system containing the node. Since the chaotic map's orbit possesses the property that any orbit will approach arbitrarily closely every point of the plane described by the chaotic map, then at some point the orbit must take it near to a desired control state. A small feedback control applied at the target point in the orbit places the dynamics of the node into a required state, since the same structure occurs at all nodes. Experimental evidence shows that if desired control state is prescribed for all nodes then eventually the lattice becomes controllable. However, it is possible for neighbouring dynamical behaviour to kick a node out of equilibrium via the coupling and so "occasional feedback control" is introduced where the feedback control is activated within the control region around the desired equilibrium for only part of the allowable time, (Mondragon 1997a).

The outline given in section 3, see (Samuel, 1997a) describes how each node site of the lattice structure can be modelled by a controlling map. The individual traffic streams entering a node can be modelled by chaotic maps and in certain models the maps can be aggregated into a single map which describes the behaviour of the traffic at the node. The next step is to provide on-line information on the Hurst parameter; this is necessary to enable the construction of an active dynamic control environment.

We propose a control scheme that actively manipulates the value of H. Our intended approach is based in manipulating H via the mean, peakedness and LRD of the traffic stream characteristics via with a local control strategy. The "philosophy" is not to destroy the chaotic behaviour of the traffic but instead to use its variability as a method of control. Chaotic systems are everywhere unstable and thus a small change in the system at any instant produces a large change at later times (this is known as "sensitive dependence on initial conditions" or, more colloquially, as "the butterfly effect") giving the controller "agility" to changes in the traffic over many timescales. Moreover as was noted in the previous section, successful control of coupled chaotic systems can be instigated with local control of each system. The implication of this to networks should be significant, since it suggests that for relatively small control actions applied locally the congestion/buffer occupancy on local and remote switches (relative to the control site) should be reduced. It is envisaged that the use of ATM mechanisms such as ABR via a chaotically initiated control sequence would provide a mechanism for the reduction of H and subsequent control of congestion.

We are proposing two different mechanisms to control traffic that is already in the network. The first mechanism seeks to reduce the variability of the traffic in a specific channel. This can be done by "careful" introduction of empty cells. The control mechanism would modify but not destroy the highly variable behaviour of the traffic. We conjecture that, individually, each of these controlled channels would change very little but that these changes would have a larger effect when the traffic is aggregated in the queue.

The second method is based in a random selector of calls in a node. The random selector would choose which call to admit by weighting dynamically the statistics of the traffic variability. The selector, modelled by a chaotic map, would assign larger probabilities to some channels than others but all the channels would have positive probabilities to be served. The first control technique is termed "Procession" and the second "Order". Conceptual views of the proposed chaotic control regime can be found in Figure. 5.

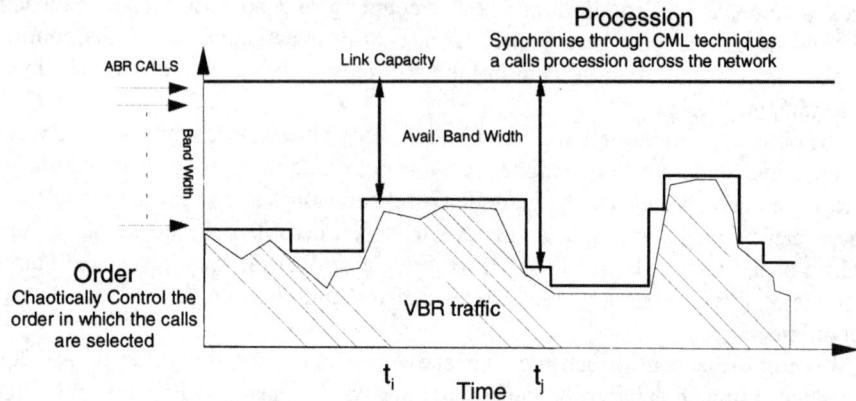

Figure 5 Conceptual view of chaotic network control as applied to ABR.

These two methods of control will be developed using chaotic maps as models of self-similar traffic because it is known that they have the correct characteristics and, moreover, that they can model high traffic rates efficiently (Samuel, 1997a and 1997b).

5. CONCLUSION

In this paper we have discussed the problems presented by self-similarity in traffic flow and summarised current views on resolving the problem of self-similar traffic within ATM networks. The weakness of conventional approaches lies in their addressing control over single timescales. We have presented results on the simulation of self-similar traffic by chaotic maps which show long range dependence varying with the map parameters. We have set out a new approach, the MAPS control paradigm, which exploits the dynamical characteristics of chaotic maps in the context of coupled map lattices to effect control over many timescales.

Acknowledgements

LGS would like to thank NORTEL and EPSRC, RJM would like to thank EPSRC and its Applied Mathematics Initiative for its support during the preparation of this work.

REFERENCES

Arrowsmith, D.K., Lansbury, A.N., and Mondragon, R.J. (1996) Control of Arnold circle map. *Int. Jour of Bif and Chaos 6:437-453.*

Chen, T. and Liu, S. (1995) *ATM Switching Systems,* Artech House, Boston.

de Prycker, M. (1991) *Asynchronous Transfer Mode: Solution for Braodband ISDN.* Ellis Horwood, London.

Duffield, N.G., Lewis, J.T., O'Connell, N., Russell, R. and Toomey, F. (1995) Entropy of ATM Traffic Streams: A Tool for Estimating QoS parameters, *IEEE Jour. on Sel. Areas in Comm.,* Vol. 13, No 6, 980-990.

Erramilli, A. Singh, R.P. and Pruthi, P. (1994a) Chaotic Maps as Models of Packet Traffic. *ITC 14, 329-338.*

Erramilli, A. Pruthi, P., and Willinger, W. (1994b) Modelling Packet Traffic with Chaotic Maps. *ISRN KTH/IT/R-94/18—SE, Stockholm-Kista, Sweden.*

Erramilli, A. Singh, R.P. and Pruthi, P. (1995) An application of deterministic chaotic maps to model packet traffic. *Queueing Systems.* Vol 20, 171-206.

Erramilli, A., Naranyan, O. and Willinger W. (1996) Experimental Queueing Analysis with Long-range Dependent Packet Traffic. *IEEE/ACM Trans on Networking,* Vol 4, No 2, 209-223.

Fan, Z. and Mars, P., (1997) The Impact of the Hurst Parameter and its Crossover effect on Long Range Dependent Traffic Engineering, *IEE 14th UKTS, 10/1-10/8.*

Fowler H.J. and Leland W.E. (1991) Local Area Network Traffic Characteristics, with Implications for Broadband Network Congestion Management. *IEEE Jour. on Sel. Areas in Comm.,* Vol 9, No 7, 1139-1149.

Krishnan, K.R. (1996) A new class of performance results for a fractional Brownian traffic model, *Queueing systems 22, 277-285.*

Leland,W.E., Taqqu, M.S., Willinger, W., and Wilson, D. (1994) On the Self-Similar Nature of Ethernet Traffic (Extended Version). *IEEE/ACM Trans on Networking,* Vol 2, No 1, 1-15.

Mandelbrot, B. and Van Ness, J.W. (1968) Fractional Brownian Motions, Fractional Noises and Applications. *SIAM Review,* Vol. 10, No 4, 422-437.

Molnár, S. and Vidács, A. (1997) On Modelling and Shaping Self-similar ATM Traffic, in *Vol 2bProc. ITC15, "Teletraffic Contributions for the information Age" (eds. V Ramaswami and P.E. Wirth), Elseveier, 1409-1420.*

Mondragon, R.J., and Arrowsmith, D.K. (1997a) Tracking unstable fixed points in parametrically dynamic systems. *Phys. Lett. A, Vol.229, No.2.*

Mondragon, R.J., and Arrowsmith, D.K., (1997b) On Control of Coupled Map Lattices: Using local dynamics to predict controllability. *Int. Jour of Bif and Chaos. 7:No 2, 383-399.*

Norros, I. (1993) Studies on a model for connectionless traffic, based on fractional Brownian motion. *Conf. On Applied Probability in Engineering, Computer and Communication Sciences, Paris.*

Ott, E., Grebogi, C., and Yorke, J. (1990) Controlling Chaos, *Phys. Rev. Lett.* 64 (11) 1196-1199, 1990.

Pruthi, P. (1995a) An Application of Chaotic Maps to Packet Traffic Modeling, *PhD Dissertation, Royal Institute of Technology, Sweden, ISRN KTH/IT/R-- 95/19--SE.*

Pruthi, P. and Erramilli, A. (1995b) Heavy-Tailed ON/OFF Source Behaviour and Self-Similar Traffic. *Proc ICC 95.*

Saito, H. (1994) *Teletraffic Technologies in ATM Networks.* Artech House, Boston.

Samuel, L.G., Pitts, J.M., and Mondragón, R.J. (1997a) Towards the Control of Communication Networks by Chaotic Maps: Source Aggregation, in *Vol 2b Proc. ITC15, "Teletraffic Contributions for the information Age"* (eds. V Ramaswami and P.E. Wirth), Elseveier, 1369-1378.

Samuel, L.G., Pitts, J.M., and Mondragón, R.J., (1997b) Fast Self-similar Traffic Generation, *14th UKTS. 8/1-8/4.*

Schormans, J.A., Scharf, E.M. and Pitts, J.M. (1993) Waiting time probabilities in a statistical multiplexer with priorities, *IEE Proc. I,* vol.140, No. 4, 301-307.

Schormans, J., Azmoodeh, M., Gordhan, S. and Davison, R. (1997) Buffer Partitioning Formula for Different Service Classes of ATM Traffic, *submitted to IEE Proc. Comms.*

Willinger, W., Taqqu, M.S., Sherman, R., Wilson, D.V. (1997) Self-Similarity through high-variability: statistical analysis of ethernet LAN traffic at the source level. *IEEE/ACM Transactions on networking, Vol. 5, No. 1, 71-86.*

30

Indicators for the assessment of congestion in TCP over ATM-UBR[1]

Mohammad A. Rahin and Mourad Kara
School of Computer Studies
The University of Leeds, LS2 UK
E-mail: {rahin, mourad}@scs.leeds.ac.uk
Telephone: +44 113 233 6590
Fax: +44 113 233 5468

Abstract

In this paper we analyse congestion in TCP over ATM-UBR, in particular the association between the congestion measures at the ATM physical and TCP transport layers. Two new indicators, protocol level congestion association factor (PLCAF) and CLR-PRR correlation coefficient are used. ATM cell loss ratio (CLR) and TCP packet retransmission ratio (PRR) are good measures of congestion at their respective layers, but do not address the relation between cell loss and its effect on TCP congestion control schemes. Both PLCAF and CLR-PRR correlation coefficient are derived to better capture the association between congestion measures at ATM and TCP levels. It is shown through simulation of multiple TCP connections over a ATM-UBR service under selected number of different operating conditions that the two new indicators can be successfully used for the analysis of congestion in TCP over ATM-UBR.

Keywords
TCP, ATM, congestion indicator, congestion association

---

[1]This work is partially funded by a UK-ADF grant (University of Leeds).

Broadband Communications   P. Kühn & R. Ulrich (Eds.)
© 1998 IFIP. Published by Chapman & Hall

1   Introduction

Performance evaluation of TCP over ATM has attracted the attention of a number of researchers [5][6][9][10]. The main causes for the performance degradation in TCP over ATM may be broadly classified into three different categories [7], *i*) due to the dynamics of TCP, *ii*) due to the behaviour of ATM and *iii*) due to the interaction between TCP and ATM layer congestion control mechanisms. A thorough knowledge about the interaction between congestions at these two distinct levels would lead to the understanding of the dynamics of TCP over ATM and would lead to a systematic appraisal of the TCP over ATM performance. A number of analytical and numerical studies e.g., [12][2] sought to estimate the burstiness of cell loss, i.e., the short overload period when cell losses occur followed by long loss-free intervals from the conditioned cell loss and string loss (loss of contiguous cells) probabilities. It is however, noted that conditional and string cell loss probabilities are not always sufficient to predict higher layer performance.

In this paper the relationship between congestion measures at TCP and underlying cell level ATM layers is investigated through simulation. Strength of association between these two measures are made using correlation coefficient of the congestion measures as well from their ratios. Of the two ATM service classes suitable for carrying TCP traffic, the UBR class is chosen in particular because, unlike ABR, UBR does not employ any congestion control mechanism at the ATM level and hence would allow the investigation of the relationship between the unhindered cell level loss to higher layer TCP level loss. The investigation is carried out with and without the presence of any intelligent packet drop policies e.g., EPD, PPD [10]. In section 2 the congestion control mechanisms in TCP and ATM-UBR are reviewed. Two congestion assessment indicators, CLR-PRR correlation coefficient and PLCAF are introduced in section 3. Section 4 describes the simulation experiments for the performance evaluation and congestion assessment of TCP over ATM-UBR under both LAN and WAN scenarios. Section 5 analyses the simulation results focusing on the congestion phenomenon. The paper is concluded in section 6 with a few remarks and direction for future works.

2   Congestion control

The congestion control mechanism in TCP is window-based providing end-to-end flow control to limit the number of packets in the network. There are two windows, one at the receiving end (receiver's window, RCVWND) as a measure of its buffer capacity and the other at the sending end (congestion window, CWND) as a measure of the network capacity. The sender can only place one window full of data at a time, but is not allowed to inject more than the minimum of RCVWND and CWND into the network.

The TCP congestion control mechanism has two distinct phases, *Slow Start* and *Congestion Avoidance* (Figure 1). During the slow start phase, either at the very

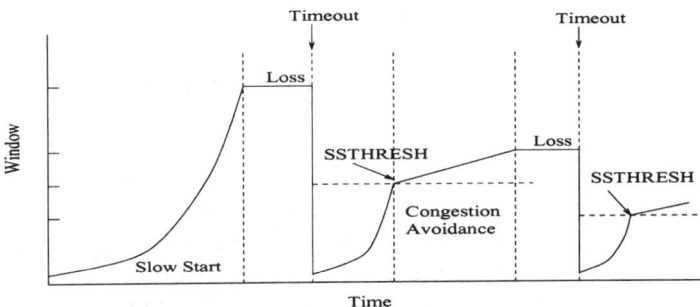

Figure 1: TCP congestion window vs. time.

beginning of the connection or when recovering from a packet loss, CWND is set to 1 TCP segment (usually 512 bytes) and is doubled once every *Round Trip Time* (RTT), i.e., on receipt of an acknowledgement. A delay-estimation algorithm is part of TCP collection of algorithms which provides a good estimate of the round-trip delay. During the congestion avoidance phase CWND is linearly increased and in doing so probes for additional available bandwidth in the network. The congestion avoidance phase starts whenever CWND reaches the value of another parameter SSTHRESH, usually initially set to 64 kB.

If a TCP packet is lost en route or received damaged, the destination informs this to the sender through duplicate acks for each out-of-order segment received. When duplicate acks are received the source refrain from sending any new segment in order not to increase the network load. At the source side, a timeout counter is maintained for the last unacknowledged segment which is reset on receipt of its ack. The source assumes congestion in the network whenever the timeout counter goes off. So as to remedy the congestion in the network the source sets SSTHRESH to half of the current value of CWND and CWND is set to one. Consequently, the slow start phase restarts. The source then retransmits the lost segment and increases its CWND by one every time a new segment is acked until CWND reaches SSTHRESH whereupon the congestion avoidance phase begins. All segments since the lost segment must be retransmitted before the source can send any new segment. This constitutes the go-back-N policy.

The TCP source continuously monitor RTT by measuring the time between the sending of a segment and when its ack is received. The retransmission timeout value is calculated from the estimates of the average and mean deviation of the RTT [4]. The fast-retransmit and fast-recovery mechanism [11] improves the TCP performance by not restarting the slow start phase for congestions indicated by the receipt of duplicate acks. However, even with fast-retransmit method the source must refrain from sending any more segments till either n (typically $n = 3$) duplicate acks are received when the lost segment is retransmitted or when the fast-retransmit timer goes off indicating that the apparently lost segment has at last reached the destination through an alternate route.

3   Assessment of congestion

Congestion in ATM network is generally specified using the metric *Cell Loss Ratio* (CLR). CLR is defined as the ratio of the number of cells that are lost en route to the total number of cells transmitted by the sources. Similarly, the TCP level congestion may be specified by the metric *Packet Retransmission Ratio* (PRR), defined as the ratio of the total number of packets retransmitted to the total number of packets sent by the sources. CLR and PRR are both long-term as opposed to instantaneous indicators of congestion at their respective layers. CLR has been found not to be a good indicator of TCP level performance [6]. In [5] TCP performance is analysed in terms of PRR in TCP over ATM. It was found that PRR is inversely related to switch buffer size, but the actual interaction between congestion measures at ATM and TCP layers under different circumstances was not established.

A single cell dropped at a switch results in the loss of one entire packet. In the worst case scenario, cells dropped are in the form of bursts of one cell long only evenly distributed in the traffic stream. This would cause very high levels of packet retransmissions and throughput loss. Ideally, unavoidable cell losses should be in bursts of a large number of cells such that at most only one or a couple of packets are affected. Both PPD and EPD aim to achieve this goal of concentrating cell losses into single packets. The interaction between cell loss and packet retransmission is difficult to predict as it depends on the nature of traffic flows as well as the cell occupancy levels at the switch. An assessment of the level of interaction, i.e., how cell losses affect TCP throughput by causing packet retransmission would lead to the understanding of effectiveness of cell drop policies, buffer allocation schemes or the source characteristics.

The association between cell loss and the consequent packet retransmission may be analysed using the correlation coefficient between CLR and PRR over a set of observations as follows.

$$Corr(C, R) = \frac{E[(C_i - \overline{C_i})(R_i - \overline{R_i})]}{\sqrt{E[(C_i - \overline{C_i})^2]}\sqrt{E[(R_i - \overline{R_i})^2]}}, \tag{1}$$

where C_i and R_i are the ith CLR and corresponding PRR observations. The correlation between CLR and PRR is a measure of the association between them. If C_i and R_i are independent, $Corr(C, R)$ is zero. On the other hand, a value close to either 1 or -1 would suggest strong association between them. Correlation between cell loss and packet retransmission may be calculated from the observed CLR and PRR of a number of TCP flows passing simultaneously through a switch or from the aggregate CLR and aggregate PRR at a switch for different values of switch buffer sizes, a factor affecting congestion. The CLR-PRR correlation may be used as a tool to compare the effectiveness of different cell drop policies or buffer allocation schemes in terms of strength of association between congestion measures at different levels in protocol stack.

The CLR-PRR correlation provides summarised information on the strength of association between ATM and TCP layer congestion measures over a set of observations. However, snapshot information on congestion correspondence for a single observation can not be readily obtained from CLR-PRR correlation. Such information may be obtained from the ratio of normalised packet retransmission to normalised cell loss, i.e. the ratio of PRR to CLR for the connection. In TCP over ATM-UBR, packet retransmission may only take place as a result of actual cell loss. The PRR to CLR ratio may thus be used as the true measure of association between cell loss and packet retransmission. We call this ratio *Protocol Level Congestion Association Factor* (PLCAF). The notion of PLCAF however, by definition is non existent in congestion free i.e., zero cell loss situations. PLCAF may be formally expressed as:

$$PLCAF = \begin{cases} \frac{PRR}{CLR} & ; CLR \neq 0, \, PRR \neq 0 \\ \textbf{notdefined} & ; \text{otherwise} \end{cases} \quad (2)$$

In non-zero cell loss situations, PLCAF is a positive number within the range $0 < PLCAF \leq (MTU/48)$. MTU is a TCP parameter specifying the largest size of TCP packets and the figure 48 is derived from the payload size of ATM cells. PLCAF may be interpreted as a measure of the number of packets affected as a result of single cell loss. For infinite TCP sources the average number of packets affected per cell loss can be obtained by multiplying PLCAF with $(48/MTU)$. A large PLCAF would indicate that cell losses are spread across the traffic stream in small bursts causing large number of TCP retransmissions. On the other hand, long bursts of cell losses causing relatively small number of retransmissions would result in smaller PLCAF, a desirable property. PLCAF plotted against switch buffer size can be used for congestion analysis in terms of buffer constraints. Also, PLCAF vs. switch buffer size plots can be used as effective tools for the comparison of different cell drop policies or traffic parameters. With an intelligent packet drop policy (e.g. PPD or EPD) active it is expected that PLCAF would decrease with increasing switch buffer size. An increase of PLCAF with increasing switch buffer size would indicate indiscriminate cell dropping and a failure to exploit relaxed buffer constraints.

Both the CLR-PRR correlation coefficient and PLCAF indicators provide information relating to the correspondence between congestion measures at ATM physical and TCP transport layers and are better indicators for the assessment and analysis of congestion in TCP over ATM as both ATM & TCP layers are considered in unison and not in isolation. However, it must be borne in mind that they alone or together do not fully specify the actual level of congestion for which either CLR and PRR must be used in conjunction.

4 Experimental setup

In this section the network configurations and parameters used in the simulations of TCP congestion behaviour over ATM-UBR are described. Fig 2 shows the network configuration used in the simulations. It consists of 2 switches, N TCP connections and $2N$ end-stations. Only one-way TCP connections (discounting the TCP ack

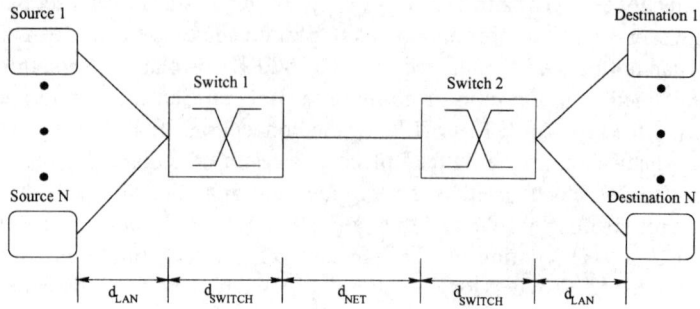

Figure 2: The N-source network configuration.

in the reverse directions) are considered. The link between the two switches is the congestion bottleneck. As study of congestion is the primary objective of the simulations, this particular configuration is selected specifically with a view to concentrate all congestions in a single link/switch. No additional background or cross-traffic sources are considered.

The N-identical TCP sources on the left side of the figure each establish a single connection with the similarly numbered destination across the two switches and the bottleneck link. The TCP sources are greedy sources with infinite supply of data and always have data ready to send as and when permitted by the TCP flow-control. The two end-station groups are considered in close proximity ($d_{LAN} = 0.5\mu s$) with their respectively connected switches. The internal delay in both switches are identical and is set to $d_{SWITCH} = 450\mu s$, the maximum such value sanctioned in ITU-T standard [1]. Two network scenarios, LAN and WAN are considered. The delay in the bottleneck link is set to a typical value $d_{NET} = 10\mu s$ equivalent to 2 km of transmission path and $d_{NET} = 10ms$ equivalent to 2000 km of transmission path for LAN and WAN scenarios respectively. Each connection in the network configuration chosen have only one path and as such the lower bound of RTT is $2*(2d_{LAN}+d_{SWITCH}+d_{NET})$ and the upper bound adds another term, a function of the buffering at the switches to it. All link bandwidths are 155 Mbps for LAN scenario and 622 Mbps for WAN scenario. The switch speeds are set accordingly, i.e., 155 Mbps and 622 Mbps for LAN and WAN respectively. Finally N, the number of sources is set to 5 and 20 for LAN and WAN respectively similar to those used in a number of published works [3][5][10] and deemed sufficient to generate traffic flows required in our study.

The STCP simulation package [8] is used in the simulation study. The retransmission timers are set to 200 ms and 500 ms for fast & slow timeouts respectively. The MTU is set to 9180 bytes as par RFC1626. The default maximum window size in STCP is 64 kB. This is sufficient to fill the network pipe in LAN, but inadequate in WAN. Despite this the maximum window size in WAN scenario is also set at 64 kB. This particular decision would present a relaxed network pipe capacity to the multiple TCP sources resulting in the facing of a much less bandwidth contention environment. The sources will be less aggressive against each other as opposed to the situation in LAN scenario.

To study the congestion behaviours at ATM and TCP layers the following three different cases for each of LAN and WAN scenarios in the N-source network configuration are simulated[2].

A. All N sources starting simultaneously. Switches do not employ any cell drop policy.

B. The N sources start at random times. Switches do not employ any cell drop policy.

C. The N sources start at random times. Both EPD & PPD cell drop policies are in force.

The above three simulation cases are particularly chosen to allow comparison of the congestion behaviour at both ATM and TCP level under different operating environments such as where TCP synchronisation plays an important role and also where intelligent cell drop policies (EPD/PPD) are employed at the switches

The switch architecture used in the simulation is a nonblocking, output-buffered type. Each input port have a private buffer of 128 cells. Each output port have additional buffer and implements a single queue on it. All VCs passing through an output buffer share this buffer. The scheduling policy at the output buffers is FIFO. In order to induce variable congestion in the switches, the size of the shared buffer in each output queue are varied for all simulation cases in both LAN and WAN scenarios. The buffer sizes arbitrarily chosen are 100, 200, 400, 600, 800, 1000, 2000, 4000, 6000, 8000 and 10000 cells. The earlier part of the range offer very tight buffer constraint, while on the other hand, the latter part of the range offer much more relaxed buffer constraint. The switches may be configured with no specific cell drop policy, PPD or EPD. The latter two may be configured either independently or in combination. For EPD, we set the threshold for discarding incoming cells according to the following formula:

$$threshold = max\left\{0.8, \frac{buffer\,size - 3.\,segment\,size}{buffer\,size}\right\}. \tag{3}$$

The threshold is thus set equal to the buffer size minus three times the TCP segment size. The first term 0.8 in the above formula avoids setting the threshold to a very low value when the buffer size is small and segment size is large.

Each simulation was run for a time equivalent to the transfer of 10^{10}bits at the link speed being simulated. This turned out to be 100.01 seconds for LAN and 24.92 seconds for WAN. Each simulation run was divided into 5 batches and average of these 5 batches are taken. Also no additional warm-time times were allowed. This was to capture the TCP behaviour during simultaneous opening of multiple TCP connections. The simulation run times are sufficiently large to capture long-term congestion behaviour.

---

[2]The notation used to refer to a particular experiment is scenario.case, e.g., LAN.A.

Table 1: Buffer requirement for zero cell loss

Number of Sources	Network Configuration	Network Capacity (Mbps)	Max. Throughput Achieved) (Mbps)	Maximum Queue (Cells)
5	LAN.A	155	133.54	4000
5	LAN.B	155	130.57	4000
5	LAN.C	155	130.57	6000
20	WAN.A	622	328.54	8000
20	WAN.B	622	207.68	800
20	WAN.C	622	207.68	1000

5   Results & evaluation

In this section the simulation results for the three cases discussed in section 4.1 under both LAN and WAN scenarios are presented[3]. The focus is mainly on the association between the congestion measures at ATM and TCP levels using the indicators CLR-PRR correlation coefficient and PLCAF as defined earlier in section 3. The switch buffer size was chosen as an artificial means to vary the level of congestion. In this study we only consider the long-term congestion behaviour of the network.

The summary of the results for zero cell loss situations are shown in Table 1. The maximum throughput achieved in all cases is smaller than the available capacity. The loss in throughout is due to various protocol (TCP, IP, AAL5, LLC, SONET) overheads, additionally for cases LAN.B, LAN.C, WAN.B & WAN.C due to some sources waiting for small periods of time before they could begin transmission. Also, in all WAN cases the decision of not using the window scaling option for TCP maximum window size resulted in severe underutilisation of the network. The last column shows the amount of buffer at which zero cell was observed for a particular network configuration. Cases LAN.C & WAN.C required slightly larger buffer compared to LAN.B & WAN.B respectively because of the EPD requirement of setting out a small overflow buffer space beyond the threshold. The maximum window size in WAN cases is kept to a value more suitable for LAN and as a result full exploitation of WAN pipe capacity is not possible by individual TCP sources. However, this allowed multiple TCP sources share the WAN pipe capacity in a less contentious manner.

The association between ATM and TCP layer congestion is first evaluated using the indicator PLCAF. PLCAF measured using the aggregate CLR and aggregate PRR for various switch buffer sizes and for all three cases are shown in Figure 3 for both LAN and WAN scenarios. In LAN.A, PLCAF is almost uniform except for a small blip at around the buffer size of 800 cells (Figure 3a). The relatively small PLCAF

---

[3]For a fuller description of the experiments and results please see:
ftp://agora.leeds.ac.uk/scs/doc/reports/1997/97_42.ps.Z.

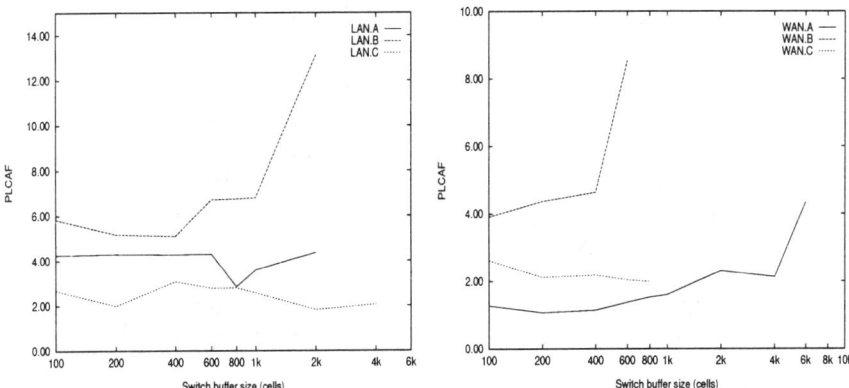

Figure 3: PLCAF as function of switch buffer size in (a) LAN (left) and (b) WAN (right) scenarios.

values in LAN.A suggest that cell losses are taking place in relatively longer bursts. This and the near uniform behaviour of PLCAF against switch buffer size in case LAN.A is due to strong source correlation. Aggregate PLCAF in LAN.B on the other hand increases with switch buffer sizes. A sharp rise in PLCAF is especially observed at buffer size of 2000 cells, the maximum buffer size beyond which no cell loss was noticed. This rise in PLCAF with switch buffer size is explained by the observation that with increasing switch buffers the cell loss phenomenon become smaller in burst size and more distributed along the traffic stream and connections. The ability of EPD to maintain strong association between CLR and PRR is seen from LAN.C data. The effectiveness of EPD is confirmed by the consistently small and fairly uniform PLCAF values in LAN.C. EPD is able to string together dispersed cell loss phenomenon spread across the traffic stream as observed in LAN.B into loss of contiguous cells affecting only a few packets. However, EPD is seen not to exploit the relaxed buffer constraints fully. It was expected that with EPD, PLCAF would show a downward trend with increasing switch buffer sizes. The greedy nature of TCP sources along with the aggressive growth of TCP window sizes are thought to be responsible for this less than ideal behaviour.

The PLCAF curves showing the correspondence between ATM and TCP layer congestion measures for WAN simulation are shown in Figure 3b. WAN.A shows slow growth of PLCAF with switch buffer size unlike the case in LAN.A where PLCAF is found to be uniform. WAN.B mirrors the LAN.B behaviour of PLCAF of abruptly rising to a high value at very relaxed switch buffer constraint. WAN.C on the other hand show decline of PLCAF with increasing switch buffer size as is expected with EPD. Under the relaxed WAN network pipe capacity considered, EPD is found to exploit the relaxed switch buffer constraints. However, unlike in LAN simulation PLCAF values for WAN.C is numerically larger than those for WAN.A. But, the numerical values of PLCAF for all three are smaller in WAN than in LAN

Table 2: CLR-PRR correlation coefficient in LAN scenario

Switch Buffer Size	LAN.A	LAN.B	LAN.C
100	0.944	0.893	0.987
200	0.980	0.974	0.996
400	0.978	0.999	0.965
600	0.998	0.819	0.857
800	0.941	0.999	0.960
1000	0.999	0.953	0.995
2000	0.951	0.957	0.969
4000	-	-	0.988
Aggregate	0.951	0.116	0.883

(Figure 3a & b). This different behaviour may be explained by the fact the sources are unable to expand their window sizes sufficiently large to fill the network pipe in the WAN scenario considered. As a result, the network is underutilised. Sources, at full speed transmission act as an on-off source, period of activity followed by period of inactivity when it is waiting to receive acknowledgement from the receivers. As a result, the sources on the whole tend to be more nicely behaved and less aggressive against each other. Because of the on-off like behaviour of the sources whose transmission start times are interleaved, when cell loss take place they are more likely to be in the form of longer bursts than before resulting in smaller PLCAF.

In the remainder of this section the interaction between ATM and TCP layer congestions using the CLR-PRR correlation coefficient is examined. This analysis for the LAN scenario is shown in Table 2. For each switch buffer size, the correlation coefficient between the set of CLRs of 5 connections and their corresponding PRRs is calculated. For all three cases, very strong CLR-PRR correlation is observed for all different buffer sizes. This is due to the sharing of common switch resources by all 5 connections, thus their behaviours are inherently related. The last row in Table 2 shows the aggregate CLR-PRR correlation coefficients. These are calculated using the set of aggregate CLRs for all buffer sizes (100, 200, 400, 600, 800, 1000 and 2000 cells & also 4000 cells for LAN.C) and the set of corresponding aggregate PRRs. LAN.A shows strong aggregate CLR-PRR correlation while in LAN.B, CLR & PRR appear to be quite independent variables. The strong correlation in LAN.A stems again from the synchronous start times of the sources and matches with the previous PLCAF analysis. While the almost independent CLR-PRR relationship in LAN.B is due to the short bursty cell loss phenomenon in indiscriminate form. Longer bursts of cell losses (maintaining the same or higher CLR) is expected to have stronger CLR-PRR correlation and smaller PLCAF. This is in evidence in LAN.C.

Finally, the association between ATM and TCP layer congestion for WAN scenario is analysed using the CLR-PRR correlation coefficients (Table 3). CLRs and PRRs of competing connections at any particular switch buffer size show strong correlation as before. However, the aggregate CLR and aggregate PRR in WAN.A

Table 3: CLR-PRR correlation coefficient in WAN scenario

Switch Buffer Size	WAN.A	WAN.B	WAN.C
100	0.999	0.960	0.964
200	0.999	0.830	0.942
400	0.999	0.955	0.908
600	0.998	1.000	0.995
800	0.996	-	1.000
1000	0.992	-	-
2000	0.961	-	-
4000	0.956	-	-
6000	0.853	-	-
Aggregate	0.873	0.997	0.983

now show slightly weaker correlation as suggested by PLCAF curve in Figure 3b. Also, very much unlike the LAN scenario, WAN.B shows very strong aggregate CLR-aggregate PRR correlation. Even though the PLCAF vs. switch buffer size plot for WAN.B appear to suggest otherwise, it was observed that the both PRR and CLR exhibit similar behaviour with changing switch buffer size and as a result show strong correlation between aggregate-CLR & aggregate-PRR in WAN.B. As expected WAN.C shows strong aggregate CLR-aggregate PRR correlation.

6   Conclusions & future work

CLR-PRR correlation coefficient and PLCAF are proposed to study the association between user perceived congestion at the TCP layer and the hidden, underlying congestion at the ATM layer. When taken an aggregate view of all connections over a number of different operating environments (changing switch buffer size), CLR and PRR are strongly correlated for sources with synchronous start times in both LAN and WAN scenarios. A similar, characteristic was also observed for sources with asynchronous start times and the ATM switches equipped with EPD & PPD cell drop policies. Using the PLCAF formulation, the association between congestion measures at ATM and TCP layers are analysed. The PLCAF analysis successfully highlighted the effectiveness of intelligent cell drop policies like EPD & PPD in terms of strength of association between CLR & PRR congestion measures.

The simulation study has shown the efficacy of both these indicators in assessing the congestion interaction. It is worthwhile to investigate effect of scaling the number of sources and also the interaction between the rate-based flow control of ABR and the window-based congestion control by TCP in TCP over ATM-ABR environment using the two indicators CLR-PRR correlation and PLCAF.

References

[1] ITU-T Recommendation I.731, Types and General Characteristics of ATM Equipment. 1996.

[2] Baumann, M., *Analysis of Cell and Frame Loss Ratios in an ATM Multiplexer*, Proc. 6th Open Workshop on High Speed Networks, pp. 31-38, Stuttgart, Germany, 8-9 Oct., 1997.

[3] Goyal, R., Jain, R., Kalyanaraman, S. and Fahmy, S., *UBR+: Improving Performance of TCP over ATM-UBR Service*, ICC'97, Montreal, Canada, 8-12 June, 1997.

[4] Jacobson, V., *Congestion Avoidance and Control*, SIGCOMM'88 Symposium, pp. 314-329, August 1988.

[5] Kalampoukas, L. and Varma, A., *Performance of TCP over Multi-hop ATM Networks: A Comparative Study of ATM Layer Congestion Control Schemes*, Technical Report - UCSC-CRL-95-13, Computer Engineering & Information Sciences, University of California, Santa Cruz, CA 95064 USA, February 1995.

[6] Kalyanaraman, S., Jain, R., Fahmy, S., Goyal, R., Lu, F. and Srinidhi, S., *Performance of TCP/IP over ABR Service on ATM Networks*, Globecom'96, London, November 1996.

[7] Kara, M. and Rahin, M.A., *Towards a Framework for Performance Evaluation of TCP Behaviour over ATM*, ICCC'97 - 13th Intl. Conf. on Computer Communication, pp. 49-60, France, Nov. 18-21, 1997.

[8] Manthorpe, S., *STCP User Manual*, Tech. Rep. 1.0, LRC-DI-EPFL, http://lrcwww.epfl.ch/~manthorp/stcp. 1996.

[9] Moldeklev, K. and Gunningberg, P., *Deadlock situations in TCP over ATM*, 4th Intl. IFIP workshop on Protocols for high speed networks, pp. 219-235, Vancouver, B.C., Canada, 10-12 August 1994.

[10] Romanow, A. and Floyd, S., *Dynamics of TCP Traffic over ATM Networks*, ACM SIGCOMM Computer Communications Review, October 1994.

[11] Stevens, R.W., *TCP/IP Illustrated, Volume 1: The Protocols*, Addison-Wesley Publishing Co., New York 1994.

[12] Takine, T., Suda, T. and Hasegawa, T., *Cell Loss and Output process Analysis of a Finite-Buffer Discrete-Time ATM Queueing System with Correlated Arrivals*, Proc. IEEE INFOCOM '93, pp. 1259-1269, San Francisco, 1993.

31

FATHOC - a rate control algorithm for HFC networks

M. M. Macedo[1,3], M. S. Nunes[1,2] , H. Duarte-Ramos[3]
1) INESC, mmacedo@inesc.pt, msn@inesc.pt 2) IST 3)
UNL/FCT
INESC R. Alves Redol 9, 1017 Lisbon, Portugal Tel. +351
13100256 Fax. +351 13145843
UNL/FCT - 2825 Monte da Caparica, Portugal Tel. +351 1
2954464 Fax +351 1 2954461

Abstract
This paper presents a rate control algorithm to use with SARP (Simple ATM based Reservation Protocol) for HFC (Hybrid Fibre Coax) networks. The SARP is a hybrid reservation and rate based protocol, based on ATM, whose main characteristics are its simplicity, high efficiency, and flexibility to support different classes of traffic. FATHOC (Fairness Achievement Through Congestion) deals with fairness achievement and congestion avoidance problems as a single problem, by trying to achieve fairness dynamically through congestion handling. FATHOC scope of application is both the ABR and nrt-VBR service categories. FATHOC algorithm was designed to work on HFC networks at the Head-end and was developed in the scope of the project ATHOC, a European ACTS research and development project.

Keywords
ATM, ABR, nrt-VBR, Hybrid Fibre Coax, medium access protocols

Broadband Communications   P. Kühn & R. Ulrich (Eds.)
© 1998 IFIP. Published by Chapman & Hall

1   INTRODUCTION

FATHOC (Fairness Achievement Through Congestion) is an algorithm for ABR rate control, and also for nrt-VBR, that achieves fairness and avoids congestion, in a new way, i.e., by trying to achieve fairness dynamically through congestion handling.

FATHOC initial design goal was to schedule the transmission grant rates of continuous type connections present on the upstream digital channel of an HFC network. Located at the Head-end of HFC networks, it implements the switch behaviour for the ABR service (ATM Forum, 1996), and can also be used on the computation of the transmission grant rates of nrt-VBR sources. FATHOC was designed to be compatible with SARP (Simple ATM based Reservation Protocol) for HFC (Hybrid Fibre Coax) networks (Sierens, 1996) (Nunes, 1996), but its scope of application may be extended to any other protocol.

SARP is a hybrid reservation and rate based protocol, based on ATM, whose main characteristics are its simplicity, high efficiency, and flexibility to support different classes of traffic (CBR, VBR, UBR, ABR), in different combinations. Two types of sources are considered for the SARP scheduling in the upstream channel: continuous sources (corresponding to CBR and VBR audio and video sources), and bursty traffic sources (UBR and ABR, e.g., LAN type traffic).

For the first kind of sources, the SARP uses a rate-based and dynamic scheduling protocol, and the adjustment of the allocated bandwidth to each terminal is based on implicit feedback of the terminal activity. The Head-end monitors the source activity based on the transmission queue piggybacked on the transmitted cells of the terminal and also by detecting empty cells (Macedo, 1997). For the second type of sources, the protocol is based on a request/credit method, using contention mini-slots.

Transmission of ATM cells on the upstream channel is allowed by the corresponding transmission grants given to each terminal on the downstream slots.

2   VIRTUAL SOURCE / VIRTUAL DESTINATION OPERATION AT THE HEAD-END

For ABR sources, full compatibility with the ABR protocol is required, even considering the particular MAC level differences between HFC networks and common ATM networks. The main difference and the additional complexity derive from the fact that in the HFC networks, cells are allowed to be transmitted to the Head-end by transmission grants issued to the stations, instead of being transmitted by the initiative of the stations, as it happens on normal switches.

As a consequence of this specific characteristic of the HFC, the Head-end must know the true transmission rate of the source in order to allow the source to transmit data and also to save bandwidth, with a minimal number of generated empty cells. This is true either when the source is transmitting below the *CCR* (or *ACR*), or when a data burst stops. Also, as the total bandwidth is limited to the bandwidth of the upstream channel, it is important to limit the bandwidth given to the ABR sources, when higher priority sources demand additional bandwidth.

In order to implement the ABR service on the HFC bus, some problems must be addressed. They are:

- Fairness achievement and congestion avoidance schemes, for dynamic scheduling of the ABR sources.

- ABR re-scheduling when other higher priority classes (namely, rt-VBR) connections demand additional bandwidth.

- Implementation of use-it-or-loose-it policies for ABR sources, when they transmit at a rate lower than CCR or when they stop transmission.

After analysis of different alternatives, it was considered that the cleaner way to solve these problems was to implement the Virtual Source/Virtual Destination feature in the Head-end.

The Virtual Source/Virtual Destination feature operation proposed by the authors for the ATHOC project can be described as following:

- Each time a FRM cell arrives to the Head-end, coming from the terminal, fairness achievement and congestion avoidance computations are done by FATHOC and an Explicit Rate is computed.

- The values of the bits *CI* (Congestion Indication) and *NI* (Not Increase) and of *ER* (Explicit Rate), are combined with the values that came in the last BRM cell from the ATM network, and with the values computed at the Head-end. The results of these computations are registered at a BRM cell sent back to the terminal.

- The Head-end does also the Source Behaviour rate computations in order to estimate the new source *ACR*, and the transmission grant rate T_g to be given to the source. The FRM cell is then transmitted to the ATM network with the computed value for ACR, and the other fields kept unchanged.

- The Head-end can also do true rate estimation, based on the activity of the source, namely in order to detect burst stops of the ABR connections.

- The Head-end can also do true rate estimation, based on the activity of the source, namely in order to adjust to the true *ACR* of the source and to detect burst stops of the ABR connections.

For non ABR sources, namely for nrt-VBR sources, the operation of the Head- is simpler. Transmission rates given to the sources are calculated exclusively based on the contracted parameters of the connection, namely *PCR* (Peak Cell Rate), and the total bandwidth available for the nrt-VBR service. However, the same basic FATHOC fairness achievement and congestion avoidance scheme are applicable.

We now describe FATHOC operation for the ABR sources and after we describe also its possible implementation for nrt-VBR sources.

3 FATHOC ALGORITHM

FATHOC basic operational principle consists in allowing sources to get progressively more bandwidth, until a congestion state is reached, and to constrain progressively the sources that are using more bandwidth until the congestion state is over.

Two different control points are defined: *CongestionStateEntering* and *CongestionStateExiting*, corresponding to two different levels of total ABR bandwidth load.

The congestion state is reached when some sources demand extra bandwidth, because they are allowed by the ATM network to rise their rates. Constrain of the sources by FATHOC is then activated, and the sources that are using larger fractions of the total bandwidth, with respect to some fairness criterion, are forced to lower progressively their rates.

Congestion state is then exited, and constraints are progressively relaxed until another congestion state happens.

The fairness criteria can be equal share, equal share of the bandwidth left after satisfaction of minimum cell rate requirements of the sources, or any other.

In those congestion/non-congestion cycles, sources using lower bandwidth, with respect to the criterion, are always rising their rates, and fairness is naturally achieved. In steady state operation, i.e., when fairness is achieved and ATM network bandwidth constraints are stable, the ABR bandwidth load will typically oscillate between non-congestion and congestion states.

In this version of FATHOC, the criterion for fairness achievement is the *equal share of ABRCapacity bandwidth left after satisfaction of minimum cell rate requirements of all sources. ABRCapacity* is the target utilisation of the ABR service bandwidth quota.

FairShare can then be defined by expression

$$FairShare_i = MCR_i + \frac{ABRSharableCapacity}{NumberOfVCs} \tag{1}$$

where

$$ABRSharableCapacity = ABRCapacity - \sum_{j=1}^{NumberOfVCs} MCR_j \tag{2}$$

However, the constraints on sources rates imposed by the ATM network (through *CI* and *NI* bits, and the *ER* field), or by low values of *PCR*, can lead to the existence of sources with bandwidth shares lower than the *FairShare* value of expression (1). When this situation occurs, it is reasonable to allow other sources, not constrained by the ATM network, to get bandwidth shares higher than that *FairShare* value of expression (1), in order to achieve higher utilisation of the bandwidth quota assigned to the ABR services.

According to the criterion chosen for fairness, sources that are not constrained by the ATM network, should be permitted to share equal fractions of *ABRCapacity* minus the sum of the *MCR* (Minimum Cell Rate) requirements of all ABR sources and minus the sum of the cell rates of all sources that are constrained by the ATM network to cell rate values lower than the *FairShare* value of the expression (1). Such a reasoning is similar to that of the MIT scheme (see Jain, 1996a) for which the value of *FairShare* is given by the expression:

$$FairShare = \frac{LinkBandwidth - \sum BandwidthOfUnderloadingVCs}{NumberOfVCs - NumberOfUnderloadingVCs}$$

and that is used by several algorithms, namely by ERICA (Jain, 1996b) (Jain, 1996c).

This new extended criterion means that in FATHOC the true value for *FairShare* for unconstrained sources, when the ATM network constrains some sources to bandwidth uses lower than that given by *FairShare* value of the expression (1), is given by expression

$$FairShare = MCR_i + \frac{ABRCapacity - \sum_{i=1}^{NumberOfVCs} MCR_i - \sum_{i=1}^{NumberOfConstrained} (CCR_i - MCR_i)}{NumberOfVCs - NumberOfConstrained} \quad (3)$$

Now, we explain the way FATHOC achieves this new fairness criterion and avoids congestion.

In FATHOC algorithm, a factor is defined for each source i, *ShareFactor_i*, whose meaning is the ratio between each fraction of the *ABRSharableCapacity* used by the source and the value of *FairShare* if no sources were constrained by the ATM network. For each source i, *ShareFactor_i* is given by

$$ShareFactor_i = \frac{(CCR_i - MCR_i)}{ABRSharableCapacity / NumberOfVCs} \quad (4)$$

Also, a global *ShareFactor* is defined, with the meaning of the maximum possible value of all *ShareFactor_i*. It is upon global *ShareFactor* that FATHOC acts to achieve fairness and avoid congestion. Therefore

$$ShareFactor = max \ \& \ bound \ (ShareFactor_i) \quad (5)$$

When a congestion state is entered, *ShareFactor* is set to the maximum of the calculated *ShareFactor*$_i$, and then progressively lowered with some decay rate. This procedure has the effect of reducing the cell rates of sources using higher bandwidth, with respect to the fairness criterion, i.e., those sources that have higher S*hareFactor*$_i$, and of reducing the total ABR bandwidth load, forcing exiting the congestion state. However, sources using less bandwidth, i.e. the sources with *ShareFactor*$_i$ lower than the global *ShareFactor*, can acquire more bandwidth, with the condition of their *ShareFactor*$_i$ not surpassing the global *ShareFactor*.

When the congestion state is exited, the global *ShareFactor* is allowed to rise, relaxing the bandwidth use. Sources that are not constrained by the ATM network, are permitted to acquire progressively higher fractions of the ABR bandwidth quota.

These procedures can be seen on figure 1, which shows the full FATHOC pseudo-code.

EnteringCongestionStateLoadFactor and *ExitingCongestionStateLoadFactor* are respectively the load factors for entering and exiting congestion state. In our simulations they were set to 91.25% and 90% of ABR bandwidth target utilisation.

The algorithm parameters *DeltaIncrShareFactor* and *DeltaDecrShareFactor* are respectively the increment and decrement to be given to ShareFactor each time a FRM cell arrives to the Head-end, when the Head-end is on a non-congestion or on a congestion state, respectively.

The decrement *DeltaDecrShareFactor* means that *ShareFactor* falls down from *MaxShareFactor*, i.e., the maximum of the *ShareFactor*$_i$ of all active ABR sources when congestion state is entered, until the (*MaxShareFactor*-1) value, in a calculated number of FRM cells, dependent on the number of the active sources. *ShareFactorIncrRate* means that *ShareFactor* increases its value by 1 (one) in a given number of FRM cells, also dependent on the number of active sources.

4 FATHOC TUNING

Tuning of FATHOC algorithm parameters *DeltaIncrShareFactor* and *DeltaDecrShareFactor* can be done taking into account the dynamic behaviour of ABR sources.

The number of active sources can vary strongly. If the *DeltaIncrShareFactor* is made independent of the number of active sources, *NumberOfVCs*, some problems could arise. For instance, if the number of sources is high, the increase of *ShareFactor* could be done after a low number of FRM cells compared with the total number of active sources, meaning that congestion could be entered quickly with few sources having their rates increased.

If we make *DeltaIncrShareFactor* dependent on the inverse of the number of active sources, we can reduce that effect, and all sources are allowed to increase progressively their rates. However, if the number of sources increase, *ShareFactor* will also increase more slowly, but it is also true that perturbations on the constraints of individual sources will have less importance on the overall bandwidth utilisation. Therefore it is reasonable to make *DeltaIncrShareFactor* dependent on the inverse of the number of sources.

FATHOC ALGORITHM

Initialisation
    ShareFactor = 1;
    congestion = false;

if FRM cell from any source i
    if not congestion
        if *LoadFactor* >= *EnteringCongestionStateLoadFactor*
            calculate *MaxShareFactor* = max of *ShareFactor*$_i$;
            ShareFactor = *MaxShareFactor*;
            calculate *DeltaDecrShareFactor*;
            congestion = true;
    else
        if *LoadFactor* < *ExitingCongestionStateLoadFactor*
        calculate *DeltaIncrShareFactor*;
        congestion = false;

    if congestion
        ShareFactor = *ShareFactor* - *DeltaDecrShareFactor* ;
    else
        ShareFactor = *ShareFactor* + *DeltaIncrShareFactor* ;
    $ER_{\text{Head-end i}}$ = MCR_i +
            ShareFactor * *ABRSharableCapacity* / *Number of VCs* ;

/* The *ER* transmitted to terminal is the minimum value of $ER_{\text{Head-end i}}$
and the *ER* that has come on a BRM cell from the ATM network */

Figure 1 Pseudo-code of the FATHOC algorithm

Similar reasoning can be made for *DeltaDecrShareFactor*.

Tuning of *DeltaIncrShareFactor* and of *DeltaDecrShareFactor* can be made such that

$$DeltaIncrShareFactor = \frac{1}{N_{FRMi}\ Number\ of\ VCs} \qquad (6)$$

$$DeltaDecrShareFactor = \frac{MaxShareFactor - 1}{N_{FRMd}\ NumberOfVCs} \qquad (7)$$

In expression (7), N_{FRMi} and N_{FRMd} are the mean number of FRM cells, received per virtual channel, calculated to produce the dimensioned *ShareFactor* changes, respectively 1 and (*MaxShareFactor*-1) for non-congestion and congestion states. These numbers of FRM cell cycles are dimensioned, taking into consideration the dynamic behaviour, but also admissible numbers.

We can limit the rise time from a non-congestion state to a congestion state to a maximum value (e.g. 150 ms), and the maximum fall time from a congestion state to a non-congestion state to another maximum value (e.g. 100 ms). Naming the rise and fall times respectively by τ_i and τ_d, we have N_{FRMi} and N_{FRMd} given by the following expressions

$$N_{FRMi} = \frac{ABRCapacity\ \tau_i}{(NRM + 1)\ C_l\ NumberOfVCs} \qquad (8)$$

$$N_{FRMd} = \frac{ABRCapacity\ \tau_d}{(NRM + 1)\ C_l\ NumberOfVCs} \qquad (9)$$

where *NRM* is the number of data cells between two FRM cells, and C_l is the cell length in bits.

However, we also limit those values to admissible values of

$$3 \leq N_{FRMi} \leq 6 \qquad (8a)$$

$$3 \leq N_{FRMd} \leq 6 \qquad (9a)$$

Doubling *ShareFactor* will be limited to a minimum of 3 FRM cycles and to a maximum of 6 FRM cycles. Halving the extra value above 1 of *ShareFactor*, will be limited to a minimum of 3 FRM cycles and to a maximum of 6 FRM cycles. For low *NumberOfVCs* that values fasten the non-congestion/congestion cycles, with respect to the cycles obtained with the former N_{FRMi} and N_{FRMd} values of expressions (8) and (9), while for high *Number of VCs* they shape the way *ShareFactor* is increased and decreased.

5 SIMULATION RESULTS

We have made several simulations of the FATHOC algorithm with ABR continuous sources. Simulations were run with a bus length of 40 km and 10 Mbps for the ABR bandwidth quota. The target utilisation was 90% of that value, meaning that *ABRCapacity* will be 9 MBps.

The load factors *EnteringCongestionStateLoadFactor* and *ExitingCongestionStateLoadFactor* were set respectively to 91.25% and 90% of *ABRCapacity*. The *DeltaIncrShareFactor* and *DeltaIncrShareFactor* values were those given by expression (8), (8a), (9), and (9a).

In the first simulation, we show how FATHOC achieves fairness and avoid congestion, and how the fairness criteria used is met. The simulation was done with six ABR sources. The first three have *MCR* = 100 Kbps and the last three have *MCR* = 500 Kbps. All sources have *PCR* = 12 Mbps, and begin at $t = 0$ ms. The duration of all simulations was 1.5 s. Figure 2 shows the results of the first simulations.

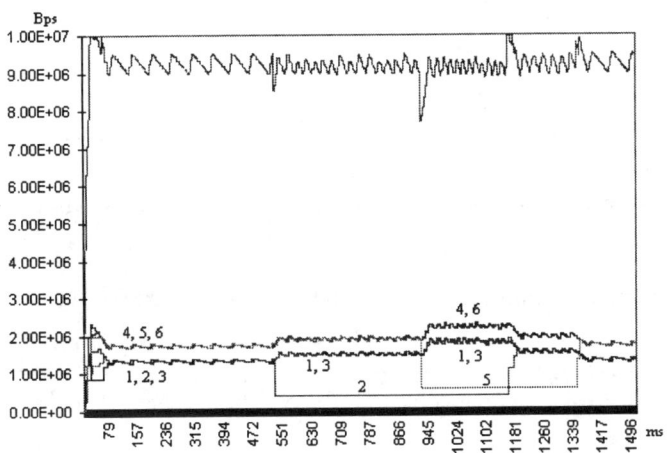

Figure 2 Simulation with different *MCR* and constraints on source rates

Note that source 2 and source 6 experienced constraints from the ATM core network. Source 2 between $t = 500$ ms and $t = 1.1$ s with *ER* = 400 Kbps, and source 5 between $t = 900$ ms and $t = 1.3$ s with an *ER* = 600 Kbps.

As it can be seen, the criterion chosen for fairness works quite well, and sources share equal fractions of bandwidth above their *MCR*. Sources 4, 5 and 6 have transmission rates that are 400 Kbps higher than sources from 1 to 3. This results from the way the *ShareFactor*ᵢ are calculated by expression (4). Note that minimum cell rate (*MCR*) requirements are not considered in the computations. It can also be seen that, when constraints on the source rates are present, the other sources rise their rates, and the fairness criteria are still met.

The other three simulations were done to test the behaviour of FATHOC in the presence of an increasing number of sources. Simulations were done with five, ten, and twenty sources. Sources had *MCR* = 100 Kbps, *PCR* = 12 Mbps, and all began at $t = 0$ ms. Several constraints, that we do not enumerate here, were imposed to the sources transmission rates.

The results of the simulations are shown on figure 3, 4, and 5.

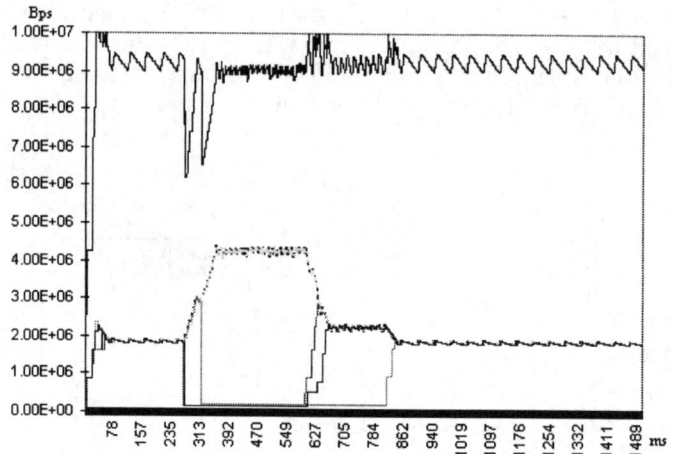

Figure 3   Simulation of FATHOC with five sources

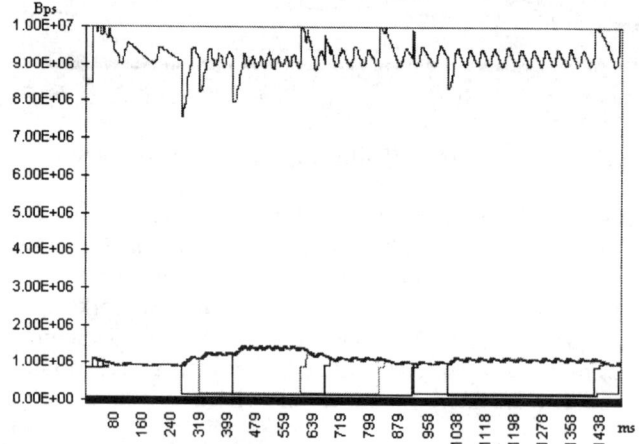

Figure 4   Simulation of FATHOC with ten sources

As can be seen, the sources that are not constrained by the ATM core network are led to share the bandwidth left available by the constrained sources. Sources transmission rates oscillations are small and decrease with the number of the active sources.

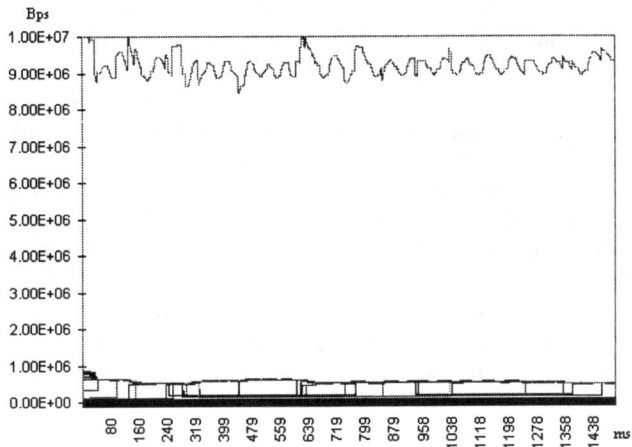

Figure 5 Simulation of FATHOC with twenty sources

6 ADAPTATION TO THE NRT-VBR CATEGORY SERVICE

Adaptation of FATHOC to nrt-VBR is simple. FRM cells does not exist, but an equivalent control interval can be easily implemented. In each control time interval, identification of non-congestion and congestion states and computation of *ShareFactor* and other algorithm parameters can be done with the same computations of the ABR version.

With nrt-VBR sources, the constraints on the transmission rates are the contracted *PCR* of the connections. When equal rate values for all sources are higher than the *PCR* values of some sources, this means that these sources are constrained, and the extra bandwidth is distributed by the sources with higher *PCR* values. It is worth noting that FATHOC was designed, not bearing in mind any specific service category, but only generic continuous type sources.

7 CONCLUSIONS AND FUTURE WORK

FATHOC is a simple algorithm for ABR fairness achievement and congestion avoidance applicable to HFC networks. Its computations are simple and very flexible to implement different fairness criteria. FATHOC can also be easily adapted to the nrt-VBR category service.

The basic FATHOC principle to achieve fairness is to let sources rise their rates until a congestion state is reached and then apply a criterion of fairness to decrease the rates of sources using more bandwidth, with respect to that criterion, until the congestion state is exited. In the congestion/non-congestion cycles, sources using more bandwidth give up bandwidth, and sources using less bandwidth acquire bandwidth, according with the developed algorithm.

FATHOC deals very well with constraints on source rates imposed by the ATM network. Unconstrained sources achieve naturally equal shares of the bandwidth left after satisfaction of minimum rate requirements, and of the constraints imposed to the source rates.

We also think that FATHOC computation framework is suitable to the application of control theory techniques. We have been working on a PID controller for FATHOC, but these investigations are still in a preliminary stage.

8 REFERENCES

ATM Forum (1996) Traffic Management Specification, Version 4.0, af-tm-0056.000, The ATM Forum, April 1996.

Sierens, C., Krijntjes, K., Nunes, M.S., et. al. (1996) A first description of the Hybrid Fibre Coax concept in the framework of the ACTS project. ATHOC IEE Meeting on Optical and Hybrid Access, BT Laboratories, Martlesham Heath, Ipswich, UK, March 1996, 1-7.

Nunes, M.S. and Macedo, M.M. (1996) A Simple Reservation MAC Protocol for HFC Networks. IEEE ATM'96 Workshop, San Francisco, USA, August 1996, T1-7.

Macedo, M.M., Nunes, M.S., Duarte-Ramos, H. (1997) SARPVBR - A VBR rate control algorithm for HFC networks. IEEE ATM'97 Workshop, Lisbon, Portugal, May 1997, 486-493.

Jain, R. (1996a) Congestion control and Traffic Management in ATM Networks: recent advances and a survey. Computer Networks & ISDN Systems, Volume 28, issue 13, October 1996, 1713-1738.

Jain, R., Kalayanaraman, S., Goyal, R., Fahmy, S., Viswanathan, R. (1996b) ERICA Switch Algorithm: A Complete Description. ATM Forum/96-1172.

Jain, R., Fahmy, S., Kalayanaraman, S., Goyal, R. (1996c) ABR Switch Algorithm Testing: A Case Study with ERICA. ATM Forum/96-1267.

Traffic Control

32

Decentralized Control Scheme For Large-Scale ATM Networks*

Lorne G. Mason, Anne Pelletier, Eric Létourneau
INRS-Télécommunications
16, Place du Commerce
Verdun, Québec, Canada, H3E 1H6
Tél: (514) 765-7836, Fax: (514) 761-8501
lorne@inrs-telecom.uquebec.ca

Abstract

Two approaches are proposed in this paper in order to facilitate network access control in large-scale ATM networks. We consider the centralized isarithmic control scheme that controls the total number of cells in the network using a permit mechanism. An implementation problem of this window-based scheme in large-scale networks is the significant control overhead associated with the flow of permits. The first approach is a partially decentralized control architecture that extends the centralized isarithmic scheme by means of an hierarchical zoning structure. An adaptive isarithmic controller is located in each zone with a distinct permit class and the zones interact through user traffic and management information flows. The decentralized architecture allows a substantial reduction of the permit flows and numerical results show that the network optimal operating point can be reached. Moreover, the level of management flows can be limited without deteriorating the network performance by using the approximate global information structure, based on mean delays. The second approach consists of decreasing the window size by letting one permit correspond to C credits. The control overhead is then reduced by a factor of C, the credit/permit ratio. Simulation results show that the network performance can even be improved compared to the centralized isarithmic scheme.

Keywords

IP/ATM, adaptive flow control

1 INTRODUCTION

A challenge in designing ATM networks is the synthesis of effective control architectures and algorithms for managing heterogeneous traffic and network resources such that quality of service constraints are met while achieving efficient use of the network resources. In this paper we address the issue of

---

*This work was supported by a grant from the Canadian Institute for Telecommunications Research under the NCE program of the Government of Canada.

Broadband Communications   P. Kühn & R. Ulrich (Eds.)
© 1998 IFIP. Published by Chapman & Hall

network configuration management in large-scale ATM networks. Network configuration management refers to the functions of virtual network topology and bandwidth allocation which enable the partition of the aggregate network capacity into its component virtual networks. The control architecture considered here and outlined in [1] supports both switched connection-oriented services and connectionless services: switched VBR connections are carried on a virtual network while delay-insensitive traffic (either connectionless or connection-oriented) is handled via the residual network component. The logical bandwidth allocated to the delay-insensitive traffic must be enforced through either rate-based or credit-based congestion control schemes (see [2] and [3] and references therein).

In this paper, we consider the adaptive isarithmic flow control scheme [4] which regulates the delay-insensitive traffic flow entering the residual network based on its optimal operating point using a permit mechanism. The resulting window size and permit distribution determine the configuration and allocated bandwidth of the residual network. To ensure that the performance of time-critical VBR traffic is not degraded, we employ the selective window control with priority scheduling [4].

Isarithmic flow control was proposed by Davies [5] for network access control in packet-switching networks. This scheme limits the total number of packets that can be in transit within the network. This is accomplished by allowing packets to enter the network only after they secure a permit. When the packet/permit combination arrives at the destination node, the permit is released and becomes available for packets entering at that node. It was found that this approach can result in degraded performance under asymmetric traffic conditions. Mason and Gu [6] proposed to employ a controller for permit distribution. In the centralized version, a controller distributes the released permits to the source queues associated with each node. Another approach is decentralized where a controller is associated with each destination node [7]. For both cases, the authors proposed several learning automata control schemes to track the optimal operating point of the network. Cotton and Mason [8] designed an adaptive central controller for fast packet-switching networks. When compared to a rate-based scheme, the isarithmic scheme has been shown to have similar performance in terms of throughput and fairness [9].

Two approaches are proposed in this paper to enable the use of the isarithmic scheme in large-scale ATM networks. The first approach is a partially decentralized control architecture which extends the centralized isarithmic approach by means of an hierarchical zoning structure. The zones are sub-networks of the given network and are connected via gateway nodes to other zones. Each zone operates a centralized adaptive isarithmic control within its boundary for a distinct permit class and it interacts with the other zones through user traffic and management information flows. The effects of zone partitioning and partial information structure on the resulting network performance are studied here.

The second approach consists of reducing the number of permits. As discussed in [10], the window size can be reduced by using one permit to control a group of cells instead of individual cell. We propose a different mechanism where we let one permit correspond to C credits. When a permit arrives at a node, the value of the counter increases by C. As in the original scheme, each admitted cell decreases the value of the counter by one. At the exit nodes each cell adds one to a counter of credits and when C are accumulated one permit is sent to the controller for dispersal. The permit flow in the network as well as the action rate of the controller are reduced by a factor of C.

We would like to mention that the control architecture described in [1] can be extended to the case of managing multiple virtual private networks which are expected to proliferate with the emergence of ATM networks – LAN interconnection being an important early example. In this concept different corporate users can be offered features similar to those of a dedicated private network while retaining the efficiency advantages of resource sharing and connectivity typical on an integrated public network. The multi-chain approach proposed in [6] allows the residual network to be subdivided into multiple such virtual networks carrying delay-insensitive traffic.

The paper is organized as follows. In Section 2, we present the performance criterion employed in this paper. Section 3 reviews the centralized adaptive isarithmic scheme and the selective windowing mechanism with priority. The decentralized control architecture is proposed in Section 4 and simulation results show that the network optimal operation point can be achieved. In Section 5, we study the effect of reduced control overhead by the introduction of the credit/permit ratio. Section 6 concludes.

2 PERFORMANCE MEASURE

A performance criterion for congestion control is the power function proposed by Kleinrock [11]. Defined as the ratio of average throughput over average delay, the power function unifies the antithetic objectives of high throughput and low delay. However, as Bharath-Kumar and Jaffe pointed out in [12], the power function is an unfair measure for network access control since it can happen that some users are denied access to the network. To prevent unfairness we employ the product of powers defined as [12]:

$$P = \prod_{o,d \mid \Lambda_{o,d} \neq 0} \frac{\lambda_{o,d}}{T_{o,d}} \qquad (1)$$

where
$\Lambda_{o,d}$: exogenous traffic from node o to node d;
$\lambda_{o,d}$: throughput of o-d traffic;

$T_{o,d}$: mean end-to-end delay of *o-d* traffic.

A proof that the product of powers is indeed a reasonable design criterion was provided by Mazumdar and *al.* [13]. The maximal product of powers point is unique and corresponds to the Nash arbitration strategy for the users: at optimum, if one user improves his power, it will adversely affect at least one other user power.

3   REVIEW OF CONTROL FRAMEWORK

In this section, we will describe in general the centralized adaptive isarithmic flow control scheme proposed by [6] [8] for high-speed data networks and the selective window control introduced by [4] to support different classes of traffic.

3.1   Centralized adaptive isarithmic flow control

In the centralized version of the adaptive isarithmic scheme, a controller is implemented to distribute the released permits as illustrated in Fig. 1. A packet is accepted to enter the network only if there is a permit available at the source queue. The packet/permit combination is routed through the network and at the destination node, the permit is sent to the central controller. The central controller is composed of two parts: the window size and the permit distribution controllers. The latter distributes adaptively the released permits while the window size controller adjusts the total number of permits, W, according to network conditions.

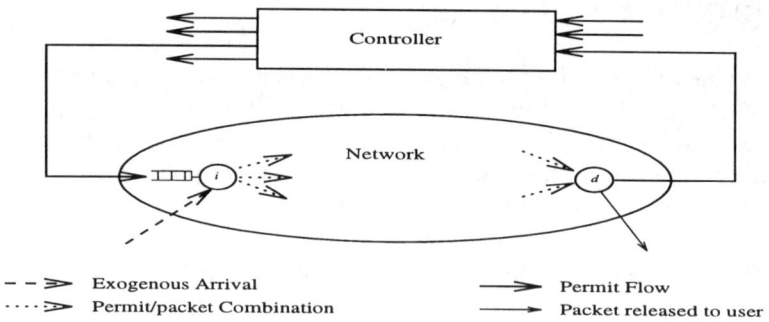

Figure 1 Centralized adaptive isarithmic flow control

The permit distribution controller is a *learning automaton* based on the structure proposed by Mason [14]: it is an L_{R-I} type automaton in an *S-*

model environment. The distribution probabilities are updated using a *learning algorithm* [8].

The second component of the central controller is the window size controller. The population of permits must also be adjusted since no value is optimum for all loads. Several methods were proposed for adaptive window size adjustment (see [15] and the references therein). A method to optimize the window size for a virtual circuit was proposed by Mitra and Seery [16]. It is based on an explicit expression relating the optimal window size to the mean round trip response time. For the isarithmic scheme, a self-optimizing method employing a cross-correlation technique for process identification was proposed by [17]. Another method to evaluate the sensitivity of the product of powers with respect to the window size is based on the phantom RPA method for estimation of derivatives [18].

3.2 Selective window control

The window size adaptation in ATM networks is more complicated than in packet-switching networks because of the frequent changes in the isochronous traffic level. To adapt the window size a fast algorithm is required which could be difficult to realize. The selective window control illustrated in Fig.2. was proposed by Liao and Mason [4] to avoid this adaptation problem.

NETWORK ACCESS STRUCTURE   LINK ACCESS STRUCTURE

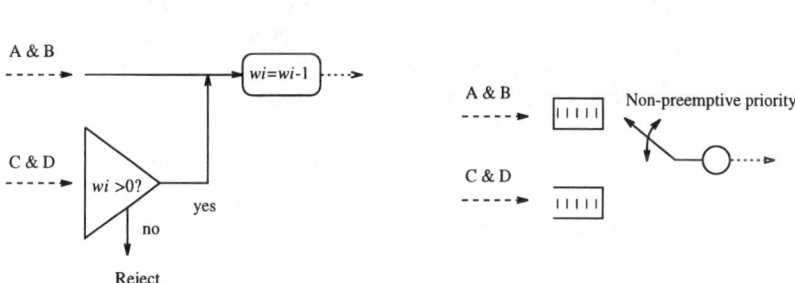

Figure 2 Selective window control with a priority mechanism

At each node, the value of counter w_i is initially set such that $\sum_i w_i = W$. It is assumed that an appropriate preventive control is implemented for isochronous traffic cells (Classes A and B in ITU-T standards). The selective window control is transparent for the isochronous traffic, but each admitted cell at node i decreases by 1 the value of the counter w_i. A delay-insensitive cell (Classes C and D) is admitted only if the counter is positive, hence this scheme adjusts the level of data traffic as a function of the level of isochronous traffic. The theoretical studies and simulation results presented in [4] have shown that the optimal window size is independent of the level of isochronous traffic. The link access structure consists of two buffers, one for each type of traffic and the service discipline gives non-preemptive priority to isochronous traffic. Simulation results have also shown that the window size has a negligible effect on the performance of the isochronous traffic because of this priority scheme. The network efficiency is further improved by using finite buffers for the isochronous cells. This prevents using the network resources to carry isochronous cells with excessive delay.

Since the selective window controls only the delay-insensitive traffic and since the priority scheme assures that the window control has a negligible effect on the isochronous traffic, we will only consider in this paper the performance of the delay-insensitive traffic as measured by the product of powers.

4   DECENTRALIZED CONTROL ARCHITECTURE

Some practical problems arise when one wants to implement a centralized control scheme in a large-scale network. On the one hand, it implies linking all the nodes to a single controller which can become fairly impossible when the number of nodes explodes. When considering a window-based control scheme, such as the centralized isarithmic control scheme described in Section 3.1 the large window size means significant control overhead. In this section we propose a partially decentralized control architecture that extends the centralized isarithmic approach and we study some design issues that are related to it.

The network management function in most existing and planned networks is either hierarchical or completely centralized. The method we proposed here assumes an hierarchical zoning structure, where each zone is a subnetwork of the given network. In each zone, one node serves as a gateway to communicate with the other zones. In the hierarchical structure considered here, a low-level zone is connected to one higher-level zone only. The number of levels as well as the number of zones connected to a higher-level zone are general. An adaptive isarithmic controller is implemented in each zone with a distinct permit class within its boundary.

The permits for a given zone must circulate in that zone only. This implies that for cells crossing zone boundaries at the gateway nodes, the attached permits are routed to the associated controller when the cells arrive at the gateway of the origin subnetwork. There is no access control for cells arriving at a gateway from another subnetwork. They are treated as endogenous arrivals at the gateway. In other words, when a cell gains access to a subnetwork it also has access to the entire network.

One attractive advantage of the centralized adaptive isarithmic control is that it automatically ensures that the global information flow, necessary for fair-efficient operation, is available without the need for further protocols and interfaces. Under the decentralized control architecture described above, management information flows must circulate between subnetworks to achieve the network optimal operating point. In each zone, the adaptive isarithmic controller will adjust the window size and the permit distribution probabilities in order to maximize the product of powers of the entire network. The permit flows in one zone provide the associated controller with the measurements (delays and throughputs) for the traffic originating and terminating in that zone. The management information flows arriving from other zones will provide the other measurements required.

We will now concentrate on the window size adjustment which requires a single number, i.e. the product of powers of all o-d pairs in the network. The permit distribution adjustment problem under the decentralized control architecture will be addressed in a further study.

4.1   Information structures

Different information structures correspond to what delay and throughput information is made available to the adaptive controllers. There are three basic cases, namely *local* (o and d in same zone), *exact global* and *approximate global* (o and d in all zones). Under the local information structure, only the o-d pairs originating and terminating in a given zone are considered by the adaptive isarithmic controller to adjust the zone's window size. The performance criterion is not the product of powers of the complete network thus it implies suboptimality (see Section 4.2 below). In the exact global information structure, all necessary information is passed across zone boundaries in order to achieve the optimal performance for the complete network. Approximations of the delay measurements are employed in the approximate global information structure to reduce the amount of management information flows circulating between the zones.

The product of powers of the complete network requires the knowledge of the throughputs and delays of all o-d pairs in the network. Each controller needs this information to adjust the number of permits in its zone. At first glance, one would think that all these parcels of information must circulate from one zone to the others but the nature of the performance criterion avoids this situation. In fact very low information overhead is required.

A partial product can be computed in a zone for all o-d pairs originating in the zone using throughput and delay information available locally while management information flows provide the delays measurements for the pairs terminating remotely. The product of powers of all o-d pairs originating in a zone is a single number per zone. The zone's controller multiplies this number with the partial products received from the other zones to obtain the network product of powers. A given zone z receives one partial product from the zone situated above. This partial product is the product of powers of all o-d pairs originating in this part of the hierarchy. Zone z also receives one partial product from each zone below. In return the given zone transmits to the higher-level zone one partial product for the *o-d* pairs originating in zone z and in the zone beneath. It also sends a partial product to each lower-level zone. In short, one number is sent by a zone to the zone above it and one number is sent to each zones below it.

Apart from this information, the management flows carry the delay measurements of the pairs that terminate remotely. Under the exact global information structure the number of delay values sent by zone z of level l to the zone of level $l+1$ is:

$$N_z(N - N_z) \tag{2}$$

where N_z is the total number of nodes in zone z and in the zones under it, and N is the total number of nodes in the network. The higher-level zone sends the same amount of information to zone z. As we move up in the hierarchy, the number of values to exchange increases which can become an implementation problem. The approximate global information structure is proposed to avoid it. Rather than carrying all the delay measurements, the management flows could carry mean delays only. Each zone computes the arithmetic mean of the delays between the gateway and each node in the zone. This value is made available to the other zones via the management flows which also carry the gateway-to-gateway delays. The amount of information grows linearly with the network's size instead of exponentially.

4.2   Simulation results

In this section, we present simulation results for an ATM network operating under the decentralized control architecture. We will compare the network performance for each information structure described above (local, exact global and approximate global). We also compared it to the performance under the centralized isarithmic control, which corresponds to one zone spanning the whole network.

The topology of the simulated networks $N1$ and $N2$ is depicted at Fig. 3. It has one level consisting of two zones (subnetworks A and B) with a gateway in each and a connection between gateways. The capacity of each unidirectional link is $1.7Gbits/s$. The two networks have the following characteristics:
$N1$: All the links cover a distance of 13 Km.
$N2$: Links $A1$-$A2$ and $B1$-$B2$: $16.7Km$, links $A1$-$A4$ and $B1$-$B4$: $9.3Km$ (all other links remaining at $13Km$).

The two traffic types simulated are the isochronous traffic (Type 1) and the delay-insensitive data traffic (Type 2). The isochronous traffic (Classes A and B in ITU-T standards) is modeled by two-state Markov Modulated Poisson Process (MMPP) which approximates the superposition process of bursty sources. The MMPP parameters are the arrival rates during states 1 and 2 ($\lambda_{1,1}$, $\lambda_{1,2}$) and the mean sojourn times in states 1 and 2 ($r_{1,1}^{-1}$, $r_{1,2}^{-1}$).

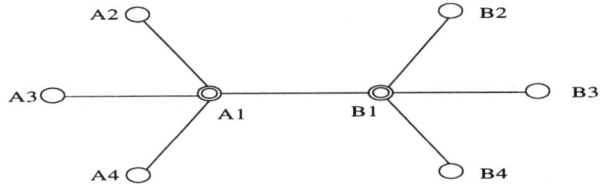

Figure 3 Simulated network topology

In the rest of this paper, the time unit is the cell transmission time. The arrival rates are:

$$\lambda_{1,1}(i,j) = \begin{cases} 0 & i = j \\ 0.16 & i,j | i,j \in subnetwork\ A \\ 0.16 & i,j | i,j \in subnetwork\ B \\ 0.08 & otherwise \end{cases}$$

$$\lambda_{1,2}(i,j) = \begin{cases} 0 & i = j \\ 0.04 & i,j | i,j \in subnetwork\ A \\ 0.04 & i,j | i,j \in subnetwork\ B \\ 0.02 & otherwise \end{cases}$$

The total offered isochronous traffic to each node is $\Lambda_1(i) = 0.5$ cell per cell transmission time and overall, $\Lambda_1 = 4$. All sources have $r_{1,1} = r_{1,2} = 0.005$. Due to the selective window control with the priority mechanism, we are assured that no isochronous cell will be refused. The residual bandwidth will be allocated to the delay-insensitive data traffic. In the simulation, the finite buffers for isochronous traffic have 100 waiting places.

The data traffic (Classes C and D in ITU-T standards) is assumed to arrive according to Poisson process. The arrival rates for Type 2 traffic are:

$$\lambda_2(i,j) = \begin{cases} 0 & i = j \\ 0.225 & i,j | i,j \in subnetwork\ A \\ 0.225 & i,j | i,j \in subnetwork\ B \\ 0.1125 & otherwise \end{cases}$$

which implies $\Lambda_2(i) = 1.125$ and $\Lambda_2 = 9$.

Table 1 presents the simulation results obtained with the decentralized implementation for each information structure (exact global, approximate global, and local). During each simulation, the window sizes in zones A and B, denoted W_A and W_B, are identical. We report the network performance, as measured by the product of powers, for a range of window sizes. Because of the magnitude of the product of powers its nth root is used, where n is the number of o-d pairs in the network (here $n = N(N-1) = 56$). The per-

Case	W_A, W_B	Exact global $\sqrt[56]{P}$	Approx. global $\sqrt[56]{\tilde{P}}$	Local $\sqrt[12]{P}_{A,\text{loc.}}$
$N1$	1200	2.42×10^{-5}	2.41×10^{-5}	5.43×10^{-5}
$N1$	1230	2.64×10^{-5}	2.63×10^{-5}	6.15×10^{-5}
$N1$	1260	2.87×10^{-5}	2.87×10^{-5}	6.83×10^{-5}
$N1$	1290	2.99×10^{-5}	2.99×10^{-5}	7.58×10^{-5}
$N1$	1320	2.79×10^{-5}	2.78×10^{-5}	8.41×10^{-5}
$N1$	1350	1.47×10^{-5}	1.47×10^{-5}	9.25×10^{-5}
$N2$	1200	2.45×10^{-5}	2.43×10^{-5}	5.51×10^{-5}
$N2$	1230	2.67×10^{-5}	2.66×10^{-5}	6.24×10^{-5}
$N2$	1260	2.90×10^{-5}	2.89×10^{-5}	6.92×10^{-5}
$N2$	1290	2.98×10^{-5}	2.97×10^{-5}	7.70×10^{-5}
$N2$	1320	2.86×10^{-5}	2.86×10^{-5}	8.58×10^{-5}
$N2$	1350	1.47×10^{-6}	1.47×10^{-6}	9.41×10^{-5}

Table 1 Simulation results with decentralized control architecture

formance measure is related to the information structure considered. For the exact global information structure, we have the exact product of powers P, whereas $\tilde{P}$ is computed under the approximate global information structure. In the last column, we report $P_{A,\text{loc.}}$ the product of powers of *o-d* pairs that originate and terminate in zone A. (Due to the symmetry, the local value for zone B is similar.)

For both $N1$ and $N2$, the optimal operating point is $W_A^* = W_B^* = 1290$ since the product of powers P is then maximized. The simulation results show that the approximate product of powers $\tilde{P}$ is a very good estimate of the exact product of powers. Under the approximate global information structure the optimal operating point is also $W_A^* = W_B^* = 1290$. We can see that the product of powers is slightly underestimated when the window sizes are under their optimal values. However the results obtained for the local information structure are not satisfactory. The local product of powers for zone A indicates that the optimal operating point would be achieved for very large window sizes. But if we consider $W_A = W_B = 1350$, we can observe that the exact global product of powers is not optimal. The local information structure considers only 12 *o-d* pairs thus it is an inadequate performance estimate.

While it was found by simulation that the optimal operating point under centralized control is $W^* = 3300$ for $N1$ and $N2$, the optimal window size for the decentralized control is $W_A^* + W_B^* = 2580$. This reduction in the total population of permits is possible because permits are returned to their controller when the cells cross zone boundaries. To conclude, these simulation results show that the decentralized approach is very attractive. The total window size is smaller and the approximate global information structure allows very low information overhead without impairing the performance.

5 CREDIT/PERMIT SCHEME

In this section we present our second approach to implement the isarithmic scheme in large-scale networks. This approach is named the credit/permit scheme. In order to reduce the flow of permits in the network, as well as to reduce the real time processing load at the adaptive controllers one could let a permit correspond to C credits. When a permit arrives to a source queue it is replaced by C credits where an admitted cell decreases by 1 the counter of credits, as previous. At the exit nodes each cell adds 1 to a counter of released credits and when C credits are accumulated a permit is sent to the adaptive controller for dispersal. The action rate of the controllers is then reduced by a factor of C as well as the signaling traffic (i.e. permits).

Because of the selective window control with priority, the credit/permit scheme does not affect the isochronous traffic and simulations were performed to assess its impact on the data traffic. The simulated network is identical to one of the subnetworks illustrated in Fig. 3 and all the links cover $13Km$. As in the previous case, the isochronous traffic is modeled by an MMPP and the data traffic by a Poisson process. The total offered traffics are: $\Lambda_1 = 2$ and $\Lambda_2 = 4.5$. Under the centralized isarithmic control, the optimal operating point of the network is W^*=2350 and $\sqrt[12]{P^*} = 4.76 \times 10^{-4}$.

Two implementations of the credit/permit scheme have been simulated. In version A, when a permit arrives at a source queue, it is immediately replaced by C credits. In version B, the credits are generated in a smooth manner that will be described below. Results for both implementation are plotted at Fig. 4.

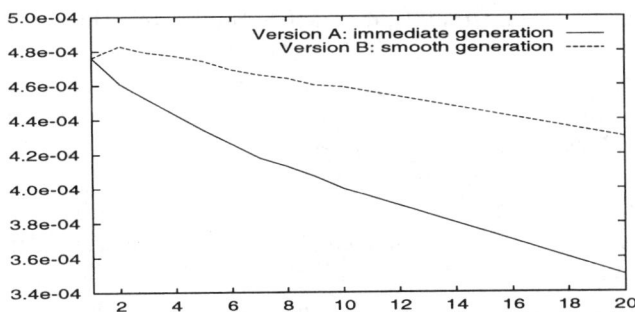

Figure 4 Version B: Product of powers versus credit/permit ratio

In this version, the network performance is even better than the original isarithmic scheme for small values of the credit/permit ratio. Since the credits are generated smoothly, the bursts of data cells are avoided. Moreover, the data traffic is shaped such that its mean delay is smaller. Again when C is large, the throughput of data traffic decreases and this degradation is more important than with version A. But in terms of the product of powers, version B outperforms version A due to the low delays involved.

6   CONCLUSIONS

This paper proposed two approaches to facilitate network access control in large-scale ATM networks. We consider the selective window control scheme with priority mechanism. It can be difficult to implement this window-based scheme in large-scale networks because of the control overhead associated with the flow of permits.

First, we proposed a partially decentralized control architecture which extends the centralized isarithmic scheme by means of an hierarchical zoning structure. An adaptive isarithmic controller is located in each zone with a distinct permit class and the zones interact through user traffic and management information flows. It was demonstrated that the decentralized architecture allows a substantial reduction of the flow of permits. Numerical results showed that the network optimal operating point can be reached under the decentralized implementation. Moreover, the level of management flows can be limited without deterioration of the global performance by employing the approximate global information structure, based on mean delays.

The second approach proposed is the credit/permit scheme that reduces the control overhead by a factor of C, the credit/permit ratio. This second approach can be used in conjunction with the first approach to further reduce the management flow overhead.

The simulation results presented here were for a hypothetical network example with a topology that lends itself naturally to a hierarchical zoning structure. The results have shown the validity of the approach in this simple case. The methodology, proposed for decentralization of access flow control is however not restricted to the simple case simulated, and results for more general topologies are currently under study and will be reported upon in due course. The merit of the proposal we believe is chiefly that it provides a potential alternative access flow control scheme to the various VC based ABR proposals for efficient transparent support of connectionless data services in a multi-service ATM network, by means of an overlay virtual network as proposed in [1] and elaborated in [19].

REFERENCES

[1] L.G. Mason, Z. Dziong, K.-Q Liao, and N. Tétreault (1990) Control Architectures and Procedures for B-ISDN, *in Proc. of* 7th *ITC Spe-*

cialists' Seminar, **Morristown, New Jersey.**

[2] H.T. Kung and R. Morris (1995) Credit-Based Flow Control for ATM Networks, *IEEE Network*, **March/April**, pp. 40–49.

[3] F. Bonomi and K.W. Fendick (1995) The Rate-Based Flow Control Framework for the Available Bit Rate ATM Service, *IEEE Network*, **March/April**, pp. 25–56.

[4] K.-Q. Liao and L.G. Mason (1994) A Congestion Control Framework for Broadband ISDN Using Selective Window Control, *IFIP Conf. on Broadband Communications (Broadband'94)*, **Paris.**

[5] D.W. Davies (1972) The Control of Congestion in Packet Switching Networks, *IEEE Trans. Commun.*, pp. 546–550.

[6] L.G. Mason and X.D. Gu (1986) Learning Automata Models for Adaptive Flow Control in Packet Switching Networks, *in Adaptive and Learning Systems – Theory and Applications*, **K.S. Narendra, New-York: Plenum Press**, pp. 213–227.

[7] F.J. Vazquez-Abad and L.G. Mason (1992) Decentralized Adaptive Isarithmic Flow Control for Packet-switched Networks, *Second ORSA Telecom. Conf., Boca Raton.*

[8] M. Cotton and L.G. Mason (1995) Adaptive Isarithmic Flow Control in Fast Packet Switching Networks, *IEEE Trans. Commun.*, **April.**

[9] E. Létourneau and L. G. Mason (1995) Comparison Study of Credit-Based and Rate-Based ABR Control Scheme, *in Proc. of the First IFIP Workshop on ATM Traffic Management, Paris*, **December.**

[10] T. Toniatti and F. Trombetta (1992) Performance simulation of end-to-end windowing in ATM networks, *in Proc. of INFOCOM*, pp. 495–502.

[11] L. Kleinrock (1979) Power and Deterministic Rules of Thumb for Probabilistic Problems in Computer Communications, *in Proc. Int. Conf. on Commun.*, pp. 43.1.1–43.1.10.

[12] K. Bharath-Kumar and J.M. Jaffe (1981) A New Approach to Performance Oriented Flow Control, *IEEE Trans. Commun.*, pp. 427–435.

[13] R. Mazumdar, L.G. Mason and C. Douligeris (1991) Fairness in Network Optimal Flow Control: Optimality of Product Forms, *IEEE Trans. Commun.*, pp. 775-782.

[14] L.G. Mason (1973) An Optimal Learning Algorithm for S-Model Environments, *IEEE Trans. on Automatic Control*, pp. 493–496.

[15] S. Pingalu, D. Tipper and J. Hammond (1990) The performance of adaptive window flow controls in a dynamic load environment, *in Proc. of INFOCOM*, pp. 55–62.

[16] D. Mitra and J.B. Seery (1991) Dynamic Adaptive Windows for High Speed Data Networks with Multiple Paths and Propagation Delays, *in Proc. of INFOCOM*, pp. 39–48.

[17] L.G. Mason and K.-Q. Liao (1993) Self-Optimizing Window Flow Control in High Speed Data Networks, *Proc. of IEEE Globecom 1992*, also *Computer Communications*, **Vol. 16, No. 11** , pp. 706–716.

[18] A. Pelletier, M. Cotton and L.G. Mason (1993) Combined Adaptive Routing and Flow Control in Fast Packet Switching Networks, *in Proc. of COMCON 4*, pp. 557–569

[19] Z. Dziong, Y. Xiong, and L.G. Mason (1996) Virtual Network Concept and its Application for Resource Management in ATM Based Networks, *in Broadband Communications, Chapman & Hall*, **Lorne Mason and Augusto Casaca, editors**, pp. 223–234.

33

Feasibility of a Software-based ATM cell-level scheduler with advanced shaping

J. Schiller
Institute of Telematics
University of Karlsruhe
Karlsruhe, Germany
j.schiller@ieee.org

P. Gunningberg
Department of Computer Systems
Uppsala University
Uppsala, Sweden
per.gunningberg@docs.uu.se

Abstract

Future servers are expected to handle a large number of connections with different Quality of Service (QoS) requirements. Networks, e.g., based on ATM technology, provide QoS via standardized traffic parameters. While the control at the network edge can handle these parameters, support for generating adequate traffic patterns (shaping) in the end-systems is limited due to hardware restrictions. This paper presents a software-based cell-level scheduler for ATM with advanced shaping mechanisms supporting priorities and fair sharing in overload situations. Furthermore, a scalable integration of the cell-level scheduling and the scheduling of DMA transfers is shown.

Keywords
ATM, shaping, scheduling, DMA, network adapter, proportional sharing, leaky bucket

1 INTRODUCTION

Future network servers are expected to handle a large number of concurrent connections with varying Quality of Service (QoS) requirements (Campbell, 1996). Examples of likely traffic sources are voice/video conferences integrated with multimedia documents, multimedia document retrieval, WWW, file transfer and real time interactive games. The bandwidth requirements will range from a few Kbit/s to

Broadband Communications   P. Kühn & R. Ulrich (Eds.)
© 1998 IFIP. Published by Chapman & Hall

several Mbit/s. Thus, servers using network technologies with link bandwidths of hundreds of Mbit/s have the potential to carry a substantial number of these connections. The problem which arises is to handle all these connections and their QoS requirements. Current WWW servers serve requests using a best-effort method, which is not sufficient when the individual connections have QoS requirements and as the number of connections increases. We assume that a future large WWW server must handle many concurrent connections with potentially highly different QoS requirements.

The network adapter of such a server is responsible for several functions in the outbound direction. It should transfer data from server memory, apply low level protocol functions, control the transmission of data according to a traffic contract and multiplex several connections. Furthermore, in our view a more advanced network adapter is needed to perform traffic shaping. Our thesis is that a CPU based approach for doing these functions for a large number of connections is both feasible and efficient. Using software and state of the art microprocessors has several benefits. The algorithms can easily be changed to suit the needs of different traffic types. In addition, a software solution benefits of new on-board processors as they become available. Our solution will scale with the increase in processing power provided the memory system keeps up. An additional issue which is investigated in this paper is if and how the traffic shaping and the scheduling of memory transactions can be combined.

The contribution of this work is a feasibility study of a software scheduler running on an on-board network adapter CPU combining the traffic shaping functionality with the scheduling of memory transactions. These operations are often decoupled, leading to hardware redundancy, double buffering of data, and the limited capacity for shaping in the I/O subsystem.

The network technology we have envisioned is ATM since it supports QoS parameters and traffic contracts for every connection. However, the proposed traffic shaping algorithms used by the network adapter are general and may be employed in any network technology with flow reservations, such as the next generation Internet protocols. In this work, we focus on how to schedule connections which have been accepted by the system. The connection acceptance phase and the QoS architecture for the applications and higher layer protocols (Campbell, 1996) which are also crucial to the server is outside the scope of this work.

2   EXISTING HARDWARE SOLUTIONS AND THEIR LIMITATIONS

Comparing our approach with the capabilities of actual shaper-chips the following topics have to be addressed:

- *Scalability*: how does the approach scale in the number of connections, where are restrictions introduced by hardware, data structures etc.
- *Efficiency*: how efficiently can the shaper utilize the outgoing bandwidth.
- *Overhead*: how large is the overhead of the implementation, e.g., how often has a host system to be interrupted, how much control information has to be exchanged between host memory and the adapter.
- *Isolation*: how well are different connections protected from each other, what happens in transient overload situations, and how individual parameters can be adjusted to the needs of a connection.

The scalability of all chip-based solutions is very limited. Values like the maximum number of supported connections (VC) are of more theoretical nature and typically only limited by the width of registers. More interesting is the number of *simultaneously* supported connections or the number of connections supported *on-chip*. These numbers of supported connections rely not only on the amount of memory to store connection state date, but rather on the algorithms, data structures, and available processing power for traversing connection information.

There are typically limits on the maximum number of *different* traffic characteristics. Additionally, the question arises what parameters can be set and if they can be used independently. Typical restrictions are 8-12 PCR (Peak Cell Rate) queues, every connection has to be in one of these queues (Fujitsu, 1997), (SIEMENS, 1997). Furthermore, the SCR (Sustainable Cell Rate) is often derived from the PCR via a per connection ratio (e.g., ½ or ¼ of the PCR). These restrictions are due to the fact, that the cell rates are generated using explicit hardware counters, only a very limited number of these fit on a chip (c.f. (ATM Forum, 1996) for further explanation of ATM traffic parameters).

Also due to space limitations, most of the data structures holding context information for a connection are located in host memory. To update these structures the host system has to be interrupted and the memory accessed. This can result in a large overhead, especially when these data structures have to be checked with full PCR (Fujitsu, 1997). Another topic is the DMA overhead, most of the approaches transfer single cells from host memory 'just-in-time' due to very small transfer buffers on chip. Typically, the receiving side has a higher priority to avoid cell losses. Thus, the sending side suffers if cells arrive or even if control information has to be updated (SIEMENS, 1997).

The isolation between different connections sharing one PCR queue is typically not addressed in any of the evaluated solutions. Some solutions provide additionally priority classes, the behavior within a priority class during transient overloads is not further determined.

2.1 UPC solutions

It is important to compare the capabilities of the UPC (Usage Parameter Control) chips located at the UNI (User Network Interface) with those of the shapers due to the fact that the UPC chips will decide if an incoming cell of a connection shaped by one of the shapers is accepted or not. The first fact one notices is the more detailed control capabilities and the variety of parameters to check. The ATM_POL3 from ATecoM (Atecom, 1997) can control up to 64k VCs checking PCR, SCR, CDV (Cell Delay Variation) and maximum burst size using a dual leaky bucket per VC. IgT has the WAC-186-B (Integrated, 1995) which uses GCRA to monitor PCR, SCR, CDV and burst tolerance for up to 16k active VCs for a bitrate up to 250 Mbit/s. Finally, the BNP2010 UPC of National Semiconductor (National, 1997) checks up to 16k VC with 3 GCRAs per VC up to a speed of 622 Mbit/s. A simple comparison shows, that the UPC can be much more precise and, hence, more restrictive than the capabilities of the best shaper chips.

3 ARCHITECTURES

The overall architectural design is illustrated in Figure 1. The adapter has a buffer memory to hold cells to be transmitted and a DMA engine that reads data from server memory into the buffer memory. Each connection has a separate queue of cells in the memory. Buffer memory is accessed from the ATM chip which does the actual transmission. The purpose of the CPU is to schedule the DMA engine for a data transfer, schedule the ATM chip to read a cell from the buffer memory according to the traffic shape and multiplexing state, and exchange control messages with the host. Note that the CPU does not touch data at all since this would be too time consuming.

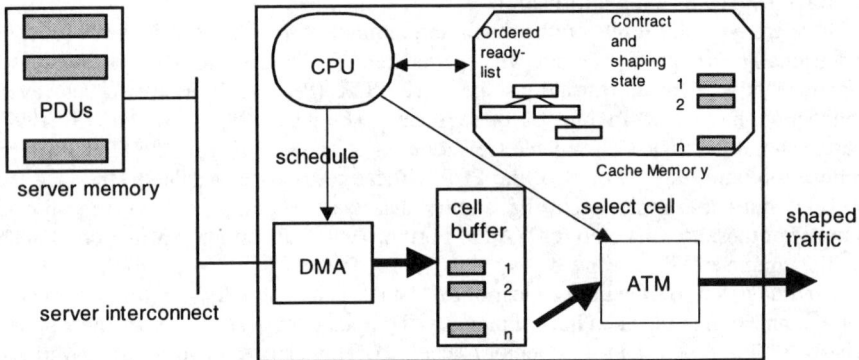

Figure 1 Network adapter architecture.

The CPU is running the following cycle for each transmitted cell.
1.   Pick the first connection identifier in the ordered ready-list of active connections, provided it is due for transmission.
2.   Initiate the ATM chip for a transfer of a cell from this connection.
3.   Shape the connection and calculate the time when the next cell of this connection should be sent according to the shaping state.
4.   Insert the connection identifier again at a new place in the ready-list according to the next transmission time.
5.   Schedule a possible DMA transfer of data from server memory for this connection.

This cycle must finish well within the time it takes to send a cell. For a 622 Mbit/s this means within 680 ns. Besides running this cycle, the CPU needs to synchronize PDU information with the server and to allocate and deallocate buffer memory. But these tasks are triggered by asynchronous events and can be done in the background.

The CPU needs a fairly large memory to hold the state of each connection and the data structure for the ready-list. The access time to this memory is crucial for the performance and it is expected that a large Level 2 (L2) cache memory is the most appropriate. The actual scheduling code is small enough to fit into Level 1 (L1) instruction cache. The adapter may have other protocol hardware on board, such as for AAL5 checksum calculation which also must be controlled by the CPU. In addition, there is some control logic that is specific to the interconnect.

4   INTEGRATED SCHEDULING, ADVANCED SHAPING, AND DMA TRANSFER

The novelty of this design is the way shaper, scheduler, and DMA transfer cooperate to fulfill the task of sending the right cell at the right time according to traffic contracts and the actual load of the adapter. The design idea is that the CPU co-schedules both the DMA and ATM functions. The CPU has enough information to bring in data from the server memory just in time for transmission since it is deterministic when the next cell in a connection is allowed to be sent according to the shaping algorithm. Furthermore, the CPU is also in full control of the multiplexing of several cells by maintaining an ordered ready-list of connections ready for transmission. With this information it is predictable when a cell will be transmitted and hence the latest time when data must be fetched from server memory. The focal point is the traffic shaper. The shaper holds the context information of a connection, i.e., PCR (Peak Cell Rate), SCR (Sustainable Cell Rate), CDVT (Cell Delay Variation Tolerance), burst size, amount of data already sent etc. The scheduler hands over a pointer to the next connection to be shaped and the shaper feeds the scheduler with timing information about the earliest time a cell from the connection can be transmitted (Figure 2). The shaper state holds information about the actual amount of data for a connection on the adapter and the PDUs stored in host memory. By interpreting this state, new data can be pre-fetched from server memory when it is necessary and viable.

4.1 The scheduler

The purpose of the scheduler is to determine when a connection is allowed to send. A connection is assigned a cell slot for transmission sometime in the future, depending on the connection state and the contention of slots between connections due to multiplexing. The scheduler maintains an ordered data structure of connections, ordered with the connection to send closest in time first and the latest at the end. The scheduler gets the earliest possible time a connection can be scheduled from the shaper. It will then try to allocate the corresponding cell slot time by checking the ready-list. If this slot is already occupied by another connection the scheduler may try to find an empty slot later in time or to reschedule the conflicting cells using the CDVT (Cell Delay Variation Tolerance) parameter or to use a static priority. An earlier slot time cannot be used by the scheduler, even if there are several empty slots since the cell will then break the traffic parameters. Such a cell would most likely be discarded by the UPC (Usage Parameter Control) mechanism as an early cell outside the contract. Scheduling later is always acceptable by UPC, but may affect the end-to-end guarantees and results in less bandwidth efficiency. Design issues for the scheduler include efficient utilization of the bandwidth, fairness at overload situations and the minimizing of the number of CPU cycles needed for the scheduling. Our fairness proposal and our measurements on CPU cycles will be discussed later.

4.2 The ready-list data structure

The ready-list has one entry for each active connection. For 64K connections, the size of this list is considerable and the time it takes to keep the list sorted is critical for the performance. The data structure chosen for this task is a binary tree, implemented as a heap. A heap can realize a binary tree efficiently without the use of pointers. All information kept in the heap has to be small to assure that most of the heap fits into the cache of a processor (e.g. 512k L2 cache for a PentiumPro). The connection state has to be fetched into the cache only when a connection is scheduled. This structure allows for a large number of connections handled simultaneously. Assuming 32 bit values for the

cell slot time and a pointer to the state the heap needs only 512kbyte for 64k active
connections.

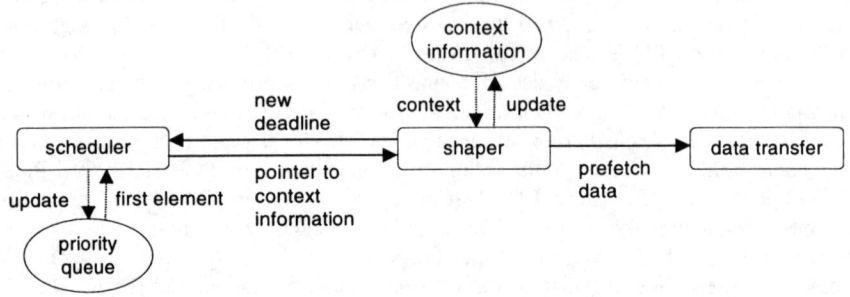

Figure 2 Interaction of shaper, scheduler, and data transfer engine.

4.3 Shaper

When called with a connection identifier, the shaper updates the shaping state and
calculates the earliest time the next cell of the connection can be scheduled without
violating the traffic contract.

The implementation of the traffic shaper uses a combination of the VSA (Virtual
Scheduling Algorithm) and LBA (Leaky Bucket Algorithm), which we derived from
the LBA and VSA specifications for the GCRA (Generic Cell Rate Algorithm) in the
ATM UNI (ATM Forum, 1996). Instead of using these algorithms for controlling
conforming cells at the UNI, we actively shape the traffic using them. This guarantees
that all cells shaped with our implementation will be accepted by the UPC. The
calculation is based on the current state of the connection (PCR, SCR, inactive) and the
mode (single/dual leaky bucket, SLB/DLB). If the connection is in PCR or SCR, the
shaper returns the new earliest transmission time for the next cell. Thus, it is guaranteed
for the SCR state that a new token will be available if this connection is scheduled the
next time. If a connection uses the DLB mode and no more tokens are left, the shaper
returns the time when the whole token bucket can be refilled completely with tokens.
Note that the shaping state of a connection will be accessed *if and only if* a cell can be
scheduled. There is no other updating necessary, e.g., filling new tokens in the bucket.
Many current implementations have to access state permanently to update the schedule
resulting in a poor performance and very limited number of connections handled at the
same time (Fujitsu, 1997), (LSI, 1997), (SIEMENS, 1997).

4.4 DMA transfer from server memory

The connection state has information about the amount of currently stored cells on the
adapter and a list of current PDUs stored in the server memory. Given the deterministic
information from the traffic shaping it is possible to keep most of the PDUs in server
memory and to move data just in time for transmission to the adapter. If all PDUs have
been sent the adapter notifies the server and the connection will enter a non-active state.
There are several potential advantages by using this deterministic information: prefetching of the right amount of data will avoid delays caused by demand fetching,
buffer size requirements can be reduced and long blocking times to other interconnnect
transactions can be avoided. By coupling the DMA transfer for a connection with the

actual sending of cells for this connection the design is simplified. The amount of cell buffer memory needed can then be decided and the shared buffer problem with asynchronous readers and writers is avoided, since the transfer is synchronized with the transmission, i.e., the consumption of data.

Two design issues must be addressed. The first one is *when* a transfer should take place. The interconnect has some access time variation that must be compensated for. This variance motivates an earlier transfer than just before the data is needed. The second issue is the *size* of the data transfer unit. The smallest unit is a cell and the largest is the PDU. A small size will cause more overhead while a big unit will consume cell memory buffers and may block other transfers. In this trade-off, the optimal transfer size of the interconnect must also be considered for efficiency.

Figure 3 shows a small example schedule generated by our prototype. 10 connections with different shapes, starting points and number of cells are scheduled together. The y-axis shows the consumption of cells, the x-axis the number of the cell slot the cell is scheduled in. Connections 2, 3, 4, 7, 8, and 9 are CBR connections, the other VBR. Connection 5 uses SLB, connections 0, 1, and 6 DLB mode. The exact shape of the plotting lines depends not only on the connections parameters but also on the current state of the other connections. This results, e.g., in slight deviations from an ideal straight line when two or more connections are scheduled for the same cell slot.

4.5 Priorities and Proportional Sharing

As soon as one implements an algorithm for scheduling one very important question is how the implementation behaves in overload situations. Overload situations can occur quite frequently, if one does not want to make only conservative reservations, i.e., allowing the sum of all PCRs never to be greater than the total capacity of the link. This would result in a very poor overall utilization if, e.g., VBR traffic sources are used. Here the PCR can be easily 10-1000 times larger than the SCR. One example is the transfer of MPEG2 coded video streams. Typical values are 1.0 to 15.0 Mbit/s for PCR, 0.2 to 4.0 Mbit/s for SCR. In addition, quite often a priority scheme is required to weight different traffic streams. One could for example give voice connections a higher priority as connections to fetch pictures from Web-pages. This would result in a higher audio quality and only minimal additional delay for the picture data transfer.

Assuming this, one can refine the question concerning the overload situation into following questions:

- *Sharing*: How is the available bandwidth shared within one priority class? What happens in overload situations caused within a priority class?
- *Isolation*: How is the interaction between different priority classes? What happens if a higher priority class already causes an overload?
- *Stability*: What happens to the system if the overload situation continues for a longer time? Does the system still provide a schedule "as good as possible"? Is the system stable?

From a users point of view the first set of questions can be answered as follows. If a user has started, e.g., several video applications that load the network completely and now starts an additional one he or she expects the available bandwidth to be shared fairly between the applications. The communication system can not make any assumptions of the importance of an application, and therefore a proportional sharing scheme is the best one can do. That means, that an application that used twice the bandwidth compared to another one still gets twice as much as the other one. But now this is less than before due to the overload. To privilege one application, one can shift the application to a higher priority. Our implementation guarantees this proportional share within one priority class independently for every connection.

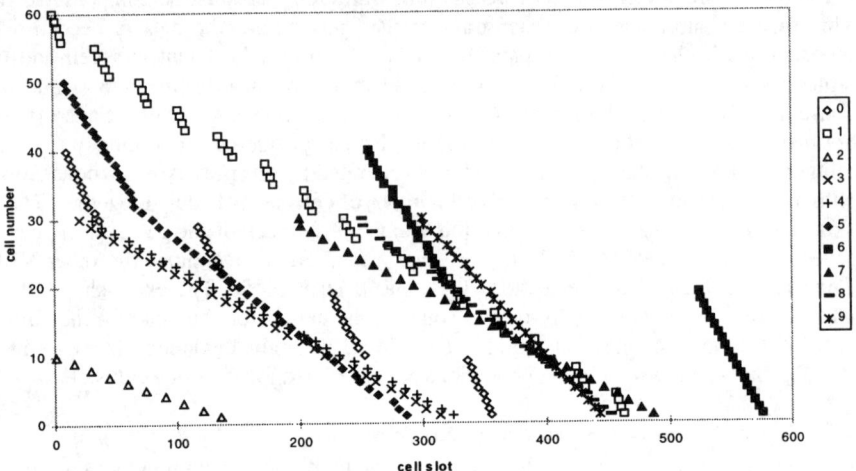

Figure 3 Example cell level schedule.

Prioritizing an application leads to the second set of questions concerning the interference between priority classes. Depending on the scheduling policy one can decide for a hard priority scheme, i.e., the scheduler tries first to satisfy connections with higher priorities and ignores lower priorities in case of an overload. This results in starvation of connections in lower priority classes. An alternative solution could provide a minimum share of the total bandwidth to avoid starvation. This is in general the better alternative due to the fact that overload situations are typically transient and common communication protocols like, e.g., TCP cannot deal properly with a total starvation but adapt well to lower bandwidth. Our implementation allows both alternatives by guaranteeing proportions of the total bandwidth in an overload situation.

All proportional sharing and handling of overload is done within the traffic shaper via adapting traffic parameters as soon as a connection is shaped and the load situation changes. This guarantees also that the prefetching of data is always harmonized with the real sending rate, so that no internal buffers can overflow in the communication system. The adaptation of the application to lower bandwidths is out of the scope of this work, but generally, this is the scheme many approaches tend to incorporate. No matter how long an overload situation exists or how strong the overload is, the system will always try to generate a schedule as close as possible to the traffic contract. The settings of the proportional sharing between different priority classes is left to the operator and depends on the policy of a service provider.

Proportional sharing and the overload behavior described above are up to now not implemented in any of the available traffic shaper chips. Either these implementations avoid overload situations by not allowing over allocation (LSI, 1997) or they throttle the total traffic via a leaky bucket (Fujitsu, 1997) (SIEMENS, 1997). Figure 4 shows an example for the effects of priority classes and proportional sharing within a priority class. Connection VC 3 has the highest priority 0, the connections 1, 4, 5, and 6 are in the same priority class 1, and finally connection 2 has the lowest priority 2. The connections 4, 5, and 6 are configured to require already 100% of the bandwidth per connection. This results in an heavy overload situation between cell slot 200 and 400. Due to the higher priority, connection 3 is not disturbed. The proportional sharing within one priority class can be seen for the connections 1, 4, 5, and 6. The slope of the graph flattens, as soon as a new connection with the same priority starts (at cell slots 200, 250, and 300). This demonstrates the proportional sharing: having the same priority and traffic parameters, two graphs must have the same slope. Finally, connection 2 has the lowest priority and starves during the heavy overload situation. If required, this could be avoided by reducing the bandwidth available for connections with higher priorities as described above.

5 PERFORMANCE EVALUATION

The algorithm was implemented using C and tested on a PentiumPro with 200MHz and 512k L2 cache (Windows NT 4.0 and Linux 2.0.28), a Digital Alpha AXP 3000/800 (Digital UNIX 3.2D-2) and a Sun Ultra I with 143 MHz (Solaris 2.5.1). The implementations on the different machines differ only in the instrumentation, not in the algorithm. The measurements were done running the complete operating system concurrently, but no other application programs. This was done on purpose to see the behavior of the algorithm in a real environment with current operating systems and not on specialized stand alone systems. The presented results give therefore an upper bound for execution times and cycles counts. The performance can only increase on dedicated systems.

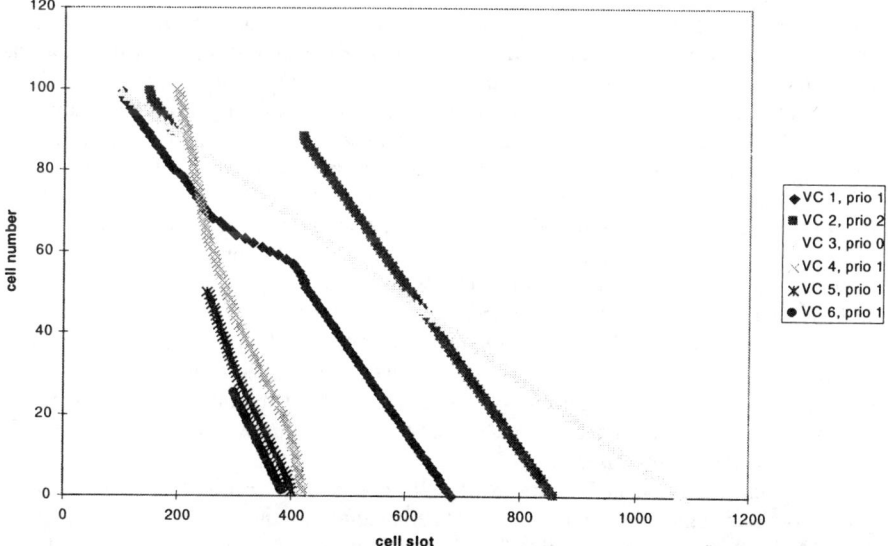

Figure 4 Effects of transient overloads, proportional sharing and priorities.

The main platform for instrumentation was the PentiumPro, Alpha and Ultra SPARC processors were used for comparison. Our main interest is in the number of CPU cycles used for scheduling of one cell. This includes the updating of all data structures, issuing of data transfer commands if necessary, and shaping of the cell stream. For counting of the CPU cycles the time stamp counter (TSC) of the PentiumPro was used. This allows for a resolution of single CPU cycles (5ns for the 200MHz CPU used). The TSC is a free running 64 bit counter not influenced by system events.

To evaluate the performance of the implementation we run worst-case scenarios that load the data structure heavily and require almost always the worst case of updating operations needed for the heap. One such scenario is for example the setup of 64k *simultaneously active* connections with identical parameters. The algorithm implemented puts no restriction on the number of connections, amount of data, or link speed. Only the actual performance of a given CPU/memory system limits this performance. To give an impression of the performance of the implementation also typical configurations were evaluated. It has to be remembered that this implementation treats every connection separately, i.e., no connections are combined or share common properties as this is the case in all existing hardware solutions due to the limited number of registers available.

Figure 5 shows cycle counts on a PentiumPro for more than 64k simultaneous active connections each sending 25000 bytes in 5 PDUs. The bandwidth chosen for the connections does not influence the performance of the algorithm, 9600 bit/s were chosen to result in a reasonable aggregated bandwidth of 616 Mbit/s (e.g., for a 622 Mbit/s adapter). Overloading an adapter using this algorithm, i.e., accepting a higher aggregated bandwidth than the total bandwidth of the adapter, does not result in a performance degradation but in an overall higher delay for cells. This is the best one can expect if the overload is done on purpose and no cells should be dropped. Most of the cells can be shaped and scheduled within 600 CPU cycles. This includes data transfer commands if necessary. Only at some points in time the cycle count goes up to about 650 cycles. It can be shown running the same algorithm on an UltraSparc that the jump in the cycle count at the beginning is a result of the PentiumPro L2 cache and not the algorithm. These UltraSparc measurements took place running a full installation of Solaris 2.5.1 but no other application programs resulting in an average time for shaping, scheduling, and data transfer of 4.5µs. To be able to handle the large size of instrumentation files values were collected and averages calculated. In addition maximum values were controlled to make sure that they are not averaged out. The binsize used for averaging is noted in the figures. 600 CPU cycles represent for the chosen processor a real-time of 3µs (200 MHz).

To stress the implementation all connections are started at exactly the same time, i.e., all data is tried to prefetch for the complete cell buffer at the beginning and then also consequently for every new PDU of a connection. The main result of these measurements is not the single number of cycles used but the fact that the number of cycles has an upper bound even under worst case conditions and has a very stable behavior for most of the cases.

Where the performance results with 64k connections show that with today's processors and standard operating systems cells cannot produced fast enough for, e.g., 155 Mbit/s adapters, this is possible for a lower number of connections. Figure 6 shows that a software solution of a shaper/scheduler can produce a cell in less than 350 cycles, i.e., 1.75 µs. This is definitely fast enough for a 155 Mbit/s adapter. Again this configuration is a worst case for 1000 connections, the cycle count per cell will drop if for example some high bandwidth connections together with a substantial number of low bandwidth connections have to be scheduled. With this lower number of connections no jump in the cycle count due to the cache behavior can be seen.

Loading the implementation with 100 connections results in a cycle count per cell of typically less than 300. If only one active connection is configured the cycle count drops to 160. The shaping takes less than 85 cycles on average and is independent of the size of the data structure.

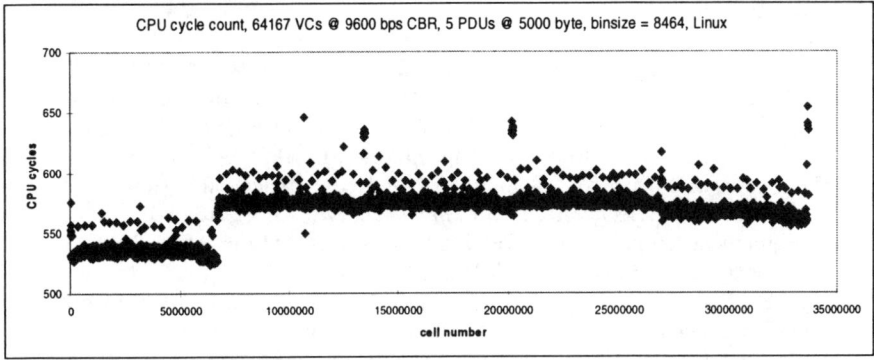

Figure 5 Cycle count on a PentiumPro/Linux sending data for 64k connections.

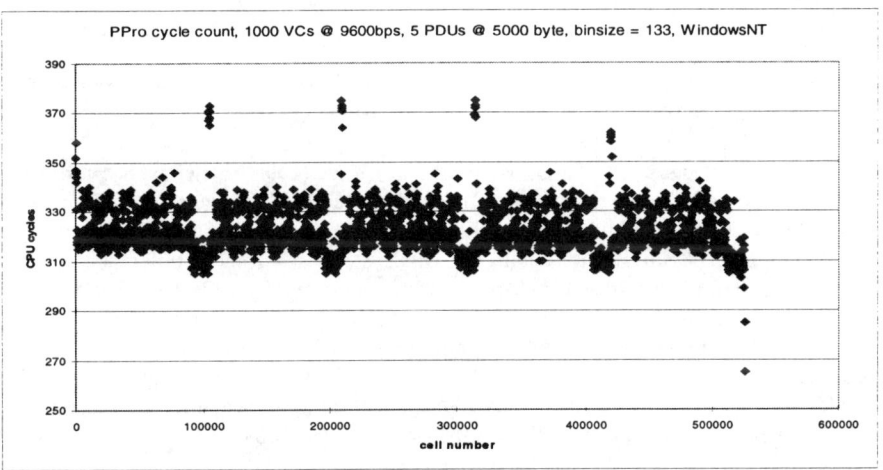

Figure 6 PentiumPro cycle count for 1000 connections.

6 CONCLUSIONS

Today, a discrepancy between the capabilities of shaping solutions, i.e., mostly chips to generate certain traffic patterns, and the UPC chips, e.g., at the UNI exists. This may lead to cell loss, although the traffic parameters agreed upon are the same for the shaper and UPC. One reason for this are the very limited capabilities of hardware solutions for shaping due to a limited chip area. Our work shows, that it is already today feasible to shape traffic for ATM networks for over 1000 connections at a rate of 155 Mbit/s using a today's general purpose CPU. Furthermore, with the proposed software solution cell-level scheduling and the scheduling of DMA transfers can be harmonized resulting in an overall higher efficiency and lower buffer requirements.

Further work will concentrate on performance measurements using high-end workstations with 600MHz CPUs and the integration of the DMA scheduling into the scheduling mechanisms of the operating system to further harmonize data transfer.

6   REFERENCES

AtecoM (1997) ATM_POL3, http://www.atecom.de/

ATM Forum (1996) Traffic management specification, version 4.0, ATM Forum

Campbell, A., Aurrecoechea, C.: Hauw, L. (1996) A Review of QoS Architectures, Proceedings of the 4. International IFIP Workshop on QoS (IWQoS), Paris

Coulson, G., Campbell, A., Robin, P., Papathomas, M., Blair, G., Sheperd, D. (1995) The design of a QoS-controlled ATM-based communication system in Chorus. *IEEE Journal on Selected Areas in Communications*, **13(4)**, 686-699

Digital Equipment Corporation (1996) Program Analysis Using Atom Tools. Digital Equipment Corporation, Maynard, Massachusetts

Druschel, P., Banga, G. (1996) Lazy Receiver Processing (LRP): A Network Sub-system Architecture for Server Systems. Proceedings of the USENIX Association Second Symposium on Operating Systems Design and Implementation, Seattle

Druschel, P., Peterson, L.L., Davie, B.S. (1994) Experiences with a high-speed network adapter: a software perspective, ACM SIGCOMM, London

Dalton, C., Watson, G., Banks, D., Calamvokis, C., Edwards, A., Lumley, J. (1993) Afterburner. *IEEE Network*, pp. 36-43

Engler, D.R., Kaashoek, M.F., O'Toole, J. (1995) Exokernel: an operating system architecture for application-level resource management. ACM SIGOPS

Fujitsu (1997) ALC (MB86687A), http://www.fmi.fujitsu.com

Georgiadis, L.; Guerin, R., Peris, V., Sivarajan, K.N. (1996) Efficient network QoS provisioning based on per node traffic shaping. *IEEE/ACM Trans. Networking*, **4**

Gopalakrishnan, R., Parulkar, G. (1995) A Framework for QoS Guarantees for Multimedia Applications within an Endsystem, 1. Joint Conference of the Gesellschaft für Informatik and the Schweizer Informatikgesellschaft, Zürich

Integrated Telecom Technology (1997) WAC-186-B, http://www.igt.com

LSI Logic (1997) ATMizer II (L64363), http://www.lsilogic.com

National Semiconductor (1997) BNP2010 UPC, http://www.national.com

Rexford, J., Bonomi, F, Greenberg, A., Wong, A (1997) A Scalable Architecture for Fair Leaky-Bucket Shaping, IEEE Infocom, pp. 1056-1064

SIEMENS (1997) SARE (PBX4110), http://www.siemens.de

Toshiba (1997) Meteor (TC35856F), http://www.toshiba.com

TranSwitch (1997) SARA II (TXC-05551), http://www.txc.com

Traw, C.B.S., Smith, J.M. (1993) Hardware/software organization of a high-performance ATM host interface, *IEEE JSAC*, **11(2)**, 240-253

Wrege, D.E, Liebeherr, J. (1997) A Near-Optimal Packet Scheduler for QoS networks. IEEE Infocom, Kobe

34

Sample-path Analysis of Queueing Systems with Leaky Bucket

Shoji Kasahara †, ‡
†*Nara Institute of Science and Technology; Graduate School of Information Science*
8916-5 Takayama, Ikoma Nara 630-01, Japan, Phone:+81-743-72-5353, Fax:+81-743-72-5359, E-mail: kasahara@is.aist-nara.ac.jp
‡*Telecommunications Advancement Organization of Japan(TAO)*
Shiba 2-31-19, Minato-Ku, Tokyo, 105 Japan

Toshiharu Hasegawa
Kyoto University; Department of Applied Systems Science, Graduate School of Engineering
Kyoto 606-01, Japan, E-mail: hasegawa@kuamp.kyoto-u.ac.jp

Abstract

This paper considers the queueing systems with leaky bucket using the sample-path argument. We consider two queueing models: one is the queueing system with leaky bucket in isolation and the other is that with multiple sources where each input is regulated by leaky bucket. In the former case, we show the sufficient condition for the rate stability of the data buffer process in terms of the token buffer process. In the latter case, supposing the burstiness constraints for the multiple inputs and the output, we give upper bounds of the buffer content process and the period during which the buffer content process is greater than a certain fixed point.

Keywords

Sample-path analysis, leaky bucket, rate stability

Broadband Communications   P. Kühn & R. Ulrich (Eds.)
© 1998 IFIP. Published by Chapman & Hall

1   INTRODUCTION

In the ATM-based Broadband ISDN (B-ISDN), it is expected that the different classes of traffic such as voice, video and data are multiplexed and carried to their destinations keeping their quality of service (QoS) satisfied the required level. In terms of the guarantee of QoS, the admission control scheme based on leaky bucket has been extensively studied and analyzed. One of the pioneering works is (Elwalid *et al.* 1991). In (Elwalid *et al.* 1991), the system with the access regulator has been analyzed in the context of the fluid queueing model.

In (Kulkarni *et al.* 1996), it turned out that the token buffer size doesn't play any role for the output process of the regulator using the leaky bucket. Our primary motivation of this paper is to investigate the relation between the token buffer size and the stability of the data buffer, and furthermore, the relation between the size of the token buffer and the behavior of the multiplexer's buffer where multiple regulated inputs are aggregated.

The queueing systems with leaky bucket have been studied assuming that the system is in equilibrium. In this context, it is not possible to capture the effects of the token buffer because it controls the bursty input from the source in the short time scale. Recently, (Cruz 1991a, Cruz 1991b) has proposed the concept of burstiness constraints. Using this constraints, we can derive the bounds of the queue length and the waiting time in the sample-path argument.

In this paper, we consider the queueing systems with leaky bucket using the sample-path argument. We consider the two queueing models: one is the queueing system with leaky bucket in isolation and the other is that with multiple sources where each input to the system is regulated by leaky bucket.

The sample-path analysis has played an increasingly important role in studying the limiting behavior of the queueing systems (see (Stidham *et al.* 1995) and the references therein). Sample-path techniques for queueing theory has been developed in (El-Taha *et al.* 1993, El-Taha *et al.* 1994, Stidham *et al.* 1993, Stidham *et al.* 1995). In (El-Taha *et al.* 1993, Stidham *et al.* 1993), the sample-path stability of an input-output process have been studied. In particular, a general sufficient condition for sample-path stability of an input-output system has been given. (El-Taha *et al.* 1994) analyzed the multi-server queueing system and derived some stability conditions. (Stidham *et al.* 1995) reviews variety kinds of application of sample-path analysis and showed how they all can be derived from the sample-path version of the renewal-reward theorem ($Y = \lambda X$).

On the other hand, Chang (Chang 1994) analyzed the queueing networks in sample-path and probabilistic arguments. In the probabilistic argument, he developed the concept of the effective bandwidth. (Altman *et al.* 1993) also considered the bounds of the performance measures like a queue length and a cycle time from both sample-path and probabilistic arguments.

Throughout this paper, we make no stochastic assumptions. The quantities and processes defined are deterministic. They may be thought of as representing a fixed sample path of a stochastic system, defined on some probability space.

The paper is organized as follows. In Section 2, we summarize the definition and some results of the stability of the general input-output process from (Stidham *et al.* 1995). In Section 3, we consider the sufficient condition of the rate stability for the queueing system with leaky bucket. In Section 4, we consider the boundaries of the fundamental quantities for the queueing system with multiple sources in which each generated input is regulated by leaky bucket.

2 DEFINITIONS AND CHARACTERIZATION OF RATE STABILITY

In this section, we summarize the definition and some results of the stability of an input-output process (Stidham *et al.* 1995).

Suppose that the state space $S = R^+$ with the Borel σ-field $\mathcal{B}(R^+)$ of subsets generated by the open sets. We consider an input-output process $\{Z(t), t \geq 0\}$ in which $Z(t)$ represents quantity in a system. We assume that $\{Z(t), t \geq 0\}$ is right continuous with left-hand limits. Let $A(t)(D(t))$ denote the cumulative input (output) to the system in $[0, t]$. Suppose that $\{A(t), t \geq 0\}$ and $\{D(t), t \geq 0\}$ are both nondecreasing, right-continuous processes. Then, $Z(t)$ is described as

$$Z(t) = Z(0) + A(t) - D(t), \quad t \geq 0. \tag{1}$$

Note that $Z(t)$ has bounded variation on finite t-intervals.

Definition 1 *An input-output process $\{Z(t), t \geq 0\}$ is said to be rate stable if*

$$\lim_{t \to \infty} t^{-1} Z(t) = 0. \tag{2}$$

Then, the following lemma is immediate from (2).

Lemma 2 *Suppose $t^{-1}A(t) \to \alpha < \infty$ as $t \to \infty$. Then the input-output process $\{Z(t), t \geq 0\}$ is stable if and only if $t^{-1}D(t) \to \alpha < \infty$ as $t \to \infty$.*

Thus, an input-output process is rate stable if the long-run input and output rate exist and are finite and equal.

The following theorem gives sufficient conditions for rate stability of the input-output process $\{Z(t), t \geq 0\}$.

Theorem 3 *Consider the input-output process* $\{Z(t), t \geq 0\}$ *defined by (1). Suppose*

$$\lim_{t \to \infty} t^{-1} A(t) = \alpha, \tag{3}$$

and there exists a real number z_0 *such that*

$$\lim_{t \to \infty} \frac{\int_0^t 1\{Z(s) > z_0\} dD(s)}{\int_0^t 1\{Z(s) > z_0\} ds} = \delta, \tag{4}$$

where $0 < \alpha < \delta$. *Then the process* $\{Z(t), t \geq 0\}$ *is rate-stable.*

Proof. See (Stidham *et al.* 1995). $\square$

Corollary 4 *Suppose the conditions (3) and (4) of Theorem 3 are satisfied with* $z_0 = 0$, *and* $0 < \alpha < \delta$. *Suppose also that*

$$\lim_{t \to \infty} t^{-1} \int_0^t 1\{Z(s) = 0\} dD(s) = 0. \tag{5}$$

Then $p(0) := \lim_{t \to \infty} t^{-1} \int_0^t 1\{Z(s) = 0\} ds$ *is well-defined and*

$$p(0) = 1 - \rho, \tag{6}$$

where $\rho = \alpha/\delta$.

Proof. See (Stidham *et al.* 1995). $\square$

3   QUEUEING SYSTEM WITH LEAKY BUCKET

In this section, we consider the rate stability of the queueing system with leaky bucket. We show our queueing model in Figure 1.

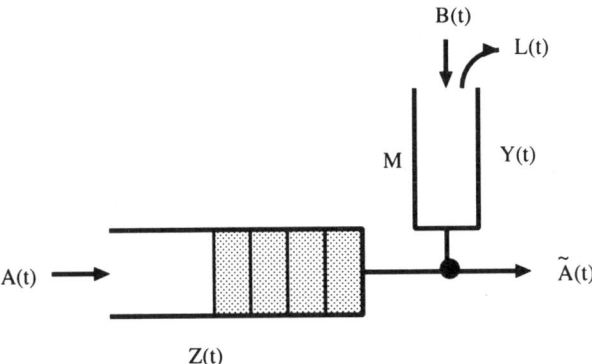

Figure 1 Queueing System with Leaky Bucket

The system has an infinite data buffer for messages and the finite token buffer whose size is M. Tokens arrive at the token buffer and arriving tokens which find the token buffer full are lost.

Arriving message is sent to the output and a token is removed from the token buffer if there is at least one token in the buffer. If there are no tokens in the token buffer, the message waits for the token arriving and is sent to the output when the token arrives.

In this system, we define the following processes;

$A(t)$: cumulative number of arriving messages at time t,

$B(t)$: cumulative number of arriving tokens at t,

$\tilde{A}(t)$: cumulative number of messages departing from the system at t,

$L(t)$: cumulative number of lost tokens at t.

Note that the output process of the token buffer is equal to $\tilde{A}(t)$.

Let $\{Y(t) : 0 \leq Y(t) \leq M, t \geq 0\}$ denote the token buffer process at time t. Then $Y(t)$ satisfies the following equation

$$Y(t) = Y(0) + B(t) - \tilde{A}(t) - L(t). \tag{7}$$

Using (7), (1) becomes

$$Z(t) = Z(0) + A(t) - Y(0) - B(t) + Y(t) + L(t). \tag{8}$$

First, we give a preliminary result that is of independent interest.

Lemma 5 *Consider the input-output process $\{Z(t), t \geq 0\}$ defined by (8). Let α and γ be non-negative constants. Suppose*

1. *the input process of the data buffer satisfies*

$$\lim_{t\to\infty} t^{-1}A(t) = \alpha, \qquad (9)$$

2. *the input process of the token buffer satisfies*

$$\lim_{t\to\infty} t^{-1}B(t) = \gamma, \qquad (10)$$

where $0 < \alpha < \gamma < \infty$.

Then the event $\{Y(t) > 0\}$ *occurs infinitely often as* $t \to \infty$. *That is, for every* $t_0 \geq 0$, *there exists a* $t \geq t_0$ *such that* $Y(t) > 0$.

Proof. Without the loss of generality, we assume that $Z(0) = Y(0) = 0$, i.e.,

$$Z(t) = A(t) - B(t) + Y(t) + L(t). \qquad (11)$$

The proof is by contradiction. It follows from (9) and (10) that, for every $\epsilon > 0$, there exists a T such that, for all $t \geq T$,

$$(\alpha - \epsilon)t \leq A(t) \leq (\alpha + \epsilon)t, \qquad (12)$$

$$(\gamma - \epsilon)t \leq B(t) \leq (\gamma + \epsilon)t, \qquad (13)$$

Now suppose $0 < \epsilon < (\gamma - \alpha)/2$. Suppose that the event $\{Y(t) > 0\}$ does not occur infinitely often as $t \to \infty$. Then there exists a t_0 such that, for all $t \geq t_0$, $Y(t) = 0$. Hence, using (11), we have for all $t \geq t_0$

$$\begin{aligned}
Z(t) &= Z(t_0) + \{A(t) - A(t_0)\} - \{B(t) - B(t_0)\} + \{L(t) - L(t_0)\} \\
&= Z(t_0) - A(t_0) + B(t_0) + A(t) - B(t) \\
&\leq Z(t_0) - A(t_0) + (\alpha - \gamma + 2\epsilon)t,
\end{aligned}$$

where the second equality follows from the fact that there are no arrival tokens lost after $Y(t) = 0$. Since $(\alpha - \gamma + 2\epsilon) < 0$, by choosing t sufficiently large, we can make the quantity on the right-hand side of the last equality negative, thus leading to a contradiction of $Z(t) \geq 0$. Therefore, for every $t_0 \geq T$, there exists at least one $t > t_0$ such that $Y(t) > 0$. That is, the event $\{Y(t) > 0\}$ occurs infinitely often as $t \to \infty$. $\qquad \square$

Now we give the following theorem for the sufficient condition of the rate stability of $Z(t)$ based on $Y(t)$.

Theorem 6 *Suppose*

1. $A(t)$ *satisfies*

$$\lim_{t \to \infty} t^{-1} A(t) = \alpha; \tag{14}$$

2. *there exists a real number c such that $0 < c \le M$ and*

$$\lim_{t \to \infty} \frac{\int_0^t 1\{Y(s) \le c\} dB(s)}{\int_0^t 1\{Y(s) \le c\} ds} = \gamma', \tag{15}$$

where $0 < \alpha < \gamma'$;
3. *The event $\{Y(t) > c\}$ occurs infinitely often as $t \to \infty$.*

Then the process $\{Z(t), t \ge 0\}$ is rate stable.

Proof. The proof is by contradiction and is similar to those of Theorem 5.1. of (El-Taha *et al.* 1993), Theorem 2.2. of (Stidham *et al.* 1993) and Theorem 2.2. of (El-Taha *et al.* 1994).

Suppose that $\{Z(t), t \ge 0\}$ is not rate stable. Then there exists a $\xi > 0$ and an increasing sequence of time points $\{\tau_n, n \ge 1\}$, with $\tau_n \to \infty$ as $n \to \infty$, such that $Z(\tau_n) \ge \xi\tau_n$ for all $n \ge 1$. Without the loss of generality, we consider the process defined in (11). We define $U(t)$ as

$$U(t) = \int_0^t 1\{Y(s) \le c\} ds.$$

It follows from (14) and (15) that, for every $\epsilon > 0$, there exists a $T < \infty$ such that

$$(\alpha - \epsilon)t \le A(t) \le (\alpha + \epsilon)t, \qquad t \ge T, \tag{16}$$

$$(\gamma' - \epsilon)U(t) \le \int_0^t 1\{Y(s) \le c\} dB(s) \le (\gamma' + \epsilon)U(t), \qquad t \ge T. \tag{17}$$

Note that from the assumption 3., $U(t) \to \infty$ as $t \to \infty$.
Let $a_n = \sup\{s : s < \tau_n, Y(s) > c\}$. Then, it follows that

$$U(\tau_n) = U(a_n) + \tau_n - a_n. \tag{18}$$

Moreover, the assumption 3. show that $a_n \to \infty$ as $n \to \infty$. Now we choose $\epsilon < (\xi - c/a_n)/5$. We have $a_n > T$ for sufficiently large n. For such n, it follows from (16), (17) and (18) that

$$Z(a_n) = Z(\tau_n) - (A(\tau_n) - A(a_n)) + (B(\tau_n) - B(a_n))$$

$$
\begin{aligned}
& -(Y(\tau_n) - Y(a_n)) - (L(\tau_n) - L(a_n)) \\
=\; & Z(\tau_n) - (A(\tau_n) - A(a_n)) + (B(\tau_n) - B(a_n)) - (Y(\tau_n) - Y(a_n)) \\
\geq\; & \xi\tau_n - (\alpha + \epsilon)\tau_n + (\alpha - \epsilon)a_n + (\gamma' - \epsilon)U(\tau_n) - (\gamma' + \epsilon)U(a_n) - c \\
=\; & \xi\tau_n + (\gamma' - \alpha - 2\epsilon)\tau_n + (\alpha - \gamma')a_n - 2\epsilon U(a_n) - c \\
=\; & \xi\tau_n - 2\epsilon a_n + (\gamma' - \alpha - 2\epsilon)(\tau_n - a_n) - 2\epsilon U(a_n) - c \\
\geq\; & \xi\tau_n - 2\epsilon a_n - 2\epsilon U(a_n) - c \\
\geq\; & \xi a_n - 2\epsilon a_n - 2\epsilon a_n - c \\
=\; & (\xi - 4\epsilon)a_n - c \\
>\; & \epsilon a_n,
\end{aligned}
$$

where the second equality is obtained from the fact that there are no lost tokens for $a_n < t < \tau_n$. On the other hand, $Z(a_n) = 0$ when $Y(a_n) > c$ for all n. Thus we have a contradiction and the proof is complete. □

Remark 7 Since ϵ is arbitrarily chosen from the positive real number, we can choose its value less than $(\xi - c/a_n)/5$. $\{a_n, n \geq 1\}$ is the increasing sequence as n is getting large and hence the effect of c/a_n becomes small and negligible. This means that the effect of the token buffer size becomes negligible if we consider the stability of the system in the long time scale.

If we restrict the condition 2. of Theorem 6 to the event $\{Y(t) = 0\}$, we obtain the following corollary.

Corollary 8 *Suppose the condition 1. of Theorem 6 is satisfied. Suppose also that*

$$
\lim_{t \to \infty} \frac{\int_0^t 1\{Y(s) = 0\}dB(s)}{\int_0^t 1\{Y(s) = 0\}ds} = \gamma', \tag{19}
$$

where $0 < \alpha < \gamma'$. Then the process $Z(t)$ is rate stable.

Proof. The condition (19) is equivalent to the following condition

$$
\lim_{t \to \infty} \frac{\int_0^t 1\{Z(s) > 0\}dB(s)}{\int_0^t 1\{Z(s) > 0\}ds} = \gamma'.
$$

Then it is immediate from Theorem 3. □

4   QUEUEING SYSTEM WITH REGULATED INPUTS

In this section, we consider the queueing system with the regulated inputs (see Figure 2). This system is a typical model of the ATM multiplexer. Here, we assume that there are K sources and that each source generates the regulated input traffic using the leaky bucket scheme considered in section 4.

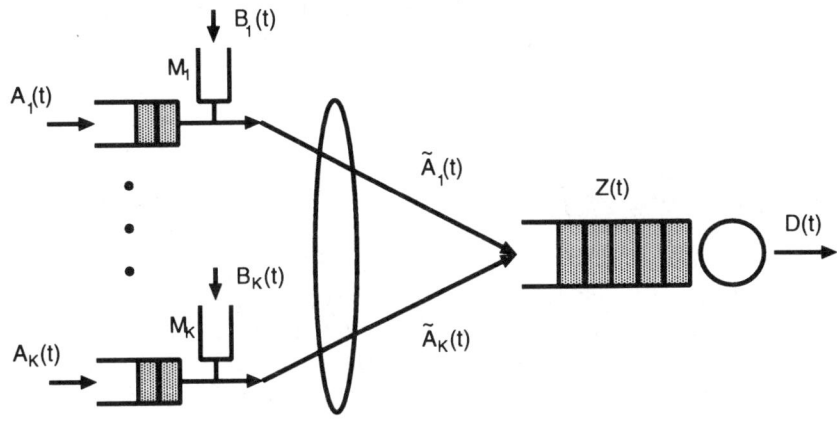

Figure 2 Queueing System with Regulated Inputs

Sources are labeled $i = 1, \cdots, K$ in sequence. The size of the token buffer of ith source is M_i. In similar to the previous section, we use the following notations:

$A_i(t)$: cumulative number of messages arriving to the ith regulator at t,

$B_i(t)$: cumulative number of arriving tokens of the ith regulator at t,

$\tilde{A}_i(t)$: cumulative number of messages departing from ith regulator at t.

We also define $\lim_{t \to \infty} A_i(t)/t = \alpha_i$ and $\lim_{t \to \infty} B_i(t)/t = \gamma_i$ for $i = 1, \cdots, K$. We assume that $\alpha_i < \gamma_i$ for $i = 1, \cdots, K$. Clearly, the message buffer process of each source is stable.

K regulated inputs are aggregated and sent to the shared buffer. Let $Z(t)$ denote the buffer contents of the system at time t. We define $D(t)$ as the total amount of the output process at t.

In this system, the total amount of inputs, denoted by $\tilde{A}(t)$, is equal to $\sum_{i=1}^{K} \tilde{A}_i(t)$. Let γ denote the average rate of the total input. Then, we have

$$\lim_{t \to \infty} \frac{\tilde{A}(t)}{t} \leq \lim_{t \to \infty} \frac{\sum_{i=1}^{K} B_i(t)}{t} = \sum_{i=1}^{K} \gamma_i \equiv \gamma.$$

Now we assume that there exists a real number z_0 such that

$$\lim_{t \to \infty} \frac{\int_0^t 1\{Z(s) > z_0\}dD(s)}{\int_0^t 1\{Z(s) > z_0\}ds} = \delta, \tag{20}$$

where $\gamma < \delta$. Then, from Theorem 3, $\{Z(t), t \geq 0\}$ is rate stable.

Following to (Altman *et al.* 1993), we define the "burstiness" of arrival and departure processes during a time interval $[s, t)$ by

$$B_{s,t}^{\tilde{A}} = \tilde{A}(t) - \tilde{A}(s) - \gamma(t - s), \tag{21}$$

$$B_{s,t}^{D} = D(t) - D(s) - \delta \int_s^t 1\{Z(\tau) > z_0\}d\tau. \tag{22}$$

That is, $B_{s,t}^{\tilde{A}}$ is the difference between the actual input and the input constantly generated with the average rate γ throughout the interval $[s, t)$. Similarly, $B_{s,t}^{D}$ is the difference between the actual output and the output constantly served with the average rate δ when the queue length is greater than z_0.

Let $b_0 = 0$ and for $n = 1, 2, \cdots$, define the following time points

$$b_n = \inf\{t > b_{n-1} \; : \; Z(t-) \leq z_0, Z(t) > z_0\},$$

$$e_n = \inf\{t > b_n \; : \; Z(t-) > z_0, Z(t) \leq z_0\}.$$

In other words, the nth event of $\{Z(t) > z_0\}$ starts at time b_n and ends at e_n. Let $T_n := e_n - b_n$.

Now we show that supposing the Cruz bounds for the burstiness of both arrival and departure processes, T_n and $Z(t)$ are bounded for $n \geq 1$ and $b_n \leq t \leq e_n$. Specifically, we assume that

$$\tilde{A}(t) - \tilde{A}(s) \leq \gamma(t - s) + M, \; 0 \leq s < t, \tag{23}$$

$$D(t) - D(s) \geq \delta \int_s^t 1\{Z(\tau) > z_0\}d\tau - \sigma_D, \; 0 \leq s < t, \tag{24}$$

where $M = \sum_{i=1}^{K} M_i$ and σ_D is non-negative constant. In (Altman *et al.* 1993), (23) and (24) are referred to as *linear burstiness bounds*. It follows from (21), (22), (23) and (24) that

$$B_{s,t}^{\tilde{A}} \leq M, \quad -B_{s,t}^{D} \leq \sigma_D, \quad 0 \leq s < t. \tag{25}$$

Remark 9 From (23), for $\epsilon > 0$, the worst case of the arrival process during $[t, t + \epsilon)$ is

$$\tilde{A}(t + \epsilon) \leq \tilde{A}(t) + M, \ 0 \leq t.$$

This means that the total amount of token buffers bounds the maximum amount of the bursty input. After the worst case occurs, the input rate is bounded by γ, the total amount of token generation rates.

Theorem 10 *For $n \geq 1$ and $b_n \leq t \leq e_n$,*

$$T_n \leq \frac{z_0 + M + \sigma_D}{\delta - \gamma}, \tag{26}$$

$$Z(t) \leq z_0 + M + \rho \sigma_D, \tag{27}$$

where $\rho = \gamma/\delta$.

Proof. First, we show (26). It follows from (22) that

$$
\begin{aligned}
\delta T_n &= D(e_n) - D(b_n) - B^D_{b_n, e_n} \\
&\leq z_0 + \gamma T_n + B^{\tilde{A}}_{b_n, e_n} - B^D_{b_n, e_n} \\
&\leq z_0 + \gamma T_n + M + \sigma_D,
\end{aligned}
$$

since the output during T_n can be no more than z_0, the work at b_n, plus the input during T_n and second inequality follows from (25).

To show (27), note that for $b_n \leq t \leq e_n$,

$$D(t) - D(b_n) = \delta(t - b_n) + B^D_{b_n, t}.$$

Then, we obtain

$$
\begin{aligned}
Z(t) &= Z(b_n) + \alpha(t - b_n) + B^{\tilde{A}}_{t_n, t} - \{D(t) - D(b_n)\} \\
&= z_0 + \rho\{D(t) - D(b_n) - B^D_{b_n, t}\} + B^{\tilde{A}}_{t_n, t} - \{D(t) - D(b_n)\} \\
&= z_0 + B^{\tilde{A}}_{t_n, t} - \rho B^D_{b_n, t} - (1 - \rho)\{D(t) - D(b_n)\} \\
&\leq z_0 + B^{\tilde{A}}_{t_n, t} - \rho B^D_{b_n, t} \\
&\leq z_0 + M + \rho \sigma_D.
\end{aligned}
$$

$\square$

Remark 11 As for the ATM multiplexer, the arriving cell is processed with the constant rate. In this case, there are no burstiness in terms of the output process $D(t)$ and hence $\sigma_D = 0$. If the condition (20) holds at $z_0 = 0$, $Z(t)$ is bounded by M. That is, if the total input rate is smaller than the output rate, the queue length of the multiplexer is no more greater than the total amount of token buffers.

REFERENCES

Altman, E., Foss, S.G., Riehl, E.R. and Stidham, Jr., S. (1993) Performance bounds and pathwise stability for generalized vacation and polling systems. Technical Report No. 93-8, Dept. of Operations Research, University of North Carolina, Chapel Hill, NC 27599.

Chang, C.S. (1994) Stability, queue length, and delay of deterministic and stochastic queueing networks. *IEEE Trans. Automa. Contr.*, **39**(5), 913–931.

Cruz, R.L. (1991) A calculus for network delay, part I: network elements in isolation. *IEEE Trans. Inform. Theory.*, **37**(1), 114–131.

Cruz, R.L. (1991) A calculus for network delay, part II: network analysis. *IEEE Trans. Inform. Theory.*, **37**(1), 132–141,

El-Taha, M. and Stidham, Jr., S. (1993) Sample-path analysis of stochastic discrete-event systems. *J. Discrete Event Dynamic Systems*, **3**, 325–346.

El-Taha, M. and Stidham, Jr., S. Sample-path conditions for multiserver input-output processes. *J. Appl. Math. Stoch. Anal.*, **7**, 437–456.

Elwalid, A.I. and Mitra, D. (1991) Analysis and design of rate-based congestion control of high speed networks, I: stochastic fluid models, access regulation. *Queueing Systems*, **9**, 29–64.

Kulkarni, V.G. and Gautam, N. (1996) Leaky buckets: sizing and admission control. Technical Report No. 95-10, Dept. of Operations Research, University of North Carolina, Chapel Hill, NC 27599.

Stidham, Jr., S. and El-Taha, M. (1993) A note on sample-path stability conditions for input-output processes. *Operations Research Letters*, **14**, 1–7.

Stidham, Jr., S. and El-Taha, M. (1995) Sample-path techniques in queueing theory. *Advances in Queueing* (ed. J.H. Dshalalow), CRC Press, Inc., 119–166.

Routing and Multicasting

35

Active Multicasting for Heterogeneous Groups

Ralph Wittmann, Martina Zitterbart
Institute of Operating Systems and Computer Networks
Technical University of Braunschweig
38106 Braunschweig, Germany
{wittmann|zit}@ibr.cs.tu-bs.de
http://ibr.cs.tu-bs.de/~wittmann/AMnet.html

Abstract

Communication environments are becoming increasingly heterogeneous. This imposes new challenges on communication support for multimedia and collaborative applications. AMnet (Active Multicasting Network) provides multi-point communication support for large-scale groups with heterogeneous receivers. Active Multicasting nodes inside the network include so-called QoS filters that remove information from continuous media streams in order to reduce data rate for low-end receivers without affecting high-end receivers. A prototype implementation of AMnet based on RSVP is presented with some first performance results.

Keywords

Group Communication, Active Multicasting, QoS Filter

1 INTRODUCTION

It is now well accepted that scalable and flexible multicasting support is increasingly needed for emerging communication systems. The demand is driven by multimedia applications, especially, by collaborative applications. The same holds for applications, such as scientific computing and distributed simulation. For many of these applications group communication is a natural paradigm. Therefore, proper support within the communication subsystem is required. This demand is acknowledged by various recent approaches that address transport protocols for reliable multicast. The focus of most current proposals is on the support of reliability. Mostly, all group members experience the same level of service, independent of their network attachment and end system equipment. Thus, all participants in the group are provided with a homogeneous quality of service (QoS).

However, heterogeneity is increasing with respect to networks and end nodes. Consider a scenario where some receivers of the group use simple PDAs, which typically are poorly equipped and often connected over error prone slow-speed wireless links (cf., Figure 1). Others may use high-end workstations that are attached through

Broadband Communications   P. Kühn & R. Ulrich (Eds.)
© 1998 IFIP. Published by Chapman & Hall

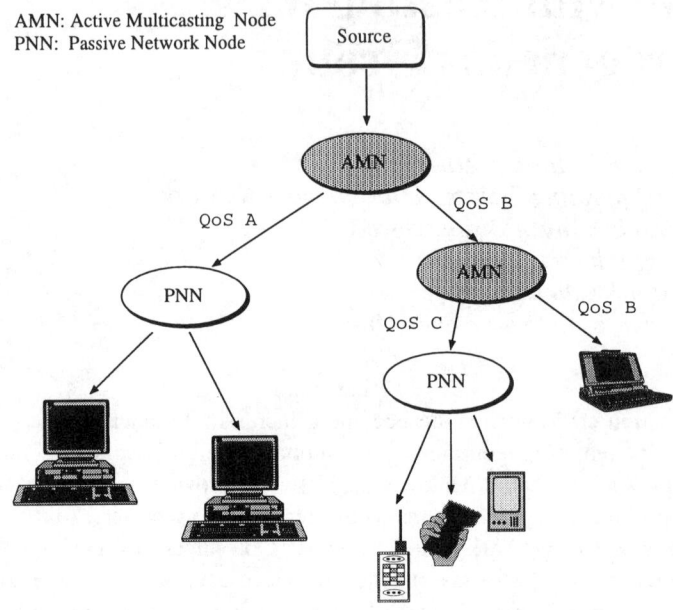

Figure 1 Multicast Tree with Active Multicasting Nodes.

a high-speed link. Due to limited processing power low-end receivers may be unable to handle the same data stream as receivers using high-end workstations. Moreover, these devices may have small displays offering low resolution and color depth. Furthermore, group members do not only differ in terms of end-system capabilities, they may also be connected to networks with different characteristics. For example, an ISDN network offers much less bandwidth than an ATM network. Hence, different participants in a multi-point communication may have different requirements with respect to quality of service. The goal of AMnet is in the provision of user-tailored services for individual group members. Data streams with different characteristics will be offered, e.g., to a high-end workstation and to a wireless attached end node. In most current approaches the service provided to individual group members is penetrated by the group member with the lowest service capabilities. Such an approach is not acceptable for multimedia and collaborative applications in heterogeneous networking environments.

AMnet especially addresses the aspect of heterogenous group communication. It is based on Active Networking (Tennenhouse & Wetherall 1996) in the sense that it uses active multicasting nodes inside the network in order to individually tailor data streams to the end users service requirements. Therefore, various service modules shall be dynamically implanted in network nodes. QoS filter and error control modules are typical examples.

An overview of the state-of-the-art in Active Networking can be found in (Tennenhouse et al. 1997).

The paper is structured as follows. Section 2 presents the AMnet approach. An early prototype is discussed in section 3. It is based on RSVP as signalling protocol for QoS filter functions. Section 4 concludes the paper and gives an outlook on future work.

2 THE AMNET APPROACH

2.1 Basic Scenario

Figure 1 depicts a multicast tree is depicted that provides heterogeneous multicasting capabilities through active network nodes, called *Active Multicasting Nodes*. Some of the network nodes included in the dissemination tree are active, others are passive with respect to enhanced multicasting tasks. Data streams with different QoS are delivered to the end users of the group, with QoS A > QoS B > QoS C. Therefore, the active multicasting nodes comprise service modules for QoS filtering and enhanced QoS signalling. Furthermore, error control capabilities are needed in the active multicasting nodes.

Thus, AMnet is based on active network nodes, that provide user-tailored services for heterogeneous group communication. Particularly, QoS filtering, enhanced signalling, and error control are considered by active multicasting nodes.

2.2 Active Multicasting Node

The structure of an active multicasting node is depicted in Figure 2. It consists of the following main components:

- service modules
- AMnet signalling
- AMnet manager
- QoS monitor
- packet filter

The service modules provide services, such as QoS filtering, error control and recovery, and the like. They can be dynamically activated and configured based on individual service requirements of participants in the group communication. These requirements are propagated through the AMnet signalling protocol. The signalling entity extracts the service requirements and forwards them to the AMnet manager. The AMnet manager is responsible for allocating resources and configuring the service modules accordingly. Information on the current load is provided by the QoS monitor. It observes the service modules and collects information about ongoing communication. The QoS monitor operates similar to the one described in (Zitterbart 1996).

AMnet signalling and the AMnet manager are not involved in the regular data path. The packet filter demultiplexes the incoming data stream and forwards signalling

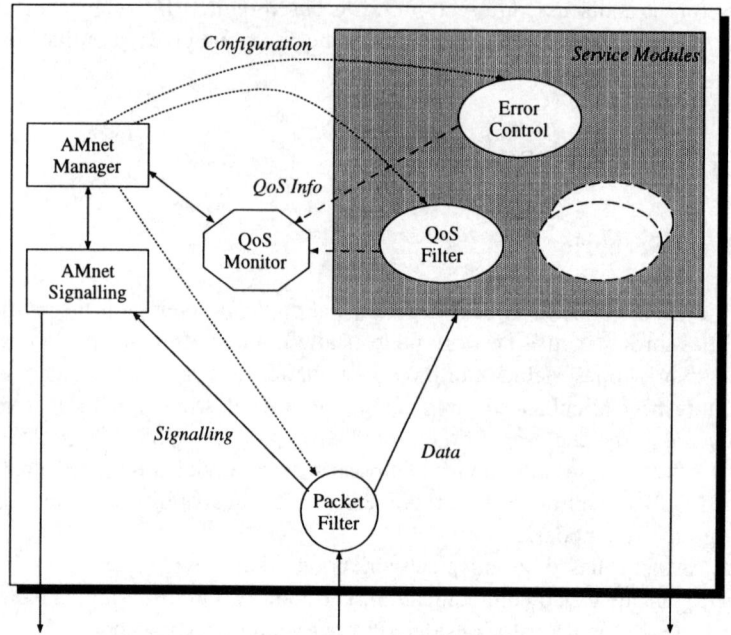

Figure 2 Structure of an Active Multicasting Node.

messages to the AMnet signalling entity and all other data to their corresponding service modules. Filtering is implemented content based, i.e., the corresponding service modules are addressed directly by the filter. Therefore, several demultiplexing steps are integrated in the packet filter.

2.3   Service Modules

The active multicasting nodes support service modules that are needed in order to support quality-based heterogeneous group communication. The service modules comprise, among others, QoS filtering, and error control and recovery.

QoS filtering
The service module QoS filtering reduces the amount of data forwarded to the next node in the multicasting tree towards the destination. QoS filtering is typically used to reduce the data rate of audio and video streams for low-end receivers. Several types of QoS filters may be active on an Active Multicasting Node (e.g., MPEG-1-filter, H.261-filter). QoS filter may perform different techniques to adapt a multimedia stream to specific QoS. As an example an MPEG-1 filter may support the following filter functions:

- Frame dropping filter
- Re-quantization filter
- Monochrome filter
- Slicing filter

The *frame dropping* filter simply drops P and B frames of the MPEG-1 stream to reduce the frame rate for less powerful receivers.

The *re-quantization filter* operates on DCT-coefficients and, thus, requires semi-decompression of the video stream. With re-quantization many near-zero coefficients may become zero which leads to a better compression ratio in the subsequent entropy encoding step. With moderate quantization steps a good trade-off between bandwidth reduction and loss of play-out quality can be achieved. However, large quantizers can result in strange artefacts (Rao & Hwang 1996).

The *Monochrome filter* removes color information. As MPEG encodes chrominance and luminance information independently, discarding of luminance components is a relatively simple operation.

Slicing filters exploit a property of MPEG encoded streams. The macro blocks of MPEG can be structured into *slices*. A slice contains information to synchronize the codec. If a slice gets lost, the codec resynchronizes on the following slice and may continue decoding. Since slices are not always included in MPEG coded data, the filter have to incorporate them. This may lead to a slightly higher data volume. However, a slicing filter can drop slices easily and, thus, reduce the bandwidth consumed on the output link.

Several QoS filters have been investigated in (Yeadon 1996). However, signalling of QoS filter has not been addressed. Filters for hierarchical encoded streams are presented in (Wolf et al. 1994). They adapt continuous media streams to a specified QoS. Hierarchically encoded streams are also used in (McCanne et al. 1996). It is designed for networks like todays Internet, i.e., networks with best effort multi-point packet delivery. Each layer is associated with a multicast group. Receivers can join and leave a group in order to change the experienced QoS. This approach is mainly targeted towards network congestion. The usage of filters to control network congestion is presented in (Wittig et al. 1994). Heterogeneity is not explicitly addressed. Application level gateways, that operate on top of RTP (Schulzrinne et al. 1996) are addressed in (Amir et al. 1995) for filtering conversion operations. Reliable multicast is not addressed in this work.

For the placement of QoS filters in the dissemination tree, several aspects have to be considered. Generally, minimal resources should be allocated with respect to bandwidth and processing power. Moreover, the end-to-end delay of user data should be minimized. Considering bandwidth, an optimal placement allocates a QoS filter as close to the root (i.e., sender) of the multicast tree as possible. The AMnet signalling protocol and the AMnet manager need to address this issue. Furthermore, dynamic join and leave operations must be supported. Therefore, dynamic allocation and re-configuration of QoS filters needs to be supported.

Error Control and Recovery

Active multicasting nodes provide also a sound bases for the implementation of error control and recovery. Thus, local group based concepts as they are discussed in the literature can be easily supported via dedicated service modules.

The problem of providing multicast communication to a large group of receivers is a vital research area in the networking community. Especially, approaches, such as LBRM (Holbrook et al. 1995) and LGC (Hofmann 1996) are similar to the concept followed by AMnet. However, these protocols do not support heterogeneous receivers. Moreover, QoS filtering of data streams may violate data integrity. Therefore, error control and recovery has to be aware of QoS filters.

Typically, mechanisms for error control and recovery are not applied in multimedia applications, such as video conferencing. However, they can still be advantageous for such applications. Due to severe time constrains retransmission of continuous media from the source to the receiver is often impossible, because of the latency incurred. Since active network nodes can provide retransmissions, the experienced retransmission delay can be significantly reduced compared to the end-to-end delay.

In AMnet, QoS filtering is supported by error control and recovery mechanisms. A negative acknowledgement scheme is applied. The active multicasting node collects the NACKs of a local group of receivers, processes them and, if necessary, passes a NACK to an active network node that is located closer to the sender of the multicast tree. The sender only receives NACKs from active network nodes of the highest hierarchy level.

Active multicasting nodes also perform error recovery. If – in case of a NACK – the requested data are still cached inside the respective network node, it retransmits the data to the receiver. If not, the NACK is passed to active network nodes at higher hierarchy levels. This releaves in many cases the burden of retransmissions from the source.

3   RSVP-BASED AMNET PROTOTYPE

A prototype of AMnet has been implemented. In the context of the Internet, signalling protocols are currently investigated with the goal of providing an integrated services architecture (Braden et al. 1994). RSVP (Zhang et al. 1993) seems to be the main candidate for deployment in the Internet. Since its design is flexible with respect to future extensions, the prototype AMnet implementation is based on RSVP. Currently the prototype consists of QoS filter modules, which are presented in the next section and an AMnet manager, which performs signaling tasks.

3.1   QoS filter

With respect to QoS filter, the focus was on compressed video streams that are MPEG-1 coded. The implementation and integration of an H.261 video is currently under way. Other filters can be found in (Yeadon et al. 1996). The filter types presented in section 2.3 have been implemented in AMnet (Kupka 1997).

These types of QoS filters are configured in the active network nodes. In addition, a set of parameters is defined which specifies the requested filter function. The parameters are *quality factor*, *skip-frames*, *B/W-mode* and *number of slices*. The quality factor corresponds to a quantization level. The filter supports the quality factors 1 to 5, with 1 representing the best play-out quality, i.e., the least bandwidth reduction. The parameter skip-frames determines the frames to be dropped. The parameter B/W-mode activates the color reduction. Finally, the slice parameter specifies the number of slices per frame that the filter has to insert into the MPEG stream.

3.2 RSVP Extensions

Since QoS filters are located in the network, signalling is needed for configuration and allocation of QoS filters and for negotiation of filter parameters. As QoS filter reside on intermediate systems, they can be viewed as network resources. Due to its flexible concept, RSVP can be extended to configure and control QoS filters. In the current AMnet prototype RSVP serves as signalling protocol.

RSVP provides the concept of classes that can be defined in order to extend RSVP to new resources. The format of the new RSVP class *QoS filter* (QF) is depicted in Figure 3(a). To assure compatibility with RSVP entities that are not extended to handle QF objects, the QF class carries the class number 208. With this class number a QF object is simply forwarded by an RSVP entity if the QF class is unknown. To select a certain type of filter class the type field is used. As an example, Figure 3(b) depicts a QF object for the implemented MPEG filter (cf., above).

The RSVP daemon has been extended with an interface to the AMnet manager to exchange QF objects (Krasnodembski 1997). In case a QF object is included in the received reservation (RESV) message, the QF object is extracted and forwarded to the AMnet manager.

3.3 Filter Merging

The AMnet manager is responsible for allocation and configuration of the QoS filter according to the requirements stated in the QF object. If the AMnet manager receives a QF object from the RSVP daemon, it needs to check whether other QoS requirements for that group have been received previously. Therefore, the AMnet manager keeps some state information concerning filter requirements of group members. The following cases can be distinguished:

1. no previous request
2. previous request for same QoS filter type
3. previous request for other QoS filter type

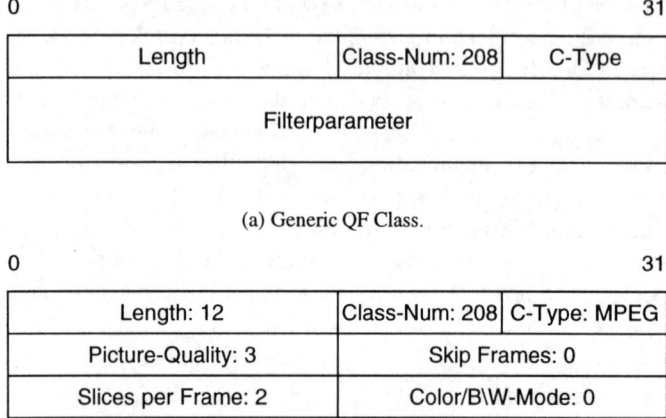

(a) Generic QF Class.

(b) Example: MPEG Filter Class.

Figure 3 RSVP Filter Class.

If no previous request is available, the RESV message is forwarded to the next node in the multicast tree.

If another filtering request for the same group and the same QoS filter is available, merging of the filter parameters is needed. The AMnet manager updates the QF object accordingly and forwards it to the RSVP daemon for further propagation towards the sender using RESV messages and standard RSVP procedures.

If a previous request for a different QoS filter type is registered at the AMnet manager, two cases need to be distinguished.

Firstly, the new request can be merged with the current request and the resulting request does not impose higher requirements than the existing request. In this case, the corresponding filter functionality is activated by the AMnet manager. The QF object that is periodically retransmitted with RSVP messages needs not to be updated. The node requires still the same amount of input data from the hierarchically higher nodes in the multicast tree.

Secondly, the newly requested QoS filter can not be merged with the active filter. In this case, different filters serve different group members in the same active network node. To receive a sufficient data stream the highest requirements for both requests have to be identified. They form the new QF object to be forwarded in the RSVP messages.

A joining group member with new requirements can also lead to the re-location of already installed QoS filters. If the data is already filtered to a lower QoS level as the requirements of the joining group member, the filter may need to be moved to an active network node that is not in the dissemination path of the new member.

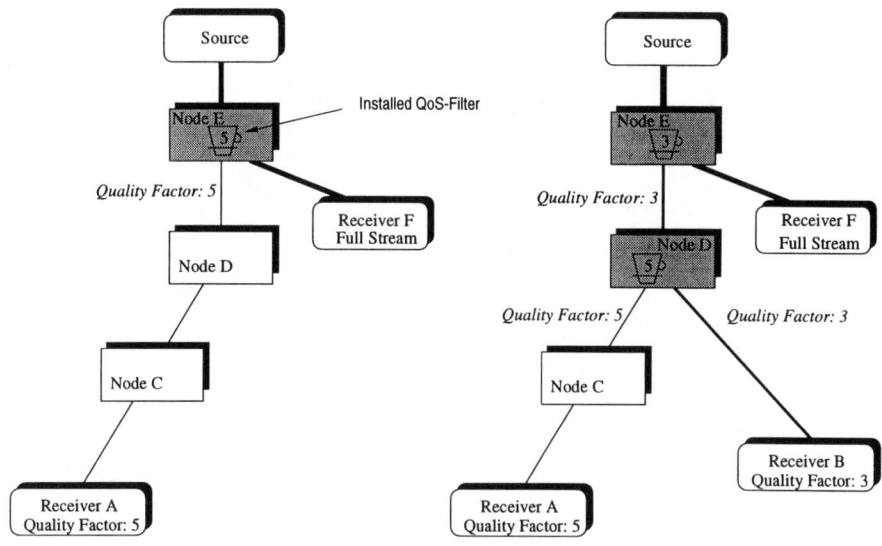

Figure 4 RSVP and QoS Filter.

QoS filters are released if group members leave the group and filtering at that node is no longer required. The same time out mechanism used by RSVP is applied for that purpose. Thus, AMnet currently only provides soft-states.

RSVP daemon and AMnet manager are involved in the control flow only. User data are forwarded directly to the corresponding QoS filter. The output of the QoS filter is then directed to the packet scheduler of the output link.

Example
In order to illustrate configuration and merging of filters, the example depicted in Figure 4 is considered. There are three receivers: A, B and F. They participate in a multicasted MPEG-1 video stream. Receiver A has severe bandwidth constraints due to a wireless link and, thus, requests a very low quality factor of 5. Receiver A issues a RESV message that includes the corresponding QF object. No other RESV messages have passed nodes C and D. This is first case mentioned in section 3.3. Thus, the RESV message is forwarded to Node E. Node E has already seen a RESV message without bandwidth constraints from receiver F. This situation corresponds to the third case in section 3.3. As a result, node E has to install a new filter that is configured to the quality factor of 5 (cf., Figure 4(a)), i.e., receiver F continues to receive the full stream and node D receives a reduced data rate.

In a subsequent step, receiver B joins the session. B requests a video stream with a moderate quality factor of 3 (cf., Figure 4(a)). The RESV message of receiver B is forwarded to node D. Node D needs to merge the different filter requests from receivers A and B. Once again, this situation corresponds to the third case in section 3.3. As a result, an RESV message with a requested video quality factor of 3

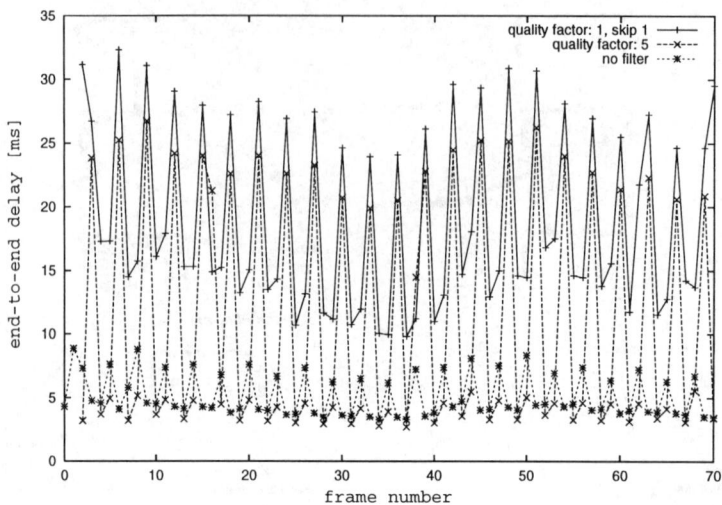

Figure 5 Filter vs. no filter.

is forwarded to node E. Moreover, a new filter is activated at Node D that performs filtering according to quality factor 5. Due to the new request at node F, the filter is re-configured by the AMnet manager to quality factor 3.

If B would have requested quality factor 5, no re-configuration would have been necessary. This corresponds to the second case listed in section 3.3.

3.4   Performance

In this section some performance results of the AMnet prototype are discussed. The end-to-end transfer delay of an MPEG-1 encoded video stream was measured with 160×120 pixels/frame. The group of pictures (GOP) was IBBPBB. Figure 5 depicts the results obtained when sender, receiver, and the active multicasting node were running on Sun Ultra 1 workstations. The workstations were connected through a lightly loaded 100 Mbit/s Fast-Ethernet. In the first experiment the video stream was transferred without filtering. In the second experiment a high quality filter (quality factor 1) was requested. In the third experiment a very low quality level (quality factor 5, dropping of B frames) was requested. Although the transfer delay depends on the requested quality level, QoS filtering introduces a substantial processing delay. Therefore, the video stream without filters leads to the best end-to-end delay. Due to the lightly loaded network with practically no data losses, QoS filters are not needed.

Figure 6 depicts results obtained in a more realistic scenario. The receiver was running on a 200MHz 586 PC, attached via a 2 Mbit/s wireless LAN (WaveLAN). The results demonstrate that QoS filtering can considerably reduce the end-to-end delay. Moreover, we obtained a reduced loss-rate of frames when QoS filter are used

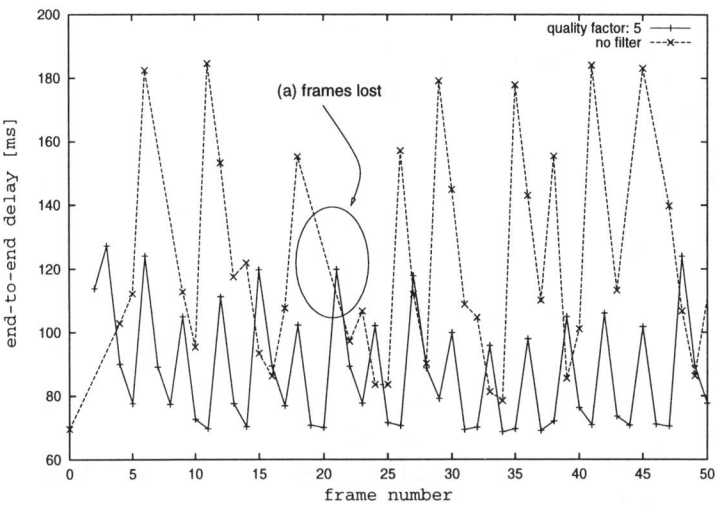

Figure 6 Wireless vs.wired transfer.

(cf., Figure 6(a)). This is especially important when the coding scheme is very sensitive to loss as in the case of MPEG. It should be noticed, that the unfiltered stream could not be decoded at the receiver. Thus, QoS filtering is highly required in such environments.

4 CONCLUSION

The AMnet approach presented in this paper provides communication support for heterogeneous group communication. Typical applications that require such services are collaborative applications and distributed simulations. AMnet is based on active networking, i.e., network nodes provide specific services to individual users. Active multicasting nodes implement QoS filters, signalling for QoS filters, error control and synchronization. The existing prototype basically considers QoS filters (MPEG-1) and RSVP-based signalling of QoS filters. The AMnet prototype is available at `http://ibr.cs.tu-bs.de/~wittmann/AMnet.html`.

Future work will focus on the impact of QoS filtering on synchronization and error control. Furthermore, merging and dynamic re-location of QoS filters will be studied in greater detail. Moreover, the prototype will be enhanced to allow for dynamically loadable service modules.

REFERENCES

Amir, E., McCanne, S. & Zhang, H. (1995), An application level video gateway, *in* 'Proc. ACM Multimedia '95', ACM, San Francisco, CA.

Braden, R., Clark, D. & Shenker, S. (1994), Integrated services in the internet architecture: an overview, RFC 1633, ISI.

Hofmann, M. (1996), A generic concept for large scale multicast, *in* 'Proceedings of International Zurich Seminar on Digital Communication (IZS'96)', Springer Verlag.

Holbrook, H., Singhal, S. & Cheriton, D. (1995), Log-based receiver-reliable multicast for distributed interactive simulation, *in* 'Proceedings of SIGCOMM '95', ACM SIGCOMM, Cambridge, MA.

Krasnodembski, K. (1997), Konzeption und Realisierung eines Signalisierungsprotokolls für Filterfunktionen, Diploma thesis, TU Braunschweig.

Kupka, T. (1997), Skalierung von MPEG-I Videoströmen, Diploma thesis, TU Braunschweig.

McCanne, S., Jacobson, V. & Vetterli, M. (1996), Receiver-driven layered multicast, *in* 'ACM SIGCOMM'96', San Francisco, USA.

Rao, K. & Hwang, J. (1996), *Techniques and Standards for Image Video and Audio Coding*, Prentice Hall PTR.

Schulzrinne, H., Casner, S., Frederick, R. & Jacobson, V. (1996), RTP: A transport protocol for real-time applications, Technical report, GMD Fokus.

Tennenhouse, D. L., Smith, J. M., Sincoskie, W. D., Wetherall, D. J. & Minden., G. J. (1997), 'A survey of active network research', *IEEE Communications Magazine* **35**(1), 80–86.

Tennenhouse, D. L. & Wetherall, D. J. (1996), 'Towards an active network architecture', *Computer Communication Review* **26**(2).

Wittig, H., Winckler, J. & Sandvoss, J. (1994), Network layer scaling: Congestion control in multimedia communication with heterogenous networks and receivers, *in* 'International COST 237 Workshop on Multimedia Transport and Teleservices', Vienna, Austria.

Wolf, L. C., Herrtwich, R. G. & Delgrossi, L. (1994), Filtering Multimedia Data in Reservation-Based Internetworks, Technical Report 43.9608, IBM European Networking Center, 69115 Heidelberg, Germany.

Yeadon, N. (1996), Quality of Service Filtring for Multimedia Communications, Phd thesis, Lancaster University.

Yeadon, N., Garcia, F., Shepherd, D. & Hutchinson, D. (1996), Continuous media filters for heterogeneous internetworking, *in* 'Proceedings of the Conference in Multimedia Computing and Networking 1996', San Jose, California.

Zhang, L., Deering, S. & Estrin, D. (1993), 'RSVP: A new reavailable ReSerVation Protocol. novel design features lead to an Internet protocol that is flexible and scalable', *IEEE network* **7**(5).

Zitterbart, M. (1996), User-to-User QoS management and monitoring, *in* 'Proceedings IFIP Workshop on Protocols for High Speed Networks', Sophia Antipolis, France.

36

Call admission and routing in ATM networks based on virtual path separation

S.A. Berezner
Department of Statistics, University of Natal, 4001 Durban, South Africa.
fax: +27 31 260 1009, e-mail: berezner@ph.und.ac.za

A.E. Krzesinski
Department of Computer Science, University of Stellenbosch, 7600 Stellenbosch, South Africa. fax: +27 21 808 4416, e-mail: aek1@cs.sun.ac.za

Abstract

Dynamic reconfiguration is a network management control which reserves transmission capacity on the communication links in order to form dedicated logical paths for each origin-destination flow. We present an efficient deterministic algorithm (XFG) to rapidly compute an optimal network configuration. We apply reconfiguration to a model of an ATM network. The XFG calculation of an optimal configuration requires a few seconds of CPU time (Pentium 100MHz) whereas a standard NLP package requires several hours to compute an optimal network configuration.

Keywords

ATM networks, call admission control, network resource management, non-linear optimization, reconfiguration, virtual paths.

1 INTRODUCTION

Resource management in ATM networks is implemented by a hierarchy of controls at the network, call and cell levels which interact to form a layered

---

*This work is supported by grants from the Australian Research Council and the South African Foundation for Research Development.

Broadband Communications   P. Kühn & R. Ulrich (Eds.)
© 1998 IFIP. Published by Chapman & Hall

traffic control. The Network Resource Manager (NRM) collects O-D traffic intensities and call blocking probabilities for the different traffic classes from the Call Admission Control (CAC). If these data exceed threshold values, the NRM may adjust the buffer sizes at the switches and compute new values for the bandwidths of the Virtual Paths (VPs) and update the VP network configuration. This is done by cross connecting groups of circuits in intermediate nodes to form (logical) direct connections between all O-D pairs to create a fully meshed network. Depending upon the traffic/network equipment variations, the update interval may vary from minutes to hours.

The CAC computes the effective bandwidth of an offered call and determines how best to route the call through the network. If a route is available, the call is connected, else the CAC may re-negotiate the call's Quality of Service (QOS) parameters or reject the call. The CAC collects cell loss and cell delay data and if these exceed threshold values, the CAC may adjust the effective bandwidth of the call or other actions may be taken. The cell level controls are responsible for detecting cell congestion and responding by traffic shaping, source rate adaptation, rate- or credit-based flow control, cell priorities and discards, and user/network parameter controls.

This paper presents a model of hierarchical resource management in ATM networks. The NRM component of the model adjusts the logical link capacities in order to optimally size the virtual path (VP) network whenever changed conditions demand such a reconfiguration. The reconfiguration maintains sufficient capacity to carry the established calls in progress. The CAC connects calls on the (logical) direct links. A call is rejected if insufficient capacity is available on the logical link. The model assumes that the effective bandwidths of the various traffic classes are known and are fixed. Cell level behaviour is assumed to be encapsulated into the effective bandwidth parameters. The model therefore contains no explicit description of cell level behaviour or cell-based resource management.

The model shows that the simplicity of reconfiguration is especially useful in dealing with the complex characteristics of multi-rate calls. We compare reconfiguration with another state independent admission/routing control (Mitra 1996) which effectively converts an ATM network into a network with fixed routing. We develop a computationally efficient algorithm for optimal network reconfiguration. We apply the algorithm to model call admission and routing in the NSF backbone network (Mitra 1996) and we demonstrate that reconfiguration achieves a comparable performance to traffic redistribution. We show that reconfiguration has both advantages and disadvantages: the advantages include conceptual simplicity, protection of the GOS for O-D pairs and much lower computational complexity. The main disadvantage of reconfiguration is that it realizes a higher overall blocking, especially for smaller networks.

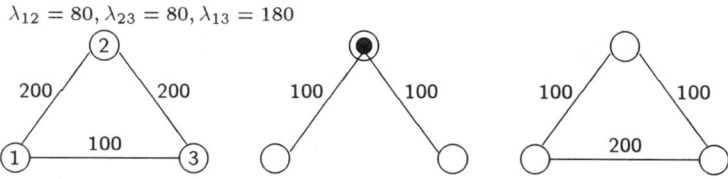

$\lambda_{12} = 80, \lambda_{23} = 80, \lambda_{13} = 180$

Figure 1 (a) Physical network before reconfiguration (b) cross connections (c) logical network after reconfiguration

2 DYNAMIC RECONFIGURATION

Consider a network which consists of J nodes and carries S services (call classes). Let d_s denote the effective bandwidth requirement of a call of class s where $s = 1, \ldots, S$. Let C_{ij} denote the capacity of a uni-directional link (i, j) such that $C_{ij} > 0$ denotes the existence of a physical link between nodes i and j. Let L denote the number of physical links. Let λ_{ij}^s denote the Poisson arrival rate of calls of class s at link (i, j) and let $1/\mu_{ij}^s$ denote the mean holding time of class s calls on link (i, j). The arrival intensity for calls of class s on link (i, j) is $\rho_{ij}^s = \lambda_{ij}^s/\mu_{ij}^s$. A call of class s connected between nodes i and j generates revenue at a rate θ_{ij}^s. The revenue generated by a call does not depend on how the call is routed, although this assumption can be relaxed if cross connection costs or other costs are taken into account. The effective bandwidth of a class s call is also independent of how the call is routed.

Consider the following network reconfiguration scheme which copes with mismatches between offered traffic and available capacity by making logical direct connections between nodes. Each such logical link is composed from several routes, where a route $r = (i, i_1, \ldots, i_k, j)$ is a sequence of nodes. All traffic classes are multiplexed onto each link and onto each route, provided the route has sufficient capacity.

Reconfiguration is performed by cross connecting circuits using configurable switches at the transit nodes. For example, consider the 3-node network presented in Figure 1. Before reconfiguration – Figure 1(a) – the network blocking probability is 24%. Figure 1(b) illustrates the cross connection of 100 circuits at node 2. After reconfiguration – Figure 1(c) – the network blocking probability is 1%.

The number x_r of circuits that should be cross connected along route r, as well as the number $x_{(i,j)}$ of circuits that should be used to carry direct traffic on the link (i, j) are obtained as a solution to the optimization problem presented below. Let $\mathcal{R}_{ij}$ denote the set of *admissible* routes between nodes i and j, including the direct route (i, j) if it exists. A route r is said to be *admissible* if circuits can be reserved on each link along this route for the sole use of calls between nodes i and j. A route is not admissible if restrictions are imposed on the route: for example there could be a limit on the number of links in the route. Routes with loops are not admissible. Let $\mathcal{R} = \cup_{i,j}\mathcal{R}_{ij}$ denote the set of all admissible routes. Let $\boldsymbol{x} = \{x_r\}_{r \in \mathcal{R}}$, $x_r \in \mathbb{I}^+$ denote the route configuration. The capacity reserved for the sole use of the traffic between nodes i and j, which is the capacity of the VP connecting nodes i and j, is given by $C_{ij}(\boldsymbol{x}) = \sum_{r \in \mathcal{R}_{ij}} x_r$.

Let $B_{ij}^s(C_{ij}(\boldsymbol{x}))$ denote the blocking probability (Ross 1995) experienced by class s calls on a link (i, j) of capacity $C_{ij}(\boldsymbol{x})$ carrying services $s = 1, \ldots, S$ with bandwidth d_s and traffic intensity ρ_{ij}^s. The total rate of earning revenue is

$$F(\boldsymbol{x}) \quad = \quad \sum_{ij} F_{ij}(C_{ij}(\boldsymbol{x})) = \sum_{ij}\sum_{s} \theta_{ij}^s \rho_{ij}^s (1 - B_{ij}^s(C_{ij}(\boldsymbol{x}))). \tag{1}$$

The Dynamic Reconfiguration and Optimization (DROP) problem is defined as: Maximize $F(\boldsymbol{x})$ subject to

$$\sum_{r \in \mathcal{A}_{ij}} x_r \leq C_{ij}, x_r \geq 0. \tag{2}$$

where $\mathcal{A}_{ij}$ denotes the set of admissible routes that use link (i, j) including the direct route (i, j). A vector $\boldsymbol{x}$ which satisfies constraint (2) is called a solution to the DROP problem. A solution $\boldsymbol{x}$ which (locally) maximizes the revenue function (1) is called a (local) optimal solution to the DROP problem.

3   TRAFFIC REDISTRIBUTION

Another method (Mitra 1996) of dealing with traffic/capacity mismatch is to redistribute the offered traffic by random sampling of the arriving flows. Let p_{ijr} denote the probability that a call arriving to node i and destined for node j is offered to route r. Thus $1 - \sum_{r \in \mathcal{R}_{ij}} p_{ijr}$ is the probability that an arriving call is not offered to the network, which is a form of call admission control. If the call is offered to the network, and if the call is blocked on route r, it is lost. This model is equivalent to a loss network with fixed flows on fixed routes and has an analytical solution (Kelly 1991).

A sequence of values is computed for the probabilities p_{ijr} which successively improve the network revenue. At each step, for the current set of

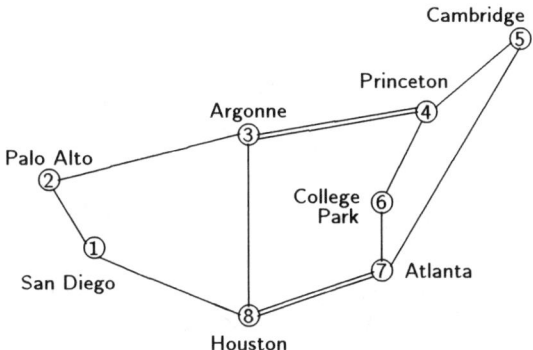

Figure 2 Core NSF network

probabilities p_{ijr}, the link blocking probabilities are calculated using the Erlang fixed point approximation. This yields a system of SL linear equations for the implied costs. The linear equations are solved to obtain the derivatives of the revenue function. The complexity of this step is $O(S^3L^3)$ which can be reduced to $O(L^3)$ if a special approximation technique is used. The derivatives are used to compute the direction of steepest ascent for the revenue function in order to obtain new values for the probabilities p_{ijr} which yield an improved revenue. This optimization procedure converges to a value for the network revenue which is not necessarily a local maximum.

Note that our reconfiguration scheme differs from the logical reconfiguration method described in Faragó (1995) which assumes that the network revenue is a convex function of the link capacities. Optimal routes are computed to carry each service class between all O-D pairs and an exact lower bound to the network revenue is calculated. This global optimum provides an initial point to compute a better estimate of the optimal network revenue. Our model assumes that the link revenues are convex functions of the link capacities so that the network revenue is not a convex function. Optimal routes are computed between all O-D pairs and many service classes are multiplexed onto each route.

4 RECONFIGURATION VERSUS REDISTRIBUTION

Consider the model (Mitra 1996) presented in Figure 2 which is a fictitious representation of the core NSF ATM backbone consisting of eight nodes

Table 1 Traffic intensity matrix

nodes	1	2	3	4	5	6	7	8
1	–	6	7	1	9	5	2	3
2	7	–	24	3	31	15	6	9
3	8	25	–	4	37	18	7	11
4	1	3	3	–	4	7	1	1
5	11	33	39	5	–	24	9	15
6	5	14	16	2	21	–	4	6
7	2	5	6	1	8	4	–	2
8	3	8	10	1	12	6	2	–

each connected by two uni-directional links. Each link carries traffic in one direction. The double lines between nodes 3 and 4 and 7 and 8 indicate that there are two links in each direction connecting these nodes. The network carries 6 services: the bandwidth requirement of the 1st service is 1 unit (16 Kbps) and the bandwidth requirements of services 2 through 6 are 3, 4, 6, 24 and 40 units respectively. The capacity of each uni-directional link is 2812 units (45 Mbps). The traffic intensity for service 3 is given in Table 1. For all other services, the values in Table 1 are multiplied by 0.5. The revenue earned by a carried call per unit of time is assumed to be independent of how the call is routed and is equal to the bandwidth of the call.

The DROP problem can be solved by NLP methods. The NSF backbone model has 332 routes and 56 VPs. Two NLP problems were defined: the first problem restricted the route lengths to 5 links (260 routes) so that the NLP has 260 variables and 56 linear constraints. The second problem does not restrict the route lengths which can be up to 7 links long. In this case the network has 332 routes and the corresponding NLP has 332 variables and 56 linear constraints. Table 2 presents the computational resources required to calculate an optimal VP configuration using the NLP package CFSQP (Lawrence 1996). The NLP terminates when the values of the objective function at the previous and current iterates differ by less than $10^{-\delta}$. An accurate CFSQP calculation of the optimum reconfiguration for this, not very large, network can take several hours (Pentium 100MHz). This suggests that general NLP solvers may not be able to compute an optimal reconfiguration within the time frame required to reconfigure the network.

Let the set of admissible routes $\mathcal{R}$ be partitioned into $\mathcal{R} = \cup_{k=0,1,\dots} \mathcal{R}^k$

Table 2 NLP

δ	5 links			7 links		
	steps	time secs	revenue	steps	time secs	revenue
0	76	760	19 970	95	1698	19 973
1	132	1048	19 995	123	2212	19 945
2	198	2618	20 003	217	3886	20 004
3	254	6256	20 004	223	5635	20 004
6	291	7375	20 004	363	18 679	20 004

where $\mathcal{R}^k$ is the set of routes for which the number of links in route $r \in \mathcal{R}_{ij}$ is no greater than $k_{ij} + k$ where k_{ij} is the length of the shortest route connecting nodes i and j. First consider the route set $\mathcal{R}^0$: traffic redistribution yields a revenue of 20 401 and the network blocking is 5.6%. Reconfiguration yields a revenue of 20 004 and a blocking of 7.4%. When applied to the route set $\mathcal{R}^1$, redistribution yields a revenue of 21 012 and reconfiguration again yields 20 004.

Overall blocking. The network revenue obtained from traffic redistribution is some 2 to 5% higher than the revenue obtained from reconfiguration. The reason for this is that redistribution employs a fully shared network where calls for many O-D pairs are multiplexed onto one route, which can result in the more efficient use of resources. Reconfiguration reserves capacity on a distinct VP for each O-D pair which results in a higher blocking. This difference will be noticeable in networks with relatively small link capacities, low connectivity where many VPs use the same links and light traffic where higher capacities are reserved for O-D pairs with light traffic.

GOS preservation. Capacity reservation, which can be considered a drawback of reconfiguration since it results in higher overall blocking, is at the same time an important instrument for GOS protection. For example, suppose the links on all routes connecting nodes i and j have 5% blocking and the length of these routes is equal to 5. Traffic redistribution will result in all calls from i to j experiencing 25% blocking, which is unacceptable. Reconfiguration does not have such a problem when calls must be carried on long routes since dedicated VPs (logical direct routes) are established for all O-D pairs.

Conceptual Simplicity. Traffic redistribution and configuration are relatively simple controls when compared to alternative routing. However, reconfiguration is simpler to implement, operate and optimize. Once the

VPs have been configured, the network transports direct traffic on (logical) direct links. The calculation of the blocking probabilities is straightforward, unlike traffic redistribution which employs the multi-rate Erlang fixed point approximation to compute the blocking. If the traffic conditions change, any resulting performance problems can be isolated and corrected. Thus if blocking for an O-D pair increases, this must be because the capacity assigned to the VP for this O-D pair is no longer adequate and must be increased.

Computational Complexity. Reconfiguration and traffic redistribution are intended to operate in near real-time. The computational resources required to calculate an optimal reconfiguration/redistribution are therefore important.

The solution of the traffic redistribution problem (Mitra 1996) has a high computational complexity since each step of the optimization requires solving a large system of linear equations, calculating the link blocking and solving the fixed point equations. In contrast, network reconfiguration is simple and requires much less computation. The multi-rate link blocking probabilities $B_{ij}^s(c)$ are calculated for $c = 0, 1, \ldots, C$ where $C = \max_{ij}(C_{ij})$ using a recursive algorithm (Ross 1995) whose computational complexity is $O(SC)$. This algorithm is adapted to work with normalized probabilities so that the calculation of the multi-rate blocking probabilities is numerically stable for large values of C. The probabilities $B_{ij}^s(c)$ are calculated once only.

The link blocking probabilities are used to compute the rate of earning revenue $F(x)$ which is given in equation (1). The dimensionality of x is equal to the number of routes in the network. For example in a network of 100 nodes, the number of shortest routes between two nodes could be $O(10^4)$. Thus even if the NLP was restricted to the shortest routes it could encounter computational difficulties. Discarding longer routes will yield a sub-optimal solution.

Bandwidth allocation within a VP. For each O-D pair the reconfiguration procedure constructs a single VP from several routes. In a multi-service environment the blocking on this VP can be higher than the blocking on a shared link of the same capacity. However, if there is a large capacity route in a VP, then high bandwidth calls can be sent over this route and low bandwidth calls can be sent to routes of smaller capacity within the VP. In this case, blocking on the VP will be almost identical to blocking on a single link of the same capacity. Thus an appropriate intra-VP routing strategy can convert this disadvantage of reconfiguration into an advantage: admission policies within each VP can be chosen to optimize the blocking/revenue for all services.

5   XFG: AN OPTIMIZATION ALGORITHM

The DROP problem can be solved using NLP methods. However, NLP methods have a high computational complexity. This section presents an efficient deterministic algorithm named XFG to compute an optimal value for the revenue function given in equation (1) subject to constraints of equation (2). The algorithm contains a mechanism to extract itself from inferior local optima.

The revenue function $F_{ij}(c)$ for each link (i, j) was defined in equation (1) for integral values of c. However, most optimization techniques require that $F_{ij}(c)$ be defined for all real nonnegative c. Therefore, instead of optimizing $F(x)$ we optimize $\widetilde{F}(x) = \sum_{ij} \widetilde{F}_{ij}(C_{ij}(x))$ where $\widetilde{F}_{ij}(c)$ is obtained by fitting a differentiable increasing convex function to $F_{ij}(c)$ where $c = C_{ij}(x)$. We also require that $\widetilde{F}'_{ij}(c) = \partial \widetilde{F}_{ij}(c)/\partial c = \theta$ at $c = 0$ for all i and j such that $C_{ij} > 0$ where $\theta = \max_{ij}(\theta^s_{ij}/d_s)$ is the maximal rate of earning revenue per unit of capacity. The meaning of this requirement is explained below.

An x that realizes an optimal revenue for $\widetilde{F}_{ij}(x)$ will realize an approximately optimal revenue for $F_{ij}(x)$. The following arguments support this statement. Although the multi-service link revenue function may be non-monotone, it is asymptotically convex and monotone and in fact it becomes convex and monotone for moderate values of c in most practical cases. We expect the optimal network configuration to have relatively large capacities for all source-destination pairs, so that the difference $\widetilde{F}_{ij}(x) - F_{ij}(x)$, which is the error in the approximation, will be small.

The optimization is based on the concept of link cost. If the route configuration is x, the cost of carrying traffic from node i to node j is defined as the derivative of $\widetilde{F}_{ij}(c)$ estimated at $c = C_{ij}(x)$. This cost is (approximately) equal to the loss in revenue when the direct traffic from node i to node j is carried on a VP whose capacity is increased (decreased) from c to $c + 1$ $(c - 1)$. In this context, the requirement that $\widetilde{F}'_{ij}(0) = \theta$ for all physical links (i, j) implies that the cost of a physical link from which all direct traffic has been displaced is equal to an overall network maximum. Since the algorithm is based on balancing link costs, this situation where a link is entirely used by nondirect traffic will not occur in an optimal solution.

The necessary condition for x to be a local optimum for $F(x)$ is that for any pair (i, j) and for any route $r \in \mathcal{R}_{ij}$ such that $x_r > 0$ we have

$$\frac{\partial F_{ij}(C_{ij}(x))}{\partial c} = \sum_{(i,j)\in r} \frac{\partial F_{ij}(C_{ij}(x))}{\partial c}. \tag{3}$$

Equation (3) can immediately be derived from the fact that all derivatives of the revenue function at the point of extremum must be equal to zero.

XFG: an algorithm for finding an optimal VP configuration.

Step $n = 0$: Construct an initial route configuration $x[0]$ which corresponds to the original network where $x_r = C_{ij}$ for $r = (i, j)$ and $x_r = 0$ for all other $r \in \mathcal{R}_{ij}$.

Step $n \to n + 1$: Compute a configuration $x[n + 1]$ which yields a larger revenue than $x[n]$. Construct a weighted graph of the physical network where each link is assigned its current cost. Select an arbitrary pair of nodes i and j and compute the least cost route $r \in \mathcal{R}_{ij}$ between i and j. The computational complexity of this step is $O(L)$. The following procedures yield a vector $x[n + 1]$ which improves the network revenue:

Cross connect. If the current cost of the VP connecting nodes i and j is greater than the cost of the shortest route $r \in \mathcal{R}_{ij}$, then cross connect $x \in \mathbf{R}^+$ units of capacity along the route r – this capacity will be subtracted from each link of the route r. The amount of capacity x to be cross connected can be found so that it maximizes the increase in revenue (line search).

Release. If the current cost of the VP connecting nodes i and j is less than the cost of the shortest route $r \in \mathcal{R}_{ij}$, then release $x \in \mathbf{R}^+$ units of capacity along the route $r' \in \mathcal{R}_{ij}$ such that $x_{r'} > 0$ and the cost of this route is maximal among all routes $r \in \mathcal{R}_{ij}$ – this capacity will be returned for the use of the direct traffic offered to each link of the route r'. Again, the amount of capacity x to be released can be found so that it maximizes the increase in revenue (line search).

Stopping rule: If for all pairs (i, j) neither cross connection nor release leads to an increase in revenue, then stop.

The order in which the cross connect and release operations are performed, and the order in which the pairs (i, j) are selected has to be specified. For example, the procedure can be cyclically applied to all pairs, or the pair that yields highest increase in revenue could be selected. The order of cross connect/release operations may influence the speed of convergence of the algorithm.

6   AN APPLICATION OF XFGI

An initial version of the algorithm has been implemented where cross connections and releases are made in integer units of capacity. A piecewise linear convex hull $\widetilde{F}_{ij}(c)$ is fitted to the revenue function $F_{ij}(c)$ and a cubic spline is used to interpolate among the values of $\widetilde{F}_{ij}(c)$. The convex hull

Table 3 XFGi VP configuration

	R^0	R^1	R^3	R^4
cross connections	17 968	3538	380	25
routes	47	29	13	7

affords a good approximation since $F_{ij}(c)$ is nearly monotone, and the cubic spline interpolation ensures that the revenue function and its first and second derivatives are continuous. We call this version of the algorithm XFGi where the "i" is an abbreviation for "integral".

The XFGi algorithm is applied to compute an optimal VP configuration for the model of the NSF backbone network presented in Figure 2. The algorithm constructs VPs in increments of 1 bandwidth unit. The algorithm converges in 21 920 steps, making 21 920 cross connections and 497 releases using 96 routes. The lengths of the cross connected routes are shown in Table 3. Note that the VPs consist primarily of circuits cross connected along the shortest routes connecting the O-D pairs. Twenty direct routes and 79 other routes are used to construct 56 VPs. The XFGi reconfiguration yields a rate of earning revenue of 20 004.2 and the NLP reconfiguration yields 20 004.3.

The computational requirements of the XFGi algorithm are determined by the search for the shortest (in terms of cost) routes which has complexity $O(JL)$. If the link capacities C_{ij} are large, the unit of bandwidth increment in the XFGi algorithm can be increased from 1 to $x > 1$. This will decrease the computational complexity of the algorithm although the computed revenue may be lower. Table 4 presents the revenue computed after reconfiguration for various values of x.

7 CONCLUSION

This paper investigates the use of capacity reconfiguration as a resource management control in ATM networks where transmission capacity is reserved on the communication links in order to form dedicated logical paths for each origin-destination flow. We present a model of dynamic reconfiguration which calculates the optimal network configuration in terms of a constrained nonlinear programming problem (NLP). An efficient algorithm is presented to reconfigure a multi-rate network and compute an optimal VP configuration. The algorithm is applied to reconfigure a model of the

Table 4 XFGi convergence

x	connects /releases	routes connected	time secs	revenue
1	21 920/497	79	19	20 004
3	9615/756	80	9	20 004
4	7963/726	79	7	20 004
6	6667/904	80	7	20 004
24	4118/783	77	4	20 004
40	3592/738	74	4	20 004
NLP: CFSQP			3886	20 004

NSF backbone network. We show that reconfiguration can be advantageous for large networks where its drawbacks namely higher overall blocking due to VP service separation, are outweighed by its advantages namely simplicity, efficiency and tractability.

REFERENCES

Faragó A., Blaabjerg S., Ast L., Gordos G. and Henk T.. A New Degree of Freedom in ATM Network Dimensioning: Optimizing the Logical Configuration. *IEEE J. on Selected Areas in Comms.*, **13:7**, 1199–1206.

Kelly F.P. (1991) Loss networks. *Annals of Applied Prob.*, **1:3**, 473–505.

Lawrence C., Zou J.L. and Tits A.L. (1996) User's Guide for CFSQP Version 2.4. Report Number TR-94-16r1, Electrical Engineering Department and Institute for Systems Research, University of Maryland, College Park, MD 20742 USA.

Mitra D. and Seery J.B. (1990) Comparative Evaluations of Randomized and Dynamic Routing Strategies for Circuit-Switched Networks. *IEEE Trans. Comm.*, **31**, 102–116.

Mitra D., Morrison J.A. and Ramakrishnan K.G. (1996) ATM network design and optimization: a multirate loss network framework. *Proc IEEE INFOCOM '96*, 994–1003.

Ross K.W. (1995) *Multiservice Loss Models for Broadband Telecommunication Networks.* Springer-Verlag, Berlin.

37

Blocking of dynamic multicast connections in a single link

Jouni Karvo, Jorma Virtamo, Samuli Aalto, Olli Martikainen
Helsinki University of Technology
P.O.Box 1100, FIN-02015 HUT, Finland
Email: {jouni.karvo,jorma.virtamo,samuli.aalto,
olli.martikainen}@hut.fi

Abstract

In this paper, a method for calculating blocking experienced by dynamic multicast connections in a single link is presented. A service center at the root of a tree-type network provides a number of channels distributed to the users by multicast trees which evolve dynamically as users join and leave the channels. We reduce this problem to a generalized Engset system with nonidentical users and generally distributed holding times, and derive the call and channel blocking probabilities as well as the link occupancy distribution.

Keywords

Multicast, blocking, queueing systems, generalized Engset system

1 INTRODUCTION

Call blocking probabilities in a circuit switched network carrying multiple traffic classes can be calculated with exact algorithms, such as the recursion of Kaufman (1981) and Roberts (1981), or with approximative methods, such as the normal type approximation (Naoumov 1995). These algorithms are applicable for point to point connections, such as telephone calls or ATM connections. They apply also for static multicast connections, where the structure of each multicast tree is fixed in advance. In a more dynamic environment, where the trees evolve with arriving and departing customers, these models are not adequate.

Multicast connections have a bandwidth saving nature. This means that a multicast connection – in taking the form of a tree where streams merge at the nodes – requires much less capacity from the network links than a bunch of separate point to point connections from the root node to the leaf nodes of the tree (see figure 1). This effect poses a problem when dimensioning a network. With multicast trees, less bandwidth is needed near the root of the tree than the sum of the bandwidths used near the leaves.

Diot *et al.* (1997) give a review on the work in the area of multicast traffic. Most

Broadband Communications P. Kühn & R. Ulrich (Eds.)
© 1998 IFIP. Published by Chapman & Hall

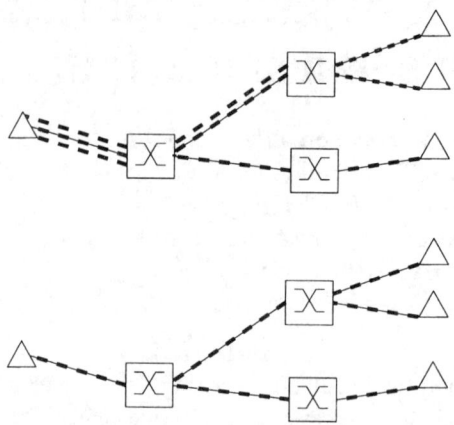

Figure 1 Point to point (top) vs. point to multipoint, or multicast connections (bottom). Data streams are presented with a thick broken line.

work in the sense of blocking in ATM networks with multicasting has been made on blocking in multicasting capable switches, see, for example, (Giacomazzi and Trecordi 1995) or (Kim 1996). Shacham and Yokota (1997) propose call admission control algorithms for real time multicast transmission. We use a similar setting and device an algorithm for calculating call blocking probabilities.

The model of dynamic multicast connections used in this paper pertains, e.g., to the case in which TV or radio station provides several programs to viewers or listeners via a telecommunication network. The model consists of a tree-type distribution network. The service center located at the root node offers the users at the leaves a set of programs delivered to the subscribing users by multicast channels. The programs run independently of their subscribers, who can join and leave the channel any time. Thus, each channel forms a dynamic tree.

A joining user is assumed to choose a channel probabilistically according to a channel preference distribution, which is the same for all the users. When joining, the user, U, creates a new branch to the tree extending from the leaf to the nearest node, A, already connected to the channel (see figure 2). Blocking may occur on any link of the new branch. On the other hand, there is no blocking on the links upstream from the connecting node A, since on that path the channel is already on. Note, however, that the joining user may extend the time the channel remains switched on.

Several multicast trees may use the same link. The required capacity in the links varies accordingly, sometimes leading to blocking when the requested capacity is not available. The probability that there is not sufficient capacity available for a channel in a specific link is called channel blocking probability on that link. By the call blocking probability we mean the probability that the user's request to subscribe to a channel is blocked. The latter is smaller, since the user's subscription is always

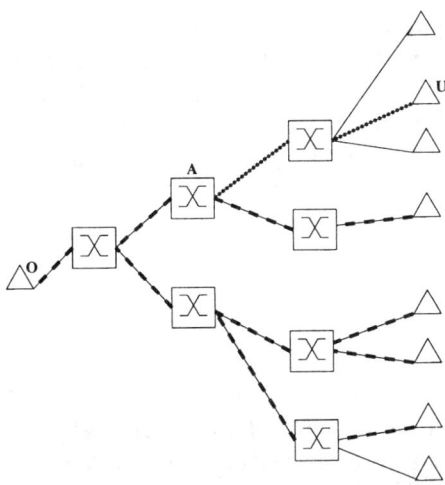

Figure 2 The tree of a channel with a new branch created by a joining user U.

accepted when the channel is already on. Both these blocking probabilities can be calculated on link or end-to-end basis. In this paper we show how to calculate the blocking probabilities of dynamic multicast connections in a specific link with finite capacity assuming that all the other links have infinite capacity.

The paper is organized as follows. In section 2 we present preliminary considerations related to a network with all the links having an infinite capacity. The case of a finite capacity link is considered in section 3. The correspondence with the generalized Engset system is given, and the required blocking probabilities are derived. An example of link occupancy and call blocking probability calculations is presented in section 4. Section 5 gives a brief summary.

2   LINK OCCUPANCY IN AN INFINITE SYSTEM

In this section we consider a link in a system, which has infinite capacity on all its links, and where, consequently, no blocking occurs. First we set the mathematical model up including the assumptions needed later on. Then we determine the mean times that an individual channel traversing the link is on and off. Finally we show how to calculate the distribution for the link capacity usage. The results will be utilized in the next section, where we focus on the blocking problem of a link with finite capacity.

Consider a link in an infinite system. The multicast channel population is denoted by I, i.e., I is the set of channels ('programs') provided by the service center. Let $c_i \in Z_+$ denote the capacity requirement of channel $i \in I$. We assume that the users downstream of the considered link subscribe to these channels according a Poisson process with intensity λ. This is a model for an infinite user population, which is a reasonable assumption in networks with a large number of users, such as TV or radio multicasting in a network (for a link not too close to the leaves of the multicast tree). Further, we assume that each user chooses the channel independently of others and from the same preference distribution, α_i being the probability that channel i is chosen. As a result, the subscriptions to channel i arrive according to a Poisson process with intensity $\lambda_i = \alpha_i \lambda$. We assume that the users' holding times are generally distributed with mean $1/\mu_i$. Finally, let a_i denote the offered traffic intensity for channel i,

$$a_i = \lambda_i/\mu_i. \tag{1}$$

Consider then the on and off times of a single channel. Let $T_{i,\text{on}}^{(\infty)}$ and $T_{i,\text{off}}^{(\infty)}$ denote their means, respectively. As mentioned above, no blocking occurs in an infinite system. Thus, if the channel is off, it is turned on every time a new subscription arrives. The channel remains in the on state (and occupies the link) as long as there are users connected to the channel. Thus, the probability p_i that channel i is on equals the probability that there is **at least one** user connected to the channel. The probability q_i that channel i is off is then the same as the probability that there are **no** users connected to the channel. On the other hand, under the assumptions made above, the number of users simultaneously connected to channel i is distributed as the number of customers in an $M/G/\infty$ queue, i.e. according to the Poisson distribution with mean a_i. Thus,

$$p_i = 1 - e^{-a_i}, \tag{2}$$

$$q_i = e^{-a_i}. \tag{3}$$

Another implication is that on and off times of the channel considered are distributed as busy and idle periods, respectively, in the corresponding $M/G/\infty$ queue. Thus,

$$T_{i,\text{on}}^{(\infty)} = \frac{e^{a_i} - 1}{\lambda_i}, \tag{4}$$

$$T_{i,\text{off}}^{(\infty)} = \lambda_i^{-1}. \tag{5}$$

The former equation follows from the fact that

$$p_i = \frac{T_{i,\text{on}}^{(\infty)}}{T_{i,\text{on}}^{(\infty)} + T_{i,\text{off}}^{(\infty)}}. \tag{6}$$

We see that the mean on time of the most popular channels as a function of the offered traffic intensity grows extremely rapidly because of the exponential term in the numerator. This indicates that there is likely to be a set of channels that are almost constantly carried on the link.

Let X denote the number of channels in use, and X_i indicate whether channel i is on ($X_i = 1$) or off ($X_i = 0$). Since

$$X = \sum_{i \in I} X_i, \tag{7}$$

where the X_i are independent Bernoulli variables with mean p_i, we have

$$E[X] = \sum_{i \in I} p_i, \tag{8}$$

$$Var[X] = \sum_{i \in I} p_i q_i. \tag{9}$$

Let then Y denote the number of capacity units simultaneously occupied in the link,

$$Y = \sum_{i \in I} c_i X_i. \tag{10}$$

Its distribution $(\pi_j)_{j=0}^{\infty}$, called the link occupancy distribution, can be calculated by the convolution algorithm (Iversen 1987), or, equivalently, from the probability generating function:

$$P(z) = \prod_{i \in I} (q_i + p_i z^{c_i}) = \sum_{j=0}^{\infty} \pi_j z^j. \tag{11}$$

As regards the mean and variance of Y, it follows from (10) and the independence of the Bernoulli variables X_i that

$$E[Y] = \sum_{i \in I} c_i p_i, \tag{12}$$

$$Var[Y] = \sum_{i \in I} c_i^2 p_i q_i. \tag{13}$$

All the results in this section are valid in a system in which all the links have infinite capacity. When multicast connections are carried on a link which has finite capacity, blocking may occur. This is studied in the next section.

3   BLOCKING IN A LINK WITH FINITE CAPACITY

In this section we show how to calculate blocking probabilities in a link with finite capacity, C, assuming that all the other links have infinite capacity.

It is important to make a distinction between various types of blocking. The **channel blocking probability** B_i^c of channel i is defined to be the probability that an attempt to turn channel i on fails due to lacking capacity, whereas the **call blocking probability** b_i^c of channel i (seen by a user subscribing to channel i) refers to the probability that a user's attempt to subscribe to channel i fails. These are different, since the user's subscription is always accepted when the channel is already on. Finally we define the **time blocking probability** B_i^t of channel i to be the probability that at least $C - c_i + 1$ capacity units of the link are occupied.

Consider a single channel $i \in I$. Denote by $T_{i,\text{on}}$ and $T_{i,\text{off}}$ the mean on and off periods, respectively, in this finite system. By considering a cycle consisting of an on period and the following off period, we deduce that the call blocking probability of channel i is

$$b_i^c = \frac{\lambda_i T_{i,\text{off}} - 1}{\lambda_i T_{i,\text{on}} + \lambda_i T_{i,\text{off}}}, \tag{14}$$

where $\lambda_i T_{i,\text{off}} - 1$ is the mean number of failed attempts to subscribe to channel i during the cycle (the last subscription arriving in the off period will be accepted), and the denominator represents the mean total number of attempts during the cycle. The frequency of accepted calls when the channel is off is clearly $\lambda_i(1 - B_i^c)$. Thus,

$$T_{i,\text{off}} = \frac{1}{\lambda_i(1 - B_i^c)}. \tag{15}$$

On the other hand, we observe that in this finite system (where the capacities of all the other links are assumed to be infinite) the on period of a channel is independent of the evolution of the other channels: once the channel is turned on all the incoming subscriptions will be accepted. This implies that the on periods are distributed as those of an infinite system. Thus,

$$T_{i,\text{on}} = T_{i,\text{on}}^{(\infty)} = \frac{e^{a_i} - 1}{\lambda_i}. \tag{16}$$

By combining equations (14), (15) and (16), we obtain the following expression for the call blocking probabilities of channel i:

$$b_i^c = \frac{B_i^c}{(1 - B_i^c)(e^{a_i} - 1) + 1}. \tag{17}$$

Thus, the only item that still remains to be determined is the channel blocking probability B_i^c. We start the derivation by observing that our finite system can be described as a generalized Engset system.

By an Engset system we refer to the well known $M/M/m/m/K$ system with a finite user population, see (Kleinrock 1975). In a generalized Engset system the users are nonidentical, that is their mean holding and interarrival times as well as the requested resources can be different. Moreover, we allow the holding times to have a general distribution.

The channels in our system represent the users in the Engset system. When the channel is on, the 'user' is active, and when the channel is off, the 'user' is idle. Thus, the holding time of user i in the generalized Engset system is generally distributed with mean $T_{i,\text{on}}$, and the interarrival time is exponentially distributed with mean λ_i^{-1}. As a consequence, we deduce that the channel blocking probability B_i^c equals the call blocking probability of user i in the corresponding generalized Engset system. Similarly, the time blocking probability B_i^t equals that of the generalized Engset system.

The time blocking probability of user i in the generalized Engset system can be calculated from the following formula:

$$B_i^t = \frac{\sum_{j=C-c_i+1}^{C} \pi_j}{\sum_{j=0}^{C} \pi_j}, \tag{18}$$

where π_j is the probability that j capacity units are occupied in an infinite system as defined in equation (11). In the special case that $c_i = 1$ for all i, this follows from the result of Cohen (1957). In the general case, it can be shown to follow from the insensitivity property of the product form probabilities of a multirate loss system (Hui 1990). It is also known that the call blocking of user i equals the time blocking (of user i) in a system where user i is removed. Thus the channel blocking probability is as follows:

$$B_i^c = \frac{\sum_{j=C-c_i+1}^{C} \pi_j^{(i)}}{\sum_{j=0}^{C} \pi_j^{(i)}}, \tag{19}$$

where $\pi_j^{(i)}$ is the probability that j capacity units are occupied in an infinite system with user i removed. These occupancy probabilities can be identified from the probability generating function

$$\sum_{j=0}^{\infty} \pi_j^{(i)} z^j = \prod_{k \in I - \{i\}} (q_k + p_k z^{c_k}), \tag{20}$$

where $I - \{i\}$ denotes the reduced set of users. Alternatively, in order to save computational effort, one may use deconvolution, which means

$$\prod_{k \in I - \{i\}} (q_k + p_k z^{c_k}) = \frac{P(z)}{q_i + p_i z^{c_i}}, \tag{21}$$

but numerical problems may arise for large systems.

To summarize, the call blocking b_i^c can be calculated from formula (17) by using (19). Note that the denominator in (17) is always greater than 1. Thus, the call blocking b_i^c seen by a user subscribing to channel i is always smaller than the corresponding channel blocking B_i^c. This reflects the fact that the users subscribing to a channel while the channel is on do not experience any blocking. We see also that, for the most popular channels, blocking seen by a user drops practically to zero, since the exponential term in the denominator grows rapidly with a_i ($b_i^c \approx B_i^c e^{-a_i}$). For a channel with $a_i \ll 1$, the channel blocking and the call blocking seen by a user are approximately the same.

Since $b_i^c \leq B_i^c \leq B_i^t$, an upper limit for the call blocking is the time blocking in a system with all channels present. No call blocking seen by a user can be higher than this, but call blocking for user approaches it for channels with channel preferences α_i near zero.

4   AN EXAMPLE

As an example, a truncated geometric distribution is used for channel choosing preferences:

$$\alpha_i \propto (1 - p)^{i-1}, \qquad i = 1, 2, \dots, |I|, \tag{22}$$

where the index i is the channel number. With this numbering the channels are in a descending order according to the usage. For numerical calculations we choose $p = 0.2$ and $|I| = 200$. We choose the average viewing time $1/\mu = 900$ s to be the same for all channels, and $\lambda/\mu = 3.5 \cdot 10^6$. The link capacity is $C = 70$ channels, and each channel has identical capacity requirement $c_i = 1$.

First, we consider an infinite capacity link. The channel usages p_i are depicted in figure 3 as a function of channel index i. The mean and variance of the number of used channels are calculated from equations (8) and (9). They are 63.4 and 3.11, respectively. In this example, where all capacity requirements are equal to 1, these same values hold for mean and variance of capacity requirement for the channels. The link occupancy distribution, π_j, equation (11), is presented in figure 4. The mean channel on times, $T_{i,\text{on}}$, equation (16), for this example are presented in figure 5. As expected, the most popular channels have practically infinite on times.

Then we consider a system with finite link capacity in a link, $C = 70$. Call and channel blocking probabilities, equations (14) and (19) respectively, for each channel

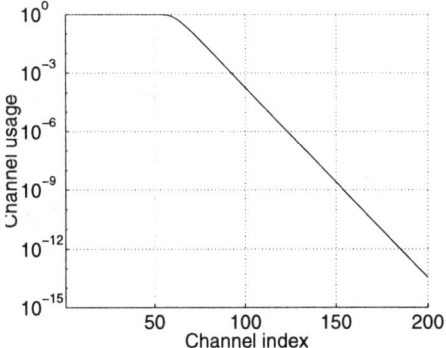

Figure 3 Channel usage versus channel index for geometrically distributed user preferences.

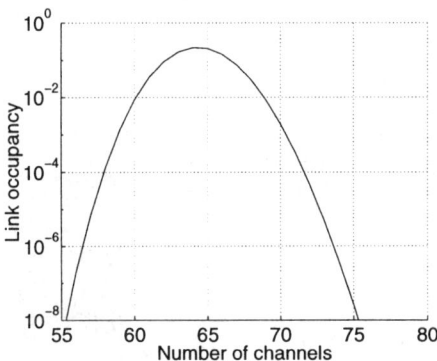

Figure 4 Link occupancy distribution for geometrically distributed user preferences.

are shown in figure 6. This shows clearly that the call and channel blocking probabilities differ for the most popular channels for which there is a high probability that the channel is already on when the user subscribes the channel.

5 SUMMARY

In this paper, we have presented a method for calculating the call and channel blocking probabilities in a link carrying multicast traffic. Multicast traffic has the property of requiring less link capacity than a set of point to point connections providing the same connectivity.

The blocking calculation presented gives us a grip of TV or radio delivery on a circuit switched multicast system, such as an ATM network with virtual circuits. We

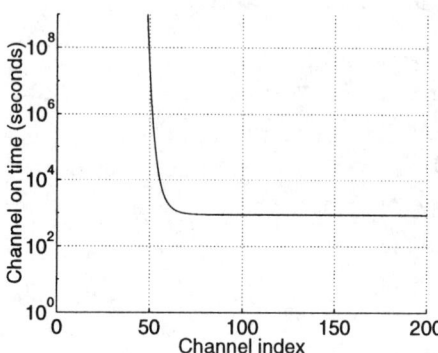

Figure 5 The mean on times $T_{i,\text{on}}$ in seconds for each channel.

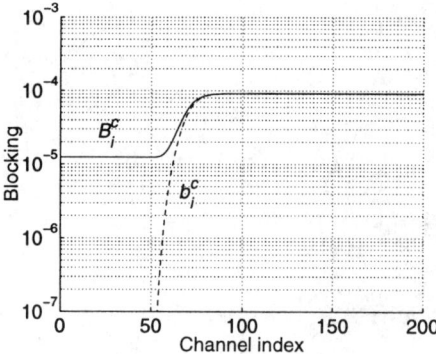

Figure 6 Channel B_i^c and call blockings b_i^c for each channel (B_i^c solid line, b_i^c dashed line).

are able to calculate the capacity needed for a link, the average channel on times, and blocking introduced by finite capacity.

The calculation started from a distribution of the users' preferences on multicast channels, from which we derived the link occupancy distribution in an infinite capacity system, and, further, the blocking probabilities in a finite system by mapping the problem to an equivalent generalized Engset system. Finally, an example was given with the geometric distribution of users' preferences.

In this paper, we have limited ourselves to the case of a single link. The case where blocking occurs on several links of a network is left for future work. Possibly methods such as Reduced Load Approximation, see for example (Ross 1995), can be used, but blockings on successive links may exhibit significant dependencies.

It is also likely that the users' actual preference distribution affects significantly the blocking introduced in the network. This part of the study would require statistics from a real-life trace, and is left for future work.

ACKNOWLEDGEMENTS

The authors wish to thank prof. Gunnar Karlsson, Petteri Koponen and Juhana Räsänen from the Helsinki University of Technology for their inspiring thoughts.

REFERENCES

Cohen, J.W. (1957) The generalized Engset formulae. *Philips Telecommunication Review* **18**(4), 158–170.

Diot, C., Dabbous, W., and Crowcroft, J. (1997) Multipoint communication: A survey of protocols, functions, and mechanisms. *IEEE Journal on Selected Areas in Communications* **15**(3), 277–290.

Giacomazzi, P., and Trecordi, V. (1995) A study of non blocking multicast switching networks. *IEEE Transactions on Communications* **43**(2–4), 1163–1168.

Hui, J. (1990) *Switching and traffic theory for integrated broadband networks.* Boston, Kluwer Academic Publishers.

Iversen, V.B. (1987) The exact evaluation of multi-service loss systems with access control. *Teleteknik, English Edition* **31**(2), 56–61.

Kaufman, J.S. (1981) Blocking in a shared resource environment. *IEEE Transactions on Communications* **com-29**(10), 1474–1481.

Kim, C.-K. (1996) Blocking probability of heterogenous traffic in a multirate multicast switch. *IEEE Journal on Selected Areas in Communications* **14**(2), 374–385.

Kleinrock, L. (1975) *Queueing systems; Volume 1: Theory.* New York, John Wiley & Sons.

Naoumov, V. (1995) Normal-type approximation for multi-service systems with trunk reservation. *Telecommunication Systems,* **4**, 113–118.

Roberts, J.W. (1981) A service system with heterogenous user requirements – application to multi-services telecommunications systems, in *Performance of data communication systems and their applications* (ed. G. Pujolle), Amsterdam, North-Holland, pp. 423–431.

Ross, K.W. (1995) *Multiservice Loss Models for Broadband Telecommunication Networks.* London, Springer Verlag.

Shacham, N., and Yokota, H. (1997) Admission control algorithms for multicast sessions with multiple streams. *IEEE Journal on Selected Areas in Communications* **15**(3), 557–566.

Traffic Flow Control

38

QoS characterization of the ATM Block Transfer mode with Instantaneous Transmission

Gérard Hébuterne
Institut National des Télécommunications
91011, Evry, France. Tel: (+33)1 60 76 45 83 gerard.hebuterne@int-evry.fr

Abstract

Using the ATM technique to carry efficiently variable bitrate connections is a challenge which gives rise to numerous studies and proposals. Aside from open loop schemes such as Deterministic Bit Rate and Statistical Bit Rate, much attention is paid on procedures able at making profit of the characteristics of "elastic traffic". Here, the source is assumed to be able to adapt its behaviour to the network state.

The *Available Bit Rate* ABR is probably the most popular solution to have been proposed. In this paper, we focuse on another one, currently under study in the ITU-T, namely the *ATM Block Transfer mode with Instantaneous Transmission*.

As for any "best effort" scheme, the network may not indefinitely allow new connections without any control. We propose to make use of the time to send a burst as the QoS criterion. We give a simple model of the procedure which captures the main characteristics of the source and network interaction, and allows an estimation of the criterion. Especially, reattempts are included in the source description, and the QoS is then specified in terms of delay and available capacity.

This work was prepared during a visit to INRS-Telecom (Montreal, Canada)

Keywords

ATM, QoS, ABT, reattempts

1 INTRODUCTION

Using the ATM technique to carry efficiently variable bitrate connections is a challenge which gives rise to numerous studies and proposals. The case where the service has stringent requirements in terms of cell delay variation and instantaneous bitrate is dealt with by *open loop schemes* – the source declaring its characteristics and requirements once and for the whole call

Broadband Communications   P. Kühn & R. Ulrich (Eds.)
© 1998 IFIP. Published by Chapman & Hall

duration. ITU-T and the ATM-Forum have defined the *Statistical Bit Rate* SBR (respectively, *Variable Bit Rate* VBR) transfer mode to take account of these classes of services (ATM Forum 1996, ITU-T 1996).

Closed loop schemes are seen as more promising in terms of network efficiency. Here, the source is assumed to have far less strict Quality of Service (QoS) requirements. Especially, the source is able to react to information messages coming from the network, giving rise to cooperative schemes. Such sources able at varying their traffic without interrupting the connection are refereed to as *elastic applications.* The dialog between the source and the network is carried out through the so-called *Resource Management* (RM) cells.

The best known example of reactive scheme is the *Available Bit Rate* mode (ABR) currently defined by the ATM-Forum (ATM Forum 1996). However, the description of the ABR mode makes it obvious that any implementation of the procedure should exhibit a high level of complexity, to say nothing of the numerous pending issues such as source behaviour control and parameter tuning. Another cooperative scheme has been proposed - see (Boyer *et al.* 1992a, Boyer *et al.* 1992b, Guillemin 1997) and is currently under consideration in the ITU-T working groups under the name *ATM Block Transfer Capability (ABT).* It is expected that ABT leads to simpler implementation while achieving a fairly efficient use of the network resources. This should be evident from the description below, where the ABT mode appears as a kind of "piecwise DBR" mode.

Whichever scheme is considered, assessing the Quality of Service (QoS) its provides is a mandatory part of its specification. Even for services described as *best-effort* services, one has to be able to give a rough characterization of the QoS level the network achieves. This may be used e.g. to assess a nominal QoS level and to help in network dimensioning, or better to serve as a basis for elementary Connection Acceptance Control (CAC) procedures.

In this study, we focus on a possible definition for the QoS attached to the ABT/IT mode. We propose a model able at predicting its performance level and at the design of the associated CAC procedure.

The paper is organized as follows. To begin with, Section 2 describes briefly the ABT/IT and its operating mode. The need for specification of the service provided and the analysis of the source requirements leads to a proposed means of describing the achieved QoS.

Section 3 presents the assumptions the model is built on. They concern the source behaviour, which is kept as elementary as possible, while taking account of the main feature of the typical services involved, and especially *reattempts.* The other assumptions concern the network behaviour and are intended at building a tractable model. The model is presented in Section 4.

The results are discussed in Section 5 through a few numerical examples. Some comments are given concerning the accuracy of the assumptions, and the use which can be made of them in order to build tentative CAC procedures.

2 THE ABT/IT TRANSFER MODE

The ABT capability is based upon the principles of the *Fast Reservation Protocol*, see (Boyer *et al.* 1992a, Boyer *et al.* 1992b). In the basic operating mode *ABT/DT*, the source opens a DBR circuit, and submits to the network requests to vary the peak cell rate allocated to it. The request, if accepted, provokes a change in the parameters (peak rate) and the source may accordingly modify its sending bitrate.

The ABT/IT capability is a variant which aims at further simplifying the procedure by eliminating the need to buffer the increase in traffic while waiting for the answer to the request. Since most often a positive answer to requests for increase is expected, the ABT management module anticipates and begins sending at the new bitrate – with the possibility of a negative answer, in which case part of the traffic will be lost.

In this study the ABT/IT variant will be further explored.

2.1 A brief description of ABT/IT

The connection is set up with some value for the allocated peak bitrate. The value of the peak rate is refereed to as *Block Cell Rate* (BCR) in the ITU-T literature. Sources willing to modify their emission rate send RM cells requesting a new value for their BCR. Immediately after they begin sending information using the new value, without waiting for any acknowledgement. The case where the bitrate is decreased raises no difficulty. In case where the bitrate is increased, the possibility exists that a node along the path has no available resource to accomodate it. A part of the flow is then discarded, and a negative acknowledgment is sent back to the node. If the upgrade request succeeds, the flow eventually reaches its destination, and a positive acknowledgment is sent to the source.

The connection makes use of some flow control mechanism. Especially, the source cannot send a new burst before the previous one is positively acknowledged. The process which manages the ABT/IT connection keeps sending unsuccessful bursts until they are correctly received. The precise way the lost bursts are corrected is beyond the scope of the present work.

The simplicity of the ABT mode can be described by stating that it can be truly described as *"piecewise DBR"* (Guillemin 1997) – between two requests, the source behaves really according to the DBR scheme.

2.2 QoS issues for ABT transfer mode

There are two levels at which QoS issues must to be examined. At the cell level, since a connection obeying the ABT mode works exactly as a DBR connection (at least, once the negotiation for the requested bandwidth has ended successfully), no special discussion is needed here and the QoS can be specified just as for the DBR mode. Moreover, guaranteeing a given level of QoS requires no specific action from the network side.

At the "burst" level, "best effort" schemes (such as ABR or ABT) rely on the possibility to lower the instantaneous bitrate offered to the sources. The maximum delay (as measured by some quantile of the end-to-end delay distribution) would yield a poor characterization: due to the intrinsic rules of the procedure, the maximum delay has to be large, so that it would be of no practical use. In fact, the main trouble sources may encounter in a best effort network is a progressive collapse of their transmission capability, due e.g. to a continuous increase in the number of open and active connections. In the ABT (for both the DT and IT variants) this happens through the rejection of the "increase bandwidth requests". Services using such a procedure as ABT/IT are likely to complete it with some error control and correction through retransmission. As a consequence, the *average time* until the request is positively acknowledged is a realistic parameter for the source to estimate the level of service. This delay gives the source an upper bound of the rate at which it can enter data into the network.

It must be emphasized at that point that not only the request rejection probability would not be an accurate QoS criterion, but also that a loss model is quite inaccurate. A model with retrials is much more effective in describing the complete behaviour of the network and the traffic sources.

3   ASSUMPTIONS AND NOTATIONS

3.1   Traffic and network models

In order to give an estimate of the network performance, we build a simple model of the traffic pattern and source behaviour.

The sources are assumed to share a set of links of the same capacity C. All sources are of the ON/OFF type and make use of the ABT/IT service. They remain silent (no allocated bandwidth) until they generate a burst at a peak rate equal to c (giving rise to a bandwidth request). This gives the simplest possible model for the process of request generation. The protocol unit in the ingress node manages the request and the acknowledgements. In order to simplify the discussion, the acknowledgments are assumed to be loss-free (one can assume they use a dedicated DBR VC).

The following events must be carefully distinguished:

● A *request* is the operation by which a source sends a data block to the destination.
● A *trial* is the event when the source makes an attempt to send data to the

destination. The trial may fail: in this case it is followed by another trial, until the source receives a positive acknowledgment.
- Each trial gives rise to *bursts* sent successively on the links of the route.

The number of nodes (or equivalently of links) between the source and the destination of the connection defines the *length* of the route. In order to keep a model as simple as possible, all connections are assumed to have the same length D_0. However, unsuccessful trials have an effective length D, $D < D_0$. Figure 1 summarizes the definitions. The path has a length $D_0 = 3$. 4 trials are necessary to process the request, and they give rise to 8 bursts.

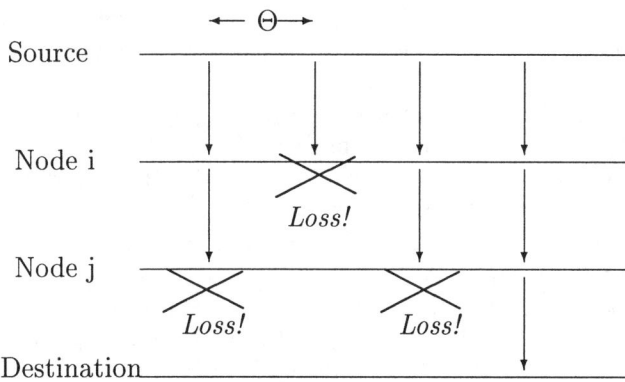

Figure 1 A *request*, made up with *trials* and *bursts*

We make the assumption that the traffic processes on the links are independent variables. The traffic is assumed to be uniformly distributed throughout the network so that all links are equally loaded. The model considers a particular link of the network. Let Λ be the number of requests per time unit sent by all connections which go through the link under study. These requests originate from nodes at distance less than or equal to D_0.

3.2 A summary of the notations

- C is the link capacity, c is the peak rate at which the sources send data, so $R = C/c$ bursts can be accomodated in a link without any loss.
- D_0 : the length of the path from the source to the destination. Actually, since a trial may end up unsuccessfully, let D denote the length of an unsuccessful trial. Let D_T be the *total number of bursts* sent through all the links of the route in order to successfully send the request (in Figure 1, the first trial has $D = 2$, etc., and $D_T = 8$).

- T is the total time to send a burst. The geographic distances in the network are assumed to be negligible so that the data is available at the receiving end without additional transmission delay, and T may be assumed as independent of D_0.

  To further simplify the expressions and notations, it is assumed that there is no additional delay in intermediate nodes. An ATM cell entering the node is immediately available for being sent if needed. To summarize, if the source begins sending a burst at time t, the burst is available in the receiver at the distance d at time $t + T$ (provided it has been successfully transmitted).

- When more than one trial is needed, the successive trials are time spaced by Θ. This delay accounts for all processing delays through the route and for any additional delay requested by the correction procedure (see discussion on Section 6).

- Λ is the arrival rate in number of requests per time unit from all the connections which seize the link under study.

- p: the probability that a burst fails while accessing the link (due to shortage of capacity on the link under consideration which already accomodates C/c bursts). With the assumptions on traffic homogeneity, p is constant throughout the network. Moreover, we make the assumption that it does not depend on the rank of the trial.

- Let π denote the probability that a trial fails. Successive links are assumed to behave independently, the relation between π and p is thus

$$\pi = 1 - (1 - p)^{D_0}. \tag{1}$$

Finally, Figure 2 displays the proposed model for performance evaluation of the ABT/IT procedure. Unsuccessful trials give rise to reattempts which are dealt with by the Management Unit. The source is assumed to wait for the acknowledgement of a request before sending another one.

4   PERFORMANCE EVALUATION

4.1   Behaviour of a request

With π as the failure probability of a trial, the probability that a request gives rise to j trials (the last one successful) is $\pi^{j-1}(1 - \pi)$. On the average $1/(1 - \pi)$ trials are needed.

Now, if a particular trial is to fail during its travel to the destination, the conditional probability that it reaches the k-th node is:

$$P(D = k \mid \text{failure}) = \frac{p(1 - p)^k}{\pi} \qquad k < D_0,$$

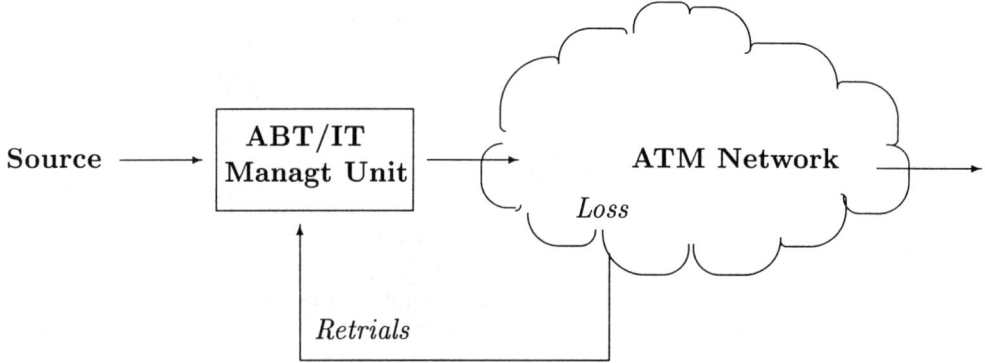

Figure 2 The traffic model with retrials

so that the mean length of failing paths is:

$$E(D \mid \text{failure}) = \frac{1-p}{p} - \frac{D_0(1-p)^{D_0}}{\pi}.$$

Finally, processing a request gives rise to a number of bursts sent in the network equal to $E(D_T)$:

$$E(D_T) = D_0 + \frac{\pi}{1-\pi}\left(\frac{1-p}{p} - \frac{D_0(1-p)^{D_0}}{\pi}\right) = \frac{\pi(1-p)}{p(1-\pi)}. \qquad (2)$$

It can be verified from eqn.(1) and (2) that $\lim_{p\to 0} E(D_T) = D_0$.

4.2 Model of the link

The "link", or rather the multiplexing stage at the node egress, receives bandwidth requests at a rate Λ per time unit from nodes at distance D_0 or less.

On the average, a request generates $E(D_T)$ bursts inside the network. Since all links are assumed equally loaded, this amounts to share this traffic among the D_0 links of the path (as if each request was contributing to the load of each link of its path to the level $E(D_T)/D_0$).

Each burst occupies a fraction of the link for a duration T. The traffic carried by the link under consideration is finally:

$$A_c = \Lambda \times \frac{E(D_T)}{D_0} \times T = \Lambda T \frac{\pi(1-p)}{p(1-\pi)D_0}. \qquad (3)$$

With $1 - p$ as probability of success, the traffic *offered* to the link is

$$A_o = \frac{\Lambda T \pi}{p(1 - \pi)D_0}. \tag{4}$$

Note that Λ should increase with D_0. In an homogeneous network the ratio Λ/D_0 should have a value roughly constant with D_0.

In the case where no rejection never occurs (this would be the case if instead of being rejected the bursts were allowed to wait in the intermediate nodes), the traffics would always have their maximum value: $A_o = C/c$, and the network would thus achieve its greater efficiency. From the result already quoted: $\lim_{p \to 0} \mathrm{E}(D_T) = D_0$ one sees that the theoretical maximum capacity would then be: $(\Lambda T)_M = C/c$ as can be expected. The *efficiency* of the procedure may be expressed as $A_o/(\Lambda T)_M$.

Now, $1/(1 - \pi)$ trials are needed to successfully transmit the request. The successfull one takes a time T while a delay Θ runs between unsussfull ones. This gives the transfer delay – or equivalently the capacity really offered to the sources:

$$\mathcal{D} = T + \frac{\pi \Theta}{1 - \pi}, \quad \mathcal{C}_{\text{eff}} = C \times (1 - \pi). \tag{5}$$

4.3   The rejection process

In order to relate the loss ratio with the traffic process, we rely on the classical simplifying assumptions made for systems with reattempts (see e.g. (Hébuterne 1987). Each link is seen as being submitted a Poisson process resulting of the superposition of the fresh offer and of all successive bursts, as if the outcome of the reattempts were independent of the initial failing. Relation 4 gives the volume of the traffic offered to the link. The instantaneous level of the fresh offered traffic is independent of the number of requests in retrial process. So, the link is modeled according to the classical Erlang model. The loss ratio is given by the Erlang formula:

$$p = \mathrm{Erl}(\frac{C}{c}, A_o). \tag{6}$$

Taken together, relations (1, 4, 6) allow to solve for p, π, A_o and the related variables.

5   NUMERICAL ANALYSIS

The efficiency of the ABT/IT procedure is measured by the available traffic the network may carry. Note that since the model assumes the rejected request can make reattempts for ever, there is no distinction between *offered* and *carried* end-to-end request traffic. The traffic carried by the links is the scarce resource which serves as a control variable.

In order to obtain significant results, the carried traffics are normalized using the maximum achievable value. That is, the parameter of interest is $A_o/(\Lambda T)_M$. Similarly, the traffic carried by the link is represented as the "link occupancy", that is the carried load, as given by equation (3), normalized to its maximum value C/c. This allows fair comparison between cases where e.g. the distance parameters are different.

The first set of curves (Figure 3) displays the influence of the relative size of the channel capacity and of the requests on the overall performance. The efficiency is the ratio $A_o/(\Lambda T)_M$ and the "link occupancy" is cA_c/C. Note that the ratio of the efficiency to the link occupancy is pD_0/π.

The higher the ratio C/c the better in terms of maximum available capacity: ratios higher than 20 are probably to be expected, since anyway no CAC would allow realistic network occupancy for lower values. Efficiencies around 60 to 70 % are achievable, with a collapse above a maximum value due to the contention. Note however that as opposed to the classical "collision" of the CSMA scheme, the contention leads to the destruction of only one of the 2 bursts.

The influence of the number of nodes is displayed in Figure 4. The performance degradate as the route lengthens, which favors networks with small diameters and accordingly large nodes. Note that however the influence of the length is not so high as the one of the ratio C/c.

At last, Figure 5 displays the variation of the additional end-to-end burst transfer delay. The delay is measured in units of Θ, for route length $D_0 = 5$. For each curve, the horizontal line indicates the point where the efficiency has the maximum value. It is clear that, as long as the network is operated for services such that C/c is not too small (say, greater than 20), the QoS, as measured by the delay, is quite satisfactory (delays due to retrials lower than 0.35Θ, here).

This analysis shows that the procedure may exhibit good performance, as long as it is operated on networks with small diameters (connections of moderate length) and for services requiring moderate bitrates (so that C/c remains high). These conditions, however, are typically the ones under which any statistical multiplexing procedure is expected to work. Provided these conditions hold, the network may achieve high efficiency (effective capacity offered to the sources higher than 60 %).

The procedure has to be complemented by some control process aiming at avoiding the operating point to move to the collapsing zone of curves (Figures 3, 4). The ABT/IT Management Unit should make use of the time out procedures which are known to provide stable operation of the CSMA/CD scheme.

6   COMMENTS AND CONCLUSIONS

The model used in the previous Sections makes a few assumptions which call for comments. Probably the strongest simplifying assumption is to consider that all bursts have the same failure probability p. Obviously, one would expect a higher loss probability for the second trial, for instance, than for the first one. To state it another way, the loss probability of a burst depends on the time elapsed since the last loss. Considering that p does not depend on the rank of the trial amounts to say that the interval between successive trials Θ is large as compared with the typical time scale of the multiplexing stage – say, as compared with the burst duration.

The derivation assumes independence between the number of pending requests in the network and the rate at which new ones arrive. In fact, since the link under study carries the traffic on a fixed number of connections, a more realistic model would be to consider the requests as being generated by a finite number of sources (Engset like traffic). The Poisson assumption corresponds to the asymptotic worst case, so that the results are conservative.

Finally, the traffic configuration deserves a few comments. The source traffic pattern consisting of bursts (On/Off source) is simplistic, and actual ATM sources can be fairly more various. Especially, the sources the ABR mode considers require a minimum bitrate (MCR), which would be quite easy to take account of in the model without degradating its accuracy. Stepwise traffic can be considered too (the source keeps on sending traffic at a given BCR and updates it now and then). While this configuration would modify the source behaviour, the model proposed still applies provided the product ΛT is given the right meaning.

The results show the ABT/IT procedure as a candidate to the provision of a "best effort" based multiplexing scheme. The dependence in the distance (length of the route measured in number of intermediate nodes) may be alleviated by organizing the network in *segments* ended by *virtual sources - destinations*, just as it has been proposed for the ABR mode. Concerning the dependence on the ratio C/c, which conditions mainly the global efficiency, the results favor large ratios, and show that values below 20 are to be avoided.

This paper has proposed a way to describe the QoS offered by a best effort service. Working on the Web gives an idea of the service provided by completely un-controlled network. This makes it obvious the need to implement some connection admission function based upon the mean transfer delay. From the results presented here such a CAC procedure would proceed by limiting the number of simultaneous connections. The full design of the procedure is however still to be undertaken. The models presented give the means to estimate its performance level.

REFERENCES

The ATM Forum: UNI Specification, Version 4.0, June 1996.

P. Boyer, F. Guillemin, M. Servel, J.P. Coudreuse (1992) Spacing Cells Protects and Enhances Utilization of ATM links. *IEEE Communications Magazine*, September 1992.

P. Boyer, D. Tranchier (1992) A reservation Principle with Application to the ATM Traffic Control. *Computer Networks and ISDN Systems*, **24**, 321-334 ,1992.

F. Guillemin: ATM Block Transfer Capability vs. Available Bit Rate Service. *European Transactions on Telecommunications*, **8**, No1, January 1997.

G. Hébuterne: *Traffic flows in switching systems*. Artech House, 1987.

ITU-T Recommendation I.371: *Traffic Control and Congestion Management*. Geneva, June 1996.

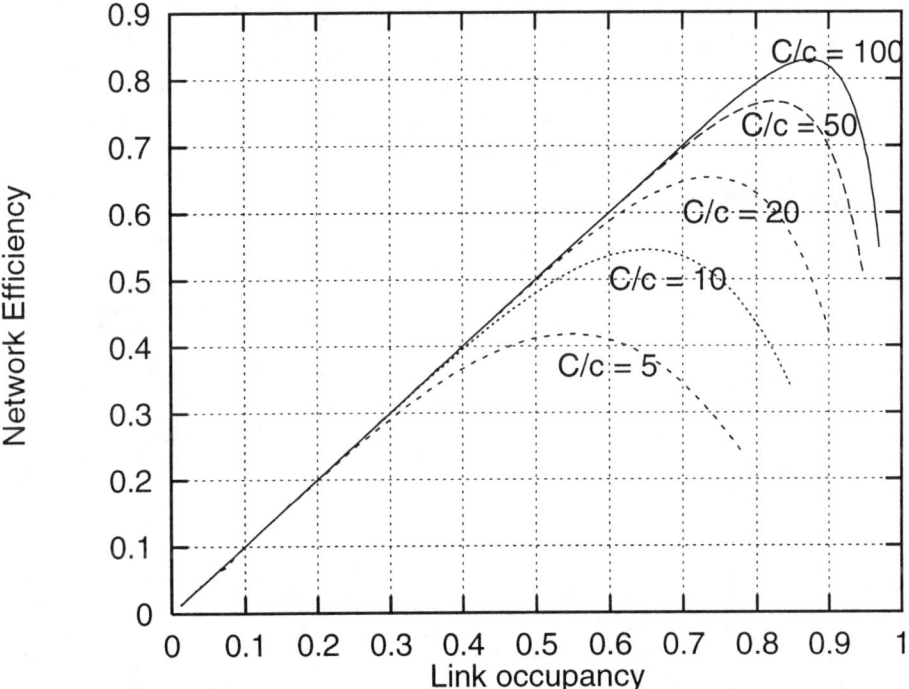

Figure 3 Relation between offered load and effective link load. Influence of the request size. The curves are drawn for a connection length $D_0 = 5$

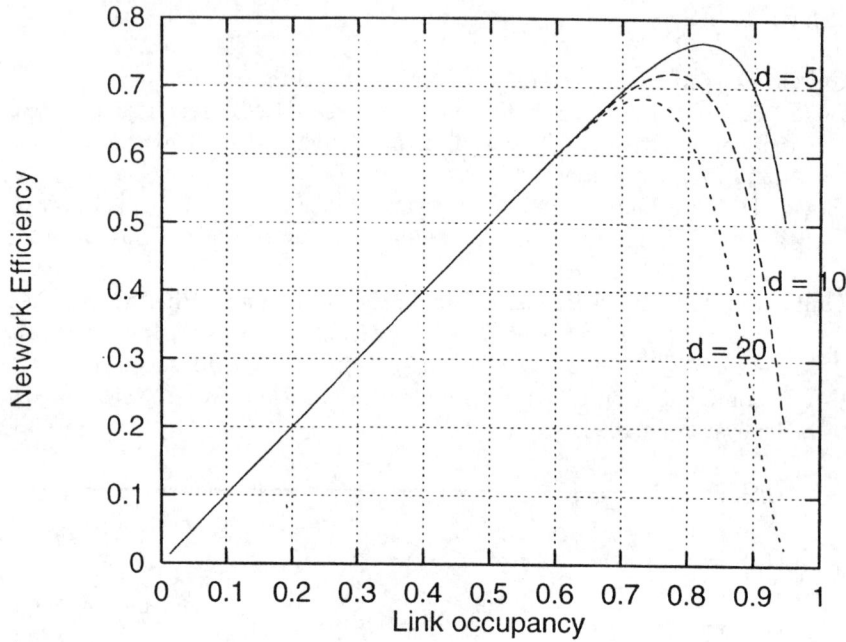

Figure 4 Offered load vs Effective link load. Influence of route length ($C/c = 50$).

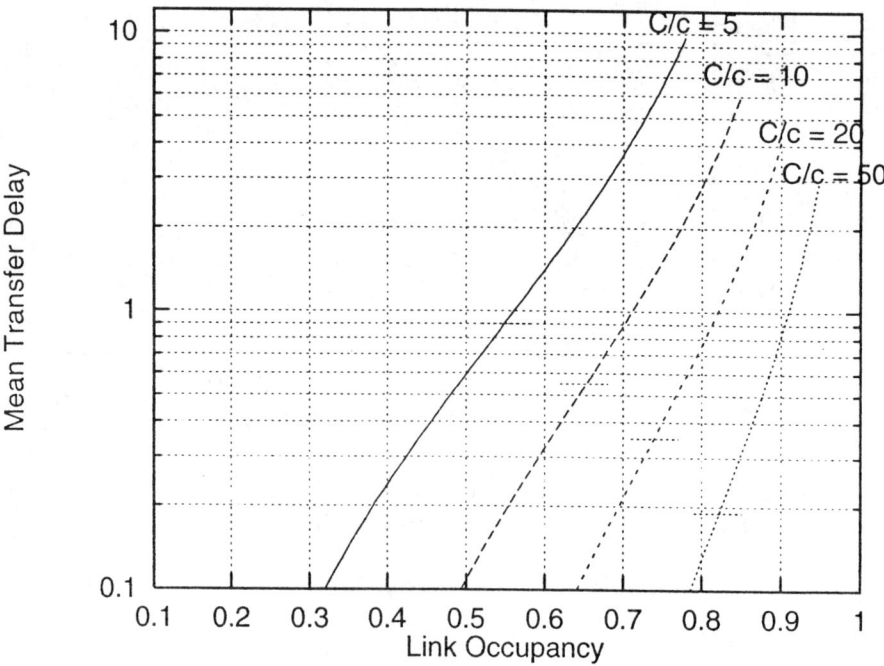

Figure 5 Mean Transfer Delay vs Effective link load ($D_0 = 5$).

39

Dynamic rate adaption for efficient use of frame relay network

A. Calveras Augé, J. Paradells
Dept. Mat. Apl. i Telemàtic. Universitat Politècnica de
Catalunya
Mod. C3 C/ Jordi Girona 1-3, 08034 Barcelona, SPAIN
Tel: +34 3 401.60.13, Fax: +34 3 401.59.81,
e-mail:{acalveras, teljpa }@mat.upc.es

Jordi Puga Antequera
Sema Group sae
Provença, 388,08025 Barcelona, SPAIN

Klaus-Jürgen Schulz
European Space Agency (ESA)
European Space Operations Centre (ESOC)
Robert-Bosch-Str. 5, D-64293 Darmstadt, GERMANY

Abstract
Modern network services like Frame Relay provide information about the load or congestion state of the network in real time to all attached end-systems. This allows the end-systems to inject traffic into the network above the agreed service contract based on committed information rate and burst excess size. This paper describes a rate control algorithm that is added to the traditional TCP transport protocol

Broadband Communications   P. Kühn & R. Ulrich (Eds.)
© 1998 IFIP. Published by Chapman & Hall

elements which exploits the congestion state information conveyed in the Forward/Backward Explicit Congestion Notification (FECN/BECN) of frame relay. The simulated network and transport protocol behaviour for typical applications of the European Space Agency[*] shows a win-win situation for the network user and the network provider.

Keywords
Frame Relay, TCP, Rate Control, Congestion Avoidance

1   INTRODUCTION

The European Space Agency (ESA) in co-operation with Sema Group sae and Polytechnic University of Catalonia   has investigated the use of frame relay network services for distribution of high rate telemetry between satellite receiving ground stations or other space data entry points and remote processing sites. The objective of this study is to connect telemetry end systems directly to the frame relay network in order to make efficient use of the frame relay inherent features of congestion avoidance, which would not have been possible by just connecting Internet Protocol routers to the frame relay network. The system shall allow a data rate of 2 Mbps and over per subscriber.

The study started from a detailed definition of the application service semantic specific to telemetry distribution applications, then identified a complete OSI like protocol stack above frame relay, thereby emphasising on the study of protocol elements in the transport layer for a high speed network scenario. In order to exploit the frame relay congestion avoidance mechanism a new protocol element for rate control was specified for the transport layer.

The complete system, i.e. 2 end-systems, a network with 3 nodes and additional background traffic was investigated in various simulation scenarios.

2   SERVICE DEFINITION

This study, based on the service definitions (volume 1 and 2, 1992), involves the transfer of information between different space agencies, in order to ensure uniformity in the Service Interface.

Telemetry Application Services require that data are relayed complete, sequence preserved, and without errors. Therefore, a protocol stack with reliable transport services.

---

[*] Part of this work is the result of a study carried out for the European Space Agency under contract: 10687/94/D/DK(SC)

3   FRAME RELAY USER NETWORK INTERFACE

Frame Relay (FR) according to Smith (1992) and Black (1992) is a network access protocol based on a fast packet switching technology, connection oriented and that preserves the transmission frame sequence. Two characteristics draw its advantages: low overhead due to minimal node processing, and the ability to support traffic bursts. Error control for incorrectly received frames and flow control processing are largely absent.

In present networks, losses are mainly due to congestion. FR has a congestion notification mechanism but a congestion control is not foreseen. So, the key point in it is the policy applied to avoid congestion, and the mechanism to notify the users when it occurs.

FR offers a range of options at subscription time: Access Rate (AR), Committed Information Rate (CIR) that provides his foreseen minimum guarantied data throughput, and Excess Information Rate (EIR) that represents the potential peak data flows.

Once congestion occurs, FR sends a notification and provides the possibility of selecting frames to be discarded (by means of FECN, BECN and DE bits respectively).

4   SIMULATION ENVIRONMENT

This section describes the models used in the simulation based study.

4.1 Application Service Modelling

CCSDS (Consultative Committee on Space Data Systems) Packets arrive at the Service Provider Application from the Space Link Services, and they are stored in the Ground Station in an infinite buffer until they are transmitted.

A general packet generation with a Poisson distribution has been assumed, with two functional modes having different inter-arrival packet mean: **Burst mode** and **Non burst mode**. The packet length is assumed to has a uniform distribution as the one given by Johnson (1991)

4.2 User Network Interface (UNI)

The User Network Interface (UNI) is the access point of the Protocol Stack to the network. It is modelled as an entity that accepts packets from higher level layers, and encapsulates them in FR frames that are injected in the network.

The UNI controls the maximum burst rate that the FR network can accept by means of the FR parameters: CIR, EIR and AR.

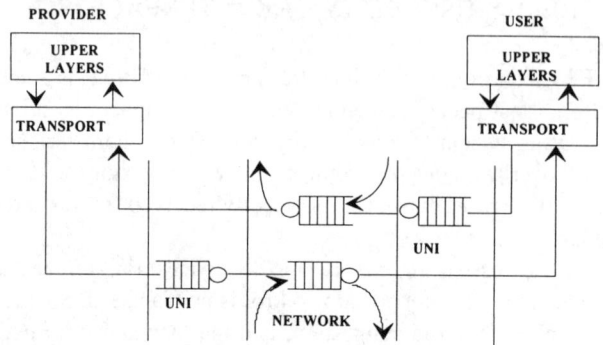

Figure 1 Layered Architecture, UNI and FR Network.

4.3 Network Modelling

FR defines an access interface to the network but not the way this network is implemented. The modelled network provides FR congestion notification (BECN and FECN bits) and a discarding strategy based in the DE bit.

A model with three nodes has been chosen for the simulation. Each node is modelled by means of a server with a finite buffer. The service time represents the packet processing time plus the transmission time, and the buffer length threshold is used to model the congestion effect.

Background traffic with exponential distribution is considered, and it also reacts to congestion. The user traffic flowing through a virtual circuit represents the 33 % of the total traffic.

The parameters that define the network model are:

- **Trunk Throughput**: 3.285 Mbps.
- **Propagation delay:** Local (5 ms), intercontinental (50 ms) and satellite link (250 ms).
- **Buffers capacity**: it defines the probability of network congestion. These buffers have been designed in the model in such way that the congestion effect appears frequently to study its consequences.

Congestion modelling

Congestion occurs when buffers are full. However, congestion notification is sent when buffers reach 80 % of its capacity.

To model the effect of congestion when a buffer reaches the 100% of its capacity, a random number of frames in the buffer with the DE bit set are discarded, and the rest of the frames are marked with the DE bit, and consequently will be discarded if necessary in next congestion situations.

5   TRANSPORT LAYER

This section presents the studied Protocol Profiles and Elements, for the particular case of telemetry applications and assuming to be used on top of a FR Network.
The eXpress Transfer Protocol (XTP) and the Transmission Control Protocol (TCP) were selected as the main candidates, and both were simulated. Although XTP is more modern no congestion avoidance mechanisms have been developed for it. Meanwhile, TCP starts from a classical protocol, but it has been studied in detail, and extensions for congestion control and high speed environments have been added.
In consequence, TCP protocol elements have been adopted as a baseline for the final solution and then adapted to FR networks. Several options within the current analysis work, including extensions and modifications to adapt it to FR networks have been considered.

5.1 Extended TCP Over Frame Relay

This section mainly describes the extended TCP Elements and the additional BECN congestion control algorithms that are most suitable in the Telemetry Delivery Applications Services over FR. The usage of BECN in the congestion control mechanism is something new, not yet totally studied.
TCP as it is decribed by Postel (1981) is a connection oriented protocol that was designed to provide end-to-end reliable data delivery. Although it uses Positive Acknowledgement with Retransmission, a Selective Repeat scheme has also been modelled as a possible extension of the protocol (Fall, 1995) (Mathis, 1996).
Accurate dynamic determination of the retransmission time-out (RTO) interval seems to be essential to TCP performance (Jacobson, 1988). RTO is determined by estimating the mean and variance of the measurement round-trip time (RTT), i.e., the time interval between sending a segment and receiving an acknowledgement for it. To avoid the ambiguity in the estimate RTO, the **Karn's Algorithm** (Jacobson, 1988) is also implemented, combined with an exponential backoff strategy for the retransmission timer. Many TCP implementations base their RTT measurements upon a sample of only one packet per window, which results in an unacceptable poor RTT estimate when the bandwidth-delay product is large. An extension to TCP can solve these problems. Using the TCP header options, the sender places a **TimeStamp** (Jacobson, 1992) in each data segment, which is sent back in acknowledgements.
A credit-based adaptive window-sliding mechanism is used for end-to-end flow control, and the algorithms known as **Slow-Start and Congestion Avoidance** (Jacobson, 1988), are used for congestion control. They operate by observing that the rate at which new packets should be injected into the network is the rate at which the acknowledgements are returned by the other end. Upon detecting congestion, the sender drastically reduces the window size and slowly starts to

increase the rate at which it transmits packets, back up to its previous levels. This approach reduces the probability of packet retransmission, and increases the effective throughput of the connection.

The **BECN Congestion Algorithm** (Jubainville, 1994) has been added as an extension, to take full advantage of FR congestion indication bits. Usually, these bits are not used by the upper layers although they are present in the protocol.

This algorithm is based in the fact that FR sets the FECN and BECN bits to indicate that congestion is occurring or has occurred in the network. The user detects these indications and assumes that the network is congested and reduces or suspends the transmission until the network recovers from the congested situation.

When a node becomes congested, FR sets these bits in each direction of the connection, in all the frames that are in transit in the node. If there is no reversing data flowing in the reverse direction of data flow, the FR Consolidated Link Layer Management (CLLM) message is used as a congestion notification message that can be forwarded to the UNI as required.

The user response to the BECN or CLLM congestion indication has been studied at the Transport Layer, introducing a **Rate Control Mechanism** at this level.

For the BECN algorithm a step count (S), is used to determine when the protocol should take congestion avoidance action.

When the UNI receives a frame with the BECN (or CLLM) bit set, data sent to the transport layer contains an indication of congestion. The procedure to be adopted by the sender on receipt of this indication is as follows:

1. If TCP's current offered data rate is greater than the agreed CIR, TCP reduces his offered rate to the CIR agreed for the circuit.

2. If a number of consecutive indications equal to the step count are received, TCP should reduce his offered rate to the next "step" rate below the current rate. This continues with a rate reduction to the next "step" rate after every S consecutive indications.

    The "step" proposed rates are:
    *   *0.675*CIR*
    *   *0.5*CIR*
    *   *0.25*CIR*

3. After having reduced the offered rate, the user may increase this rate by 0.125 times the throughput, after any S/2 consecutive received data without congestion indications.

Other studies based on TCP and Explicit Congestion Notification are of interest. Floyd (1994) proposes the guidelines for TCP's response to Explicit Congestion Notification Mechanisms, Calhoun (1995) discusses Congestion Management in the IPV6 Internetworks, and Schulzrine (1996) proposes an enhancement to the TCP's congestion control mechanisms using binary congestion notification.

Another extension is added to the TCP, the **Fast Recovery and Fast Retransmit algorithm** (Stevens, 1997). In the TCP error recovery mechanism, a received out of sequence segment generates immediately a duplicated ACK. Since FR does not disorder packets, a duplicated ACK means that data is lost. Nevertheless, it has to be pointed out that the RTO expiration of a packet that has only been delayed but not lost, may cause a received out of sequence segment and consequently a duplicated ACK to be received. In this case the reception of a duplicated ACK does not mean that a packet has been lost.

Nevertheless, if a number of duplicated ACK's are received in a row, it is a strong indication that a segment has been lost. A retransmission of what appears to be the missing segment is performed then, without waiting for the RTO to expire. In this case, the TCP congestion avoidance algorithm is also performed, but not the slow start. In the specific case of the study the number of duplicates that force retransmission is two.

6 DISCUSSION

The selection of the protocol stack has been validated by means of computer based simulation techniques. Results and conclusions are valid for the particular case of Telemetry Applications on top of FR, and a particular decision about each Protocol Element is provided.

6.1 Environment

The obtained results for the evaluation are obtained using the following parameters:
- *Source Parameters:*
  - *Burst Packet inter-arrival: 4,87 msec*
  - *Non Burst Packet inter-arrival:19,46msec*
  - *Burst Rate Duration: 1 sec*
  - *Non Burst rate Duration: 2 sec*
  - *Packet Length:*
    - *Maximum: 2000 octets, Minimum: 500 octets, Average: 1250 octets*
- *Transport Layer:*
  - *MSS: 1560 octets*
- *Network:*
  - *Transmission time: 3,9 msec, Node Buffer Length: 50 frames*
- *UNI:*
  - *AR: 2Mbps, CIR: 1Mbps*

The performance measurements are:
- *Application efficiency (%):*
  It is the ratio of amount of data generated by the application to that injected in the network.
- *Frame Losses (%):*
  It is the ratio of bits transmitted through the network to those injected in it.
- *End to end rate (bps)*

6.2 Fast Retransmit And Fast Recovery

TCP extensions for high performance protocols, propose another error control and recovery, to act quickly when losses occur at the network.

One advantage of this algorithm is to transform the Go-Back-N retransmission scheme to a "quasi-selective" retransmission mechanism, avoiding a large number of unnecessary retransmitted packets.

Consequently, it is **recommended** to use the fast retransmission and fast recovery option, mainly, when bandwidth-delay product is high. Nevertheless, when a fixed retransmission timer is considered as the best solution to avoid unnecessary retransmissions (this concept will be discussed in following sections), the **Fast Retransmission and Fast Recovery Algorithm are then mandatory**.

Table 1 shows these ideas. The parameters and protocol profiles are the same as before but the BECN Algorithm that is not implemented.

Table 1   Fast Retransmit and Fast Recovery effect on application efficiency, frame losses and end-to-end rate.

Mechanism used	Application efficiency	Frame losses (%)	End-to-end rate (Mbps)
Without Fast ret. and fast reco.	89,23	1,7	0,973
With Fast ret. and fast reco.	91,97	0,65	0,962

6.3 Congestion Control

Congestion control mechanisms in the Transport Layer are **mandatory**, since FR only notifies and discards packets in case of congestion, and this is the main cause of data losses in networks with very low bit error rates.

It has been shown how each type of congestion control affects the protocol performance: TCP incorporated Congestion Control Mechanisms and BECN Congestion Control Mechanism. Table 2 shows the results.

Both congestion control mechanisms are recommended since:

- BECN congestion algorithm prevents the network from congestion situations to occur, without reducing drastically the offered traffic.
- When frames are lost due to congestion, the TCP congestion control reduces drastically the offered rate.

Table 2 Congestion Control Mechanisms effect on the performance parameters used

Mechanism used	Application efficiency	Frame losses (%)	End-to-end rate (Mbps)
None	82,22	2,718	0,94
BECN	91,09	0,07	1,021
TCP	91,97	0,645	0,962
Both	93,25	0,046	1,033

It can be seen that the combination of both mechanisms improves the results. The usage of BECN alone offers a significant performance.

6.4 RTO And RTT Measurement

The RTT Measurement algorithm updates the Retransmission timer, adjusting it dynamically to the state of the network. This measure is adequate when the bandwidth-delay product is low, and the measure really shows the actual state of the network, but in case a high bandwidth-delay product, a more accurate measure has to be made to avoid unnecessary retransmissions. The **TimeStamp** facility may be optionally used to increase the accuracy of that measure.

Another important point is that unnecessary retransmissions have been shown to affect the fast retransmission and fast recovery algorithm, as well the TCP congestion control algorithm. Therefore, it is important to avoid these unnecessary retransmissions.

After the simulation studies taking into account the bursty traffic scheme and the network topology, the conclusions about the RTT measurements and the RTO dynamic adaptation are:

- As the network delay increases, the dynamic measurement of the retransmission timer is not adequate, since the actual network congestion situation in which the new timer acts, is rather different from the situation in which the timer was measured.
- Although the TimeStamp option gives a more accurate RTT, this optional mechanism is not sufficient to avoid unnecessary retransmissions.

In conclusion, the use of **the RTT measurement algorithm is not recommended**, specially when bandwidth-delay product is significant. Therefore, a solution in this case is to consider a fixed RTO that should be greater than the maximum expected RTT, being long enough to avoid the unnecessary retransmissions.

Nevertheless, the use of a fixed RTO has to be complemented with the use of the **Fast Retransmission and Fast Recovery Algorithm**. In this case, the algorithm forces the retransmission of the lost packet before the timer expiration.

The results are shown in the following tables for **fixed** and **dynamic** RTO.

Table 3   The effect on the performance of the RTT measurement and the RTO adjustment.

Mechanism used	Application efficiency	Frame losses (%)	End-to-end rate (Mbps)
Dynamic adjust of RTO	91,03	0,027	0,124
No adjust of RTO.	95,30	0,206	0,776

The topology and the type of network (based of frame relay) lead to this recommendation about the usage of a fixed RTO. In a more complex network with the possibility to change routes it is not clear if the solution with fixed RTO could be implemented.

7   CONCLUSIONS

The main conclusion of the study is that using the identified Protocol Elements of the extended TCP, and the BECN congestion control algorithm at the Transport Layer is the best option for the delivery of telemetry over FR networks.

The simulation results have allowed to select the most suitable TCP protocol elements and parameters that show the best efficiency, specially in congested networks.

Following graphs show the comparison between two protocol profiles:

Case 1. This protocol profile includes the TCP extensions, the Fast Recovery and Fast Retransmission algorithm, the Go-Back-N error recovery algorithm, considers the BECN Algorithm against congestion, and a fixed RTO. This is the recommended protocol profile.

Case 2. This protocol profile neither considers the TCP extensions for high performance nor the BECN algorithm against congestion situations, and a dynamic RTO.

The percentages of improvement [*] with the recommended protocol profile are shown in tables 4 and 5.

---

[*]**Percentage of Improvement (%):** percentage the recommended case adds to the other, referred to this case. In the Frame Losses a negative value means an improvement (with a maximum value of - 100%) and a positive one a worsening.

Table 4 Improvement obtained using the proposed profile

Application Efficiency	Frame Losses	End-to-End Rate
17,637%	-98,823%	8,955%

Table 5 Performance of the protocol with the two cases C1 (case 1) and C2 (case 2) commented above.

Mechanism used	Application efficiency	Frame losses (%)	End-to-end rate (Mbps)
Case 1	96,72	0,03	1,0237
Case 2	82,22	2,72	0,9396

8   FUTURE WORK

The ATM Forum has specified 5 ATM service classes (ATM Forum, 1996), one of them being the so called Available Bitrate (ABR) service, which includes an explicit rate control mechanism in concept similar to the frame relay FECN/BECN information to convey information about the network load and congestion state from the network to the user. Future work will have to address how the ATM ABR service will interface to the transport layer which would then take care of proper rate adaptation.

The results of the study will now be verified in a real network scenario. ESA has let a contract to Sema Group (Number: 11772/96/D/DK(SC)) for development of the end systems, which will be connected to a frame relay test network.

9   REFERENCES

(1992) Description of CCSDS Ground Infrastructure Cross-Support Services. Volume 1: Service Concept (DRAFT). 20th April 1992 (CCSDS-P3V1).

(1992) Description of CCSDS Ground Infrastructure Cross-Support Services. Volume 2: Space Data Services. 4th April 1992 (CCSDS-P3V2).

ATM Forum (1996) *Traffic Management Specification, Version 4.0*, April 1996.

Black, U. D. (1992) *Frame Relay Networks: specifications and implementations*. McGraw-Hill.

Calhoun, (1995) *Congestion control in IPv6 Internetworks*, internet draft, May 1995.

Fall, K.and Floyd, S. (1995) *Comparisons of Tahoe, Reno, and Sack TCP*. A Postscript version of this document is available from ftp://ftp.ee.lbl.gov/papers/sacks.ps.Z, December 1995.

Floyd, S. (1994) *TCP and explicit congestion notification.* A Postscript version of this document is available from `ftp://ftp.ee.lbl.gov/papers/tcp_ecn.4.ps.z`, 1994.

Jacobson, V (1992) *TCP Extensions for High Performance.* RFC 1323. May 1992.

Jacobson, V. (1988). Congestion Avoidance and Control. *Computer Communication Review,* vol. 18, no. 4, 314-329.

Johnson, M. J. (1991) Coping with data from Space Station Freedom. *Computer Networks and ISDN Systems,* 22, 131-142.

Jubainville, R. (1994) Congestion control for frame relay. *Telecommunications,* March 1994, 77-80.

Mathis, M., Mahdavi, J., Floyd, S. and Romanov, A. (1996) *TCP Selective Acknowledgements Options.* Internet Draft. 1996.

Postel, J. (1981) *Transmission control protocol,* RFC 793, January 1981.

Schulzrinne, S. H. (1996) Binary Congestion Notification in TCP, IEEE June 25 1996,772-776.

Smith, P. (1992) *Frame Relay. Principles and Applications.* Addison-Wesley.

Stevens, W (1997) *TCP slow start, congestion avoidance, fast retransmit and fast recovery,* RFC 2001, Janury 1997.

10   BIOGRAPHIES

Anna Calveras obtained the electrical engineering degree from the Polytechnic University of Catalonia. She worked at Sema Group Spain where she participated in several research projects. Later she returned to the University where she holds a position as an assistant professor and currently she is working in her Ph.D. in the optimisation of the TCP protocol on heterogeneous network environments.

Josep Paradells obtained the Ph.D. degree from the Polytechnic University of Catalonia. He is associate professor at the mentioned University.

Jordi Puga obtained the electrical engineering degree from the Polytechnic University of Catalonia. Currently he is working at Sema Group Spain where he leads the satellite group.

Klaus-Jürgen Schulz obtained the Ph.D. in informatics from the University of Zurich. Currently he is working at European Space Agency (ESA) in the European Space Operations Centre (ESOC) where he participates in projects related with the communication systems at the ground segment.

40

Stop & Go ABR: A Simple Algorithm for the Implementation of Best Effort Services in ATM LANs

M. Ajmone Marsan[1], K. Begain[2], R. Lo Cigno[1], M. Munafò[1]

[1] Dipartimento di Elettronica, Politecnico di Torino
Corso Duca degli Abruzzi, 24 – 10129 Torino – Italy
e-mail: {ajmone,locigno,munafo}@polito.it

[2] Department of Computer Science, Mu'tah University
Mu'tah 61710 – Jordan e-mail: begain@science.sci.mutah.edu.jo

Abstract

In this paper we propose a novel algorithm for the implementation of best effort services in ATM LANs. The algorithm is a peculiar version of ABR in which sources can transmit only at two different cell rates, the Peak Cell Rate (PCR) and Minimum Cell Rate (MCR); for this reason, the proposed algorithm is called Stop & Go ABR. The Stop & Go ABR algorithm is first described, and then evaluated by detailed simulation of a simple ATM LAN configuration with best effort TCP connections as well as interfering VBR connections, showing that it is capable of providing very good performance and fairness.

Keywords

ATM, traffic management, ABR, performance

1 INTRODUCTION

The data communication services typical of computer networks, such as Local Area Networks (LANs), Metropolitan Area Networks (MANs), and, most important, the Internet, are based on a datagram, or connectionless, packet switching approach, and generally do not guarantee the success in the delivery of the information, thus being often termed Best-Effort.

This is in contrast with the traditional approaches adopted in telecommunication networks, that are based on either circuit switching or packet switching with virtual circuits, also called connections, and that try to guarantee the delivery at the destination of the information generated at the source.

The original conception of the Asynchronous Transfer Mode (ATM) was due to researchers in the telecommunications field, and as a result ATM is based on connection-oriented cell switching, and is naturally matched to guaranteed services, not to best effort services.

The selection of the most adequate approach for the provision of best effort

Broadband Communications   P. Kühn & R. Ulrich (Eds.)
© 1998 IFIP. Published by Chapman & Hall

services in ATM networks has been the subject of many debates within the technical literature as well as standardization committees.

Today, two techniques are standardized for the provision of best effort services in ATM networks, known as the UBR and ABR service categories or transfer capabilities [6, 5].

UBR stands for Unspecified Bit Rate; UBR provides very simple means for the transfer of the data resulting from best effort services through ATM networks. The problem of UBR is that it can be quite inefficient [13, 11, 2], depending on the network configuration and load.

ABR stands for Available Bit Rate; ABR provides algorithms with variable degree of sophistication and efficiency to exploit the bandwidth not used by guaranteed services for the transfer of the data resulting from best effort services through ATM networks. The problem of ABR is that the algorithms that permit good performance to be obtained are rather complex to implement.

In particular, three ABR operating modes are specified (named EFCI, RRM and ERM respectively, see [5] for their description); all are based on the use of special flow control cells, called Resource Management (RM) cells, to notify sources about the congestion of ATM switches along the path followed by the connection. ABR sources are required to react to the information arriving from the network by adequately modifying (increasing or reducing) their cell transmission rates.

The key to success of an ABR implementation is threefold: i) high performance; ii) robustness and great resilience; iii) low implementation cost both for the end-user and for the network nodes. Several proposals have appeared in literature, based either on theoretical approaches [10, 15, 9], or on more heuristic and empiric considerations [7, 12, 3]. Most of them try to optimize performance, while robustness and cost received less attention.

In this paper we propose a novel algorithm for the implementation of best effort services in ATM LANs that is a peculiar version of ABR in which sources can transmit only at two different cell rates, the Peak Cell Rate (PCR) and Minimum Cell Rate (MCR). For this reason, the proposed algorithm is called Stop & Go ABR.

2   STOP & GO ABR

Stop & Go ABR is a version of ABR that was designed with the following main objectives in mind: i) be extremely simple to implement, both at end user ATM terminals and within ATM switches; ii) be acceptably efficient and fair in typical LAN environments; iii) be compliant with the ATM Forum Traffic Management Specification [5].

Stop & Go ABR can be considered a particular version of either the Relative Rate Marking (RRM) or the Explicit Rate (ER) ABR operating modes, that allows only two values for the transmission speed of ABR sources: the Minimum Cell Rate (MCR) and the Peak Cell Rate (PCR).

Parameter	Value	Parameter	Value	Parameter	Value
Nrm	32	Mrm	2	TCR	10
RIF	1	RDF	1	CDF	1
TBE	ID	ATDF	ID	Trm	ID
MCR	ID	PCR	ID	ICR	PCR/MCR

Table 1 Parameter vector for Stop & Go ABR implementation, the label I.D. means that values are implementation dependent

In this paper we shall consider Stop & Go ABR as deriving from RRM ABR by setting both the Rate Increase Factor (RIF) and the Rate Decrease Factor (RDF) to 1.

Table 1 gives the ABR parameter vector that must be used in order to implement Stop & Go ABR. Most parameter values are fixed, hence reducing the complexity of the negotiation at connection setup; the others, identified in the table as Implementation Dependent (ID), can be either negotiated at connection setup or set by the network manager to a suitable value.

For what the Initial Cell Rate (ICR) is concerned, it is assumed that ICR = PCR; however, if the network is not able to support a value of the Transient Buffer Exposure (TBE) large enough to ensure that $\frac{TBE}{FRTT} \geq PCR$ where FRTT is the round trip time measured at connection setup, then ICR must be set to MCR.

The behavior of the source and destination ATM equipment in Stop & Go ABR is exactly as specified in Sections 5.10.4 and 5.10.5 in [5].

The switch behaviors are not standardized in [5], in order to leave space for the competition among equipment manufacturers. The study of Stop & Go ABR in this paper assumes that ATM switches use two separate buffers for each output interface: one buffer is used for high priority connections (CBR and/or VBR) and the second for ABR connections (that are considered to be of lower priority); moreover, switches are assumed to implement RRM ABR with a control algorithm based only on the ABR buffer occupancy.

More precisely, the control algorithm within ATM switches is based on the comparison of the ABR queue length Q_l with a threshold t. The NI (No Increase) and CI (Congestion Indication) bits within backward RM cells are set as follows for all ABR connections:

$$
\begin{aligned}
Q_l &< t \implies NI = 0 \quad CI = 0 \\
Q_l &\geq t \implies NI = 1 \quad CI = 1
\end{aligned}
\tag{1}
$$

with no specific algorithm for the enforcement of fairness among different connections; in practice, the NI bit is not used.

In general, in RRM ABR, the NI and CI bits are used to form a 3-state feedback (when CI=1, NI has no meaning) with the semantic "increase the cell rate" (NI=CI=0), "keep the actual cell rate" (NI=1,CI=0), and "decrease the cell rate" (NI=-,CI=1), but with just one threshold in the buffer it is impossible to discriminate more than two cases.

The use of two thresholds in order to create an hysteresis cycle, and also use the NI bit, was shortly investigated, but was discarded for the following two reasons: i) a "Stop & Go" system has intrinsically only two states, so that the (NI=1,CI=0) combination has little meaning; ii) traffic configurations may arise where Q_l remains between the two thresholds while some connections are transmitting at their PCR and others are transmitting at their MCR, thus creating stable, unfair situations.

Different ATM switch control algorithms, derived either from RRM ABR or from ER ABR, are being considered for further study on Stop & Go ABR.

3   PERFORMANCE EVALUATION WITH TCP CONNECTIONS

Stop & Go ABR was designed for the implementation of best effort services in LAN environments, where round trip delays are short, and congestion is not likely to arise in more than one node at a time.

The performance analysis of Stop & Go ABR in such a setup must take into account the interactions between the control algorithms of Stop & Go ABR and those of TCP, that is by far the most widely used transport protocol for the provision of best-effort services in LANs.

For this reason, in the characterization of the performance of Stop & Go ABR we studied its behavior with TCP connections in a rather simple LAN setup. We restrict the analysis to greedy TCP connections transmitting only maximum size segments whose dimension is 1460 bytes (this value is the one presently used on Ethernet LANs) excluding all overheads; to those we add a 48-byte overhead (20 bytes for the TCP header, 20 for the IP header, and 8 for AAL5 overheads). The TCP implementation used in the simulations is the officially distributed BSD TCP-reno 4.3, that was adapted to run on top of CLASS* [1, 4], a cell level ATM network simulator developed at Politecnico di Torino in co-operation with CSELT, the research centre of Telecom Italia. The only major modification introduced into the TCP code concerns the timer granularity that was set to 50 μs, in order to adapt it to large bandwidth-delay product networks.

---

*CLASS stands for Cell Level ATM network Services Simulator; it is entirely written in C language and portable on most computing platforms. More information on CLASS are available at the URL http://www.tlc.polito.it/class.html.

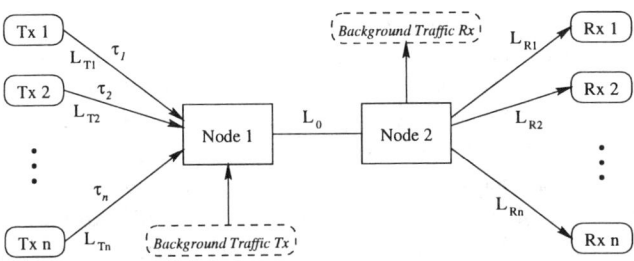

Figure 1 Bottleneck topology

Parameter	Value	Parameter	Value	Parameter	Value
TBE	500	ATDF	10.0	Trm	100
MCR	1 Mbit/s	ICR		PCR	

Table 2 Implementation dependent ABR parameter values used in the simulations

3.1 The simulation scenario

Fig. 1 presents the "bottleneck" topology, that can be considered a simplified representation of any LAN scenario, where only the congested node (node 1) is considered, together with the sources and destinations that load the congested output link.

Several TCP transmitters (say n) are connected to node 1, and the corresponding TCP receivers are connected to node 2. The n TCP transmitters contend for the transmission resources on the congested link connecting nodes 1 and 2. Both nodes implement Stop & Go ABR as described in Sect. 2; however, only node 1 is congested, due to the demultiplexing in node 2 of all the TCP connections that cross the only link connecting the two nodes. The same congested link is also used by some background traffic that is not controlled with the ABR mechanism, since it is supposed to be generated by CBR or VBR sources, hence to have higher priority. TCP connections are unidirectional, and TCP receivers return only ACKs to the corresponding transmitters.

Table 2 reports the values of the implementation dependent Stop & Go ABR parameters used in the simulations. The PCR value is one of the variables of our study. The other variables in the study are the number of TCP connections, the background traffic characteristics, and the network span.

In the scenario taken as a reference for the simulation experiments we assume that all links in the network have capacity 150 Mbit/s, that the link between the two nodes is 10 km long, and that the distance of transmitters from node 1 and that of receivers from node 2 are also 10 km, that is,

$L_0 = L_{Ti} = L_{Ri} = 10 \ km$ in Fig. 1. Such distances correspond to a round trip delay approximately equal to 0.15 ms, which can be assumed to be typical for medium size LANs, if processing delays are taken into account. The number of TCP connections is set to 3, and the background traffic is assumed to be generated by one On-Off VBR source with exponentially distributed On and Off periods that transmits at constant bit rate C_b during On periods. The average durations of the On and Off periods are 20,000 and 10,000 slots, respectively (roughly 56 and 28 ms), while C_b varies with the background traffic load; namely C_b is 1.5 times the value of the background traffic load values in the horizontal axis of the plots shown in the following Subsection. The buffer sizes within the switch are 100 cells for the higher priority traffic (that does not need a large buffer), and 1,000 cells for the lower priority ABR traffic. This latter value corresponds to a buffer size of 53 kbytes, that is fairly small for the equipments available on the market. The threshold is set to 25% of the ABR buffer size; namely $t = 250$.

If not otherwise stated, numerical results are obtained with the values just described for the reference simulation scenario.

In order to measure the network performance, we use three different metrics.

TCP goodput – this is the throughput seen by the TCP end user, after all faulty and duplicated TCP segments have been discarded.

TCP efficiency – this is the ratio between the goodput and the load offered by the TCP transmitter.

Link Waste (LW) – this is the bandwidth that remains unused on the bottleneck link, expressed as a fraction of the link capacity.

3.2   Numerical results

Let us first of all analyse the impact of the PCR value on the network performance in the reference simulation scenario with Stop & Go ABR.

Fig. 2 presents 4 plots where the value of the PCR of the three TCP connections is varied.

All results are plotted against the background offered load. The left vertical scale measures the goodput of the three TCP connection, that are reported with solid lines; the dot-dashed line represent the capacity theoretically available for each TCP connection (thus, ideally, all solid lines should lie on the dot-dashed line). The left vertical scale measures the efficiency of the three TCP connections (reported with dashed lines, that ideally should be equal to 1), as well as the link waste (reported as a dotted line, that hopefully should be as near as possible to 0).

The 4 plots refer to the values PCR = 25, 50, 75, and 100 Mbit/s, as indicated on each plot. The overall network performance is quite satisfactory, showing a good exploitation of the network resources with high efficiency,

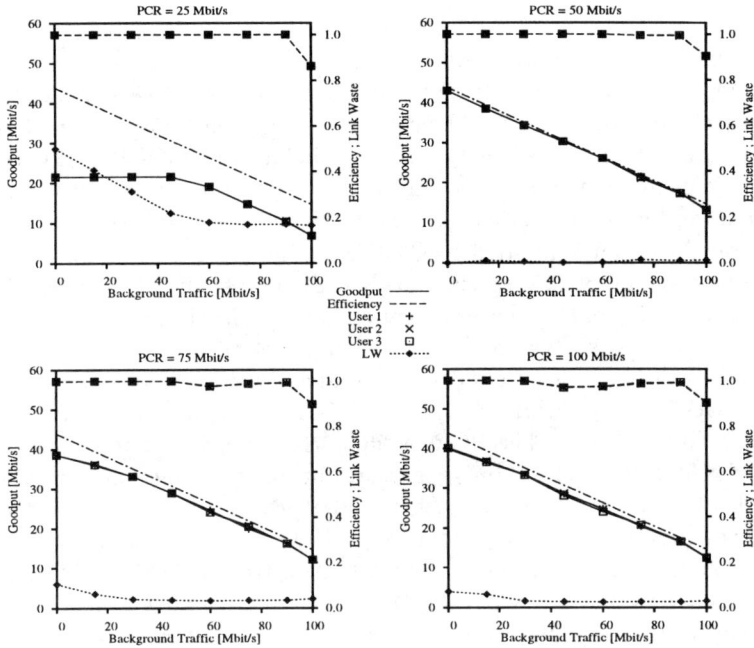

Figure 2 Performance with 3 TCP connections with equal length and PCR

together with a surprising degree of fairness, given the simplicity of the implementation.

It is remarkable that also when the peak load offered by the ABR connections (PCR = 100 Mbit/s – lower right plot) is twice the link capacity, the behavior of the network is fair and quite smooth, while it has been shown [13, 11, 8, 2] that plain TCP over ATM generally shows a poor behavior.

Considering the plot for PCR = 25 Mbit/s, one might expect that, as the background load increases and reaches 75 Mbit/s, the goodput of the TCP connections, that at lower loads were rate limited, reaches the available capacity. In fact, the characteristics of the background traffic prevent this behavior: since the background traffic is generated by one On-Off source, the spare capacity mainly results from silence periods of the On-Off source, that the TCP connections can not exploit due to their PCR limitation.

One final remark on the PCR = 50 Mbit/s plot: the performance in this case is so much better than in other cases because the sum of the PCRs is exactly equal to the link capacity and the background traffic is On-Off – a situation that is probably not realistic.

A specific comment is needed for the efficiency curves. It is in general expected that the efficiency drops below 1 when cells are lost in the network and, as a consequence, faulty segments are discarded at the receiver, triggering the retransmission of the lost segment either for a timeout expiration or, more probably, for repeated ACKs reception at the source. However, no

TCP cells are dropped either from the ABR buffer in the congested node, or anywhere else in the network during the simulation runs, hence the efficiency reduction is due to some different cause. As a matter of fact, the expiration of a retransmission timer is not necessarily triggered by a segment loss, but it can be due to an exceedingly long time before the ACK comes back. The TCP retransmission timer τ_o are set according to the formula $\tau_o = \hat{\mu}_t + 4\hat{\sigma}_t$ where $\hat{\mu}_t$ and $\hat{\sigma}_t$ are smoothed estimations of the round trip time average and standard deviation computed by TCP with the algorithm proposed by Van Jacobson (see [14] pp. 300). When the background is Off for a reasonably long time $\hat{\mu}_t$ is very close to the propagation round trip time, i.e., $\simeq 0.15$ ms in our case, and $\hat{\sigma}_t$ is very small, so that τ_o is set to a very small value, say < 1 ms. When the background traffic source switches to the On state, it is served with priority within the node, and TCP traffic must wait until capacity is available to serve it. The PCR of the background traffic for a background load of 100 Mbit/s is 150 Mbit/s, so that during the whole duration of an On period, whose average is roughly 56 ms, TCP traffic is not served: under these circumstances it is common that a timeout expires even if no cells have been lost in the network. It might be argued that setting the TCP timer granularity to a larger value can avoid this phenomenon, but indeed this is true only to a very limited extent. If the granularity is increased to 100 or 500 μs, the qualitative behavior described above is not modified; if the timer granularity is instead increased to a value much larger than the propagation delay, say 50 or 100 ms, that are values pretty similar to those used in today implementations, the above phenomenon is probably avoided, but TCP becomes completely unable to adapt to network changes since its sensitivity is too low to detect congestion.

The efficiency reduction is due to this phenomenon not only in the results presented in Fig. 2, but also in all other results presented throughout the paper, since cell losses were never observed in simulation runs.

Let us now consider a slightly different situation, where each TCP source is at a different distance from the congested ATM switch.

Fig. 3 reports the results for the same PCR values of Fig. 2, but with link lengths $L_1 = 1$ km, $L_2 = 5$ km and $L_3 = 10$ km, respectively.

The behavior of TCP is known to be biased against longer connections; however, if the ABR mechanism is able to prevent cell losses, the TCP congestion control mechanism is not triggered, and the performance is dictated by the ABR control loop, whose behavior is influenced by the loop delay. Fig. 3, however, shows that the overall network performance is satisfactory, and even fairness does not suffer appreciably form the different source distances from the control point, even if they vary over an order of magnitude. The reason for this quite good behavior lies in the fact that the ABR feedback is intrinsically sampled: neglecting delay jitters, it is sent once every $t_f = \dfrac{1}{\text{Nrm}} \cdot R_e \cdot t_s$ seconds,

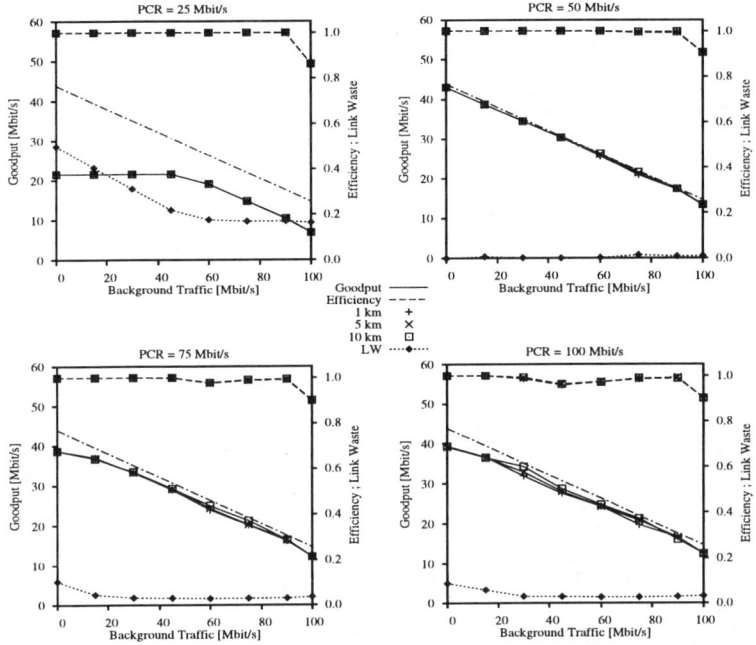

Figure 3 Performance with 3 TCP connections with different length and equal PCR

where R_e is the effective rate of the source*, and t_s is the slot duration. In the case under consideration t_f varies between 136 μs when $R_e = 100$ Mbit/s and 13.6 ms, when $R_e = 1$ Mbit/s: both values are larger than the control loop delay of the 10 km connection that is roughly 50 μs. In practice this means that the sampled closed loop system has a transfer functions whose response delay is equal to 1 sample independently from the source distance.

Fig. 4 shows performance results in the case when TCP connections have different PCRs; both cases of equal length connections and of different length connections are considered.

The goodput of TCP connections is proportional to their PCR, as it might be expected, and as it is probably perceived as "fair" from both the users and the network point of view. Once again, connection lengths have a marginal impact on performance; in particular, if we observe the lower plots, that refer to the case of different length connections, it is clear that the source with higher PCR gets the most out of the network, independently from being the nearest or the farthest from the congested node.

Up to now we have considered a rather small LAN scenario, where Stop & Go ABR proved to perform quite well; this scenario was considered because

---

*If the connection is persistent, as in our study, R_e is equal to the Current Cell Rate (CCR) read in the backward RM cells; however in the general case the source can transmit at lower rates, or even be silent.

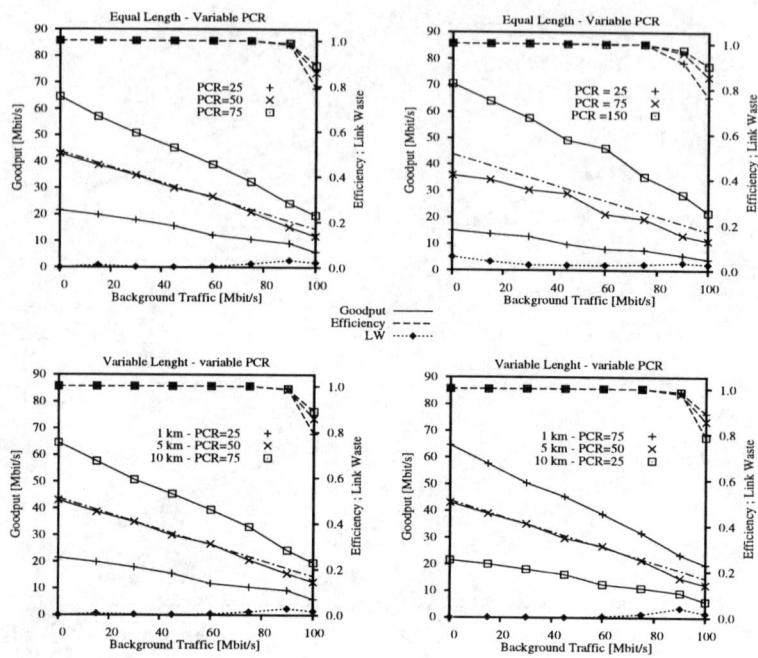

Figure 4 Performance with 3 TCP connections with different lengths and PCRs

we believe it can be the most reasonable for the application of Stop & Go ABR; however, the reader might think that we restricted our analysis to such scenario because in more complex environments Stop & Go ABR provides poor performance. In order to show that this is not the case, we also provide a glimpse to the performance of Stop & Go ABR under different operating conditions.

Fig. 5 presents performance results for Stop & Go ABR in a scenario with 10 TCP connections in a network that spans 300 km (i.e., $L_0 = L_{Ti} = L_{Ri} = 100\ km$). The high-priority background traffic is now obtained by the superposition of 10 On-Off sources with burstiness 10. The On and Off periods are exponentially distributed with average equal to 3000 and 27000 slots, respectively, and the PCR of each connection is equal to the global offered background traffic load. The ABR buffer size is 5,000 cells and $t = 1250$ cells, keeping the same ratio to the buffer size as for the previous cases; the buffer size for the high-priority background traffic is increased to 10,000 cells, since for high loads the sum of the background traffic PCR is greater than the link capacity.

From the curves in Fig. 5 we see that the performance of Stop & Go ABR is still rather good, even if the bandwidth-delay product has been increased by one order of magnitude. By comparing these results with those discussed before, it can be seen that the efficiency of TCP connections deteriorates more

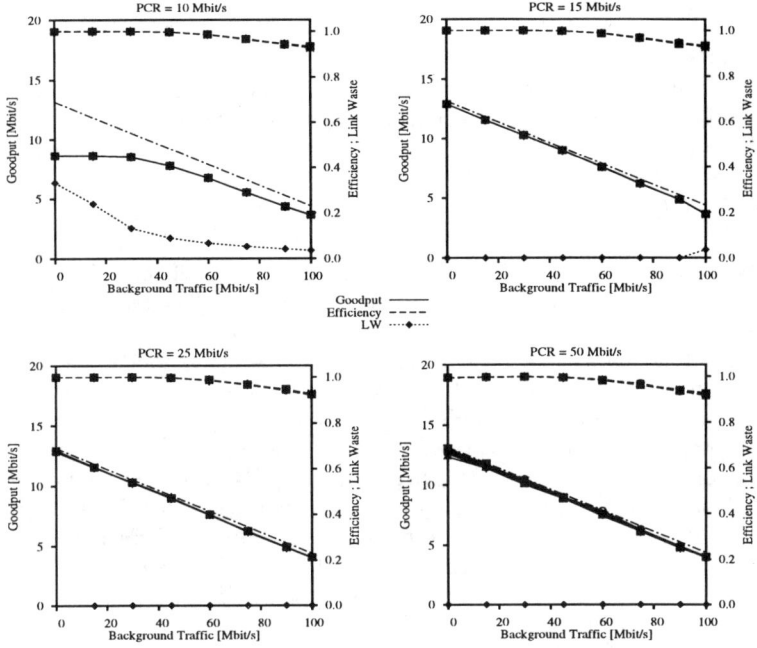

Figure 5 Performance with 10 TCP connections with equal length and PCR

smoothly as the background traffic load increases: this is not due to the larger network span, but to the different characteristics of the background traffic that, being produced by 10 sources, shows a greater variability. It is worthwhile noticing that even in the lower right plot, where the sum of the ABR PCRs is more than 8 times the link capacity, the behavior is still very smooth and fair, and the Stop & Go algorithm prevents the network from losing any cell of the ABR connections; the same is not true for the background traffic that, even with a 10,000 cell buffer and priority over ABR traffic, suffers from cell losses (not shown in the plots) for loads greater than 60 Mbit/s.

4 CONCLUSIONS

In this paper we described the Stop & Go ABR algorithm, that is a peculiar version of ABR in which sources can transmit only at two different cell rates, called peak cell rate (PCR) and minimum cell rate (MCR).

The performance and fairness of Stop & Go ABR were evaluated by detailed simulation of a simple ATM LAN configuration with best effort TCP connections as well as interfering VBR connections.

Numerical results showed that Stop & Go ABR is capable of providing very good performance and fairness for a range of different parameter values.

Rather surprisingly, Stop & Go ABR proved to perform quite well even in the case of network spans of the order of hundreds of km, and very bursty interfering VBR connections.

Further studies on Stop & Go ABR will consider different switch control algorithms deriving either from RRM ABR or ER ABR.

REFERENCES

[1] M.Ajmone Marsan, A.Bianco, T.V.Do, L.Jereb, R.Lo Cigno, M.Munafò, "ATM Simulation with CLASS", *'Performance Modeling Tools'*, Performance Evaluation 24 1995, pp.137–159

[2] M.Ajmone Marsan, A.Bianco, R.Lo Cigno, M.Munafò, "Some Simulation Results about Shaped TCP Connections in ATM Networks", in: D.Kouvatsos (editor), Performance Modeling and Evaluation of ATM Networks – Vol.2, Chapman and Hall, London, 1996

[3] M.Ajmone Marsan, A.Bianco, R.Lo Cigno, M.Munafò, "Four Standard Control Theory Approaches for the Implementation of RRM ABR Services", in: D.Kouvatsos (editor), *Performance Modelling and Evaluation of ATM Networks – Vol.3*, Chapman and Hall, London, July 1997

[4] M.Ajmone Marsan, A.Bianco, C.Casetti, C.F.Chiasserini, A.Francini, R.Lo cigno, M.Munafò, "An Integrated Simulation Environment for the Analysis of ATM Networks at Multiple Time Scales", to appear on *International Journal of Computer Networks and ISDN Systems*

[5] ATM Forum AF-TM-0056.000, "ATM Forum Traffic Management Specification", Version 4.0, April 1996

[6] ITU-TSS Study Group 13, Recommendation I.371 "Traffic Control and Congestion Control in B-ISDN", Geneve, Switzerland, July 1995

[7] R.Jain, S.Kalyanaraman, R.Viswanathan, R.Goyal, "A Sample Switch Algorithm", *ATM Forum-TM 95-0178R1*, February 1995

[8] S.Keung, K.Siu, "Degradation in TCP Performance under Cell Loss", ATM FORUM 94-0490, April 1994

[9] A.Kolarov, G.Ramamurthy, "A Control Theoretic Approach to the Design of Closed Loop Rate Based Control for High Speed ATM Networks", *proceedings of IEEE INFOCOM'97*, Kobe, Japan, April 1997

[10] S.Mascolo, D.Cavendish, M.Gerla, "ATM Rate Based Congestion Control Using a Smith Predictor: an EPRCA Implementation", *proceedings of IEEE INFOCOM'96*, S.Francisco, CA, USA, March 1996

[11] H.Li, K.Siu, H.Tzeng, "TCP over ATM with ABR service versus UBR+EPD service", ATM FORUM 95-0718, June 1995

[12] L.Roberts et al. "Enhanced PRCA (Proportional rate-control algorithm), ATM FORUM 94-0735R1, August 1994

[13] A.Romanow, S.Floyd, "Dynamics of TCP Traffic over ATM Networks", *IEEE JSAC*, v.13, no.4, pp.633-641, May 1995

[14] W.R.Stevens, "TCP/IP Illustrated: The Protocols", Vol.1, Addison Wesley, 1994

[15] Y.Zhao, S.Q.Li, S.Sigarto, " A Linear Dynamic Model for Design of Stable Explicit-Rate ABR Control Schemes", *proceedings of IEEE INFOCOM'97*, Kobe, Japan, April 1997

Source and Traffic Modelling

41

A Locally Stationary Semi-Markovian Representation for Ethernet LAN Traffic Data

S. Vaton and E. Moulines
Ecole Nationale Supérieure des Télécommunications
46 rue Barrault, 75634 Paris Cedex 13, France.

Abstract

In the past few years a large number of teletraffic measurements have been extensively studied by many contributors. Most of the authors agree on saying that the traffic measured on today's broadband networks is long-range dependent. Oddly enough these contributors do not question the stationarity of the traffic on these long time scales when the hypothesis of stationarity is essential to speak of long-range dependence. Concurrently it has been demonstrated that some kind of non stationarities in a short-range dependent process can lead, if they are not detected, to the untrue conclusion of long-range dependence.

We prove on the basis of different tests of stationarity that the hypothesis of short range dependence and the hypothesis of stationarity are contradictory on long time-scales. Contrary to many authors who decide in favor of the long-range dependence we propose to model the measured traffic as a locally stationary and markovian process. We exhibit a new markovian model and we show how one can track the varying parameters of this model by means of a recursive maximum likelihood algorithm.

We then generate a non stationary markovian traffic whose varying parameters are matching the parameters of the measured traffic. We verify that the use of a classical visual index of long-range dependence brings to the same conclusion of long-range dependence for the non stationary and markovian model than for the measured traffic.

Keywords

Long-range dependence, long memory, non-stationarity, tests, Hidden Markov Model.

1 INTRODUCTION

In the past few years a large number of contributions have been devoted to the statistical study of different traffics measured on today's broadband networks.

Broadband Communications   P. Kühn & R. Ulrich (Eds.)
© 1998 IFIP. Published by Chapman & Hall

There is now a consensus to say that this traffic is long-range dependent. These contributions are very significant. They are indeed at variance with markovian models of traffic such as the Poisson process or the Markov Modulated Poisson Process for which queuing results have been established.

It has what is more been established that some quality measures such as the overflow probability or the average packet delay are strongly underestimated by markovian models if the traffic is long-range dependent. The only queuing results that can be established for long-range dependent processes are overestimations of the overflow probability of a queue fed by a long-range dependent process. These results appeal to the difficult theory of great deviations.

We wonder if the traffic might not be a stationary and long-range dependent process but a non stationary and markovian process. It is known ([1],[2],[3]) that deterministic jumps or trends in the mean of a time series without long-range dependence can mislead to the conclusion of long-range dependence if one relies on visual indexes of long-range dependence such as the variance time plot. In the above-mentioned contributions the authors investigate hours of traffic and oddly enough they do not test the stationarity of the traffic on these time-scales.

To support our intuition we investigate a traffic stream that is commonly studied in the litterature (LBL-PKT3). This trace was originally investigated by Paxson and Floyd [4] who conclude that it exhibits a high degree of burstiness that can not be explained by markovian models and who discuss how this burstiness might mesh with self-similar models of traffic.

The rest of the paper is organised as follows. In Section 2 we expose different tests of stationarity for mixing processes and their application to the LBL-PKT3 stream. We conclude that the hypotheses of stationarity and of mixing are incompatible on long time scales which proves that stationary markovian models are inadequate on long time-scales. In Section 3 we propose a new model, the Shifted Exponential Hidden Markov Model (SEHMM) and we briefly recall how one can track the parameters of this model. We then simulate a non stationary SEHMM. We show that for this synthetic non stationary and markovian model the variance time plot index leads to the same conclusions as those obtained by Paxson and Floyd for the real traffic. These findings question the consensus according to which the traffic measured on modern broadband networks is long-range dependent.

2   TESTS OF STATIONARITY

2.1   Theory of the tests

(a)   General Framework
Basically the tests of stationarity that we propose rely on the comparison of different empirical statistics calculated on two neighbour segments of finite

length of the stream. The hypothesis of stationarity is rejected if the empirical statistics for the two neighbour segments are significantly different.

Denote by $\{X_t\}$ the sequence of the inter arrival times (IAT) from which a set of finite length $\{X_t\}_{1 \leq t \leq T}$ is observed. Suppose that one aims at testing if this finite observation is strict sense stationary. Denote by τ_1 the presumed change point. For the sake of homogeneity we also define $\tau_0 = 0$ and $\tau_2 = T$. We do as if $\{X_t\}_{1 \leq t \leq \tau_1}$ and $\{X_t\}_{\tau_1+1 \leq t \leq T}$ were two realizations of finite length $T_i = \tau_i - \tau_{i-1}$ of two processes $\{X_t^1\}$ and $\{X_t^2\}$.

In what follows we test the stationarity of the IAT process in the sense of (i) the mean of the process (ii) the sampled cumulative distribution function $\mathbb{E}(g(X_t))$ where $g(x) = (\mathbb{I}_{\Delta_1}(x), \cdots, \mathbb{I}_{\Delta_N}(x))'$, $(\Delta_i)_{1 \leq i \leq N}$ being a partition of $\mathbb{R}^+$ and (iii) the first covariance coefficients $\mathbb{E}((X_t^2, X_t X_{t+1}, \cdots, X_t X_{t+N-1})')$. We introduce a new time series $\{Z_t\}$ that is defined as (i) $Z_t = X_t$ (ii) $Z_t = g(X_t)$ or (iii) $Z_t = (X_t^2, X_t X_{t+1}, \cdots, X_t X_{t+N-1})'$ depending of the non stationarities that we want to detect.

(b)   Central Limit Theorems

Different Assumptions are needed to establish a Central Limit Theorem for the vector of the empirical statistics.

Assumption 1 $\{X_t\}$ *is strict sense stationary.*

Assumption 2 $\frac{T_i}{T} \overset{T \to \infty}{\to} c_i > 0.$

Assumption 3 $\{X_t\}$ *is α-mixing with an α-mixing coefficient that verifies* $\sum_{n=0}^{+\infty} \alpha_n^{\delta/(2+\delta)} < +\infty$ *and with* $\mathbb{E}(|X_t|^{2+\delta}) < +\infty.$

Assumption 4 $q_t \overset{t \to \infty}{\to} +\infty$ *and* $q_t = o(t).$

Let us recall that the α-mixing coefficient of the process $\{X_t\}$ is defined as $\alpha_n = \sup_{A,B} |\mathbb{P}(A \cap B) - \mathbb{P}(A)\mathbb{P}(B)|$ the supremum being taken on all sets A in $\mathcal{M}_{-\infty}^t$ and B in $\mathcal{M}_{t+n}^{+\infty}$ where $\mathcal{M}_a^b = \sigma(X_t, a \leq t \leq b)$.

The Assumption 3 is verified by many usual processes and in particular by a large class of Markov processes. It is in particular verified by any ARMA process if the density of the innovation is strictly positive on $\mathbb{R}$ [5] and by any finite state irreducible Hidden Markov Chain. For a survey about mixing processes and about the Central Limit Theorem for such processes we refer the reader to [6] and the references therein.

Theorem 1 *Assume (A1-A2-A3). Then it holds that*

$$\sqrt{T}(\frac{1}{T_i} \sum_{t=\tau_{i-1}+1}^{\tau_i} Z_t - \mathbb{E}(Z_t)) \sim \mathcal{AN}(0, c_i^{-1}\Gamma_0), \quad 1 \leq i \leq 2$$

where $\Gamma_0 = \sum_{\tau=-\infty}^{+\infty} \gamma_Z(\tau)$ *with* $\gamma_Z(\tau) = \mathbb{E}(Z_t Z'_{t+\tau}) - \mathbb{E}(Z_t)\mathbb{E}(Z'_t)$

Note that Γ_0 is equal to the spectral density matrix of $\{Z_t\}$ at zero frequency. This remark permits the construction of a consistent estimator

$$\hat{\Gamma}_0 = \frac{1}{T} \sum_{0}^{+m_T} w(k) \Re((\sum_{1}^{T}(Y_t - \hat{\rho})e^{j\frac{k+1}{T}t})'(\sum_{1}^{T}(Y_t - \hat{\rho})e^{j\frac{k+1}{T}t}))$$

where $m_T = \sqrt{T}$ and $w(k) = \mathbb{I}_{k=0} + \frac{2}{2m_T+1}\mathbb{I}_{1 \leq k \leq m_T}$.

Denote by $\hat{Z}^i_T = \frac{1}{T_i}\sum_{t=\tau_{i-1}+1}^{\tau_i} Z^i_t$ the empirical statistics for the segment of index i and denote by $\hat{Z}_T = ((\hat{Z}^1_T)'(\hat{Z}^2_T)')'$ the vector of the empirical statistics for the two segments.

Theorem 2 *Assume (A1-A2-A3). Then it holds that*

$$\sqrt{T}(\hat{Z}_T - (11)' \otimes \mathbb{E}(Z_t)) \sim \mathcal{AN}(0, \Gamma) \quad \text{where} \quad \Gamma = \begin{pmatrix} c_1^{-1}\Gamma_0 & 0 \\ 0 & c_2^{-1}\Gamma_0 \end{pmatrix}$$

and where $A \otimes B$ *denotes the Kronecker product of A and B.*

To demonstrate the Theorem 2 we mimic the approach of Epps in [7]. We define a new estimator $\tilde{Z}^i_T$ where the first q_T terms are removed

$$\tilde{Z}^i_T = \frac{1}{T_i} \sum_{\tau_{i-1}+q_T+1}^{\tau_i} Z_t$$

The basic idea consists in proving that $\sqrt{T}\hat{Z}^i_T$ and $\sqrt{T}\tilde{Z}^i_T$ converge in distribution to the same normal distribution and in proving that $\sqrt{T}\tilde{Z}^1_T$ and $\sqrt{T}\tilde{Z}^2_T$ are asymptotically independent, in the sense that

$$T(\mathbb{E}(\exp(i\tilde{Z}^1_T u^H + i\tilde{Z}^2_T v^H)) - \mathbb{E}(\exp(i\tilde{Z}^1_T u^H))\mathbb{E}(\exp(i\tilde{Z}^2_T v^H))) \overset{T\to\infty}{\to} 0$$

It results from the Davydov Theorem [8] that

$$|\mathbb{E}(\exp(i(\tilde{Z}^1_T u^H + \tilde{Z}^2_T v^H))) - \mathbb{E}(\exp(i\tilde{Z}^1_T)u^H)\mathbb{E}(\exp(i\tilde{Z}^2_T)v^H)| \leq \alpha(q_T) \overset{T\to+\infty}{\to} 0$$

and though $\tilde{Z}^1_T$ and $\tilde{Z}^2_T$ are asymptotically independent.

Denote by $D^i_T = \sqrt{T}(\hat{Z}^i_T - \tilde{Z}^i_T) = \frac{\sqrt{T}}{T_i}\sum_{t=\tau_{i-1}+1}^{\tau_{i-1}+q_T} Z_t$ the difference between $\sqrt{T}\hat{Z}^i_T$ and $\sqrt{T}\tilde{Z}^i_T$. It results from (A2-A4) that the covariance matrix of D^i_T tends to zero as T tends to infinity. It then results from the Theorem 1 and from the Slubtski Theorem [9] that $\sqrt{T}(\tilde{Z}_T - (11)' \otimes \mathbb{E}(Z_T)) \sim \mathcal{AN}(0, \Gamma)$ which concludes the proof of Theorem 2.

(c)   Tests
● **Stationarity of the mean and of the marginal distribution**

As stated above the test of stationarity consists in comparing $\hat{Z}_T^1$ and $\hat{Z}_T^2$. For the mean and for the marginal distribution of the process we consider the difference between $\hat{Z}_T^1$ and $\hat{Z}_T^2$ on the two neighbour segments $\hat{Z}_T^1 - \hat{Z}_T^2 = U\hat{Z}_T$, where $U = (1 - 1)$ in the test of stationarity of the mean of $\{X_t\}$ and $U = (1 - 1) \otimes I_N$ in the test of stationarity of the marginal distribution of $\{X_t\}$.

Theorem 3 *Assume (A1-A2-A3). Then it holds that*

$$\sqrt{T}U\hat{Z}_T \sim \mathcal{AN}(0, U\Gamma_0 U')$$

Theorem 4 *Assume (A1-A2-A3). Then it holds that*

$$T\hat{Z}_T(\Gamma^{-1/2})^H\Gamma^{-1/2}\hat{Z}_T \sim \chi^2(N)$$

where $\Gamma^{\frac{1}{2}}$ denotes the square root of Γ, $\Gamma = \Gamma^{\frac{1}{2}}(\Gamma^{\frac{1}{2}})'$.

- **Stationarity of the first correlations**
  In [10] Mauchly introduces the sphericity statistics to test whether two gaussian random vectors have the same covariance matrix. Drouiche and Mokkadem ([11],[12],[13],[14]) generalize this test to the case of time series to test whether two processes have proportional spectra. Vaton [15] proposes to use this measure of spectral similarity to test whether a process is second order stationary. We briefly recall the principles of the test of second order stationarity proposed by Vaton [15].
  For any positive sequence $\rho = (\rho_0, \rho_1, \cdots, \rho_{N-1})'$ denote by $T_N(\rho)$ the Toeplitz matrix $T_N(\rho) = \sum_{\tau=0}^{N-1} \rho_\tau M_\tau$ where M_τ is the matrix whose entry (i,j) is equal to $M_\tau(i,j) = \delta_\tau(|i - j|)$. Denote by μ and ν two positive sequences and define

  $$S(\mu, \nu) = \frac{(det(T_N(\mu)T_N^{-1}(\nu)))^{1/N}}{\frac{1}{N}Tr(T_N(\mu)T_N^{-1}(\nu))}$$

  It results from the arithmetico-geometric inequality that $S(\mu, \nu) \leq 1$ with equality when $\mu = \alpha\nu$ where α is a proportionality constant.
  Our idea is to derive the asymptotic distribution of the ratio $S(\hat{Z}_T^1, \hat{Z}_T^2)$ normalized by a factor that depends on the length T of the observation and to reject the hypothesis of stationarity if the obtained value is lower than a prescribed threshold determined by the false alarm probability.
  Note that $S(\hat{Z}_T^1, \hat{Z}_T^2)$ is a deterministic function of $\hat{Z}_T$. This permits the derivation of an asymptotic result for $S(\hat{Z}_T^1, \hat{Z}_T^2)$. The demonstration of this result is based on a Taylor development of S at point $(\mathbb{E}(Z_t), \mathbb{E}(Z_t))$. As S is maximum at point $(\mathbb{E}(Z_t), \mathbb{E}(Z_t))$ a second order Taylor development is needed.

Theorem 5 *Assume (A1-A2-A3). Then it holds that*

$$2TS^*(\hat{Z}_T) \xrightarrow[(d)]{T \to \infty} Z^H \nabla^2 S((11)' \otimes \mathbb{E}(Z_t))Z \quad \text{with} \quad Z \sim \mathcal{N}(0, \Gamma)$$

where $\nabla^2 S^*((11)' \otimes \mathbb{E}(Z_t))$ denotes the Hessian of S at point $(11)' \otimes \mathbb{E}(Z_t)$.

The only technical points that are needed to establish the expression of $\nabla^2 S$ are the second order differential of the determinant and of the inverse of any matrix M. These expressions can be obtained by differential calculus :

$$\log|M + \Delta M| = \log|M| + Tr(M^{-1}\Delta M) - Tr(M^{-1}\Delta M M^{-1}\Delta M) + o(||\Delta M||^2$$
$$(M + \Delta M)^{-1} = M^{-1} - M^{-1}\Delta M M^{-1} + 2M^{-1}\Delta M M^{-1}\Delta M M^{-1} + o(||\Delta M||^2)$$

- **Thresholds**

The Theorems 4 and 5 permit to reject the set of Assumptions (A1-A3) with a false alarm probability of α. If the obtained statistics is superior to the $(1 - \alpha)$ quantile of the asymptotic distribution one concludes that (A1-A3) is wrong which means that A1 and A3 are mutually exclusive.

Note that in the Theorem 5 the asymptotic distribution is a quadratic form in a multidimensional Gaussian random variable and that the prescribed threshold is obtained by Monte-Carlo simulation.

2.2   Results

The simulations are replicated for thirteen time-scales ranging from six seconds to one hour and thirty minutes and for ten pairs of neighbour segments for each time-scale.

On the Figures 1 and 2 we plot $T\hat{Z}_T^H(\Gamma^{-1/2})^H\Gamma^{-1/2}\hat{Z}_T$ for all the pairs of neighbour segments. The 90% and 99% quantiles of $\chi^2(N)$ are represented in dotted lines. (A1-A3) is rejected when $T\hat{Z}_T^H(\Gamma^{-1/2})^H\Gamma^{-1/2}\hat{Z}_T$ is superior to the $(1 - \alpha)$ quantile of $\chi^2(N)$.

On the Figure 3 we plot the cumulative distribution function $P(X \leq 2TS(\hat{Z}_T^1, \hat{Z}_T^2))$ for the asymptotic distribution $X = Z^H\nabla^2 S|_{(\mathbb{E}(Z_t),\mathbb{E}(Z_t))} Z$ where $Z \sim \mathcal{N}(0, \Gamma)$. The 90% and 99% fractiles for the distribution of X are represented in dotted lines. ((A1-A3) is rejected if $P(X \leq 2TS(\hat{Z}_T^1, \hat{Z}_T^2))$ is superior to $(1 - \alpha)$.

The conclusions of our simulations is that (A1-A3) is wrong for most pairs of neighbour segments for long time-scales. Classical models such as the stationary Poisson process or the stationary Markov Modulated Poisson Process are consequently not adapted to the traffic that we investigate on these long time-scales.

Note that the Assumption A3 is wrong for long-range dependent processes such as the Fractional Gaussian Noise or the fractionally integrated autoregressive moving average process. Consequently the tests developed do not permit to reject A1 for long-range dependent processes. The difficulty to decide between long-range dependence and non stationarities has already been discussed by Duffield *et al.*in [2].

It is thus difficult to decide if the evidences of auto-similarity mentioned by many authors result from a real auto-similarity of the traffic or from some non-stationarities that might have mislead to the conclusion of auto-similarity ([1],[2],[3]) or from the coexistence of both phenomena.

3 A NON STATIONARY AND SEMI-MARKOVIAN MODEL

3.1 The Shifted Exponential Hidden Markov Model

As mentioned in Section 2 it is difficult to decide between a real auto-similarity of the traffic and some non stationarities. Our intuition is that the hypothesis of local stationarity is as plausible as the hypothesis of long-range dependence. Contrary to many authors who suggest modeling the BISDN traffic as a long-range dependent process we propose to model the measured traffic as a locally stationary and markovian process.

One way of modeling time series that are suspected to be locally stationary consists in using a parametric model whose parameters are jumping from time to time or are drifting with time at a rate that is sufficiently fast for the non stationarities to be perceptible and sufficient slow for parameters tracking to be possible.

We propose to model the observed process as a locally stationary Hidden Markov Chain with conditional laws that are shifted exponential. We call this model the Shifted Exponential Hidden Markov Model (SEHMM).

Denote by $\{O_t\}$ the successive inter-arrival times and denote by $\{Q_t\}$ a finite state Markov Chain whose transition matrix is denoted by P and whose initial distribution is denoted by π. The parameters of the distribution of O_t conditionally to $Q_t = i$ are the shift s_i and the intensity λ_i of the shifted exponential distribution. Denote by $\mathcal{F}_t = \sigma(o_t, o_{t-1}, \cdots)$ and by $\mathcal{G}_t = \sigma(q_t, q_{t-1}, \cdots)$ the filtrations associated to the processes $\{O_t\}$ and $\{Q_t\}$. Then

$$P(Q_t = i \mid \mathcal{F}_{t-1}, \mathcal{G}_{t-1}) = P(Q_t = i \mid q_{t-1})$$
$$\forall A \in \mathcal{B}, \quad P(O_t \in A \mid Q_t = i, \mathcal{F}_{t-1}, \mathcal{G}_{t-1}) = P(O_t \in A \mid Q_t = i)$$
$$= \int_A \mathbb{I}_{[s_i, +\infty[}(u)\lambda_i exp(-\lambda_i(u - s_i))du$$

The SEHMM is a new model proposed by Vaton *et al.*[16] from the analysis of the traffic measured by Paxson and Floyd [4]. This model has many attractive features, among which the existence of simple on-line and off-line algorithms of estimation and control for such models.

One should remark that the SEHMM is a generalization of the model of Kofman *et al.*. According to Kofman *et al.*who analyzed the traffic measured by Jain and Routhier [17] the distribution of the interarrival times is a mixture of exponentials. The SEHMM is also close to the Markov Modulated Poisson

Process (MMPP). In both cases the model is semi-markovian and the marginal distribution of is a mixture of exponential distributions.

The vindication of this new model as well as off-line and on-line procedures of estimation of the parameters of this model are detailed in [16]. Note that the estimation of the shifts s_i is particularly involved. This estimation can not be performed in a maximum likelihood sense since the likelihood has many discontinuity points. The Cramer-Rao variance lower bound that justifies the use of the maximum likelihood estimator is not even defined; the conditions under which this bound is derived are indeed not fulfilled. Vaton and Chonavel [18] propose an algorithm of estimation of several shifts in the case of incomplete data. This algorithm is based on a Fourier transform of the marginal distribution of the process; it exploits the fact that the shift of a distribution is equivalent to a modulation by a complex exponential function of the Fourier transform of this distribution. In what follows the shifts s_i are supposed to be constant and known. This point has been verified on the traffic that we investigate in this contribution.

3.2   A recursive estimation procedure

In our context it is of major interest to derive a recursive algorithm of estimation of the model that we propose since we wish to cope with varying parameters. The recursive estimation of the parameters of a HMM has been studied by several authors (see Elliott [19] for a review). In this contibution we briefly recall the original procedure developed by Mevel [20].

Denote by $\theta(t)$ the value of the parameters at time t. Contrary to the procedure developed by Elliott [19] the procedure developed by Mevel is not based on the EM paradigm but it exploits directly the particular structure of the log-likelihood derivative. Denote by

$$b_i(o_t; \theta(t)) = \mathbb{1}_{[s_i(t),+\infty[}(o_t)\lambda_i(t)exp(-\lambda_i(t)(o_t - s_i(t)))$$

the probability density function of O_t conditionally to $Q_t = i$ and to $\Theta = \theta(t)$ and denote by α_t the one step ahead prediction filter at time t

$$\alpha_t(i) = P(Q_{t+1} = i \mid o_t, o_{t-1}, o_{t-2}).$$

Because of the semi-Markov property the one step ahead prediction filter, its gradient $g_t = \nabla\alpha_t$ and its Hessian $h_t = \nabla^2\alpha_t$ can be computed recursively

$$\alpha_{t+1} = F(\alpha_t, o_{t+1}) \quad \text{and} \quad g_{t+1} = G(\alpha_t, g_t, o_{t+1}) \quad \text{and} \quad h_{t+1} = H(\alpha_t, g_t, h_t, o_{t+1}) \tag{1}$$

The computation of the log-likelihood of $\{o_1, \cdots, o_t\}$ and of its gradient is

$$\log p(o_{1:t}; \theta) = \sum_{\tau=0}^{t-1} \log p(o_\tau \mid o_{1:\tau-1}) = \sum_{\tau=0}^{t-1} \log(\sum_i \alpha_\tau(i)b_i(o_{\tau+1})) \tag{2}$$

which permits a stochastic approximation procedure

$$\theta(t+1) = \theta(t) + \gamma_t \frac{\partial \log p(o_t \mid o_{1:t-1}\theta_t)}{\partial \theta} \tag{3}$$

where γ_t is a sequence of step-size (in the tracking context, we set $\gamma_t = \gamma$). The idea is to exploit the recursive formulae 1 to compute recursively the term of excitation in this stochastic gradient algorithm. Mevel [20] demonstrates that the stationary points of this algorithms are the extrema of the Kullback information and he demonstrates the asymptotic normality of the estimator under the hypothesis of stationarity.

3.3 A visual index of long-range dependence

A time series $\{X_t\}$ is long-range dependent if its autocovariance function $r(j)$ is $r(j) \sim Cj^{2H-2}$ as $j \to +\infty$ where $1/2 < H < 1$. It is well known [21] that if $\{X_t^{(m)}\}$ denotes the aggregated series

$$X_t^{(m)} = \frac{1}{m} \sum_{k=(t-1)m+1}^{tm} X_k$$

the sample variance var$X^{(m)}$ of the aggregated series is var$X^{(m)} \sim \sigma_0^2 m^{2H-2}$ as $m \to +\infty$.

This permits the construction of a visual index of long-range dependence. One plots $\log \text{var} X^{(m)}$ versus $\log m$ for various aggregation levels. If the series is long-range dependent the graphic fits a straight line with a slope $-1 < \beta = 2H - 2 < 0$. The slope of this straight line provides an estimate of the Hurst parameter H.

This visual index of autosimilarity is the main evidence of many contributors to sustain that the traffic measured on modern broadband networks is long-range dependent.

As we suspect that some non stationarities might have mislead to the conclusion of long-range dependence we mean to exhibit the same visual index for a non stationary SEHMM that we simulate. We compare our conclusions with the conclusions of these authors. The parameters of the non stationary SEHMM are matching the varying parameters of the real traffic that we estimate with the stochastic gradient algorithm exposed above.

The logarithm of the sample variance of the aggregated process is plotted versus the logarithm of the aggregation level on Figure 4 for both the real traffic and the traffic that we simulate. The full line is a reference that corresponds to a time series that is not correlated. The graphic resembles a straight line in the case of the locally stationary SEHMM as well as in the case of the real traffic. The estimates of H deduced from the slope of the straight lines are close if one takes into account the bad quality of this estimate.

4   CONCLUSION

In this contribution we have developed different tests of stationarity for mixing processes. Thanks to these tests we have established by intensive simulation that any stationary and semi markovian model is inadequate on long time scales for the traffic measured on modern broadband networks. The hypothesis of stationarity and the hypothesis of mixing that these models postulate are indeed incompatible on these time scales.

We suspect that the apparent auto-similarity revealed by many contributors results at least partly from some non stationarities of the traffic. We have proved by simulation that a locally stationary and markovian model exhibits the same evidence of autosimilarity as the real process. Our findings question the consensus that has become established around the long range dependence of the traffic measured on modern broadband networks. They lead the way to some new realistic and tractable models.

Acknowledgments

This work was supported by France Telecom research center under contract number PE95-7633. This work was accomplished while S. Vaton was an invited student at the Ecole Nationale Supérieure des Télécommunications de Bretagne.

REFERENCES

[1] V. Klemes, "The hurst phenomenon: a puzzle?," *Water Resour. Res.*, vol. 10, pp. 675–678, 1974.

[2] N. Duffield, G. Lewis, N. O'Connel, and F. Toomey, "Statistical issues raised by the bellcore data," Preprint.

[3] V. Teverovski and M. Taqqu, "Testing for long-range dependence in the presence of shifting means or a slowly declining trends, using a variance type estimator," *J. of Time Series Analysis*, vol. 18, no. 3, pp. 279–304, 1997.

[4] V. Paxson and S. Floyd, "Wide-area traffic: The failure of poisson modeling," in *SIGCOMM'94*, London, England, aug 1994, pp. 257–268.

[5] A. Mokkadem, "Entropie des processus linéaires," *Probability and mathematical statistics*, vol. 11, no. 1, pp. 79, 1990.

[6] P. Doukhan, *Mixing: properties and examples*, Lecture Notes in Statistics. Springer-Verlag, 1994.

[7] T.W. Epps, "Testing that a gaussian process is stationary," *The Annals of Statistics*, vol. 16, no. 4, pp. 1667, dec 1988.

[8] P. Billingsley, *Probability and measure*, Wiley series in Probability and Mathematical Statistics. Wiley, 1979.

[9] P.J. Brockwell and R.A. Davies, *Time Series: Theory and Methods; Second Edition*, Springer Series in Statistics. Springer-Verlag, 1991.

[10] A. Mauchly, "Significance test for sphericity of a normal n-variate distribution," *Ann. Math. Stat.*, vol. 11, pp. 204–209, 1940.

[11] A. Mokkadem and K. Drouiche, "A new test for time series," Prepublications d'Orsay.

[12] A. Mokkadem, "On some new tests for time series," *CRAS Paris, Series I*, vol. 318, pp. 755–758, 1994.

[13] K. Drouiche, "Measuring randomness," Preprint. To appear in the Journal of the Royal Society B.

[14] K. Drouiche, "A new test for whiteness," Preprint. To appear in the IEEE Trans. on Signal Processing.

[15] S. Vaton, "A new test of stationarity and its application to teletraffic data," in *ICASSP'98*, Seattle, USA, may 1998.

[16] S. Vaton, E. Moulines, H. Korezlioglu, and D. Kofman, "Statistical identification of lan traffic data," in *ATM'97*, Bradford, England, july 1997, pp. 15/1–10.

[17] R.Jain and S.A. Routhier, "Packet trains: and a new model for computer network traffic," *IEEE-JSAC*, vol. 4, no. 6, pp. 986–995, sep 1986.

[18] S. Vaton and T. Chonavel, "A fourier transform estimation of shifts in the case of incomplete data," in *SSAP'98*, Portland, USA, sep. 1998.

[19] R.J. Elliott, L. Aggoun, and J.B. Moore, *Hidden Markov Models: Estimation and Control*, Applications of Mathematics. Springer-Verlag, 1995.

[20] L. Mevel, *Statistique Asymptotique pour les Modèles de Markov Cachés*, Ph.D. thesis, Université de Rennes 1, 1997.

[21] J. Beran, *Statistics for long memory processes*, Chapmann and Hall, 1994.

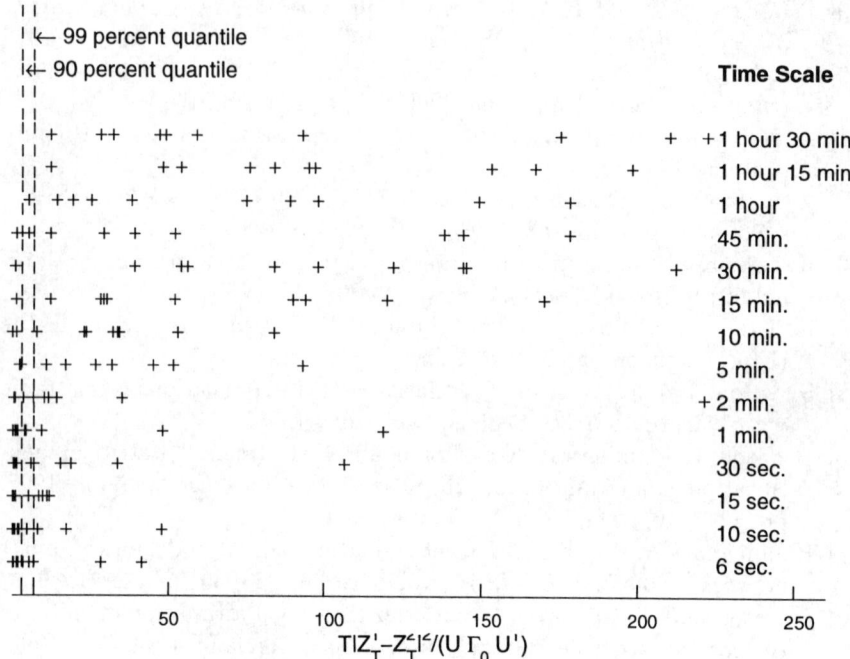

Figure 1 Stationarity of the mean

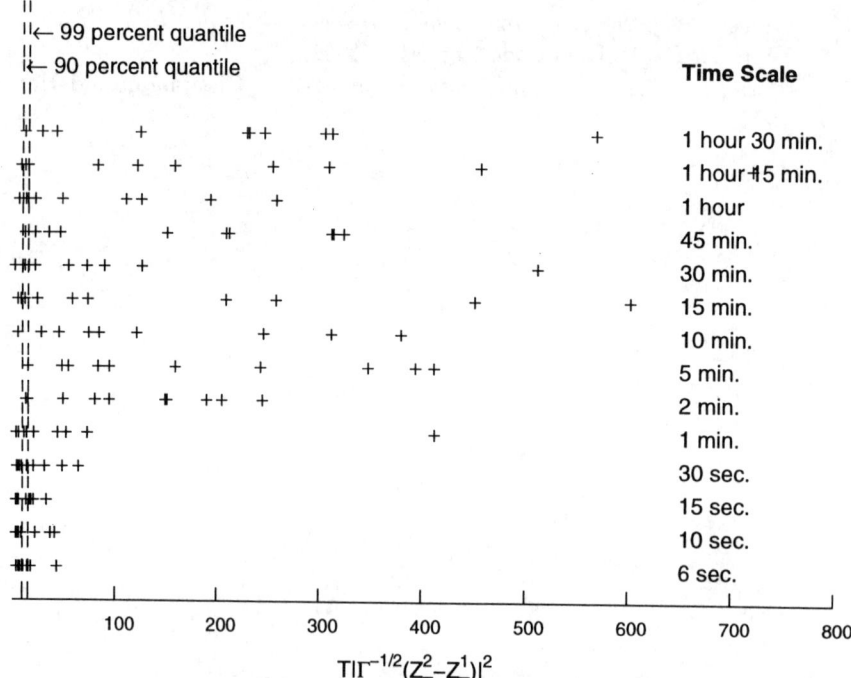

Figure 2 Stationarity of the marginal distribution

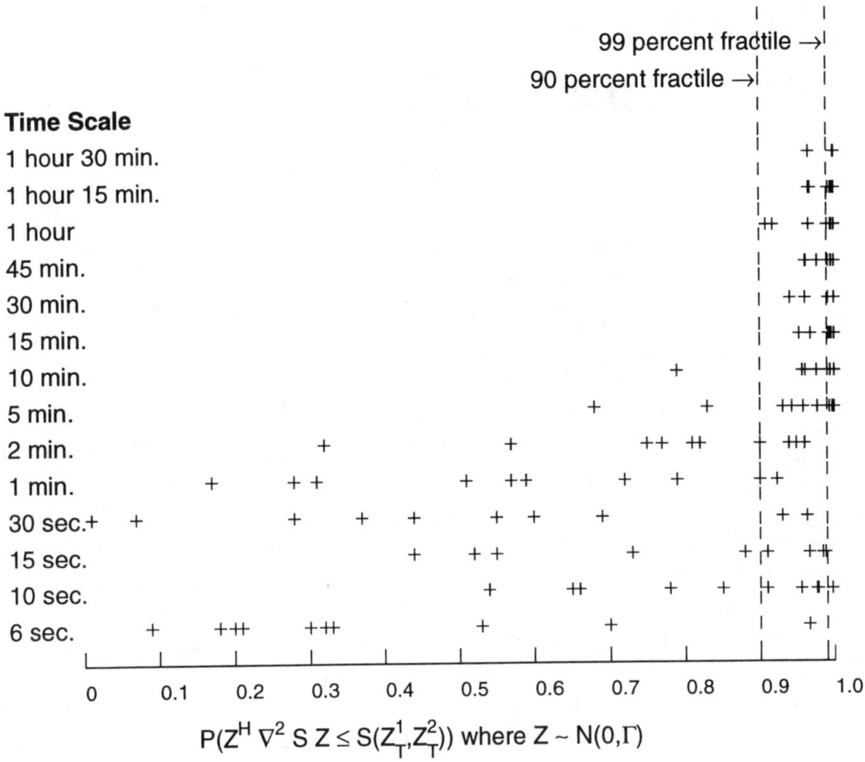

99 percent fractile →
90 percent fractile →
Time Scale
1 hour 30 min.
1 hour 15 min.
1 hour
45 min.
30 min.
15 min.
10 min.
5 min.
2 min.
1 min.
30 sec.
15 sec.
10 sec.
6 sec.
0
0.1
0.2
0.3
0.4
0.5
0.6
0.7
0.8
0.9
1.0
P(Z^H ∇^2 S Z ≤ S(Z_T^1,Z_T^2)) where Z ~ N(0,Γ)

Figure 3 Stationarity of the first five correlations

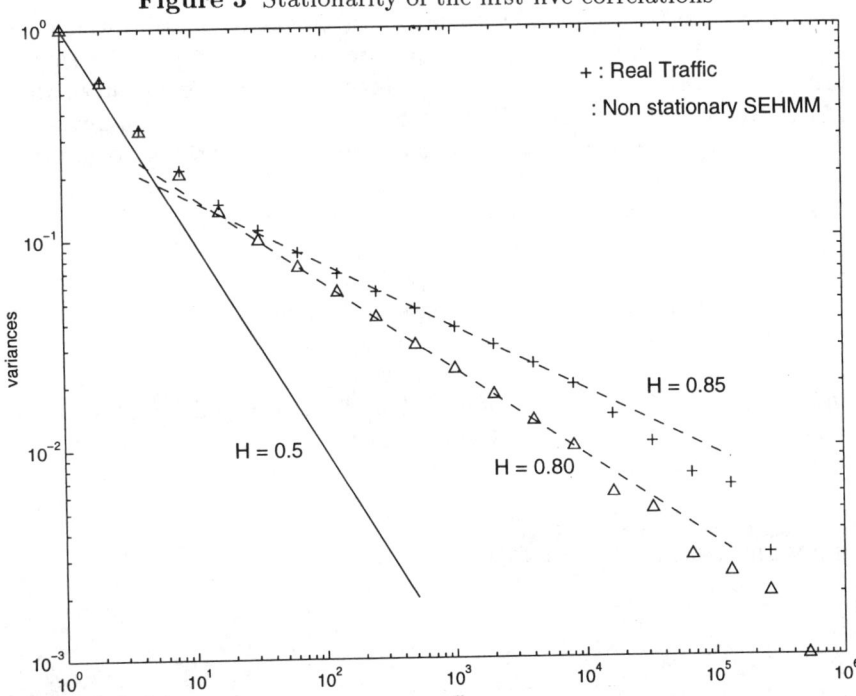

+ : Real Traffic
: Non stationary SEHMM
H = 0.85
H = 0.80
H = 0.5
variances
m
10^0
10^-1
10^-2
10^-3
10^0
10^1
10^2
10^3
10^4
10^5
10^6

Figure 4 Variance of the aggregated process versus the aggregation level

42

Bidimensional fluid model for VBR MPEG video traffic

L. de la Cruz, M. Fernández, J. Alins and J. Mata

Department of Applied Mathematics and Telematics
Polytechnic University of Catalonia
C/ Jordi Girona, 1 i 3. Mòdul C-3, Campus Nord
08034 - Barcelona, SPAIN
E-mail : {ljcruz,marcos,juanjo,jmata}@mat.upc.es
Phone : +34-[3] 401 60 14 - Fax : +34-[3] 401 59 81

Abstract

The broadband bandwidth of ATM networks allows to develop new video services. To reduce the resource allocation in these networks, the video services employ compression techniques. The most common used technique is the MPEG algorithm. In order to maintain the image quality the MPEG coders generate a variable bit rate. Usually, a smoothing system is inserted between the video coder and the user network interface. The purpose of this system is to improve the performance of ATM networks. In this way, the traffic delivered to the network maximize the multiplexing gain.

To evaluate and design ATM networks a characterization of the supported video traffic is needed. In this sense, this paper presents an analysis of the smoothed MPEG video traffic. Likewise, a new procedure to adjust a binomial Markov Modulated Fluid Process (MMFP) is presented. The goodness-fit-tests reveal some deficiencies on the autocovariance and probability density functions on the model. In order to correct these failures a new bidimensional MMFP model is elaborated. This new model is basically the aggregated process of two ON/OFF minisources types. For this point of view, the model can be dimensioned using each minisource type to capture the short and long range dependencies of the video traffic. The results show a well fit of the above mentioned functions.

Keywords
ATM Networks, VBR MPEG video traffic, Markov Modulated Fluid Models, Short and Long Range dependencies

Broadband Communications   P. Kühn & R. Ulrich (Eds.)
© 1998 IFIP. Published by Chapman & Hall

1 INTRODUCTION

Broadband Networks based on the Asynchronous Transfer Mode (ATM) will support, among others, traffic coming from variable bit rate video (VBR) coders, which are capable of maintaining a constant picture quality of the decoded image. The characterization of such VBR video sources becomes important in the analysis and design of Broadband Integrated Services Digital Networks (B-ISDN). The network architecture and its characteristics, such as cell-loss probabilities, transmission delay, statistical multiplexing gain, buffering, etc., are strongly related to the statistical properties of the sources and the coding schemes involved. Therefore, source models are useful to analyze and to dimension the network components (Nikolaidis, 1992)(Mata, 1996)(Mata, 1996).

The MPEG coding algorithm was developed to achieve a high compression ratio with a good picture quality (Le Gall, 1991). MPEG can be used to transmit real-time variable bit rate broadcast video and it is suitable for video-on-demand in ATM networks (Verbiest, 1988)(Pancha, 1994).

A MPEG video sequence of pictures (SoP) is divided into groups of N pictures (GoP). A GoP consists of subgroups of M pictures where the first is a reference picture, intra (I) or predicted (P), and the rest are bidirectionally-predicted (B). The image quality depends on the values M, N and the selected quantizer step size (q). MPEG coder can be set in an open-loop mode to maintain the subjective quality with a fixed q. In this case, the output bit rate is variable. A suitable choice of q, M, N parameters is important to minimize the traffic bit rate for a fixed subjective quality or for a constant signal-to-noise ratio (SNR) (Mata, 1994)(Mata, 1996).

The variations of the bit rate generated in the codification are produced by intrinsic and extrinsic reasons. The intrinsic reasons are related, fundamentally, to the codification modes applied on the frames. Thus, the I frames need a higher number of bits than the frames P or B because the I frames only exploit the spatial redundancy using the DCT transform technique. In addition, the P frames tend to generate greater number of bits than B ones, since only motion compensation is applied respect to the previous reference image. Within the codification of the frames, another factor that give rise to the bit rate variations is the exploitation of the entropy using run-length codes. The extrinsic reasons that produce fluctuations in the bit rate depend on the content of the frames to code. The frames with high grade of detail or complex texture reduce the efficiency of the spatial redundancy exploitation. The high activity scenes with fast camera movements, zooms and plane changes difficult the use of predictive compression techniques. Therefore, the output bit rate suffers an important increment in both cases.

In general, the coders do not deliver directly the traffic to the user interface because, usually, a smoothing system is enabled to reduce the resource allocation on the ATM networks. The smoothing is achieved inserting a small storage buffer between the coder and the user network interface. The storage process introduces an additional delay on the cells delivered to the network. This system allows to maintain the bit rate approximately constant during a time interval. In this way, the variability and the peak rate of the traffic is decreased. Another basic function of the VBR MPEG traffic smoothing is to remove the intrinsic periodic fluctuations produced by the coding algorithm. These fluctuations reduce the performance of multiplexers and switch fabrics. This reduction appears when , at their input ports, there are several video sequences whose coding modes are in phase. Therefore, the main advantage of the VBR MPEG traffic shaping is the improvement achieved on the statistical multiplexing gain. Recent studies propose the best time interval to smooth the VBR MPEG video traffic is the GoP interval (De la Cruz, 1997).

In previous works, several models are proposed to capture the behavior of the traffic generated for the VBR MPEG coders (Maglaris, 1988)(Sen, 1989)(Heyman, 1992)(Grunenfelder, 1991)(Melamed, 1992). These models try to fit the main statistical descriptors of the actual traffic. Usually, the mean, the standard deviation, the peak rate, the autocorrelation function and the probability distribution function are fitted for different temporal series generated at slice, frame or GoP time intervals.

This work is focused on the characterization of the VBR MPEG traffic delivered to the network. To accomplish this study, the GoP traces of a long MPEG video sequence has been employed. This sequence is a recorded concert of the America music band. The sequence is composed by several scene whose contents are different in their textures and activity levels. The MPEG coder parameters are chosen to the values N=6, M=2 and q=9. Therefore, the defined GoP pattern is IBPBPB. The spatial resolution of the frames is 352x288 pixels per frame for luminance and 176x144 for the chroma components, with a frequency of 25 frames per second. The sequence is 41695 frames (6950 GoPs) long, that correspond to approximately 30 minutes of video. The image quality achieved on the coding process is 37.8 dB.

The actual data analysis has allowed to propose two models. In section 2 a binomial Markov Modulated Fluid Process (MMFP) is fitted to synthesize the statistical behavior of the real VBR MPEG traffic. The results show an inaccurate capture of the autocovariance function for high lags. In addition, this model do not fit well the probability density function for high generation levels. Observing these deficiencies, a new bidimensional MMFP is introduced in section 3 to improve the goodness fit. This model presents a perfect adjustment of the short and long terms of the autocovariance function and also fit well the real data histogram. Finally, the conclusions of this work are presented in section 4.

2   BINOMIAL FLUID MODEL FOR VBR MPEG TRAFFIC SOURCES AT GOP LEVEL

In this section, a simple fluid model based on a monodimensional Markov Chain is analyzed (Mata, 1994)(Maglaris, 1988). The objective is to elaborate a new method to adjust the model to capture the statistical behavior of the real video traffic. In addition, the deficiencies of this type of models will be clearly stated.

The Markov Modulated Fluid Processes (MMFP) presented in Figure 1 are characterized by means of an exponential autocovarianze function and a binomial probability distribution function. The study and adjustment of this model is achieved through the analysis of the $M/M/\infty//S$ queuing system. In this process, the state probabilities follow a binomial distribution (Maglaris, 1988).

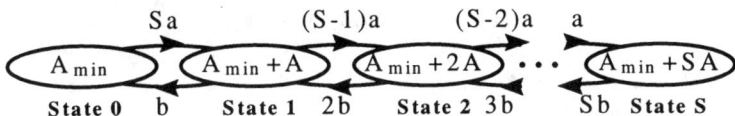

Figure 1.- Binomial MMFP.

Characterizing the process as a fluid model, each state in the Markov Chain represents a constant bit rate. These models are useful when the number of basic units generated during the reference interval is very large. In this cases, the unit generation is not a relevant event and the analysis can be focused on the global volume of transferred information. In our case, the reference interval is as the needed time to transmit a GoP. In this period the number of bits or cells to transmit is very high. Therefore, the performance evaluation of ATM networks supporting VBR MPEG traffic is suitable via fluid models.

The presented binomial MMFP model is composed by S+1 states. Since the real data never reach the null bit rate, in this model a minimum bit rate (A_{min}) is set at state 0. The difference between the generated bit rate for consecutive states has been fixed to a constant value (A). Regarding this difference, the bit rate is discretized in a set of S+1 states. Thus, on state $i \in [0,S]$ the output bit rate is $A_{min}+iA$ bits per GoP. The transition rates are exponentially distributed and their values proportional to the state number. In this way, on the state $i \in [0,S]$ the birth rate is $(S-i)a$ and the death rate is ib.

The binomial MMFP have the property to be decomposed through an aggregation of ON/OFF elemental processes, called minisources. For these minisources, a non null minimum bit rate (A_l) and a maximum bit rate (A_m) have been set to OFF and ON states respectively. The difference between A_l and A_m is the step value A. For these minisources, the exponential birth and death rate are a and b respectively. With this construction, the binomial MMFP is equivalent to the aggregation of the S independent minisources, where: $A_{min} = SA_l$; $A_m = A+A_l$ and $A = A_m - A_l > 0$.

To relate the model parameters S, A_l, A_m, a and b to the real data, it is necessary to derive a new equations set. Let be Z a random variable associated to the number of people in the system $M/M/\infty//S$ and let be X a random variable connected to Z by X = AZ.

Since temporal series of real data do not reach null generation values, an offset value A_{min} must be considered on the synthesized temporal series. This effect is captured by a new binomial random variable Y, linked to X as : $Y = X + A_{min}$. Under these definitions, the Y random variable follows the specified binomial MMFP. Therefore, the Y random variable will have to fit well the statistical behavior of real data. The binomial MMFP behavior is obtained by the addition of the S independent random variables associated to the S minisources. Thus, the mean rate, the variance and the covariance function are:

$$m_Y = S\eta = S(A_m \pi_{ON} + A_1 \pi_{OFF})$$
$$\sigma_Y^2 = S(A_m - A_1)^2 \pi_{ON} \pi_{OFF} = S A^2 \pi_{ON} \pi_{OFF} \qquad (1)$$
$$c_{YY}(\tau) = S(A_m - A_1)^2 \pi_{ON} \pi_{OFF} e^{-(a+b)\tau}$$

where π_{OFF} and π_{ON} are the steady-state probabilities to find the minisource in states OFF and ON respectively.

The procedure explained in (Mata, 1994) to determine the model parameters is based on the random variables Y and X. The former has to match the real process characteristics and the X random variable will be auxiliary used. The statistical properties of both random variables are the same, but their means are SA_1 units different. To fit the model, the sequence "Live in Central Park" has been used. Applying this technique, the resulting parameters are: S=9 minisources; A_1=7619.38 bits/GoP; A=34890.51 bits/GoP; a =0.0338 s^{-1}; b=0.0903 s^{-1}.

To evaluate the goodness fit of the model, the real data and the model autocovariance functions are compared in Figure 2(a). The calculated function for the real series presents, approximately, an exponential decay only for lags lower than a few tens of frames. This interval is well fitted by the model. Nevertheless, this behavior is inaccurate for higher lag values because the real traffic function presents a slower decay in that region. The slow decay of the real data autocovariance function is related to the long range dependence of the video traffic (Garret, 1993).

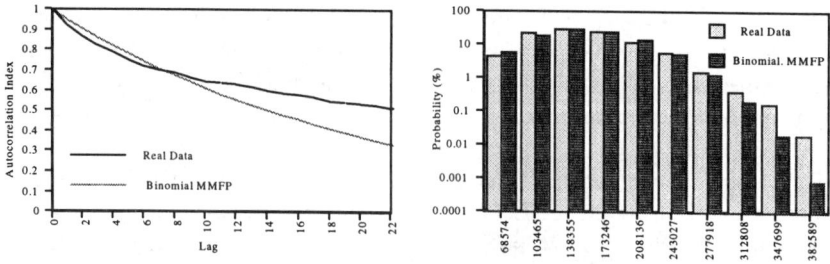

Figure 2.- (a) Data and model autocovariance functions (b) Histograms

In Figure 2(b), the probability density functions are compared using the histogram bars. The shape presented by the real temporal series can be approached by a binomial distribution. The main difference between the model and the real data appears on the highest values. In this border area, the model do not reach the suitable probability values.

3   VBR MPEG VIDEO MODELING USING A BIDIMENSIONAL MMFP

In order to improve the model proposed in the last section a bidimensional MMFP is introduced. The structure of the model is presented in Figure 3. This two-dimensional structure is interpretable as the aggregation of two basic minisources types. Let be S_1 and S_2 the number of minisources of each type respectively.

The discussed bidimensional MMFP model has been denoted $M/M/\infty//S_1+S_2$ model. This model was proposed in (Sen, 1989) The technique presented in this pioneering paper are not applicable in the present work since new characteristics of the VBR video traffic are considered.

To develop the $M/M/\infty//S_1+S_2$ model, the bit rate for the minisources of type 2 has been set to a null value in OFF state, that is $J_1=0$. This implies $J = J_m-J_1 = J_m$.

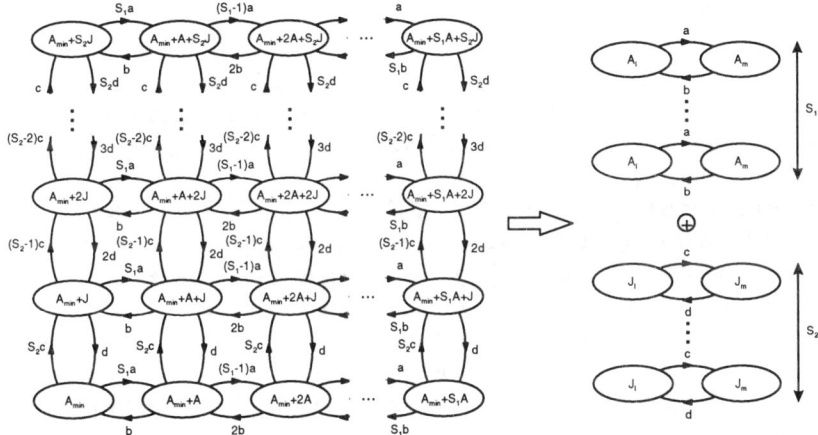

Figure 3.- Bidimensional MMFP model and its break down in two types of minisources

This consideration does not produce loss of generality, since the minimal constant value can be concentrated in a type of minisources or can be distributed between the two types. In both case, the resulting model is the same MMFP. The analysis of the model $M/M/\infty//S1+S2$ can be carried out using the results obtained for the binomial $M/M/\infty//S$ model. To develop the model, two binomial random variables, called $M/M/\infty//S_1$ and $M/M/\infty//S_2$, are associated to the minisources of type 1 and 2 respectively. These random variables are independent mutually. In this way, the behavior of the model $M/M/\infty//S_1+S_2$ is related to the random variable composed by the addition of the corresponding random variables to each set of minisources.

Let be the following random variables:

X = bit rate generated in the $M/M/\infty//S_1+S_2$ model
Y = bit rate generated in the $M/M/\infty//S_1$ model
Z = bit rate generated in the $M/M/\infty//S_2$ model

then:

$$X = Y + Z \qquad \text{and} \qquad m_X = m_Y + m_Z$$

$$\sigma_X^2 = \sigma_Y^2 + \sigma_Z^2 \qquad \text{and} \qquad c_{XX}(\tau) = c_{YY}(\tau) + c_{ZZ}(\tau) \tag{2}$$

Let be p and q the steady-state probability in ON state for each minisource type respectively. Using the previously obtained results for the bit rate generated by the binomial MMFP, the mean, variance and autocovariance can be expressed as:

$$m_X = S_1 pA + S_2 qJ$$
$$\sigma_X^2 = S_1 p(1-p)A^2 + S_2 q(1-q)J^2$$
$$c_{XX}(\tau) = S_1 p(1-p)A^2 e^{-\gamma_1\tau} + S_2 q(1-q)J^2 e^{-\gamma_2\tau} \tag{3}$$

Since the generated values for the coder are associated to the random variable X, the values of γ_1 and γ_2 are deduced by means of the real data autocovariance function. In this way, the relation between the variances of the random variables Y and Z must be determined.

In the adjustment of the binomial $M/M/\infty//S$ model at GoP level a correct capture of the autocovariance function was achieved for the first lags. However, the long range dependence was practically ignored. Therefore, a separate analysis of the short and long terms seems suitable. In this way, each sources types can be synthesized to capture both temporal behaviors. Thus, the sources of type 2 will be employed to fit the long range dependence and the sources of type 1 for the short range of the autocovariance function. This implies that the relation of the correlation coefficients is:

$$\gamma_1 \gg \gamma_2 > 0 \tag{4}$$

Since for lags of more than 100 GoP intervals the influence of the short term is negligible, it can be considered:

$$c_{XX}(\tau) \cong \sigma_Z^2 e^{-\gamma_2\tau} \qquad \tau > 100 \tag{5}$$

The least square estimation technique has been applied to determine the value of the autocovariance function for lags between 100 and 250 GOP intervals. In Figure 4(a) the resulting adjustment is shown for the value $\gamma_2 = 0.02541$ rad/seg.

In order to obtain the short term dependence of the autocovariance function, it is necessary to extract the long term dependence of the real data autocovariance function. According to the expression (5), the variance of the Z random variable can be approximate:

$$\sigma_Z^2 \cong \left. \frac{c_{XX}(\tau)}{e^{-\gamma_2 \tau}} \right|_{\tau=100} \tag{6}$$

and, therefore:

$$c_{YY}(\tau) \cong c_{XX}(\tau) - \sigma_Z^2 e^{-\gamma_2 \tau} . \tag{7}$$

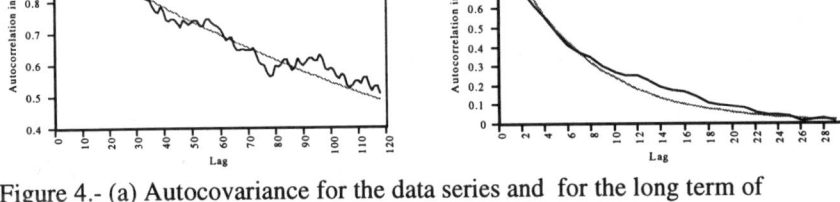

Figure 4.- (a) Autocovariance for the data series and for the long term of $M/M/\infty//S_1+S_2$ model (b) Autocovariance for the data series and for the short term of $M/M/\infty//S_1+S_2$ model

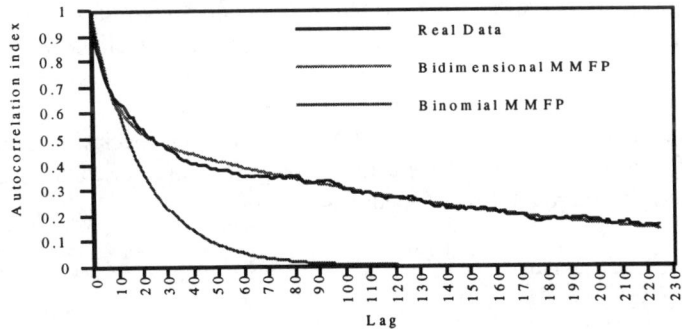

Figure 5.- Autocovariance of the actual data series and the binomial and bidimensional MMFP

Applying the same estimation technique as the long range term, the obtained autocovariance index is $\gamma_1 = 0.5929$ rad/seg, with $\sigma_z^{\,2} = 1.217 \cdot 10^9$.

These values allows to verify the previous hypothesis $\gamma_1 \gg \gamma_2 > 0$. The goodness-fit-test is showed in figure 4(b). Finally, the variance of the Y random variable is derived as:

$$\sigma_Y^2 = \sigma_X^2 - \sigma_Z^2 = 9.63 \ 10^8.$$

(8)

The results of the discussed break down technique are presented in Figure 5. In this figure, the accurate capture of the temporal data is observed for the different time scales. Likewise, the unfitted behavior of the binomial MMFP is shown.

The analysis of real traffic allows to state the set of equations to find the model parameters. This equations determine a finite number of solutions. In this case, using the measured values of the actual time series:

$$\min(X) = 67527 \qquad ; \quad \max(X) = 382589$$
$$m_X = E[X] = 154072.07 \quad ; \quad \sigma_X^2 = 2.17 \ 10^9$$

(9)

and the derived variances using the break down of the autocovariance function:

$$\sigma_Y^2 = 9.63 \ 10^8 \quad ; \quad \sigma_Z^2 = 1.217 \ 10^9$$

(10)

the MMFP bidimensional model can be fitted.

The synthesized MMFP model attempts to capture the smoothing traffic properties of the VBR MPEG video sources. In this synthesis, the probability to surpassed the peak rate of the real traffic has been considered very low. Thus, the resulting model can be modified to reach a higher peak rate increasing the number of minisources S_1. Another possibility is to fix on the equations set the maximum rate (A_{max}) to a higher value than the real data maximum.

In this case, the solved equations force the maximum generation state of the model to the maximum value of the actual series. Therefore, the model parameters can be determined through the set of equations:

$$J = \frac{A_{max} - S_1 A - A_{min}}{S_2}$$
$$S_1 p A + S_2 q J = m - A_{min}$$
$$S_1 p (1 - p) A^2 = \sigma_y^2$$
$$S_2 q (1 - q) J^2 = \sigma_z^2$$

(11)

Note that the system is undetermined. In this case, all the solutions have been found for the different integer values that verify $S_1>0$ and $S_2>0$. Using numerical methods, the system only presents solution for values which satisfy $S_1+S_2\leq13$. The obtained values from p and q (with p>q) are very insensitive to the increment of S_2. Hence, the increment of the number of minisources of type 2 produce a set of states with a extraordinarily low probability of visit when S_2 is greater than 2. To synthesize a non overconfigured model only the cases $S_2=1$ and $S_2=2$ has been considered.

Analyzing the solutions for $S_2=1$, the results are unsuitable. In Figure 6(a) the relative minimum of the probability distribution function of the bidimensional MMFP model is observed for $S_2=1$ and $S_1=7$. This minimum reflects the strong transition between the maximum bit rate when the minisource of type 2 is inactive and the minimum bit rate when this minisource is in ON state. This model underestimates the probability of the intermediate generation bit rates. Therefore, the appropriated value of S_2 is 2 to capture the real traffic characteristics.

Table 1 Minisource parameters of the bidimensional MMFP

Model	$A_1 (bits/GoP)$	$A(bits/GoP)$	$a (sec^{-1})$	$b (sec^{-1})$
S1=5	13710	27757	0.3	0.2929
S2=2	0	87614	0.0022	0.02321
S1=6	11500	25468.6	0.2663	0.3266
S2=2	0	80601.47	0.00266	0.002275

The next step in the model synthesis is to establish the number of type 1 minisources. In order to avoid an overconfigured model, the probability of a lot number of type 1 minisources are in the active state can not be very low. Analyzing the results, the values of S_1 higher than 6 present a very reduced sojourn probabilities to higher states. In order to reduce the effects of the discretized bit rate generated by the model, the highest significant value of S_1 must be set. In this way, the suitable choices are $S_1=5$ or $S_1=6$.

For the accurate adjustment of the MMFP model, the study of the minimum value generated by the model is necessary. In order to fit well the probability density function of the real data a small correction in this bit rate can be applied.

After this consideration, the obtained result is presented in table 1 and the goodness-fit-test of the probability density function is shown in figure 6(b).

4   CONCLUSIONS

This paper is focused on the characterization of the video traffic generated by a VBR MPEG coder at GoP level. Regarding a smoothing system allocated between the MPEG coder and the user network interface, a long real video traffic is analyzed at this level. To capture the statistical behavior of the actual data two MMFP models are proposed. This type of model has been selected by its analytical tractability on ATM networks.

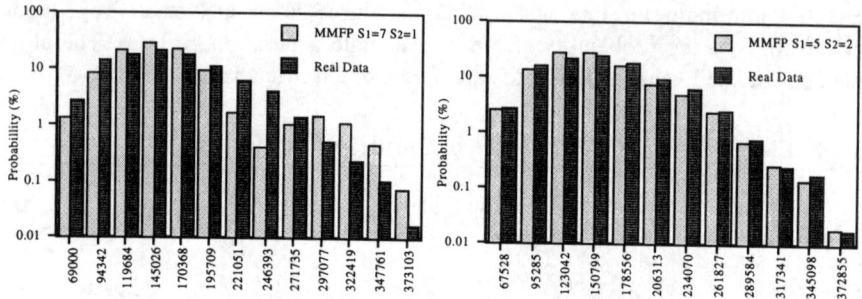

Figure 6.- Histogram of the real traffic and the bidimensional MMFP for (a) S1=7, S2=1 and (b) S1=5, S2=2

The first model considered is the binomial MMFP. In order to adjust this model a new technique is introduced. The goodness-fit-tests applied to validate the model has been the autocovariance and the probability density functions. The results exhibit some deficiencies on both functions. On the autocovariance function a faster decay is noticed and on the probability density function the high activity levels of the model do not reach the actual values.

Starting from these effects a new bidimensional MMFP is introduced. This fluid model can be obtained through the aggregation of two types of independent minisources. This composition give rise to a two-dimensional structure of the MMFP model. The model synthesis has been developed by means of the autocovariance function. This function can be decomposed in a short term and a long term of decay. The long term shows the long range dependence or self-similarity of the video traffic. Assigning each type of minisources to both decay terms the model fits well the autocovariance function. A detailed study of the number of minisources of each type allows to adjust the probability density function too. In this way, this work demonstrate the heterogeneous MMFP are suitable to characterize the VBR MPEG traffic.

5   REFERENCES

De la Cruz, L., Alins, J. and Mata, J. (1997) Prediction Techniques for VBR MPEG Traffic Shaping. *Proceedings of IEEE GLOBECOM'97*, **3**, 1434,39.

Garret, M.W. (1993) Contributions Toward Real-Time Services on Packet-Switched Networks. Ph.D. Dissertation CU/CTR/TR 340-93-20, Columbia University, New York.

Grunenfelder, Cosmas, Manthrope and Odinma-Okafor (1991) Characterization of Video Codecs as Autoregressive Moving Average Processes and Related Queueing System Performance. *IEEE J. on Selected Areas in Communications,* **9**, no. 3, 284-92.

Heyman, D.P., Tabatabai, A. and Lakshman, T.V. (1992) Statistical Analysis and Simulation Study of Video Teleconference Traffic in ATM Networks. *IEEE Trans. On Circuits and Video Techn.,* **2**, no. 1, 49-59.

Le Gall, D. (1991) MPEG: A Video Compression Standard for Multimedia Applications. *Communications of the ACM,* **34(4)**, 305-13.

Maglaris, B., Anastassiou, D., Sen, P., Karlsson, G. and Robbins, J.D. (1988) Performance Models of Statistical Multiplexing in Packet Video Communications. *IEEE Trans. On Commun.,* COM-36, no. 7, 834-43.

Mata, J., Sallent, S., Balsells, J., Zamora, J. and Van der Kolk, A. (1994) Statistical Models for MPEG Video Standard. *Proceedings of IEE EUSIPCO'94,* **1**, 624-7.

Mata, J. and Sallent, S. (1996) Source Traffic Descriptor for VBR MPEG in ATM Netwoks, in *Multimedia Communications and Video Coding* (ed. Plennum Press).

Mata, J., Pagán, G. and Sallent, S. (1996) Multiplexing and Resource Allocation of VBR MPEG Traffic on ATM Netwoks. *Proceedings of the IEEE ICC'96,* **3**, 1401–5.

Melamed et al. (1992) TES-Based Traffic Modelling For Performance Evaluation Of Integrated Networks. *Proceedings of the IEEE INFOCOM '92,* 75-84.

Nikolaidis, I. and Akyildiz, I.F. (1992) Source Characterization and Statistical Multiplexing in ATM Networks. *Paper of the College of Computing.* Georgia Institute of Technology, Atlanta, USA.

Pancha,P., and El Zarki, M. (1994) MPEG Coding For Variable Bit Rate Video Transmission. *IEEE Communications Magazine,* **32**, no.5, 54-66.

Sen, P., Maglaris, B., Rikli, N., Anastassiou, D. (1989) Models for Packet Switching of Variable Bit Rate Video Sources. *IEEE J. on Selected Areas in Commun.,* SAC-7, 865-9.

Verbiest, W., Pinnoo, L. and Vosten, B. (1988) The Impact of the ATD Concept on Video Coding. *IEEE J. on Selected Areas in Communications,* SAC-6, **9**, 1623-32.

6   ACKNOWLEDGEMENTS

This work has been developed using the Global ATM Standard Simulator (GLASS) and supported by the SIGLA (GLASS) Project [Spain CICYT, TEL 96-1452].

43

Superposition of Markov Sources and Long Range Dependence

F. Geerts and C. Blondia
Performance Analysis and
Telecommunication Systems Research Group
Dept. Mathematics and Computer Science
University of Antwerp
Universiteitsplein, 1, B-2610 Antwerp– Belgium
`fgeerts@uia.ua.ac.be-blondia@uia.ua.ac.be`

Abstract

This paper introduces a model to study the phenomenon of long range dependence. This model consists of an infinite superposition of independent Markovian ON/OFF–sources. A condition for assuring long range dependence is given and the Hurst parameter together with the correlation decay is derived for a specific example. We also give a physical interpretation of the existing long range dependence by means of the Ising model.

Keywords

ATM, Ising Model, Long Range Dependence, Markov Sources, Phase Transitions

1 INTRODUCTION

Recent measurements on Ethernet traffic, e.g. (Leland et al. 1993*a*, Leland et al. 1993*b*, Leland et al. 1994) show that its profile exhibits Long Range Dependent (LRD) characteristics. Also for variable bit rate video traffic a similar behaviour has been observed (Beran et al. 1995). LRD means that correlations extend to an infinite time scale and the correlation decay follows a power law. Traditional finite state Markovian traffic models, such as Markov Modulated Poission Processes (MMPP), Markovian Arrival Processes (MAP), etc..., have an exponential correlation decay and can therefore not adequately model this type of ATM data traffic.

These observations have triggered new research activities on models which are able to capture LRD characteristics. Several approaches have been proposed in literature. The theories of Fractional Brownian Motion (Norros 1994), chaotic maps (Pruthi 1995) and regularly varying functions (Boxma 1996) have have been successfuly applied to study LRD properties. An alternative approach consists of modeling LRD over a chosen time scale by using a Markovian approximation (Andersen et al. 1995). Recently, several authors (Daniels & Blondia 1997, Likhanov et al. 1995, etc...) use the superposition of an infinite number of ON/OFF–sources to characterize LRD

Broadband Communications   P. Kühn & R. Ulrich (Eds.)
© 1998 IFIP. Published by Chapman & Hall

traffic.

This paper follows the last approach. We consider a class of processes consisting of the superposition of an infinite number of ON/OFF–sources. Through the characterisation of the sum of the covariances, it is possible to establish a simple explicit necessary and sufficient condition for the process to be LRD. This condition expresses an asymptotical eigenvalue degeneracy of the transition matrix. The simplicity of the model allows to derive explicit formulas for the powerlaw correlation decay and the Hurst parameter. It is also shown that under LRD conditions the mean queue length is infinite.

In physics, long range dependence occurs in turbulence, quantum field theory, $1/f$ noises, and critical phenomena. For instance, in the two–dimensional Ising model, a critical phenomenon appears. The Ising model is a model for ferromagnetism, and describes phase transitions. A phase transition is e.g. a transition from non–aligned spins to aligned spins (magnetism). This transition from an unordered system to an ordered one, happens at the critical point (e.g. a certain temperature). At this point, the correlation decay between spin regions goes from an exponential decay to a powerlaw decay. The eigenvalue degeneracy of the transfer matrices (see later) is also here a necessary condition for the existence of long range dependence. It appears that our model can be incorporated in the Ising model.

More research is needed to study the possibility of applying known techniques from statistical physics to our model, and the question of a physical interpretation of the queueing process in this context remains open.

This paper is structured as follows. In the next section, the single ON/OFF–source is described and the eigenvalue structure of the transition matrix is related to the correlation structure. This is then generalized to a finite superposition of sources of this type. At the end of this section we introduce our model. In Section 3 we prove a necessary and sufficient condition for the long range dependence of this model, and two examples are given. As a direct consequence, we prove in Section 4 the infiniteness of the mean queue length, and discuss some problems concerning the queueing behaviour. In Section 5, the exponent of the power decay of the correlation is derived, and an explicit formula is given for the Hurst parameter. We introduce a physical counterpart, namely the Ising model, of our model in Section 6 and we observe the relationship between a phase transition and long range dependence. Conclusions are drawn in Section 8.

2   CORRELATION AND EIGENVALUE STRUCTURE

An ON/OFF–source is defined to be a two–state discrete time Markov chain. In the OFF state the process generates 0 cells/slot, and in the ON state the process generates 1 cell/slot. The duration of the ON state is geometrically distributed with parameter β. Similarly the duration of the OFF state is geometrically distributed with parameter α. Let $X = \{X_i\}_{i \in \mathbb{N}}$ be the two–state discrete time Markov chain with irreducible and aperiodic transition matrix P, $P = \begin{pmatrix} \alpha & 1 - \alpha \\ 1 - \beta & \beta \end{pmatrix}$. The stationary probability vector is given by $\pi = (\pi_0, \pi_1) = \left(\frac{1-\beta}{2-\alpha-\beta}, \frac{1-\alpha}{2-\alpha-\beta}\right)$, and the arrival rate of this process is $\lambda = \frac{1-\alpha}{2-\alpha-\beta}$. The covariance $\gamma(k) = E[X_i, X_{i+k}] - E[X_i]E[X_{i+k}]$ is given by $\gamma(k) = \gamma^k \pi_0 \pi_1 = \gamma^k \frac{(1-\alpha)(1-\beta)}{(2-\alpha-\beta)^2}$, where $\gamma = \alpha + \beta - 1$ is the second largest eigenvalue of P. As P is stochastic and irreducible, it follows that $\gamma < 1$, and hence $\lim_{k\to\infty} \gamma(k) = 0$.

Let $X^{(\ell)}$ be N such Markovian ON/OFF–source with transition matrix $P^{(\ell)}$ and arrival rate $\lambda^{(\ell)}$, $\ell = 1, 2, \ldots, N$. We consider the superposition of these N sources, $Y_i^{(N)} = \sum_{\ell=1}^{N} X_i^{(\ell)}$. Since each ON/OFF–source can be viewed as a D-MAP, and the superposition of a finite number of D-MAP's is a D-BMAP (Blondia 1992, Blondia & Geerts 1997), we conclude that the corresponding transition matrix P_N, is given by the Kronecker product $P_N = \bigotimes_{\ell=1}^{N} P^{(\ell)}$. Because the ON/OFF–sources are independent, the covariance $\gamma_N(k)$ of the superposed sources equals $\gamma_N(k) = \sum_{\ell=1}^{N} \left(\gamma^{(\ell)}\right)^k \frac{(1-\alpha_\ell)(1-\beta_\ell)}{(2-\alpha_\ell-\beta_\ell)^2}$. The queueing model used in what follows is the D-BMAP/D/1–queue. The service time is assumed to be constant and chosen as time unit. The average number of cells arriving in the queue is given by $\lambda^{(N)} = \sum_{\ell=1}^{N} \pi_1^{(\ell)}$. We assume that $\lambda^{(N)} < 1$ to ensure the existence of a stochastic equilibrium for the queueing system.

Let us now consider an infinite superposition of ON/OFF–sources. Denote $Y_i^{(\infty)} = \sum_{\ell=1}^{\infty} X_i^{(\ell)}$, $P_\infty = \lim_{N\to\infty} P_N$, $\gamma_\infty(k) = \lim_{N\to\infty} \gamma_N(k)$, and the arrival rate $\lambda^{(\infty)} = \sum_{\ell=1}^{\infty} \lambda^{(\ell)}$. We shall derive some properties of $Y^{(\infty)} = \{Y_i^{(\infty)}\}_{i\in\mathbb{N}}$ in the next sections.

3   LONG RANGE DEPENDENCE PROPERTIES

The sequence $Y_1^{(\infty)}, Y_2^{(\infty)}, \ldots$ of stationary random variables is called long range dependent if $\sum_{k=1}^{\infty} \text{Cov}\left(Y_1^{(\infty)}, Y_k^{(\infty)}\right) = \infty$ (Beran 1994, Roberts et al. 1996). For our model we have to ensure that $\sum_{k=1}^{\infty} \gamma_\infty(k) = \infty$.

Proposition 1 *A superposition of an infinite number of Markovian ON/OFF–sources $Y^{(\infty)}$ is long range dependent if and only if*

$$\sum_{\ell=1}^{\infty} \frac{1 - \alpha_\ell}{(1 - \beta_\ell)^2} = \infty, \tag{1}$$

where α_ℓ and β_ℓ are the elements of $P^{(\ell)}$.

Proof. We must proof that the series $\sum_{\ell=1}^{\infty} \sum_{k=1}^{\infty} \left(\gamma^{(\ell)}\right)^k \frac{(1-\alpha_\ell)(1-\beta_\ell)}{(2-\alpha_\ell-\beta_\ell)^2}$ diverges. We shall prove that this series has the same divergent behaviour as the series of condition (1). For this we need two observations

Firstly the stability condition $\lambda^{(\infty)} < 1$ implies that $\lim_{\ell\to\infty} \frac{1-\alpha_\ell}{2-\alpha_\ell-\beta_\ell} = 0$. This means that there exist an M' such that for $\ell \gg M'$, $1 - \epsilon < 1 - \frac{1-\alpha_\ell}{2-\alpha_\ell-\beta_\ell} < 1$, for a fixed ϵ.

Secondly the inequality $\sum_{\ell=1}^{\infty} \frac{\gamma^{(\ell)}}{1-\gamma^{(\ell)}} \frac{(1-\alpha_\ell)(1-\beta_\ell)}{(2-\alpha_\ell-\beta_\ell)^2} < \max_\ell \frac{\gamma^{(\ell)}}{1-\gamma^{(\ell)}}$ implies that long range dependency exists only if $\sup_\ell \gamma^{(\ell)} = 1$. If $\gamma^{(m)} = 1$ for a finite $m < \infty$, then the model consists of two seperate and identical submodels. Hence, we assume that $\lim_{\ell \to \infty} \gamma^{(\ell)} = 1$. This implies that $\exists M''$ such that for $\ell \gg M''$, $1 - \epsilon < \gamma^{(\ell)} < 1$ for the same ϵ. We let $M = \max\{M', M''\}$. For $\ell \gg M$, we have $(1-\epsilon)^2 \frac{(1-\alpha_\ell)}{(2-\alpha_\ell-\beta_\ell)(1-\gamma^{(\ell)})} < \frac{\gamma^{(\ell)}}{1-\gamma^{(\ell)}} \frac{(1-\alpha_\ell)(1-\beta_\ell)}{(2-\alpha_\ell-\beta_\ell)^2} < \frac{(1-\alpha_\ell)}{(2-\alpha_\ell-\beta_\ell)(1-\gamma^{(\ell)})}$. After rewriting these expressions in function of $\frac{1-\alpha_\ell}{1-\beta_\ell}$, and due to the fact that, $\lim_{\ell \to \infty} \frac{\frac{1-\alpha_\ell}{1-\beta_\ell}}{\frac{1-\alpha_\ell}{1-\beta_\ell}+1} = 0 \Leftrightarrow \lim_{\ell \to \infty} \frac{1-\alpha_\ell}{1-\beta_\ell} = 0$, the following bounds are obtained, $(1-\epsilon)^2 \frac{1}{2} \frac{1-\alpha_\ell}{(1-\beta_\ell)^2} \leq \frac{1-\alpha_\ell}{(1-\beta_\ell)^2(\frac{1-\alpha_\ell}{1-\beta_\ell}+1)^2} \leq \frac{1-\alpha_\ell}{(1-\beta_\ell)^2}$, for $\ell \gg M$. This shows that the two series have the same divergent behaviour, and concludes the proof. ∎

From this Proposition it follows that for appropriate choices of α_ℓ and β_ℓ, the resulting superposition of heterogeneous Markovian ON/OFF–sources is LRD.

We now give two examples.

Example 1 (Daniels & Blondia 1997)

Let $P_\ell = \begin{pmatrix} 1 - (1/a)^\ell & (1/a)^\ell \\ (b/a)^\ell & 1 - (b/a)^\ell \end{pmatrix}$, with $1 < b < a$. It is clear that $\gamma^{(\ell)} = 1 - (1/a)^\ell - (b/a)^\ell$ converges to 1 as ℓ tends to infinity. We see that the model P_∞ is LRD iff the series $\sum_{\ell=1}^{\infty} \left(\frac{a}{b^2}\right)^\ell$ diverges, or iff $b^2 \leq a$.

Example 2

Let $P_\ell = \begin{pmatrix} 1 - 1/\ell^p & 1/\ell^p \\ 1/\ell^q & 1 - 1/\ell^q \end{pmatrix}$, for $\ell = 2, 3, \ldots$. To ensure $\lambda < 1$ we need $p > q + 2$. It is clear that $\gamma^{(\ell)} = 1 - 1/\ell^p - 1/\ell^q$ converges to 1 as ℓ tends to infinity. We see that the model P_∞ is LRD iff the series $\sum_{\ell=2}^{\infty} \left(\frac{1}{\ell^{p-2q}}\right)$ diverges, or iff $p \leq 2q + 1$.

4 QUEUEING BEHAVIOUR

The mean queue length $\bar{L}$ of an infinite superposition of Markovian ON/OFF–sources is given by (Neuts 1989, Chapter 6)

$$\bar{L} = \lambda + \frac{1}{1-\lambda} \sum_{\ell=1}^{\infty} \sum_{k>\ell} \pi_1^{(\ell)} \pi_1^{(k)} \left(1 + \frac{\gamma^{(\ell)}}{1-\gamma^{(\ell)}} + \frac{\gamma^{(k)}}{1-\gamma^{(k)}}\right). \tag{2}$$

The following Proposition is a direct consequence of Proposition 1.

Proposition 2 *The mean queue length $\bar{L}$ of an infinite superposition of Markovian ON/OFF–sources is infinite if and only if it is long range dependent.*

Pr oof. First assume that the arrival process $Y^{(\infty)}$ is LRD and consider the following term of (2), $\sum_{\ell=1}^{\infty} \sum_{k>\ell} \pi_1^{(\ell)} \pi_1^{(k)} \frac{\gamma^{(k)}}{1-\gamma^{(k)}}$. Interchanging the summation indices and using similar bounds as in Proposition 1, it is clear that $\sum_{\ell=2}^{\infty} \frac{1-\alpha_\ell}{(1-\beta_\ell)^2} \pi_1^{(1)} \leq \sum_{\ell=2}^{\infty} \sum_{k=1}^{\ell} \pi_1^{(k)} \pi_1^{(\ell)} \frac{\gamma^{(\ell)}}{1-\gamma^{(\ell)}} \leq \lambda \sum_{\ell=2}^{\infty} \frac{1-\alpha_\ell}{(1-\beta_\ell)^2}$, and hence by Proposition 1, $\bar{L} = \infty$. Now, let $\bar{L} = \infty$. Because the term in expression (2) $\sum_{\ell=1}^{\infty} \sum_{k>i} \pi_1^{(\ell)} \pi_1^{(k)} \frac{\gamma^{(\ell)}}{1-\gamma^{(\ell)}} \leq \lambda \sum_{\ell=1}^{\infty} \frac{1-\alpha_\ell}{(1-\beta_\ell)^2}$, is bounded, it immediately follows that $\sum_{\ell=1}^{\infty} \sum_{k>i} \pi_1^{(\ell)} \pi_1^{(k)} \frac{\gamma^{(\ell)}}{1-\gamma^{(\ell)}} = \infty$, implying long range dependence of $Y^{(\infty)}$. This proves the Proposition. ∎

More interesting properties of the queue distribution, like e.g. the tail of the queue length distribution, have not yet been derived for our model. The eigenvalue degeneracy of P_∞ induces severe difficulties. In the absence of this degeneracy, one can rely on the dominant pole approximation(Laevens & Bruneel 1997). In this case there is an unique isolated dominant pole which governs the asymptotic behaviour of the queue (Abate et al. 1994, Falkenberg 1994, Mieghem 1996). The eigenvalue degeneracy transforms the isolated pole into an accumulation point in the complex plane. As a consequence there is not a single dominating pole, but an infinite number of poles which have to be taken into account.

Using large deviation techniques, Buffet and Duffield (Buffet & Duffield 1992, Dembo & Zeitouni 1993, Duffield 1992, Duffield 1993) derived a bound for the loss probability. This method does not seem to be applicable to our model.

Boxma (Boxma 1996) has shown that in a special case, $P(\tau_A = m) \sim m^{-\beta} \Rightarrow Pr(U > m) \sim m^{-(\beta-2)}$ holds, using the Fluid Flow approach. We currently believe that this is also true for our model, but have been unable to establish this result (Laevens & Bruneel 1997) for the superposition of Markovian ON/OFF–sources.

5   THE CORRELATION DECAY, HURST PARAMETER AND THE INDEX OF DISPERSION FOR COUNTS

In this section we give a more detailed study of example 2. The method follows a similar reasoning as for example 1 (Daniels & Blondia 1997).

5.1   Correlation Decay and Hurst Parameter

Proposition 3 *The correlation decay of the arrival process $Y^{(\infty)}$ of example 2 is given by*

$$Cov(Y_1^{(\infty)}, Y_k^{(\infty)}) \sim k^{\frac{q-p+1}{q}}, \tag{3}$$

for large k.

PROOF. We need to find the decay of the series $\sum_{i=2}^{\infty}(1 - \frac{1}{i^p} - \frac{1}{i^q})^k \frac{i^{(q+p)}}{(i^p+i^q)^2}$. Observe that the second factor can be bounded by $\frac{1}{3}\frac{i^{q+p}}{i^{2p}} \leq \frac{i^{q+p}}{(i^p+i^q)^2} \leq \frac{i^{q+p}}{i^{2p}}$. A second simplification of (3) is is done by replacing $(1 - \frac{1}{i^p} - \frac{1}{i^q})^k$ by $(1 - \frac{1}{i^q})^k$, i.e.

$$\sum_{i=2}^{\infty}(1 - \frac{1}{i^p} - \frac{1}{i^q})^k i^{(q-p)} \leq \sum_{i=2}^{\infty}(1 - \frac{1}{i^q})^k i^{(q-p)}.$$

Because $(1 - \frac{1}{i^q})^k i^{(q-p)}$ is a nonnegative descending function, we use a continuous variable x instead of i and we apply Cauchy's integral test,

$$\int_2^{\infty}(1 - \frac{1}{x^q})^k x^{(q-p)}dx \leq \sum_{i=2}^{\infty}(1 - \frac{1}{i^q})^k i^{(q-p)}$$

$$\leq (1 - \frac{1}{2^q})^k 2^{(q-p)} + \int_2^{\infty}(1 - \frac{1}{x^q})^k x^{(q-p)}dx.$$

To evaluate this integral, we use the inequalities $(1 - \frac{1}{x^q})^k \leq e^{-\frac{k}{x^q}} = \frac{1}{e^{\frac{k}{x^q}}} \leq \frac{x^q}{k}$. Furthermore, it is clear that $1 - k\frac{1}{x^q} \leq (1 - \frac{1}{x^q})^k \leq 1 - k\frac{1}{x^q} + \frac{k^2}{2}\left(\frac{1}{x^q}\right)^2$. We want that $1 - k\frac{1}{x^q} > 0$, so it is sufficient that $x > k^{\frac{1}{q}}$. We can now give an upper bound,

$$\int_2^{\infty}(1 - \frac{1}{x^q})^k x^{(q-p)}dx \leq \int_2^{k^{1/q}} \frac{x^{2q-p}}{k}dx$$

$$+ \int_{k^{1/q}}^{\infty}\left(1 - k\frac{1}{x^q} + \frac{k^2}{2}\left(\frac{1}{x^q}\right)^2\right) x^{q-p}dx.$$

The right hand side is clearly proportional to $k^{\frac{q-p+1}{q}}$.
Using $1 - k\frac{1}{x^q}$ as under bound and after some similar calculations we find the bounds

$$C_1 k^{\frac{q-p+1}{q}} \leq \sum_{i=2}^{\infty}(1 - \frac{1}{i^q})^k i^{(q-p)} \leq C_2 k^{\frac{q-p+1}{q}},$$

where C_1 and C_2 are some constants. We now have to assure that these bounds are also valid for the original sum. For this it is neccesary to bound

$$\sum_{i=2}^{\infty}\left((1 - \frac{1}{i^q})^k - (1 - \frac{1}{i^p} - \frac{1}{i^q})^k\right) i^{(q-p)}. \tag{4}$$

We have that $(1 - \frac{1}{i^q})^k - (1 - \frac{1}{i^p} - \frac{1}{i^q})^k \leq k\left(1 - \frac{1}{i^q}\right)^{k-1}\frac{1}{i^p}$. Using similar techniques as above, it is possible to show that the difference (4) is bounded by $Ck^{\frac{q-p+1}{q}-\delta}$, with

C a constant and $\delta > 0$. The resulting powerlaw decay is $k^{\frac{q-p+1}{q}}$. This proves the Proposition. ∎

The degree of long range dependence is often expressed by means of the Hurst parameter.

If $\text{Var}\{X_1 + \cdots + X_n\}$ of long range dependent sequence grows at speed n^{2H}, where $H \in (\frac{1}{2}, 1]$, then the number H is called the Hurst parameter of the sequence.

Proposition 4 *The Hurst parameter for the discrete time arrival process $Y^{(\infty)}$ of example 2 is given by*

$$H = \frac{3q - p + 1}{q} \tag{5}$$

PROOF. The Hurst parameter can be derived from the power decay of the covariance (Roberts et al. 1996). If the power of the covariance decay is $k^{-\beta}$, then the Hurst parameter is given by $H = \frac{2-\beta}{2}$. It directly follows from Proposition 3 that $\beta = -\frac{q-p+1}{q}$, hence the Hurst parameter for example 2 is $H = \frac{3q-p+1}{2q}$. From the conditions $p > q + 1$ and $p < 2q + 1$, it follows that $H \in (\frac{1}{2}, 1]$. ∎

For completeness, we mention that for example 1, the powerlaw decay is given by $k^{-\frac{\log b}{\log b - \log a}}$, and hence $H = \frac{\left(2 - \frac{\log b}{\log b - \log a}\right)}{2}$ (Daniels & Blondia 1997).

5.2   The Index of Dispersion for Counts

In this section we derive an expression for the limit of the Index of Dispersion for Counts (IDC) of the process $Y^{(\infty)}$.

Denote N_k the number of arrivals in an interval of length k. The *Index of Dispersion for Counts* (IDC) at time k is defined to be the variance of the number of arrivals in an interval of length k divided by the the mean number of arrivals in this interval, i.e. $I(k) = \frac{\text{Var}(N_k)}{\text{E}(N_k)}$. Denote $I^{(\ell)}(k)$ the IDC of the process $X^{(\ell)}$ with $\lim_{k \to \infty} I^{(\ell)}(k) = J^{(\ell)}$ and $I^{(\infty)}(k)$ the IDC of the process $Y^{(\infty)}$, with $\lim_{k \to \infty} I^{(\infty)}(k) = J^{(\infty)}$. From (Blondia & Geerts 1997), we know that

$$J^{(\ell)} = \frac{\pi^{(\ell)} P_1^{(\ell)} e - 3[\pi^{(\ell)} P_1^{(\ell)} e]^2 + 2\pi^{(\ell)} P_1^{(\ell)} Z^{(\ell)} P_1^{(\ell)} e}{\pi^{(\ell)} P_1^{(\ell)} e},$$

with $Z^{(\ell)}$ the fundamental matrix of the Markov chain $P^{(\ell)}$, given by $Z^{(\ell)} = [I - (P^{(\ell)} - e\pi^{(\ell)})]^{-1}$, and $P_1^{(\ell)}$ given by $P_1^{(\ell)} = \begin{pmatrix} 0 & 0 \\ 1/\ell^q & 1 - 1/\ell^q \end{pmatrix}$. Furthermore,

$$\lim_{k \to \infty} I^{(\infty)}(k) = \frac{\sum_{\ell=1}^{\infty}[\lambda^{(\ell)} - 3(\lambda^{(\ell)})^2 + 2\pi^{(\ell)} P_1^{(\ell)} Z^{(\ell)} P_1^{(\ell)} e]}{\sum_{\ell=1}^{\infty} \lambda^{(\ell)}}. \tag{6}$$

It is easy to show that

$$Z^{(\ell)} = \frac{1}{(\ell^p + \ell^q)^2} \begin{pmatrix} \ell^p(\ell^p + \ell^q + \ell^{2q}) & \ell^q(\ell^p + \ell^q - \ell^{p+q}) \\ \ell^p(\ell^p + \ell^q - \ell^{p+q}) & \ell^q(\ell^p + \ell^q + \ell^{2p}) \end{pmatrix}.$$

Hence,

$$\pi^{(\ell)} P_1^{(\ell)} Z^{(\ell)} P_1^{(\ell)} e = \frac{\ell^q}{(\ell^p + \ell^q)^3} [\ell^{2q} - \ell^{2p} + \ell^{2p+q}].$$

Using this expression in (6), we obtain that

$$J^{(\infty)} = \frac{\lambda^{(\infty)} - 3 \sum_{\ell=1}^{\infty}(\lambda^{(\ell)})^2 + 2 \sum_{\ell=1}^{\infty} \frac{\ell^q[\ell^{2q} - \ell^{2p} + \ell^{2p+q}]}{(\ell^p + \ell^q)^3}}{\lambda^{(\infty)}}. \tag{7}$$

From equation (7) it follows that the limit of the IDC of the process $Y^{(\infty)}$ is infinite if $p \leqslant 2q + 1$, which is exactly the condition under which the process has the long range dependence property. This is in agreement with the criterion that a process is long range dependent if its IDC is diverging.

6 CORRESPONDENCES BETWEEN LRD IN TELECOMMUNICATIONS AND PHASE TRANSITIONS IN STATISTICAL PHYSICS

There is an important similarity between our model and a model of phase transitions in statistical physics, namely the Ising model. This model was introduced in 1925 by Ising (Ising 1924) as a model for ferromagnetism, and is solved analytically by Onsager in 1944 (Onsager 1944, Kaufman 1949, Kaufman & Onsager 1949, T.D. Schultz & Lieb 1964).

We consider electrons, located on a rectangular lattice, who can have two different spins, spin up or spin down. With each microscopic configuration $\mathcal{O} = \{\omega(i, j) = \text{up}(+1)/\text{down}(-1) \mid 1 \leq i \leq n, 1 \leq j \leq m\}$, one associates a probability $P(\mathcal{O}) = \frac{1}{Z}e^{E(\mathcal{O})}$, where $E(\mathcal{O})$ is a function, called the interaction energy, and where $Z = \sum_{\mathcal{O}} \exp E(\mathcal{O})$ is called the partition function. In the classical Ising model the interaction is nearest–neigbour, i.e. only adjacent electrons interact. The theory of Markov chains can be incorporated in this formalism (Georgii 1988, Kemeny et al. 1976, Prum & Fort 1991). From this local information, one whishes to deduce macroscopic statistical properties (magnetization, ...), by taking the thermodynamic limit, i.e. expanding the lattice to the whole plane.

The calculation of the macroscopic properties can be done in an elegant way using transfer matrices. The principle is to put the values of the interaction energy in a (transfer)matrix (see Figure 1).

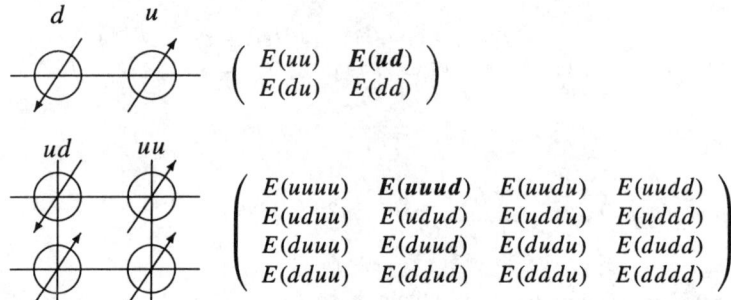

Figure 1 Transfer matrices for the one– and two–dimensional Ising model. The bold entries correspond to the shown spin configuration

It can now be proven that in the thermodynamic limit, i.e. for an infinite transfer matrix, the correlation function of two spins in different columns, decays to zero according to a power law, if the transfer matrix is asymptotically degenerate, i.e. the second greatest eigenvalue equals the greatest eigenvalue!

We now simplify the Ising model, by assuming no vertical interactions. The corresponding transfer matrix is then the Kronecker product of the 2×2 transfer matrices, corresponding to the single rows. It is now clear that Figure 2 establishes the link with our model.

Of course the Ising model is far more complex, admitting vertical interactions (dependent sources). The eigenvalue structure of the infinite transfer matrix depends on the temperature. The temperature at which the system undergoes a phase transition, i.e. transition from short to long range dependence, is called the critical temperatute T_c. For our model, we constructed the transfer (transition) matrices of the rows (sources) in such a way that the resulting infinite transfer matrix is always asymptotic degenerate (see Proposition 1). Nevertheless, we can view condition (1) as a way of determining an abstract critical temperature. If we take e.g. example 2, we can fix q and take p as 'temperature'. The critical value is then $p_c = 2q + 1$. For a more accurate description of the Ising model see (Domb & Green 1972, Thompson 1972))

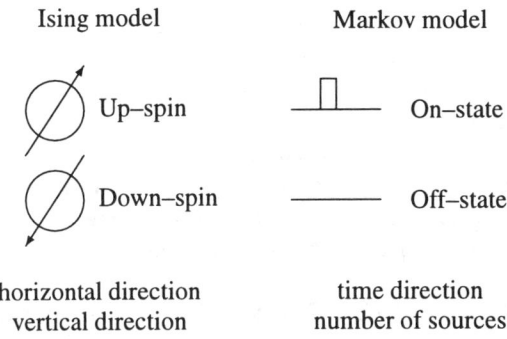

Ising model   Markov model

⌀) Up–spin   ⌐Π⌐   On–state

⌀) Down–spin   ─────   Off–state

horizontal direction   time direction
vertical direction   number of sources

Figure 2 Correspondences between the Ising Model and the Arrival Process $Y^{(\infty)}$

7 CONCLUSIONS AND FUTURE WORK

In this paper we proved a necessary and sufficient condition for long range dependence of an infinite superposition of heterogeneous Markovian ON/OFF–sources. Two simple examples are given and the corresponding Hurst parameters are derived. We are currently investigating the existence of a general method for calculating the Hurst parameter of our model. It is also shown that for our model, long range dependence directly implies infinite mean queue length. The characterization of the queue length distribution remains open. We give a physical interpretation of our model by means of the Ising model. The existence of long range dependence is a consequence of the asymptotical degeneracy of the transition matrix. Further research is needed to study the possibility of applying known techniques of statistical physics to the context of ATM modelling. The admittance of dependent sources (vertical interactions) seems a first step. Also an interpretation of the queueing process in terms of the Ising model, could be of great help in deriving the tail of the queue length distribution.

REFERENCES

Abate, J., Choudry, G. & Whitt, W. (1994), 'Asymptotics for steady–state tail probabilities in structured Markov queueing models', *Comm. Stat. Stoch. Models* **10**, 99–143.

Andersen, A., Jensen, A. & Nielsen, B. F. (1995), Modelling and performance study of packet–traffic with self–similar characteristics over several timescales with Markovian arrival processes (MAP), *in* I. Norros & J. Virtamo, eds, 'Proceedings of the 12th Nordic Teletraffic Seminar, NTS12 Symposium 154', VTT, Finland, pp. 269–293.

Beran, J. (1994), *Statistics for Long–Memory Processes*, Vol. 61 of *Monographs on Statistics and Applied Probability*, Chapman & Hall.

Beran, J., Sherman, R., Taqqu, M. & Willinger, W. (1995), 'Long range dependence in VBR video traffic', *IEEE Trans. Comm.* **COM-43**.

Blondia, C. (1992), 'A discrete–time batch Markovian arrival process as B-ISDN traffic model', *Belgian Journal of Operations Research, Statistics and Computer Science* **32**.

Blondia, C. & Geerts, F. (1997), The correlation structure of the output of an ATM multiplexer, *in* 'Proceedings of the fifth IFIP Workshop on Performance Modelling and Evaluation of ATM Networks', Ilkley.

Boxma, O. (1996), 'Fluid queues and regular variation', *Performance Evaluation* **27&28**, 699–712.

Buffet, E. & Duffield, N. (1992), Exponential upper bounds via martingales for multiplexers with Markovian arrivals, Technical Report DIAS-APG-92-16, Dublin Institute for Advanced Sciences.

Daniels, T. & Blondia, C. (1997), A discrete–time ATM traffic model with long range dependence characteristics, *in* 'Proceedings of PMCCN'97', Tsukuba, Japan.

Dembo, A. & Zeitouni, O. (1993), *Large Deviations Techniques and Applications*, Jones and Barlett.

Domb, C. & Green, M., eds (1972), *Phase Transitions and Critical Phenomena*, Vol. 1, Academic Press.

Duffield, N. (1992), Rigorous bounds for loss probabilities in multiplexers of discrete heterogeneous Markovian sources, Technical Report DIAS-APG-92-31, Dublin Institute for Advanced Sciences.

Duffield, N. (1993), Exponential bounds for Markovian queues, Technical Report DIAS-APG-93-01, Dublin Institute for Advanced Sciences.

Falkenberg, E. (1994), 'On the asymptotic behaviour of the stationary distribution of Markov chains of M/G/1–type', *Comm. Stat. Stoch. Models* **10**(1), 75–97.

Georgii, H. (1988), *Gibbs Measures and Phase Transitions*, Vol. 9 of *De Gruyter Studies in Mathematics*, Walter de Gruyter.

Ising, E. (1924), 'Beitrag zur Theorie des Ferromagnetismus', *Z. Physik* **31**, 253.

Kaufman, B. (1949), 'Crystal statistics, II. Partition function evaluated by spinor analysis', *Phys. Rev.* **76**, 1232.

Kaufman, B. & Onsager, L. (1949), 'Crystal statistics, III. Short–range order in a binary Ising lattice', *Phys. Rev.* **76**, 1244.

Kemeny, J., Snell, J. & Knapp, A. (1976), *Denumerable Markov Chain*, Vol. 40 of *Graduate Texts in Mathematics*, Springer–Verlag.

Laevens, K. & Bruneel, H. (1997), Some preliminary resuls on traffic characteristics and queueing behaviour of discrete–time ON/OFF–sources, *in* 'Proceedings of the fifth IFIP Workshop on Performance Modelling and Evaluation of ATM Networks', Ilkley. to appear.

Leland, W., Taqqu, M., Willinger, W. & Wilson, D. (1993*a*), Ethernet traffic is self–similar: stochastic modelling of packet traffic data.

Leland, W., Taqqu, M., Willinger, W. & Wilson, D. (1993*b*), Statistical analysis of high time–resolution Ethernet LAN traffic measurements, *in* 'Proceedings of INTERFACE'.

Leland, W., Taqqu, M., Willinger, W. & Wilson, D. (1994), 'On the self–similar nature of Ethernet traffic, extended version', *IEEE/ACM Trans. Networking* **2**(1).

Likhanov, N., Tsybakov, B. & Georganas, N. (1995), Analysis of an ATM buffer with self–similar ("fractal") input traffic, *in* 'Proceedings of INFOCOM'95', IEEE.

Mieghem, P. V. (1996), 'The asymptotic behaviour of queueing systems: large deviations theory and dominant pole approximations', *Queueing Systems* **23**, 27–55.

Neuts, M. (1989), *Structured Stochastic Matrices of M/G/1 Type and Their Applications*, Marcel Dekker.

Norros, I. (1994), 'A storage model with self–similar input', *Queueing Systems* **16**, 387–396.

Onsager, L. (1944), 'Crystal statistics, I. A two–dimensional model with an order–disorder transition', *Phys. Rev.* **65**, 117.

Prum, B. & Fort, J. (1991), *Stochastic Processes on a Lattice and Gibbs Measures*, Vol. 11 of *Mathematical Physics Studies*, Kluwer Academic Publishers.

Pruthi, P. (1995), An Application of Chaotic Maps to Packet Traffic Modelling, PhD thesis, Royal Institute of Technology, Dept. of Teleinformatics.

Roberts, J., Mocci, U. & Vitramo, J., eds (1996), *Broadband Network Teletraffic*, Springer.

T.D. Schultz, D. M. & Lieb, E. (1964), 'Two–dimensional Ising model as a soluble problem of many fermions', *Rev. Mod. Phys.* **36**, 856.

Thompson, C. (1972), *Mathematical Statistical Physics*, The Macmillan Company, New York.

Performance Modelling and Dimensioning

44

Multi-layer modelling of a multimedia application

M. Baumann[1], T. Müller[1], W. Ooghe[2],
A. Santos[3], S.B. Winstanley[4], and M. Zeller[5]

[1] Dresden University of Technology, Chair for Telecommunications
[2] University of Ghent, Department Electrical Engineering
[3] Telefónica de España, TID
[4] Queen Mary and Westfield College, Electronic Engineering
[5] Alcatel Switzerland, Customer Service

Contact:
Dresden University of Technology, Chair for Telecommunications
D-01062 Dresden, Germany
tel. +49 351 463-3942, fax +49 351 463-7163
e-mail {baumann,muellert}@ifn.et.tu-dresden.de

Abstract

With the increasing deployment of ATM, application oriented QoS measures become more and more important. In order to investigate relationships between QoS measures on different protocol layers, end system models describing these layers have to be developed. In this paper, a case study for the modelling of a multi-media application comprising video and audio components is presented. The model covers ATM, AAL, and application layer processes. For the application level, a stochastic process describing distribution and correlation of video frame sizes is proposed. Different simplifications of the rather complicated compound model are discussed and assessed w.r.t. correlation structure and behaviour in traffic control functions. Furthermore, the detailed model is used to investigate relationships between loss measures on ATM, AAL and video frame level.

Keywords
QoS, Multimedia, ATM, Simulation

1 INTRODUCTION

ATM based broadband ISDN have to support a variety of traffic classes in an integrated framework allowing a flexible and efficient usage of network resources. A very important application class including e.g. real-time multi-media applications

Broadband Communications   P. Kühn & R. Ulrich (Eds.)
© 1998 IFIP. Published by Chapman & Hall

is characterized by traffic streams depending both on bounded transfer delays and certain transmission guarantees. For this type of traffic, only open-loop traffic control strategies can be applied. Both ITU–T (ITU–T 1996) and ATM Forum (ATM Forum 1996) have foreseen service classes to support delay-sensitive CBR and VBR traffic. The most important traffic control functions associated to these classes are usage parameter control UPC and connection admission control CAC. From a user's point of view, two main questions are related to these control functions: How does the ATM layer QoS ensured by CAC functions map onto application layer, user-related QoS measures? Which UPC parameters are appropriate for the application, i.e. which bandwidth and buffer resources have to be reserved in the network? Only if both questions can be answered satisfactory, a trade-off between connection price and user-perceived QoS can be found. Of course it is not possible to perform an in-depth analysis for each particular application possibly transmitted over ATM. Instead it is necessary to derive guidelines as general as possible. As a starting point, however, real applications have to be considered. As part of the ACTS project EXPERT (ACTS 1996), the addressed interactions between traffic control functions and applications are considered by means of user trials and theoretical investigations. The application model described and assessed in this paper is based on traffic measurements carried out in the EXPERT test bed in Basel. It covers the most important protocol layers in a typical computer based multi-media application.

The paper is organized as follows. Section 2 describes background and general operation of the investigated multi-media application. Section 3 establishes the complete source model and derives different simplifications. In section 4, the stochastic process modelling the video frame generation is developed. Section 5 investigates the correlation properties of the proposed models and evaluates the ability to reflect cell loss ratios in a peak rate shaper and a policing function. The most detailed source model is used in section 6 to examine relationships between loss measures on ATM, AAL and video frame level. Section 7 draws conclusions from the presented study.

2   GENERAL OPERATION OF THE APPLICATION

The considered multi-media application ISABEL (Quemeda *et al* 1996) is based on standard UNIX workstations with ATM adapter cards. It comprises video and audio components. The video part applies a Motion JPEG codec board, for the audio part a constant frame rate is transmitted. The normal UNIX network protocol stack is used to hide network specific details. In order to fulfill the real-time requirements of a dialog-oriented application, UDP had to be chosen as transport layer protocol. Due to the given ATM adapter card, IP frames are transmitted via AAL type 5. It has to be noted, that the generated ATM cells cannot be transmitted over low-cost service classes like UBR (ATM Forum 1996). Instead, the service classes CBR or VBR-rt have to be used. Only these classes can provide both delay and loss commitments necessary for a proper operation.

The multi-media workstations have been used mainly for two purposes. First, the traffic characteristics generated during a session of approximately 12 minutes have been stored in a trace file. The trace contains interarrival times of all cells sent by one party, together with AAL end-of-frame marks. Thus, it was possible to derive AAL frame boundaries. Additional heuristics and some knowledge about the internal operation of ISABEL even allowed to derive video and audio frame boundaries from the trace. Secondly, the traffic generated by one end system has been policed with a police function unit realizing a leaky bucket algorithm. In section 5, results from these experiments and corresponding simulation runs are compared.

UDP itself cannot give any delay guarantees. Thus, the only way to adapt the overall transmission delay to the application's requirements consists in optimizing the segmentation process inside of the application. ISABEL subdivides the audio bit stream of 256 kbit/s (16 kHz sampling frequency, 16 bits per sample) into packets of $L_A = 1600$ bytes each. Taking into account the protocol overhead of UDP/IP and AAL, an AAL frame consisting of $N_A = 35$ cells is generated every $T_A = 50ms$ by the audio stream. Given the appropriate hardware and software, video images should experience the same overall delay as audio signals. Hence, it is possible to perform additional segmentation of the video bit stream. ISABEL segments the video stream into parts of $L_V = 4096$ bytes each which are preceded by a header of 28 bytes during UDP/IP processing. This eventually leads to $N_V = 87$ cells per AAL frame, except for the last frame of a video image which will be shorter in most cases. The modelling of video image sizes is described in detail in section 4.

3 EXACT AND SIMPLIFIED SIMULATION MODELS

Based on the implementation details given in the last section, a simulation model has been developed which tries to cover all processes as exactly as possible (figure 1). Since all sub-streams of the multi-media application are transmitted using one

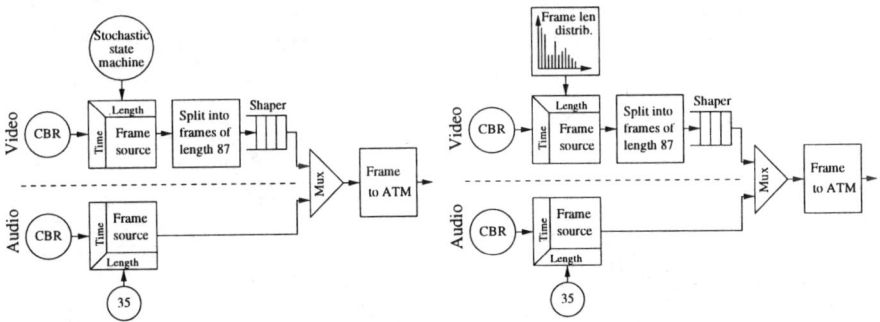

Figure 1 Exact model (SSM-FRLEN)   **Figure 2** Model H-FRLEN

virtual channel on ATM level, video and audio streams are multiplexed already on frame level (block Mux). The very right block converts incoming frames into appropriate strings of cells which are sent with full line speed. Video and audio frames are generated by the parts above and below the dashed line, respectively. Due to limitations of the experimental set-up, only a video image frequency of $f_V = 9.4/s$ could be reached. The CBR source of the video part therefore generates video cycles with a duration of $T_V = 1/f_V = 106ms$. For each video image, a stochastic state machine described in section 4 draws the current video frame length in cells, already taking into account the overheads added by the subsequent segmentation processes. The next two blocks of the video branch model the image segmentation on application level ($N_V = 87$ cells per UDP frame, see section 2). Investigations of the trace file showed, that the application spaces the generation of UDP packets with a distance of $S_V = T_V/5$. The mean number of UDP frames per video image evaluates to 3.95, see section 4. Thus, a video image can be transmitted in average during one video cycle. The packet spacing mechanism has been modelled by means of a shaper equivalent to the device depicted further below in figure 8, but operating at frame level. The frame size is not taken into account, the buffer capacity is unlimited. The audio part consists of a source generating frames (length $N_A = 35$ cells) with constant distances of $T_A = 50ms$. This most detailed source model is referred to as model SSM-FRLEN (generation of frame lengths by a stochastic state machine).

   The source model of figure 1 is relatively complex. This leads to memory and time consuming simulation models, and an analytical treatment becomes almost impossible. Therefore, simplified source models have been derived. As a first step, the synchronized stochastic state machine is replaced by a discrete frame length distribution derived from the measurement trace, see figure 2. This reduces the complexity remarkably, but it has disadvantages, too. During a video scene, the sizes of successive frames are similar. By using only the frame length distribution, the correlation between these sizes is lost. The model is referred to as H-FRLEN (histogram of video frame lengths).

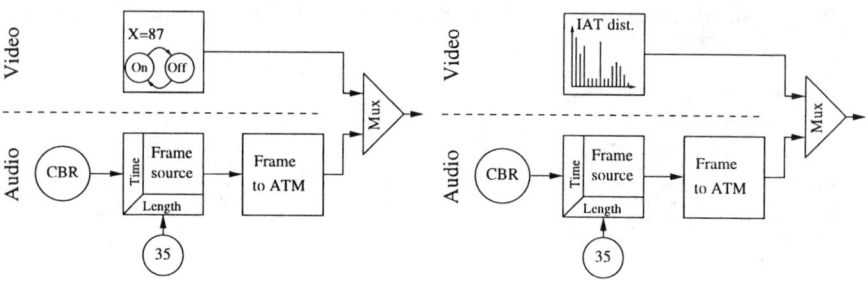

Figure 3 Model ON(D)-OFF   **Figure 4** Model H-IAT

   The models SSM-FRLEN and H-FRLEN cover the traffic behaviour on video and AAL frame level. A further reduction of complexity can be achieved if the video frame level is skipped entirely. In figure 3, the video part is modeled directly on ATM level. Hence, the multiplexing of video and audio traffic streams is shifted to the ATM level. The video frames are generated by an ON-OFF process with deterministic number of 87 cells per ON state and geometrically distributed duration of the OFF period (model designated by ON(D)-OFF). During the ON period, cells are sent back to back. The very regular structure of the interarrival times for AAL and video frames is completely lost. In the video part of the last model, all knowledge about the operation of higher layers is removed, see figure 4. The model is only based on the interarrival time distribution on cell level. The IAT distribution is directly derived from a measured trace where the audio part of the application has been turned off. The model is referred to as H-IAT (histogram of interarrival times on cell level).

4   STATE MACHINE FOR GENERATION OF VIDEO IMAGE SIZES

A number of contributions dealing with the modelling of MPEG and Motion-JPEG streams is already available, e.g. (Melamed *et al* 1994, Rose 1995). The approach adopted here resembles the scheme proposed in (Rose 1995).

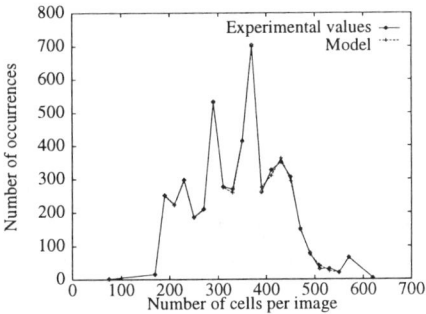

State	Mean	St.Dev.	Weight
1	360.2	76.7	0.539
2	196.7	6.6	0.065
3	230.2	12.7	0.080
4	289.3	8.6	0.092
5	362.8	4.0	0.111
6	436.7	20.2	0.098
7	571.1	8.2	0.015

Figure 5 Comparison of experimental and source model histograms

Table 1 Moments and weights of the Normal distributions

As starting point, the distribution of video image sizes in cells has been derived from the measurement trace. This distribution is depicted in figure 5. Although the exact results do highly depend on the actual video scenes during a session, the general behaviour has been found to be typical also for other video sequences. A superposition of weighted Normal distributions is appropriate to fit the overall image size distribution. The number of distributions has been determined by a "best guess", followed by a fitting procedure and an assessment of the compound distribution using a Chi-square test. Table 1 contains the results obtained from the fitting to the distribution given by figure 5. The number $N = 7$ of Normal distributions corresponds to the 6 obvious peaks of the distribution, complemented by an underlying distribution with relatively high standard deviation (state number 1). In order to construct the state machine, it is first necessary to associate each possible image size to one of the N states. Given an image size of X cells, state number i is chosen, if the measure $Z_i = (X - \mu_i)/(\sigma_i \cdot w_i)$ becomes minimal for this state. For the ith distribution, μ_i, σ_i, and w_i designate the mean value, the standard deviation, and the weight, respectively. Application of this criterion translates the list of consecutive video sizes derived from the measured trace into a list of consecutive video state numbers. From this sequence, conditioned relative frequencies p_{ij} of changing to state j, provided that the current state is i, can be derived. The evaluation of the transition matrix $\mathbf{Q} = (p_{ij})$ of the stochastic automaton revealed a special property. Except for state 1, the probabilities for changing to other states than the current one, or to state 1, vanish. Thus, state 1 can be interpreted as transition state between all other states. Once being on a certain level of activity, the correlation between subsequent image sizes remains high. A change of the video scene causes a transition to state 1 from where, in turn, the next level of activity is reached. Equation 1 gives the actual values derived from the trace. The matrix structure seems interesting for solving the problem analytically (Wuyts *et al* 1997).

$$\mathbf{Q} = \begin{pmatrix} 0.954 & 0.003 & 0.009 & 0.007 & 0.012 & 0.014 & 0.001 \\ 0.029 & 0.971 & 0 & 0 & 0 & 0 & 0 \\ 0.062 & 0 & 0.938 & 0 & 0 & 0 & 0 \\ 0.041 & 0 & 0 & 0.959 & 0 & 0 & 0 \\ 0.057 & 0 & 0 & 0 & 0.943 & 0 & 0 \\ 0.073 & 0 & 0 & 0 & 0 & 0.927 & 0 \\ 0.03 & 0 & 0 & 0 & 0 & 0 & 0.97 \end{pmatrix}. \tag{1}$$

To summarize, video image sizes are generated as follows. With each request, a random number according to the normal distribution of the current state is drawn. Then, the current state is changed applying transition matrix $\mathbf{Q}$. The random number drawn in the first step is returned.

5   EVALUATION OF MODEL PERFORMANCE

In this section, the traffic source models described in sections 3 and 4 are compared with respect to their influence on the performance of subsequent ATM networks. This is done by evaluating the autocorrelation functions of the traffic streams. Furthermore, the CLR in a peak rate shaper and the acceptance region of a leaky bucket policing function are compared. Different results available in the literature (Takine *et al* 1993, Grünenfelder *et al* 1994) show that the correlation structure of a traffic stream strongly influences the performance of network elements like statistical multiplexers. The autocorrelation function of the counting process therefore is an important descriptor of the characteristics of a traffic stream. The counting process is constructed by subdividing the time axis into intervals of width w and determining the number of cell arrivals per interval. The autocorrelation function ACF then is defined by $ACF(l) = cov(x_i, x_{i-l})/var(x_i)$, where x_i is the number of cell arrivals in the i-th interval of width w. In the following $l \cdot w$ is referred to as lag. By changing w the dependencies in different time scales can be observed.

In figure 6, the ACF for $w = 10$ slots is given. For the measured trace a nearly linear behaviour for lag $= 0$ to lag $= 135$ can be observed. This is due to the regular size of 87 cells per AAL frame stemming from the video part. The multi-media end system was attached to the measurement setup via a 100 Mbit/s TAXI interface.

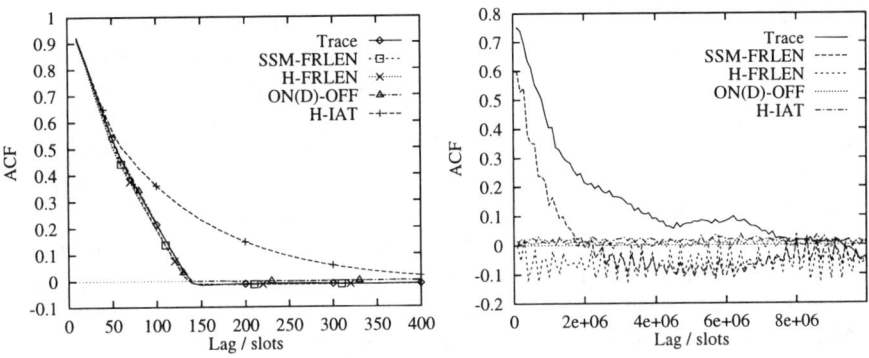

Figure 6 Short range autocorrelation, interval width $w = 10$ slots

Figure 7 Long range autocorrelation, interval width $w = 10^5$ slots

Hence, a burst of 87 cells is stretched over a time period of 135 slots. Except H-IAT, all models can emulate the short range autocorrelation very well. All these models realize the deterministic AAL frame size of 87 cells. The model H-IAT only uses the histogram of the cell interarrival time. Therefore it is not able to reproduce the deterministic frame size. In figure 7 the ACF for $w = 10^5$ slots is given (long range autocorrelation). The observed lag ranges up to an equivalent duration of 27 s. It can be seen, that the ACF of the measured trace stays positive over a long period of time. This long range correlation is caused by the slow picture (and bit rate) changes during a scene. Only SSM-FRLEN is capable of modelling this behaviour to some extent. All other models do not deal with the application layer correlations and therefore fail to model the long range dependencies.

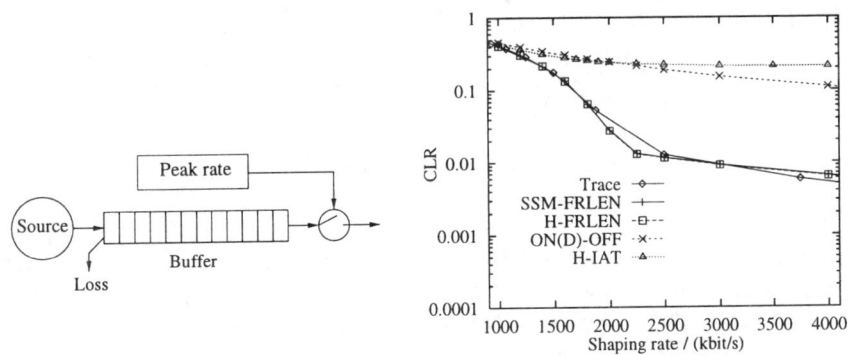

Figure 8 Peak rate shaper

Figure 9 CLR in the shaper, buffer size 100 cells

As a next step of model assessment, the CLR in a peak rate shaper has been investigated. Normally, the user will choose the parameters in such a way that shaper buffer overflows do not occur. In order to better evaluate the models, the rather hard requirement of zero loss has been relaxed. Shaping and further below policing have been used to assess the models, since only the characteristics of one single traffic stream influence the system performance. As has been outlined in (Baumann *et al* 1996), the results for multiplexing are far less sensitive against small changes of traffic characteristics.

The traffic streams produced by the different models and the trace have been fed into the shaper which is depicted in principle in figure 8. The CLR has been measured for a varying shaping rate. The CLR for a buffer size of 100 cells is shown in figure 9. The results of H-IAT and ON(D)-OFF are not satisfactory. The streams produced by SSM-FRLEN and H-FRLEN, however, cause nearly the same CLR as the trace. For this small buffer size the different modelling of the application layer and therefore the different properties w.r.t. large time scales seem to be less important. This behaviour changes, if larger shaper buffers are used. Figure 10 shows the CLR for a buffer size of 1000 cells. As in the case of smaller buffers, H-IAT and ON(D)-OFF overestimate the CLR. But for this buffer size H-FRLEN underestimates the CLR while the curve for SSM-FRLEN is relatively exact. This demonstrates that the correct modelling of application layer processes with their long-time memories becomes more important with increasing buffer sizes.

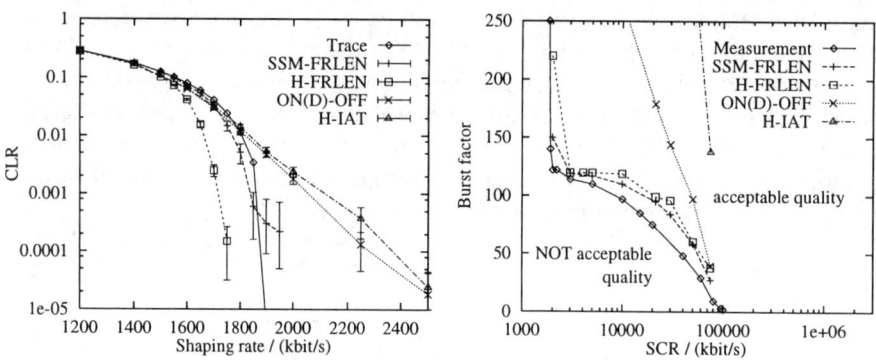

Figure 10 CLR in the shaper, buffer size 1000 cells

Figure 11 Acceptance region for policing

Finally, the influence of leaky bucket policing on the application has been examined. Subjective tests showed that the considered multi-media application can accept a CLR of up to 10^{-4}. Again, it should be noted that a policing function normally is dimensioned with a target of no loss. The free parameters of a leaky bucket are increment I, and limit L (ATM Forum 1996). The SCR (sustainable cell rate) then is $SCR = 1/I$, the burst factor BF follows with $BF = L/I$. Figure 11 shows the acceptance region for policing, if the allowed CLR is 10^{-4}. The values produced by the different models are compared to measurement results obtained with a policing unit available in the EXPERT testbed in Basel (EXPERT 1997). The curves for H-IAT and ON(D)-OFF extremely differ from the measurement results. If these models would be used to estimate the values necessary for traffic contract negotiation, then this eventually would lead to higher costs for the user. SSM-FRLEN and H-FRLEN can provide a good estimation of the acceptance region. The difference between these models is small. This again confirms that long range correlation properties are less important for the investigation of systems with rather short system memories.

6 MULTIPLEXING EXPERIMENTS

The source model has been used to investigate relationships between QoS measures on ATM, AAL, and application level. In order to avoid a superposition of too many effects, only the video part of the detailed model has been applied. Thus, the source model comprises the upper half of the structure depicted in figure 1. On ATM level, the CLR is considered. An AAL frame is lost, if at least one corresponding cell is dropped by the ATM network. This is described by the frame loss ratio (FLR). If the ratio of FLR and CLR is smaller than the mean number of cells per frame, then cell losses tend to "clump". On application level, the ratio of corrupted and emitted video frames is considered. With ISABEL, a video frame received only partially does not lead to the loss of the entire picture. Nevertheless, the resulting QoS degradation always is perceivable. Therefore, the term video frame loss ratio (VFLR) is used to describe the subjective QoS on application level.

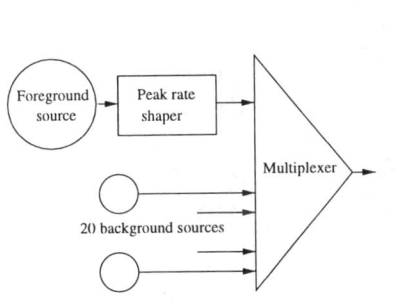

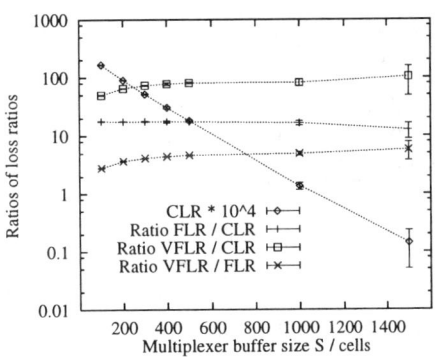

Figure 12 Investigated model configuration

Figure 13 Ratios of loss ratios for different buffer sizes S

For all simulation experiments, the configuration depicted in figure 12 has been used. The video traffic stream is superimposed by 20 background traffic streams generated by ON/OFF sources with geometrically distributed durations of ON and OFF phases. During the ON phase, a background source sends cells with a constant distance of $\Delta = 10$, the mean phase durations are $\overline{T}_{ON}$ and $\overline{T}_{OFF}$. All values are normalized to the duration of one ATM time slot. The multiplexer is a simple FIFO multiplexer with a buffer size of S cells. A cell (if available) is served during every time slot. The traffic of the reference source is shaped by a peak rate shaper which ensures a peak cell rate of B_{peak} (unit: Mbit/s). The first experiment considers ratios between CLR, FLR, and VFLR, if the buffer size of the multiplexer is changed. The question to be answered was, whether higher-layer loss ratios improve with the same speed as the CLR. In table 2, all parameters of the configuration are collected. The total traffic load in the multiplexer and the load generated by the background traffic are designated by ρ_{Tot} and ρ_{BG}, respectively. The traffic load of the reference traffic stream evaluates to $9.2 \cdot 10^{-3}$. Figure 13 shows the known fact, that the logarithm of the CLR decreases linearly with the buffer size (note the CLR scaled up by 10^4). A comparison between the ratio FLR/CLR and the mean number of cells per AAL frame reveals that an AAL frame loss normally is caused by more than one cell loss. The ratio FLR/CLR only falls slightly with falling CLR. This indicates, that the degree of cell loss clumping does not increase substantially with falling absolute CLR. Compared to the results e.g. in (Blondia *et al* 1991), this is somewhat surprising. Often it has been observed that the probability of losing successive cells increases with decreasing absolute loss probability. The ratio VFLR/FLR which describes the clumping of AAL frame losses, also is only slightly dependent from the absolute CLR. Here, the ratio VFLR/FLR is almost equal to or even higher than the mean number of AAL frames per video frame. The loss process is slightly modulated by the foreground traffic. Therefore losses of foreground cells mostly occur during long video frames. The almost constant ratio VFLR/CLR confirms that all loss ratios improve nearly in the same extent.

Parameter	Value	$\overline{T}_{ON}$	$\overline{T}_{OFF}$	ρ_{BG}	CLR
$\overline{T}_{ON}$	1000 time slots	100	104.5	0.98	$(1.4 \pm 0.5) \cdot 10^{-4}$
$\overline{T}_{OFF}$	1400 time slots	500	600	0.91	$(1.2 \pm 0.3) \cdot 10^{-4}$
B_{peak}	4 Mbit/s	1000	1400	0.83	$(1.1 \pm 0.2) \cdot 10^{-4}$
S	100 ... 1500	2000	3600	0.71	$(1.0 \pm 0.2) \cdot 10^{-4}$
ρ_{BG}, ρ_{Tot}	0.83, 0.84	3000	6200	0.65	$(1.2 \pm 0.3) \cdot 10^{-4}$

Table 2 Parameters experiment 1   **Table 3** Source parameters of experiment 2

For a second experiment, the buffer size of the multiplexer has been fixed to 1000 cells. As grade of freedom, the characteristics of the background traffic have been altered by varying the burst lengths. In parallel, the silence durations have been adapted in such a way, that the resulting CLR remained constant at approximately 10^{-4}. The number of background sources again was 20. Table 3 shows the traffic parameters, traffic loads and measured CLR. In figure 14, the results of this experiment are depicted. Again, the clumping of AAL frame losses is more or less independent from the traffic characteristics. The slightly falling ratio FLR/CLR indicates that cell loss clumping increases, if the traffic becomes more "unfriendly" for statistical multiplexing. Here, increasing burst lengths reduced the possible system utilisation (see table 3). This general tendency is confirmed by the last experiment. The multiplexer buffer size has been set to 1000 cells, and the background traffic characteristics are

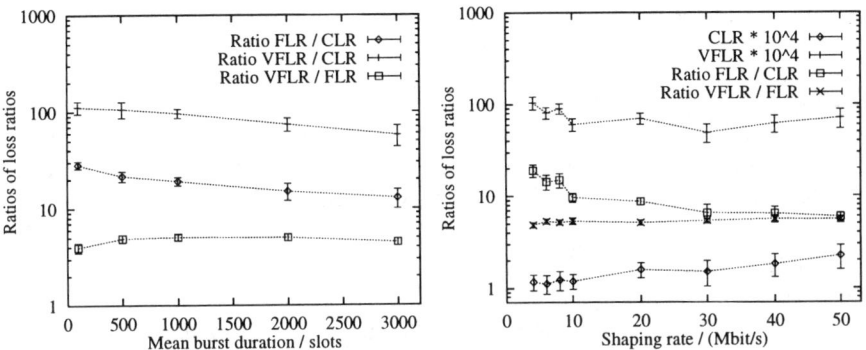

Figure 14 Ratios of loss ratios for differ- **Figure 15** Ratios of loss ratios for different burst durations $\overline{T}_{ON}$ ent shaping rates B_{peak}

the same as given in table 2. The shaping rate of the shaper limiting the peak cell rate
of the foreground traffic, has been varied from 4 to 50 Mbit/s. Figure 15 shows a CLR
rising together with the peak bit rate of the foreground traffic. The ratio FLR/CLR,
however, becomes more friendly for the application, if the shaping is performed at
a rate not too close to the mean bit rate. This may be explained as follows. Suppose
constant lengths and distances of overload periods in the multiplexer. Then the probability that a frame can be transmitted without hitting a lossy period, is higher for
frames with shorter duration. Eventually a better subjective QoS is achieved (note
the VFLR scaled up by 10^4 in figure 15), since the mapping between AAL and video
frame loss ratio again stays more or less constant.

To summarize, an improvement of the CLR is directly translated into better loss
measures on AAL and application level. The clumping of both cell and AAL frame
losses is remarkably independent from i) the absolute CLR value, ii) background
traffic characteristics, and iii) buffer sizes. Corrupted application layer frames usually are only subject to at most one AAL frame loss. The ratio VFLR / FLR thus
approaches the mean number of AAL frames per application layer frame. The ratio
FLR / CLR normally is lower than the mean number of cells per AAL frame.

7   CONCLUSION

In this paper, results of a case study concerned with the modelling of a multi-media
application have been presented. The crucial video traffic stream model covers interarrival times and sizes of video frames, AAL frames, and cells. The autocorrelation
properties of the cell counting process therefore are reflected with good accuracy.
Comparisons between traffic experiments and simulations with policing and shaping
devices turned out, that correlation properties in the time scale of tens of seconds
are less critical, if systems with realistic buffer sizes shall be investigated. Nevertheless, the UDP/IP and AAL segmentation processes should be modelled with high

accuracy. Using the most accurate traffic model, relations between loss measures on ATM, IP frame, and video frame level have been investigated. The results indicate that these relations are remarkably independent from ATM background traffic characteristics and system buffer sizes.

Acknowledgement. This work has been carried out as part of the ACTS project AC094 EXPERT 'Platform for Engineering Research and Trials'.

REFERENCES

ACTS project AC094 (1996) Platform for Engineering Research and Trials. WWW home page http://www.elec.qmw.ac.uk/expert/.

ATM Forum (1996) Traffic Management Specification Version 4.0. ftp://ftp.atmforum.com/pub/approved-specs/af-tm-0056.000.ps.

Baumann, M., and Müller, T. (1996) Simulation und Verifikation von ATM–Quellen-modellen (in German). *Proc. of: 10. ASIM Symposium – Simulationstechnik'*, Dresden, September 1996, pp. 195–200.

Blondia, C., and Casals, O. (1991) Cell Loss Probabilities in a Statistical Multiplexer in an ATM Network. *Proc. of the 6th GI/ITG Fachtagung: Messung, Modellierung und Bewertung von Rechensystemen*, München, 1991, Springer–Verlag, pp. 121–136.

EXPERT, ACTS project AC094 (1997) *First results from Trials of Optimized Traffic Features*. Deliverable 10, March 1997.

Grünenfelder, R., and Robert, S. (1994) Which Arrival Law Parameters are Decisive for Queueing System Performance? *Proc. ITC-14*, Antibes Juan Les Pins, France, June 1994, pp. 377–386.

ITU–T (1996) Recommendation I.371, Traffic Control and Congestion Control in B-ISDN. Geneva, May 1996.

Melamed, B., Reininger, D., and Raychaudhuri, D. (1994) Variable Bit Rate MPEG Video: Characteristics, Modeling andMultiplexing. *Proc. ITC-14*, Antibes Juan Les Pins, France, June 1994, pp. 295–306.

Quemada, J., de Miguel, T.P., Azcorra, A., Pavon, S., Salvachua, J., Petit, M., Robles, T., and Huecas, G. (1996) ISABEL: a CSCW application for the distribution of events. *Proc. COST 237 Workshop on Multimedia Networks and Systems*, Barcelona.

Rose, O. (1995) *Statistical properties of MPEG video traffic and their impact on traffic modeling in ATM systems*. Technical report no. 101, University of Würzburg, Department of Computer Science, February 1995.

Takine, T., Suda, T., and Hasegawa, T. (1993) Cell Loss and Output Process Analysis of a Finite-Buffer Discrete-Time ATM Queueing System with Correlated Arrivals. *Proc. IEEE INFOCOM '93*, San Francisco, pp. 1259–1269.

Wuyts, K., and Boel, R.K. (1997) Efficient performance analysis of ATM buffer systems by using the spectral analysis of rate matrices. *Proc. of the fifth IFIP workshop on performance modelling and evaluation of ATM networks*, Ilkley, pp. 33/1–33/9.

45

Multiplexing periodic sources in a tree network of ATM multiplexers[*]

J.M. Barceló, J. García
Polytechnic University of Catalonia
Computer Architecture Department, c/ Gran Capitan, Modulo C6-E105, Barcelona E-08071, Spain
tel : + 34 3 4016798, fax : + 34 3 4017055,
e-mail: joseb@ac.upc.es, jorge@ac.upc.es

Abstract

We obtain the queue length distribution in a tree of discrete time queues with constant service time whose input is periodic traffic. In the context of ATM the study could be applied to CBR sources. The tree consists of M-stages. To solve this system, we first solve a 2-stage tree network. Given this configuration, a more complex tree network can be easily solved making use of the properties of the discrete time queues with identical service times. We also give closed formulas for the average waiting time and average number of cells in any queue of the tree network.

Keywords

ATM, tree topology, periodic sources

1 INTRODUCTION.

In this paper we derive closed-form formulas for the queue length distributions in a discrete-time M-stage tree queueing network loaded with periodic traffic sources. In this type of network, the queues can be grouped in M groups or stages. Every queue of a stage is fed by all the exit traffic of any given number (which could be 0) of queues from the previous stage as well as by certain number (which could be 0) of external sources of traffic. All the entrance traffic in the network is routed to the root queue, which occupies the first stage (see in Figure 5 an example for the case $M = 3$). We consider the discrete-time case in which all the queues have a single server with constant service time. As is shown in Morrison (1978) only two configurations are relevant to solve this system: the case $M = 1$, which corresponds to a single server queue with a constant service time and the case $M = 2$. The solution of the case $M > 2$ can be found solving systems with $M = 1$ and $M = 2$.

We are interested in tree networks of discrete time queues with constant ser-

---

[*]This work was supported by project TIC95-0982-CO2-01 and XUGA 10503A96

Broadband Communications   P. Kühn & R. Ulrich (Eds.)
© 1998 IFIP. Published by Chapman & Hall

vice time whose input is periodic traffic (e.g. CBR sources or periodic sources with back-to-back cells). For this kind of traffic, we compute the probability distribution function of the virtual waiting time in any queue of the tree.

In Modiano *et al* (1996) a solution to the average queuing delay in a tree network of discrete time queues with constant service time where the arrival process is Poisson is presented. Their solution is based on an equivalent network where priority is given to customers in transit. From this priority model it is easy to derive the average waiting time and average number of clients in the system in any queue of the tree. We will make use of this result to give a closed formula for the average waiting time when the input traffic is periodic. We think that although this kind of topology is limited, its solution can lead to give us more insight to more general queue systems such as queues in tandem with cross traffic.

The paper is organized as follows: in section 2, we compute the CPDF (Complementary Probability Distribution Function) of the virtual waiting time in this kind of topologies with periodic traffic. We first calculate the case of two multiplexing stages, to later extend the solution to M-stage networks. In section 3, we apply the equivalent priority model proposed in Modiano *et al* (1996) to obtain the average waiting time and average number of cells in the system in any queue of the tree network. Finally we give some results in section 4.

2   MULTIPLEXING PERIODIC SOURCES IN AN M-STAGE TREE QUEUEING NETWORK.

2.1   Two-stage tree network.

In this section we compute the CPDF for the multiplexing of periodic sources in two-stage tree networks as a first step to later extend it to M-stage networks. Each source is periodic and emits a constant number, b, of back-to-back cells remaining silent during the rest of the period, T-b slots. The time at which the burst becomes active is uniformly and independently distributed within the period, see Figure 1. By choosing $b = 1$, we have CBR sources. We consider the pooling of K buffers into a single queue, the root queue, which receives also N_r sources as input (called exogenous traffic). Each one of the i (i=1...K) buffers multiplexes N_i sources. We consider every source in the network to have identical characteristics, burst b and period T, and every queue to have a constant service time of one ATM slot. In fact the hypothesis that every source has the same characteristics can be relaxed to the hypothesis that the external sources which feed a given queue have the same characteristics and all the sources which enter the network are of equal period. We assume stability at each queue of the tree considering that $N_i b < T$ for all $i = 1, ... K$ and $(\sum_{i=1}^{K} N_i + N_r) \cdot b < T$.

We are interested in the distribution of the queue length at the root queue. The distribution of the Virtual Waiting time of a single queue system fed by periodic sources can be computed, for instance, by means of the Benes bound, see Roberts and Virtamo (1991) for the multiplexing of periodic sources emitting one cell and see García *et al* (1995) for the multiplexing of periodic sources emitting bursts of b cells. For simplicity's sake we substitute the burst arrivals of the sources which feed the second stage queues by batch arrivals of size b. This does not suppose any change in the results for the root queue, see Figure 2. Therefore, when we calculate the queue length distribution in the root queue we assume that each source which enters the queues of the second stage generates the following pattern: a batch of b cells followed by a silence of $T-1$ slots, see Figure 1. The arrival epoch of the batches also is independently and uniformly distributed within the period T.

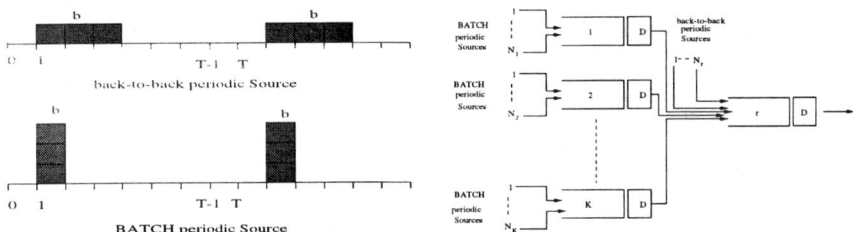

Figure 1 Back-to-back and Batch periodic sources.

Figure 2 Two-stage tree network.

We define $N^r(t,0)$ as the number of cell arrivals to the root queue r in the interval [-t,0), and L^r_{-t} as its queue length at time -t. We can express $P\{L^r_0 > x\}$ as:

$$P\{L^r_0 > x\} = \sum_{t=1}^{T} P\{N^r(t,0) = t + x \ \ and \ \ N^r(u,0) < u + x, t < u \le T\} \quad (1)$$

Note that events that occur in the queues of the second stage at any time will influence the root queue one slot later. We define the following functions:

- $B^r(t_1, t_2)$ is the number of exogenous sources that become active in an interval $[-t_1, -t_2)$ in the root queue.
- $B^i(t_1, t_2)$ is the number of batch arrivals to the i-th queue in an interval $[-t_1, -t_2)$, with $i = 1, \cdots, K$.
- $G(t, t + x, n_1, ..., n_K, n_r) = P\{N^r(t,0) = t + x \ \ | \ \ N^r(u,0) < u + x, t < u \le T, B^1(t+1,1) = n_1 ..., B^K(t+1,1) = n_K, B^r(t,0) = n_r\}$
- $L(t, n_1, ..., n_K, n_r) = P\{N^r(u,0) < u+x, t < u \le T \ \ | \ \ B^1(t+1,1) = n_1 ..., B^K(t+1,1) = n_K, B^r(t,0) = n_r\}$

- $B(t, n_1, ..., n_K) = P\{B^1(t+1,1) = n_1, ..., B^K(t+1,1) = n_K\}$.

Equation (1) can be transformed into:

$$P\{L_0^r > x\} = \sum_{t=1}^{T} \sum_{n_1=0}^{N_1} \cdot\cdot \sum_{n_K=0}^{N_K} \sum_{n_r=0}^{N_r} G(t, t+x, n_1, ..., n_K, n_r) \cdot$$
$$\cdot L(t, n_1, ..., n_K, n_r) \cdot B(t, n_1, ..., n_K) \cdot P\{B^r(t,0) = n_r\} \quad (2)$$

(a)   Term $G(t, t+x, n_1, ..., n_K, n_r)$:

In order to obtain this term we should bear in mind that the event $\{N^r(u,0) < u+x, t < u \le T\}$ implies that the K queues are empty at time -(t+1). Therefore the contribution which each queue of the second stage makes to the root queue during the interval $[-t,0)$ can only have as its origin the cells which arrive at the said queue during the interval $[-(t+1),-1)$. We introduce now the following notation:

- $N^{ri}(t,0)$ is the number of cells that arrive at the root queue r from the i-th queue in the interval $[-t,0)$.
- $R^{ri}(t, k_i, n_i)$ is the probability $P\{N^{ri}(t,0) = k_i \mid queue\ i\ is\ empty\ at\ -(t+1), B^i(t+1,1) = n_i\}$.
- $N^{rr}(t,0)$ is the number of cells that arrive at queue r from the exogenous sources.
- $R^{rr}(t, k_r, n_r)$ is $P\{N^{rr}(t,0) = k_r \mid queue\ r\ is\ empty\ at\ -t, B^r(t,0) = n_r\}$.

Probability $G(t, t+x, n_1, ..., n_K, n_r)$ will be the convolution of R^{ri} ($i = 1, ...K$) and R^{rr}:

$$G(t, t+x, n_1, ..., n_K, n_r) = \sum_{\substack{\sum_{j=1}^{K} k_i + k_r = t+x}} \prod_{i=1}^{K} R^{ri}(t, k_i, n_i) \cdot R^{rr}(t, k_r, n_r) \quad (3)$$

(b)   Terms $R^{rr}(t, k_r, n_r)$ and $R^{ri}(t, k_i, n_i)$:

The term $R^{rr}(t, k_r, n_r)$ can be obtained from a partial result obtained in García *et al* (1995). In that system, three regions were considered: $t < b$, $b \le t < T - b$ and $T - b \le t < T$. For a proof of this result we refer to García *et al* (1995). Term $R^{rr}(t, k_r, n_r)$ is then expressed as:

$$\begin{cases} \frac{1}{t^{n_r}} q_t^{(n_r)}(k_r) & if \quad t < b, \quad n_r \le N_r \\ \sum_{j=0}^{n_r} \binom{n_r}{j} \frac{(t-b+1)^j}{t^{n_r}} q_{(b-1)}^{(n_r-j)}(k_r - bj) & if \quad b \le t < T - b, \quad n_r \le N_r \\ \sum_{j=0}^{N} \binom{N_r}{j} \frac{(t-b+1)^j}{t^{N_r}} q_{(b-1)}^{(N_r-j)}(k_r - bj) & if \quad T - b \le t < T, \quad n_r = N_r \end{cases} \quad (4)$$

Where $q_a(y)$ is the discrete-time unitary pulse in the interval $[1,a]$ and $q_a^{(k)}(y)$ is its k-th discrete-time convolution. A simple expression for $q_a^{(k)}(y)$ is derived in García *et al* (1995) (see also Feller):

$$q_a^{(k)}(y) = \sum_{s=0}^{\lfloor \frac{y-k}{a} \rfloor} (-1)^s \binom{k}{s} \binom{y - s \cdot a - 1}{k - 1} \quad if\ k > 0\ \ for\ y = k, \ldots, ka \quad (5)$$

And $q_a^{(k)}(y) = 0$ for other values of y. For $k = 0$ we define $q_a^{(0)}(y) = \delta(y)$, where $\delta(y)$ is the discrete-time impulse function.

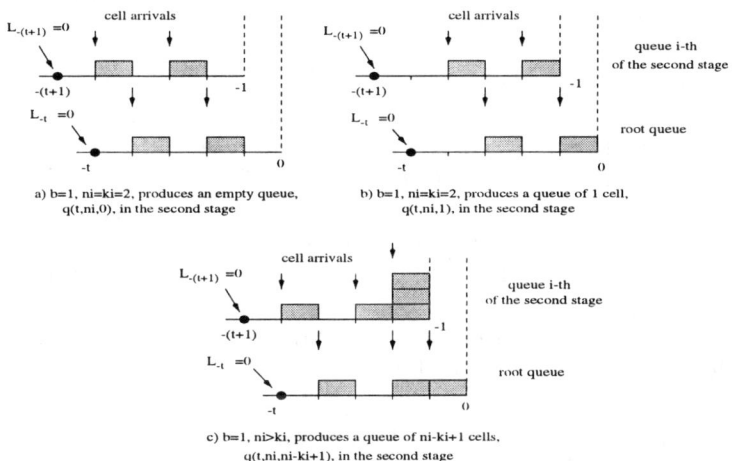

a) b=1, ni=ki=2, produces an empty queue, q(t,ni,0), in the second stage

b) b=1, ni=ki=2, produces a queue of 1 cell, q(t,ni,1), in the second stage

c) b=1, ni>ki, produces a queue of ni-ki+1 cells, q(t,ni,ni-ki+1), in the second stage

Figure 3 Cases to calculate the term $R^{ri}(t, k_i, n_i)$, (with b=1).

To calculate the term $R^{ri}(t, k_i, n_i)$, we must take into account the fact that the temporal axis in the queues of the second stage is shifted one time slot (the arrivals which are relevant to the root queue in the time instant $t = 0$ are produced in the queues of the second stage during the interval $[-(T+1),-1)$). Since each queue begins empty, we can substitute queue i for a queue of period t in the interval $[-(t+1),-1)$, to which there arrive n_i batches of size b uniformly and independently distributed. It follows that, if $n_i b$ cells arrive at queue i and k_i cells are emitted, at time instant $t = -1$ there will be either a queue of length $n_i b - k_i + 1$ (in the case $n_i b > k_i$, see Figure 3.c) or a queue

of length 0 or 1 (in the case $n_i b = k_i$, see Figures 3.a,b). In consequence we
will have the following cases:

$$R^{ri}(t, k_i, n_i) = \begin{cases} 1 & if \quad k_i = 0 \quad and \quad n_i = 0 \\ 0 & if \quad k_i = 0 \quad and \quad n_i > 0 \\ 0 & if \quad k_i > 0 \quad and \quad n_i \cdot b < k_i \\ q_b(t, n_i, 0) + q_b(t, n_i, 1) & if \quad k_i > 0 \quad and \quad n_i \cdot b = k_i \\ q_b(t, n_i, n_i b - k_i + 1) & if \quad k_i > 0 \quad and \quad n_i \cdot b > k_i \end{cases} \quad (6)$$

Where $q_b(T, N, x)$ is the probability that the queue length of an $ND/D/1$
with period T and N batch arrivals of length b is x. As it is easily demonstrated, the expression $q_b(T, N, x)$ is equal to $Q_b(T, N, x - 1) - Q_b(T, N, x)$,
where for $T + x > Nb$:

$$Q_b(T, N, x) = \sum_{i=\lceil \frac{x}{b} \rceil}^{N} \binom{N}{i} (\frac{ib - x}{T})^i (1 - \frac{ib - x}{T})^{N-i} (1 - \frac{(N - i)b}{T - ib + x}) \quad (7)$$

(c)   Term $L(t, n_1, ..., n_K, n_r)$:
To calculate $L(t, n_1, ..., n_K, n_r) = P\{N^r(u, 0) < u + x, t < u \le T \mid B^1(t + 1, 1) = n_1, ..., B^K(t+1, 1) = n_K, B^r(t, 0) = n_r\}$ we use similar arguments as in
Roberts and Virtamo (1991) or García *et al* (1995). The event $P\{N^r(u, 0) < u + x, t < u \le T \mid B^1(t + 1, 1) = n_1, ..., B^K(t + 1, 1) = n_K, B^r(t, 0) = n_r\}$
corresponds to the situation in which the root queue of an auxiliary network
of queues loaded with periodic arrivals of period $T - t$ in the interval [-(T-t),0)
is empty at time 0.

During a period, the root queue of the said auxiliary network receives cells
with different origins, see Figure 4.b:

- cells which come from the exits of the second stage queues,
- cells which come from bursts generated by the external sources which directly feed the root queue and that began during the actual period, and
- cells which come from bursts generated by the external sources which directly feed the root queue and that began in a previous period but that in
  part contribute to the actual period.

The cells which enter the root queue from the second stage queues in the
auxiliary system are those which arrived at the second stage queues in $N_i - n_i$
batches of b cells uniformly distributed in the interval [-(T-t+1),-1). At this
point it is important to note that for the auxiliary root queue to be empty in
the time instant $t = 0$ it is necessary for the second stage queues to be empty
too. This means that all cells which arrive at the second stage queues between
[-(T-t+1),-1) must be served beforehand and routed to the root queue at time
instant $t = 0$.

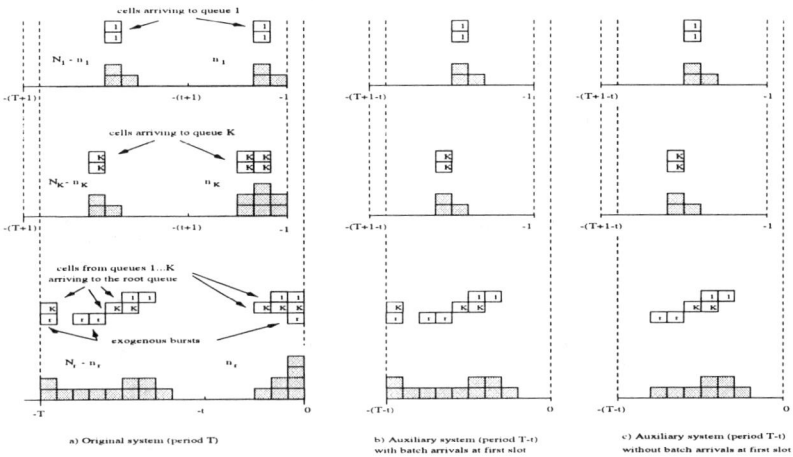

Figure 4 Equivalent system to calculate term $L(t, n_1, ..., n_K, n_r)$. a) Original system. b) Auxiliary system, where bursts from the auxiliary second stage queues and exogenous bursts plus cells belonging to bursts that began in the previous period arrive at the auxiliary root queue. c) Auxiliary system, where only bursts from the auxiliary second stage queues and exogenous bursts arrive at the auxiliary root queue

Arrivals which proceed directly from external sources and which feed the root queue in the auxiliary system consist of $N_r - n_r$ bursts of b cells uniformly distributed in the interval $[-(T-t),0)$.

Lastly, in the root queue of the auxiliary system we have an entrance in batch at the beginning of interval $[-(T-t),0)$ with which we model the contribution of the bursts which (in the original system) began to be emitted in the previous period. This batch consists of $\sum_{i=1}^{K} n_i b + n_r b - t - x$ cells.

Since the original root systems is stable, $\sum N_i b + N_r b < T$, the auxiliary system will be so too: $\sum_{i=1}^{K}(N_i - n_i)b + (N_r - n_r)b + \sum_{i=1}^{K} n_i b + n_r b - t - x < T - t$. Therefore the batch arrival at the beginning of the interval $[-(T-t),0)$ will be served before the end of the interval, and will not contribute to the length of the queue in the instant $t = 0$. We have then a situation in which the only arrivals which have to be counted in order to obtain an empty queue in the auxiliary root queue at time 0 are those which proceed from the exit of the $N_i - n_i$ batches and the $N_r - n_r$ exogenous sources (see Figure 4.c). Therefore:

$$L(t, n_1, ..., n_K, n_r) = 1 - \frac{(\sum_{i=1}^{K}(N_i - n_i) \cdot b + (N_r - n_r) \cdot b)}{T - t} \tag{8}$$

(d)   Terms $P\{B^r(t,0) = n_r\}$ and $B(t, n_1, ..., n_K)$:
$P\{B^r(t,0) = n_r\}$ has a binomial distribution.

$$P\{B^r(t,0) = n_r\} = \left(\begin{array}{c} N_r \\ n_r \end{array} \right) (\frac{t}{T})^{n_r} \cdot (1 - \frac{t}{T})^{N_r - n_r} \qquad (9)$$

Since each queue is independent and the batches are uniformly distributed, the term $B(t, n_1, ..., n_K)$ can be calculated as:

$$B(t, n_1, ..., n_K) = \prod_{i=1}^{K} P\{B^i(t+1, 1) = n_i\} = \prod_{i=1}^{K} \left(\begin{array}{c} N_i \\ n_i \end{array} \right) (\frac{t}{T})^{n_i} \cdot (1 - \frac{t}{T})^{N_i - n_i} \quad (10)$$

2.2   M-stage tree network.

Once the two stage system is solved, we can generalize the results for an M-stage tree network. Multiplexing stage i is pooled by K_{i-1} queues and N_{pq} exogenous sources (with q=1...K). We can make use of a property of the rooted tree networks with discrete-time single server queues presented by Morrison (1978). Morrison showed that in such networks, the rooted tree network can be reduced to a two-stage network with a prescribed input. Specifically, to calculate the queue length distribution at the root queue, we can assume that the external sources that enter the queues which hang on each branch of a second stage queue, enter the second stage queue directly.

Using this argument, and the fact that each queue of the network is a root queue of a corresponding sub-tree, the analysis of the network with M-stages can be done from the analysis of equivalent systems with $M = 1$ and $M = 2$.

For example, consider a three multiplexing stage network, see Figure 5. Each queue is fed by exogenous sources. In order to obtain the queue length distribution at queue r (the root queue), we can substitute this system for an equivalent two-stage network (Figure 6). In this equivalent system, only the sources which enter the root queue directly are back-to-back periodic sources. The rest behave as Batch periodic sources.

3   AVERAGE DELAYS AND AVERAGE NUMBER IN THE QUEUE IN THE TREE NETWORK.

To give close-form formulas for the average queueing delay and average waiting time in any queue of the tree network, we will make use of an equivalent network based on priorities proposed by Modiano et al (1996). In this proposal, they first consider a simple case with two queues in tandem and Poisson traffic as input. High-priority is given to customers in transit from one queue to the other, while low-priority is given to the exogenous traffic that enters directly

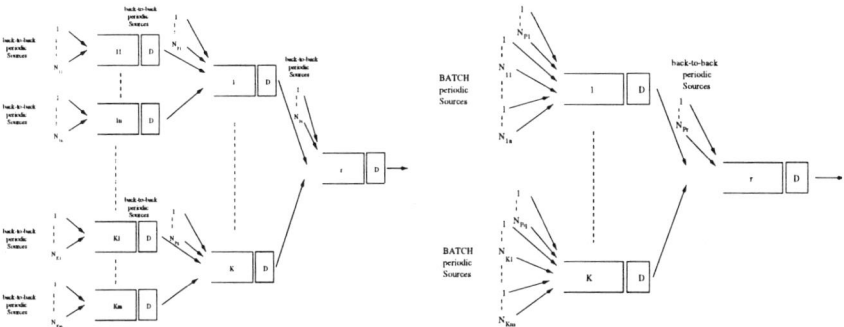

Figure 5 Three stage network.   **Figure 6** Equivalent network.

the first queue. If $\overline{Q}$ is the average overall queue size (low and high priority traffic), $\overline{Q_1}$ is the average queue size at the high priority queue and $\overline{Q_2}$ is the average queue size at the low priority queue, then it holds that $\overline{Q_2} = \overline{Q} - \overline{Q_1}$.

Solving this system, an extension to larger systems can be easily derived. If we consider a two stage system in which the second stage consists of K queues whose output traffic is routed to the first stage as in Figure 2, the average number of customers in the root queue will be $\overline{Q_r} = \overline{Q} - \sum_{i \in K} \overline{Q_i}$.

Where $\overline{Q_i}$ is the average number of customers in queue $i = 1, ..., K$ including queue room and service room and $\overline{Q}$ is the average number of customers in the total system except for those being serviced in the root queue.

We will apply this formula to calculate average delays in any queue system of the tree whose input is homogeneous periodic traffic of period T and burst b. Dron *et al* (1991) show that the average waiting time in the queue, $\overline{W}$, of a queue system multiplexing N CBR periodic sources of period T_{cbr} with load $\rho = \frac{N}{T_{cbr}}$ is the following:

$$\overline{W} = \frac{(N-1)!}{2} \sum_{k=1}^{(N-1)} \frac{1}{T_{cbr}^k} \frac{1}{(N-1-k)!} \tag{11}$$

The number of cells in the queue $\overline{Q}$ can be calculated using Little's formula: in periodic systems this can be interpreted as following, see Humblet *et al* (1993):

$$P\{W = x\} = P\{a \text{ departure leaves } x - 1 \text{ cells in the queue}\} =$$
$$\frac{P\{Q=x\}}{P\{Q>0\}} = \frac{T_{cbr}}{N} P\{Q = x\} \text{ with } x > 0 \tag{12}$$

Therefore applying this relation and taking into account that the service time is the time unit:

$$\overline{Q} = \rho \cdot \overline{W} = \frac{N!}{2T_{cbr}} \sum_{k=1}^{(N-1)} \frac{1}{T_{cbr}^k \cdot (N-1-k)!} \tag{13}$$

In the general case of multiplexing back-to-back periodic sources with burst size b and period $T = b\, T_{cbr}$, the average waiting time in the queue can be expressed as b times the average waiting time in the queue for an $ND/D/1$ system with N CBR sources of period T_{cbr}. Thus:

$$\overline{Q} = b\overline{Q}_{cbr} = \frac{N!\, b}{2T_{cbr}} \sum_{k=1}^{(N-1)} \frac{1}{T_{cbr}^k} \frac{1}{(N-1-k)!} = \frac{N!\, b^2}{2T} \sum_{k=1}^{(N-1)} \left(\frac{b}{T}\right)^k \frac{1}{(N-1-k)!} \tag{14}$$

We consider the same tree topology as in the former section. The second stage has K queues fed each one by N_i periodic sources of period $T = b \cdot T_{cbr}$ being $i = 1,...,K$ and b the burst size. Let ρ_i be the load at each queue ($\rho_i = \frac{N_i\, b}{T}$). The output of each of these queue systems is multiplexed with N_r exogenous periodic sources with load $\rho_r = \frac{N_r\, b}{T}$, see Figure 2. The total load in the root queue is $\rho = \sum_i \rho_i + \rho_r$. We define N as the total number of sources in the system: $N = \sum_i N_i + N_r$. Therefore, we can express the average number of cells in the root queue as:

$$\overline{Q}_r = \frac{N!\, b^2}{2T} \sum_{k=1}^{(N-1)} \frac{b^k}{T^k(N-1-k)!} - \sum_{i=1}^{K} \frac{N_i!\, b^2}{2T} \sum_{k=1}^{(N_i-1)} \frac{b^k}{T^k(N_i-1-k)!} \tag{15}$$

The average waiting time in the queue is given by Little's formula $\overline{W}_r = \frac{T}{Nb}\overline{Q}_r$ and the average waiting time and average number of cells in the system can be expressed as: $\overline{W}_{r_s} = \overline{W}_r + 1$ and $\overline{Q}_{r_s} = \overline{Q}_r + \rho$.

4   RESULTS.

In order to see how the number of second-stage queues influences the root queue, we have studied various configurations with different loads. In Figure 7 we have drawn the required buffer length in a root queue so that the probability of cell losses is below 10^{-10} as a function of b when the second-stage has two queues. We call the number of sources which enter the second-stage queues N_1 and N_2 respectively. As can be seen in Figure 7, the worst situation occurs with balanced loads (e.g. $N_1 = N_2 = 6$ over a period $T = 15b$).

Finally in Figure 8, we show the queue length distribution when the second stage has $K = 2, 3, 4, 12$ queues. The curves are calculated with a total load

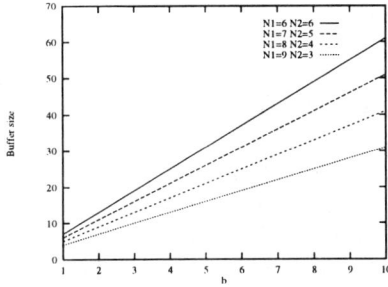
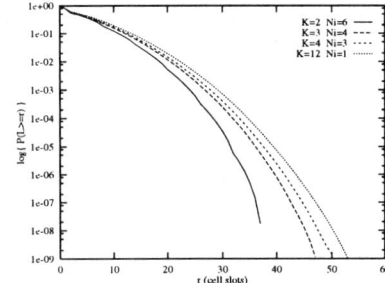

Figure 7 Buffer size in the root queue versus b. Two stages. **Figure 8** Queue length distribution for 2,3,4,12 queues in the second stage.

	two stages, K=2			two stages, K=4		
ρ	$T_{cbr} = 100$	$T_{cbr} = 1000$	$M/D/1$	$T_{cbr} = 100$	$T_{cbr} = 1000$	$M/D/1$
0.90	2.365	3.695	4.090	2.612	3.957	4.354
0.80	1.286	1.610	1.666	1.482	1.817	1.875

Table 1 Average waiting time in the queue for CBR sources and Poisson.

of $\rho = 0.8$ and $b = 6$. The period of the sources is $T = 90$ and the load is balanced among the K queues ($N_k = \frac{12}{K}$ sources in each queue). The worst situation is when each queue is fed by only one source ($K = 12$). In that case, the same results would be obtained as if 12 sources were multiplexed in one stage.

In table 1 we compare the average waiting time in the queue, $\overline{W}$, for a root queue that is pooled by $K = 2$ and $K = 4$ buffers. In this case the load is balanced among the buffers. The average waiting time in the root queue is compared for the different periods and for the case in which Poisson traffic of load equal to the periodic traffic enters at each queue. We can observe what it is known: Poisson is pessimistic for high loads as in the single queue system. For $K = 2$ the average delay decreases slightly respect to the case of $K = 4$, but not too much. We also observe that if the number of buffers are increased it looks more like a single queue system.

5 CONCLUSIONS.

We have obtained the probability distribution function of the virtual waiting time in the root queue of a tree network of discrete time queues with constant service time whose input is periodic traffic. Using the Benes approach it is possible to compute the Virtual Waiting time CPDF in any queue of the tree network. For that purpose, we first solve a tree of two stages. The second stage consists of K queues fed by periodic sources. The output of each of these queues is routed to the second stage and is multiplexed with exogenous periodic traffic. We assume that each queue either in the first stage or the second is stable.

Once the two-stage system is solved, we can make use of a property of the rooted tree networks with discrete-time single server queues presented by Morrison (1978) in to solve the M-stage queueing network. In such networks the external sources that feed queues that hang on a branch of a second stage queue, can be connected directly to the corresponding second stage queue root. In this way, we have an equivalent system of two stages.

To compute the average waiting time and the average number in the queue, we apply a model, presented in Modiano *et al* for Poisson traffic, based on an equivalent network where high priority is given to traffic in transit and low priority is given to exogenous traffic. We apply this method considering now that the input traffic is periodic and give closed formulas for the former average parameters.

REFERENCES

[1] ITU-T Recommendation I-371 (1995) "Traffic Control and Congestion Control in B-ISDN", Draft issue, Geneva, 1995.

[2] ATM Forum (1996) "ATM Forum Traffic Management Working Specification", Version 4.0, 1996

[3] Dron, L., Ramamurthy, G. and Sengupta, B. (1991) "Delay analysis of continuous Bit Rate Traffic over an ATM network", *IEEE JSAC*, Vol. 9, No 3, April 1991

[4] Feller, W. "An Introduction to Probability theory and its Applications", Vol 2, Ed. John Willey.

[5] García Vidal, J., Barceló, J.M. and Casals, O. (1995) "An Exact Model for the Multiplexing of Worst Case Traffic Sources", *IFIP Conference on Performance on Computer Networks*, PCN'95, Instambul, Turkey, October 1995.

[6] Humblet, P., Bhargava, A., and Hluichyi, M. (1993) "Ballot Theorem applied to the Transient Analysis of $nD/D/1$ queues", *IEEE/ACM Transactions on Networking*, Vol 1, No 1, February 1993.

[7] Modiano, E., Wieselthier, J.E. and Ephemerides, A. (1996) "A simple analysis of average queueing delay in tree networks", *IEEE Transactions on Information Theory*, Vol. 42, No 2, March 1996

[8] Morrison, J.A. (1978) "A Combinatorial Lemma and its Application to Concentrating Trees of Discrete-Time Queues", *The Bell System Technical Journal*, Vol. 57, No 5, May-June 1978.

[9] Roberts, J.W. and Virtamo, J. (1991) "The Superposition of Periodic Cell Arrival Streams in an ATM Multiplexer", *IEEE Trans. on Comm.*, Vol. 39, No. 2, Feb. 1991.

[10] Roberts, J.W., Mocci, U. and Virtamo, J. Eds. (1996) "Broadband Network Teletraffic" - *Final Report of Action COST 242*. Springer Verlag

46

Dimensioning of ATM Networks with Finite Buffers under Call-Level and Cell-Level QoS Constraints

A. Girard
INRS-Télécommunications
16 Place du Commerce, Verdun, (Que) H3E 1H6 Canada. email:
andre@inrs-telecom.uquebec.ca, Ph: (514)-765-7832, Fax: (514)-761-
8501.

C. Rosenberg
École Polytechnique de Montréal
Département de Génie Electrique et Génie Informatique, PO Box
6079, Succ. "Downtown", Montréal, H3C 3A7 Canada. email:
cath@comm.polymtl.ca, Ph: (514)-340-4123, Fax: (514)-340-4562.

Abstract

In broadband networks with guaranteed Quality of Service (QoS), dimensioning becomes an even harder task than in traditional data or circuit networks. The network operator has two resources to manage — the trunk capacity and the switch memory (buffers) — and multiple QoS constraints to respect — call blocking, cell loss ratio and cell delays. We investigate how this new context can affect the design, cost and dimensioning of networks. In particular, we study how much can be gained if we do not restrict ourselves to the two limit cases of a bufferless or an infinite buffer system. We deal first with the single-service case for a single link. We then investigate the multi-service case on a single link. We study the cost of integration when service classes have very different delay constraints. The main results obtained are that finite buffers can have a very significant effect on the optimal cost whenever buffer costs are high, as in satellite systems, especially when a delay constraint is imposed on the problem and that services integration can be very costly when different delay requirements are involved.

Keywords

Broadband Networks, dimensioning, design of networks with finite buffers, quality of service, satellite systems

Broadband Communications   P. Kühn & R. Ulrich (Eds.)
© 1998 IFIP. Published by Chapman & Hall

1   INTRODUCTION

Designing and dimensioning ATM networks present some new challenges to the network operators. Some of them are related to the fact that these networks have to support multiple services with very different characteristics and others to the fact that these services require commitments on QoS (rather than simply objectives) at the call and cell level (other challenges are dealt with in (Girard & Rosenberg 1997)). The context in which this study has been done is the one of guaranteed bandwidth ATM Transfer Capabilities (ATCs) (ITU-T 1996), i.e., the Deterministic Bit Rate (DBR) and the Statistical Bit rate (SBR). We focus on the SBR ATC which is used to attain reasonable efficiency through statistical multiplexing while providing the stringent QoS guarantees demanded by some VBR applications.

The QoS requirements at the cell level is on Cell Loss Ratio (CLR) and may also be on cell delay (mean delay, maximum delay and/or Cell Delay Variation (CDV)). These requirements put very stringent constraints on the buffer dimensioning. Hence, the network operator has to manage two resources from QoS and cost standpoints, namely the trunk capacity and the switch memory (buffers). This is particularly important for systems where buffers are very expensive or where buffer size is severely limited, as in satellite switches.

A large literature exists on design methods for broadband networks (see (Hui 1988, Ash, Chang & Labourdette 1994, Girard & Lessard 1992, Girard & Zidane 1995) for some references). All these methods are based on the same technique. First, an effective bandwidth is defined for each service class. The multi-service design problem is then equivalent to a multi-rate circuit-switched network design problem. Standard design methods, coupled to fast approximation to the multi-rate blocking formulas, can then be used to obtain a good near-optimal solution.

The three main weaknesses of this approach are 1) that most of the currently proposed effective bandwidth models are based on approximations assuming either no buffer or an infinite buffer, 2) that usually a single cell QoS constraint, namely the cell loss probability is considered and 3) that the underlying source model is not usually in line with the ITU-T traffic descriptors specification (ITU-T 1996).

We use simple models to investigate how finite buffers can affect the design, cost and dimensioning of this new type of networks. We have decomposed the paper in 2 parts. The first one deals with the single-service case and a single buffer. Section 2 formulates the problem in this case and gives some numerical results showing that finite buffers and delay constraints can have a significant cost impact. In the second part, corresponding to Section 3, we study a multi-service model in the single-buffer case. Section 4 presents our conclusions and future work.

2 SINGLE-SERVICE SINGLE LINK CASE

In this section, we consider a single multiplexer or a switch output including buffer and line in which all sources (e.g., connections) are identical. In this way, we can study the effect of the multi-QoS commitments and multi-resource context with a simple call admission control procedure without the added complexity of effective bandwidth (see Section 3). No assumptions are yet made on the source model and the approach that we propose below is general. We define the following parameters:

w the revenue generated by each connected call.
A the parameter λ/μ of the Poisson process describing the arrival of connection requests to the server.
L the call loss probability.
$\overline{L}$ the maximum call loss probability. This is the call-level QoS constraint.
B the call blocking probability. This is the probability that a call cannot be connected on the link at the time of its arrival. In the single-buffer case, this is identical with the call loss probability L. In a network, this may or may not be the case, depending on the call admission and routing techniques used.
C the capacity of the output line, i.e., the rate of the server.
D the maximum transit delay for each cell going through the switch or multiplexer.
$\overline{D}$ the maximum transit delay allowed for cells in the switch. This is one of the 2 cell-level QoS constraints.
P the cell loss probability.
$\overline{P}$ the maximum cell loss probability. This is one of the 2 cell-level QoS constraints.
K the buffer size, expressed in number of cells.
$E(A, N)$ the Erlang-B function for traffic intensity A and N servers.
β a vector that describes the source and various QoS parameters.

2.1 Call admission control

Since we want to examine the trade-off between buffer size and capacity (i.e., server rate), we determine the optimal buffer size and rate by solving the problem

$$\min_{K,C} \ z(K,C) \text{ subject to } B(A, K, C, \beta) \leq \overline{L} \text{ and } K, C \geq 0 \qquad (1)$$

where $N(K, C, \beta)$ is the largest number of calls with parameter vector β that can be accepted such that all the cell QoS constraints are met and $z(K, C)$ is some yet unspecified objective function that depends on K and C.

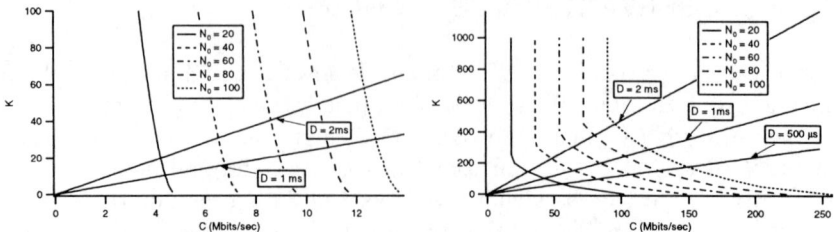

Figure 1 Contours of $N(K,C,\overline{P})$ at different levels N_0. Sources are Class 1

Figure 2 Contours of $N(K,C,\overline{P})$ at different levels N_0. Sources are Class 2

We have an explicit form for the blocking function if we assume that the call arrival process is Poisson.

$$B(A,K,C,\beta) = E[A, N(K,C,\beta)]. \tag{2}$$

The blocking constraints (1) can then be rewritten as $E[A, N(K,C,\beta)] \leq \overline{L}$. which can be further simplified if we define N_0 as the solution of the equation $E(A,N) = \overline{L}$. The constraint (1) on call loss probability can then be expressed, for a given pair $(A, \overline{L})$, simply as

$$N(K,C,\beta) \geq N_0. \tag{3}$$

Eq. (3) takes care of the call loss probability constraint while the set of all cell QoS constraints are present through the parameter β of $N(K,C,\beta)$.

2.2   The problem domain

We can obtain further insight on the problem structure by taking advantage of the particular cell QoS constraints that we have chosen, $\overline{P}$ and $\overline{D}$. We can replace the constraint $N(K,C,\overline{P},\overline{D}) \geq N_0$ by two decoupled constraints. These constraints still depend implicitly on the source parameters that have not been indicated to simplify the notation. The first is of the form $N(K,C,\overline{P}) \geq N_0$ and deals only with the cell loss QoS constraint. It states that the acceptance region, defined by $N(K,C,\overline{P})$, should be large enough so that the call-level QoS constraints can be met. Since D is the largest delay that each cell can experience going through the switch, we can write $D = K/C$ and the QoS delay constraint can be expressed as $D \leq \overline{D}$ which gives, in terms of buffer size, $K \leq \overline{D}C$. Note that this decoupling of the two cell QoS constraints is possible only because the maximum cell delay is independent of the number of connections in the system.

We now examine the structure of the problem. We have selected two service

Class	PCR (Mbits/sec)	SCR (Mbits/sec)	MBS (cells)	$\overline{P}$
1	0.320	0.064	50	10^{-7}
2	9.	0.9	20	10^{-7}

Table 1 Source parameters for the Worst-Case Traffic (WCT) model

classes characterized by the parameters indicated in Table 1. We do not claim that those values are representative of real services. Instead, they were chosen simply to have very distinct behavior but still within a reasonable range.

First we look at the contours of $N(K, C, \overline{P}) = N_0$ in the (K, C) space (i.e., the isocontour $K(C)$ at level N_0). This is shown on Fig. 1 for Class 1 services. The curves are labeled with values of N_0, each value corresponding to a set of pairs $(A, \overline{L})$. In addition, we have indicated the lines corresponding to the cell delay constraint labeled in milli- or micro-seconds.

We can examine the effect of the source rate on the domain by looking at Fig. 2 where we have assumed that each source operates at a higher rate (Class 2 services). On these graphs, the vertical segment of the curve corresponds to the condition $\rho = 1$, since we do not expect that real networks will operate with utilization higher than 1. As we can see, there is no qualitative difference between the two figures. We have also examined the effect of the cell loss QoS when the value of $\overline{P} = 10^{-9}$. We could observe little qualitative difference in the form of the domains.

2.3 Optimal dimensioning

The optimization problem is then

$$\min_{K,C} z(K, C) \text{ subject to } N(K, C, \overline{P}) \geq N_0 \text{ and } K \leq \overline{D}C \text{ with } K, C \geq 0$$

where N_0 has been calculated from the values of A and $\overline{L}$. We choose to minimize the dimensioning cost, in which case we have $z = C_K K + C_C C$ and the optimization problem is

$$\min_{K,C} z = C_K K + C_C C \text{ subject to } N(K, C, \overline{P}) \geq N_0 \ (\lambda) \text{ and } K \leq \overline{D}C \ (\mu)$$

where the dual variables are indicated with their corresponding constraints. Note that the variables λ and μ defined here have nothing to do with the λ and μ that are used to represent the arrival and service rate for the call process, as used in the definition of A.

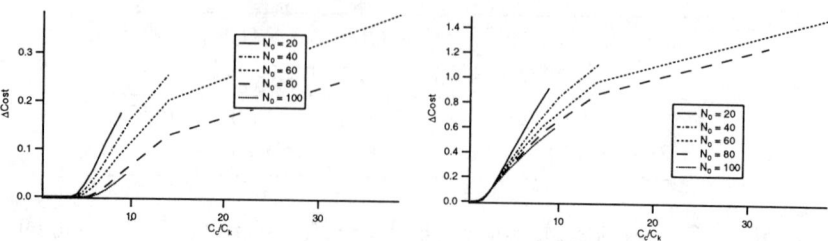

Figure 3 Dimensioning cost of the delay constraint, sources are Class 2, $\overline{D}_2 = 1$ ms

Figure 4 Dimensioning cost of the delay constraint, sources are Class 2, $\overline{D} = 500$ μs

2.4   Cost of the delay constraint

We can gather from Figs. 1 and 2 some information on the structure of the optimal solution as a function of the cost ratio $a \stackrel{\triangle}{=} C_C/C_K$ (here expressed in (\$/Mbits)/(\$/cell buffer space)). When a is small, the transmission cost is small and the optimal solution is at $K = 0$. In that case, the delay constraint is not binding. If a is large, the optimal solution is to choose C as small as possible consistent with the condition $\rho \leq 1$. We call the *trade-off region* the range of values of a where these two conditions are not met. In that range of values, it is possible to trade-off buffers for capacities. The point is that the size of this trade-off region is determined by the variation of the slope of the contour over the range of capacities. We see that for low-bandwidth sources, the trade-off region occurs at relatively high values of a while this region is in a much lower range of values for the high-bandwidth source. This notion will be useful again when we discuss multi-service systems.

We are now in a position to quantify the cost of imposing a maximum delay constraint QoS in addition to the traditional cell loss constraint. The results are presented as a function of the unit cost ratio a. For a given value of a, we can compute C_0, the optimal dimensioning cost without the delay constraint and C_1,the optimal dimensioning cost when the delay constraint is present. We use the ratio

$$\Delta = \begin{cases} (C_1 - C_0)/C_0 & \text{if the delay constraint is active} \\ 0 & \text{otherwise.} \end{cases}$$

We present the case corresponding to Fig. 2 since the other cases are not qualitatively different. The value of Δ is plotted on Fig. 3 as a function of a when the delay is $\overline{D} = 1$ msec. We can note three things on this graph.

First, the trade-off region is small. Near the origin, a is small and the optimal solution is $K = 0$. In that case, the delay constraint is not binding. At the other end, a is large and the optimal solution is to choose C as small as possible. In that case, the buffer would be chosen $K = \infty$ but this is prevented by the delay constraint and the optimal solution is always at the point where the delay constraint is tight. For values above that point, increasing a does not induce a corresponding change in the buffer size and the cost increment is linear in a.

Second, the cost increase is more important for small call traffic values, i.e., small N_0. Recall that the value of N_0 increases as the call traffic increases, for a given call loss probability value. The third is that the cost increase gets larger as the buffer cost gets relatively less expensive (high a). This result could have been deduced from Fig. 2 directly. What is not so immediate is the size of Δ, in this case as high as 20%.

Similar results are shown on Fig. 4 for a delay of 500 μsec. We see here a substantial increase in cost due to the delay constraint, as high as 140%.

The obvious conclusion is that for homogeneous traffic, adding the delay constraint can have a significant impact on the dimensioning cost of the system. We can also speculate that in a network dimensioning model, this will tend to regroup traffic on large links and eliminate small capacity links, since the delay constraint is much more costly for these small system. This would produce networks with a low-connectivity topology and tend to favor long paths to connect some origin-destination pairs, something that should generally be avoided.

3 MULTIPLE-SERVICE SINGLE LINK CASE

While the results of Sections 2 are interesting in their own right, we are really interested in the effect of finite buffers in the case of multi-service networks, (i.e., in the case of multiple connection types). For multiple call types, we define the additional parameters:

m the number of call types.

A a vector of offered call traffics whose component A_i, $i = 1, m$ represents the arrival rate of calls of type i.

x a vector $[x_1, x_2 \ldots x_m]$ where x_i is the number of calls of type i present in the system.

$Q_i(\mathbf{x})$ the ith cell QoS function (CLR, average delay, CDV, etc).

$\overline{Q}_i$ the largest valued permitted for the ith QoS function.

We now assume that we can model the cell process for each call type by a single parameter W_i called the *effective bandwidth* (Ahmadi & Guérin 1990, Rege 1994, Bean 1994, Dziong, Liao & Mason 1991, Choudhury, Lucantoni & Whitt 1994, Gibbens & Hunt 1991, Kelly 1991, Elwalid & Mitra 1993). With this technique, we define

W the vector of effective bandwidths for all call classes.

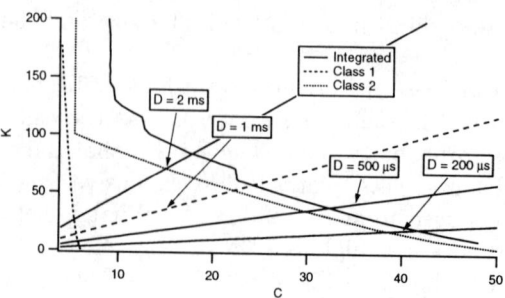

Figure 5 Problem domains for the multi-service system, Class 1 and 2, low call traffic

If we can assume that all effective bandwidths are a multiple of a given base rate C_0, then we can define $N(C)$ the number of circuits for a server rate C and we have $N(C) = \lfloor C/C_0 \rfloor$. In this model, we may have an integer or a fractional number of servers. We may also assume that the effective bandwidths can take any relative values, in which case the number of servers is a real number. This is not really a problem as long as we have continuous approximations (in N) for the blocking function. This is in effect similar to the extension of the Erlang B function to real values of the trunk group. The blocking probability is then defined as

$$B^i = \Pr\left\{\sum_{i=1}^m x_i W_i > N(C)\right\} \tag{4}$$

which can be evaluated, for Poisson arrival processes, exactly (Kaufman 1981, Roberts 1981) or approximately (Labourdette & Hart 1990, Gazdzicki, Lambadaris & Mazumdar 1993). Note that to be used in our context, the effective bandwidths should depend now both on the buffer size and the server rate, in addition to the source parameters, which are not indicated here. This is an additional difficulty since most currently known models for effective bandwidth are calculated for a *single* cell QoS constraint, generally the cell loss probability and usually for K assumed either very large or very small. Nevertheless, if we assume that we do have a suitable effective bandwidth model, the trunk group dimensioning problem becomes, for the cost minimization version

$$\min_{C,K} z = C_C C + C_K K \text{ subject to } B^i(\mathbf{A}, \mathbf{W}(C, K), N(C)) \le \overline{L}^i.$$

3.1   Problem domain

We can see the effect of integration and the interaction with the delay constraints on Fig. 5. First, we have shown the problem domain (i.e., the function

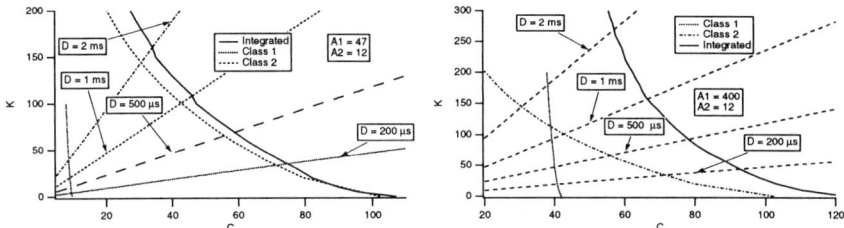

Figure 6 Problem domains, separate and integrated systems, Class 1 and 2, intermediate traffic

Figure 7 Problem domains, separate and integrated systems, Class 1 and 2, high traffic

$K(C)$) for the two call types separately and for the integrated model. We can see that the wideband service clearly dominates the shape of the integrated domain. Also note the steps in this domain while the corresponding curves for the isolated systems are smooth. This effect is real and probably depends on the discrete changes in values for the effective bandwidths caused by the fact that only integer values of connection numbers are possible.

A more interesting effect is seen when we consider the delay constraints indicated on the figure. Suppose that the two classes have distinct delay constraints $\overline{D}_1 = 200$ μs and $\overline{D}_2 = 2$ ms. In the integrated system, we have no choice but design for the more stringent value of these constraints. The feasible region is then given by the intersection of the $\overline{D} = 200$ μs line and the integrated admission curve. This value will in turn force a high value of bandwidth with a corresponding high cost as soon as the transmission cost is significant with respect to the buffer cost. If, on the other hand, we design each system independently, we can use the intersection of the $\overline{D} = 2$ ms curve with the Class 2 admission curve and the $\overline{D} = 200$ μs intersection with the Class 1 admission curve. Because these curves are very different, there will be substantial savings in having two separate systems instead of a single integrated one.

If, on the other hand, the values were $\overline{D}_1 = 2$ ms and $\overline{D}_2 = 200$ μs, then we would expect that the integrated system would provide savings as compared with the separate systems.

We have also investigated the effect of the call arrival rate on the admission regions. This is shown on Figs. 6–7. On Fig. 6, we see the effect of increasing A_2 while maintaining A_1 constant. The slope of the Class 2 curve varies over a wider range that for the case $A_2 = 2$ which means that the trade-off region will be larger. Also, in that case, the effect of the narrow-band traffic is not very important and the integrated system is driven by the wideband class almost exclusively.

We have then fixed A_2 to 12 and increased the value of A_1 to have a load comparable to the Class 2 load. Figure 7 clearly shows now that *both* classes have an effect on the domain of the integrated system. The narrow-band class

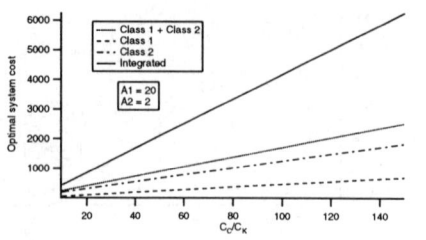

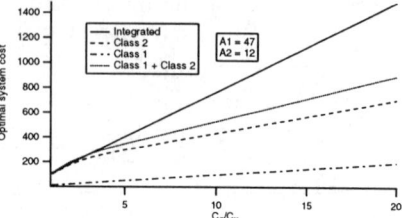

Figure 8 Cost of integration with delay constraints. Low-traffic system, $\overline{D}_1 = 2$ ms, $\overline{D}_2 = 200$ μs

Figure 9 Cost of integration with delay constraints. Medium-traffic system, $\overline{D}_1 = 2$ ms, $\overline{D}_2 = 200$ μs

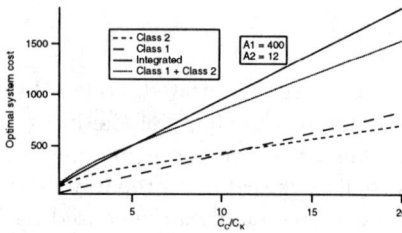

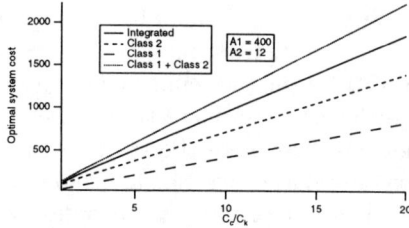

Figure 10 Cost of integration with delay constraints. High-traffic system, $\overline{D}_1 = 2$ ms, $\overline{D}_2 = 200$ μs

Figure 11 Cost of integration with delay constraints. High-traffic system, $\overline{D}_1 = 200$ μs, $\overline{D}_2 = 2$ ms

seems to act as a limiting factor preventing low values of C while the wideband class pushes up the buffer size.

3.2   Cost of integration

We now investigate the cost of integration when the delay constraints are very different for the two classes. We give on Figs. 8–10 the cost of separate systems designed for the delay constraint of each class separately. We also give the cost of the integrated system designed for the more stringent delay constraint, and the sum of the costs of the individual systems. There are substantial costs incurred by integration for the values of the delays chosen here, although in the separate systems, we would need to build two transmission systems with twice the termination cost.

The second is that the suitability of integration depends, in addition to the cost parameters, also on the arrival rate of call traffic. This is seen on Figs. 9 and 10. In the first case, the advantage of integration occurs for a very small range of values of a and is barely noticeable. In the second case, this range is larger and the benefit of integration is larger over that range. This is another example of the importance of taking into account the interaction of the call and cell processes when deciding on such issues as integration vs segregation.

The higher cost of the integrated system is also heavily dependent on which of the two services has the tightest delay constraint. If we interchange the values of the constraints so that we have $\overline{D}_1 = 200$ μs and $\overline{D}_2 = 2$ ms, we see from Fig. 11 that the multiplexing gain is sufficient to offset the fact that the integrated system must be designed for the smallest delay.

4 CONCLUSIONS

We have investigated here the effect of finite buffers on the optimization of call-level and cell-level QoS-constrained multi-service networks.

We have found that, when there is no delay constraint, the positive cost of the buffer yields a small increase on the total cost which is nearly linear with the cost of the transmission facility. When the presence of the buffer manifests itself in the form of a maximum delay constraint, however, we have shown that this can lead to large increases in the optimal cost (in some cases, up to 100%). We have also noted that this effect is more important for small call traffic values and speculated that this would lead to networks with concentrated traffic and hence sparse topologies.

For the multiple-service case, the main conclusion is that integrating sources on a single server may be more costly than serving them on separate facilities. This effect becomes more important as buffers get more expensive, as is the case for satellites systems. This conclusion, however, is strongly dependent on the delay constraints, the source parameters as well as the call arrival process.

Acknowledgment

This work was partly funded by NSERC Strategic grant No. STR0166996.

REFERENCES

Ahmadi, H. & Guérin, R. (1990), Bandwidth allocation in high-speed networks based on the concept of equivalent capacity, *in* 'Proc. 7th ITC Specialist Seminar'.

Ash, G., Chang, K. & Labourdette, J.-F. (1994), Analysis and design of fully shared networks, *in* Labetoulle & Roberts (1994), pp. 1311–1320.

Bean, N. (1994), Effective bandwidth with different quality of service requirements, *in* V. Iversen, ed., 'Proc. IFIP'94: Integrated broadband communication networks and services', Elsevier Science B.V., pp. 241–252.

Choudhury, G., Lucantoni, D. & Whitt, W. (1994), On the effectiveness of effective bandwidths for admission control in ATM networks, *in* Labetoulle & Roberts (1994), pp. 411–420.

Dziong, Z., Liao, K. & Mason, L. (1991), Buffer dimensioning and effective bandwidth allocation in ATM-based networks with priorities, *in* 'Proc. ITC Specialist Seminar', pp. 154–165.

Elwalid, A. & Mitra, D. (1993), 'Effective bandwidth of general markovian traffic sources and admission control of high-speed networks', *IEEE/ACM Transactions on Networking* **1**(3), 329–343.

Gazdzicki, P., Lambadaris, I. & Mazumdar, R. (1993), 'Blocking probabilities for large multirate Erlang loss systems', *Adv. in Appl. Prob.* **25**, 997–1009.

Gibbens, R. & Hunt, P. (1991), 'Effective bandwidth for multi-type UAS channels', *Questa* **9**, 17–28.

Girard, A. & Lessard, N. (1992), Revenue optimization of virtual circuit ATM networks, *in* 'Proc. Networks'92'.

Girard, A. & Rosenberg, C. (1997), A unified framework for network design with generalized connections, *in* V. Ramaswami & P. Wirth, eds, 'Teletraffic contributions for the information age: Proc. ITC 15', Vol. 2 of *Teletraffic science and engineering*, Elsevier, pp. 319–328.

Girard, A. & Zidane, R. (1995), 'Revenue optimization of B-ISDN networks', *IEEE Transactions on Communications* **43**(5), 1992–1997.

Hui, J. (1988), 'Resource allocation for broadband networks', *IEEE Journal on Selected Areas in Communications* **6**(9), 1598–1608.

ITU-T (1996), 'Recommendation I.371: Traffic and congestion control in B-ISDN'.

Kaufman, J. (1981), 'Blocking in a shared resource environment', *IEEE Transactions on Communications* **29**(10), 1474–1481.

Kelly, F. (1991), 'Effective bandwidths at multi-class queues', *Queueing Systems* **9**, 5–16.

Labetoulle, J. & Roberts, J., eds (1994), *Proc. 14th International Teletraffic Congress*, Elsevier.

Labourdette, J. & Hart, G. (1990), 'Link access blocking in very large multimedia networks', *Computer Communication Review* **20**(4), 108–117.

Mignault, J. (In preparation), A reference resource allocation method for ATM networks, PhD thesis, Ecole Polytechnique de Montréal, 2900 Edouard-Montpetit, CP 6079, Succ A Montréal, Qué. Canada H3C 3A7.

Mignault, J., Gravey, A. & Rosenberg, C. (1996), 'A survey of straightforward statistical multiplexing models for call access control in ATM networks', *Telecommunication Systems* **5**(1–3), 177–208.

Rege, K. (1994), 'Equivalent bandwidth and related admission criteria for ATM systems—a performance study', *International Journal of Communication Systems* **7**, 181–197.

Roberts, J. (1981), A service system with heterogeneous user requirements: Application to multi-services telecommunications systems, *in* G. Pujolle, ed., 'Performance of Data Communication Systems and their Applications', North-Holland Publishing Co., pp. 423–431.

INDEX OF CONTRIBUTORS

KEYWORD INDEX